PESTS, DISEASES, AILMENTS and ALLIES of AUSTRALIAN PLANTS

An introduction to some of the good, bad and interesting
creatures that you might find in your garden

With aids to their identification, symptoms and
recommendations for control

DAVID L. JONES, B.AGR. SCI, DIP. HORT,
W. RODGER ELLIOT and SANDRA R. JONES, PHD
LINE DRAWINGS BY TREVOR L. BLAKE and DAVID L. JONES

First published in 2015 by Reed New Holland Publishers
Sydney

Level 1, 178 Fox Valley Road, Wahroonga, NSW 2076, Australia

newhollandpublishers.com

A record of this book is held at the National Library of Australia.

ISBN 978 1 87706 997 0

Managing Director: Fiona Schultz
Publisher: Diane Ward
Project Editor: Simon Papps
Designer: Thomas Casey
Photographs: David Beardsell, Angus Carnegie, Mark Clements, Pierre Cochard, Ted Edwards/CSIRO Ecosystems
Sciences, Rodger Elliot, Alan Everett, John Fanning, Roger Farrow, Chris French, Chris Goudey, Bill Hall, David L. Jones,
Sandra R. Jones, Tim Jones, Emily Noble, Peter Norris, Alan Stephenson, Max Sutcliffe, Ashley Walmsley, Annie Wapstra,
Hans Wapstra, Mark Wapstra, Gary Wilson, Michael Wood, Tony Wood
Line drawings: Trevor L. Blake and David L. Jones
Production Director: Arlene Gippert
Printed in China

10 9 8 7 6 5 4 3 2

Keep up with New Holland Publishers:
 NewHollandPublishers
 @newhollandpublishers

PESTS, DISEASES, AILMENTS and ALLIES of AUSTRALIAN PLANTS

An introduction to some of the good, bad and interesting
creatures that you might find in your garden

With aids to their identification, symptoms and
recommendations for control

DAVID L. JONES, B.AGR. SCI, DIP. HORT,
W. RODGER ELLIOT and SANDRA R. JONES, PHD

LINE DRAWINGS BY TREVOR L. BLAKE and DAVID L. JONES

THEMES

- To learn the name of that creature before you take the steps to kill it, because you might be doing yourself a favour by leaving it alone. Accurate identification may save you and the creature a lot of anguish.

- A garden full of insects and other small creatures is much more stimulating and interesting than one sterilised by synthetic chemicals.

- There is a misapprehension that natural chemical compounds are safer than manufactured chemicals. Some of the most toxic chemicals are to be found in nature. Consider the toxins Atropine, Cerberin, Nicotine and Ricin for example.

- Treat chemicals with the respect they deserve, don't see them as a first response and only use them as a last resort.

DISCLAIMER

The material provided in this book is general in nature and has been sourced from research, guidelines and recommendations of third parties. Such material is assembled in good faith, but does not necessarily indicate a commitment to a particular course of action. Laws governing pesticide use are different in each Australian state and change over time. It is essential that the user follow instructions on the product label and only use registered products. The decision to use a particular product rests solely with the reader. Before relying on the material in any important matter, users should carefully evaluate its accuracy, currency, completeness and relevance for their purposes, and should obtain any appropriate professional advice relevant to their particular circumstances.

DEDICATION

To those people who are against the indiscriminate use of pesticides in our environment.

ABBREVIATIONS

cm	centimetre	sp.	species singular	alt.	altitude
mm	millimetre	spp.	species plural	eucalypts – we use this term to
m	metre	subsp.	subspecies		denote species of *Angophora*,
		Bt	*Bacillus thuringiensis*		*Corymbia* and *Eucalyptus*.

ACKNOWLEDGEMENTS

It takes much support and input from many people to bring a book of this size and complexity together and we thank all who have been involved.

We are extremely grateful to Tony Wood for generously allowing us to use his fantastic photos of insects. These photos actually provided impetus in the early stages of the revision. Max Sutcliffe generously allowed us to use some of his magnificent bird photos which add greatly to the book. Chris French took many photos on our behalf during his travels in the Western Australian bush. David Beardsell was happy for us to use his photos from the first book. We also thank Roger Farrow and Pierre Cochard for providing photos of grasshoppers and the Emperor Gum Moth, Peter Norris for photos of Red-legged Earth Mite, Ted Edwards for organising photos of wood moth damage from CSIRO Ecosystems Sciences, Angus Carnegie for photos of Myrtle Rust, Ashley Walmsley, Editor of *Good Fruit & Vegetables* magazine, and Alan Everett for the photo of the green snail, Tim Jones for photos of phasmids and Alan Stephenson for the photo of Dendrobium Beetle. Others who provided photos include Chris Goudey, Mark Clements, John Fanning, Bill Hall, Emily Noble, Hans, Mark and Annie Wapstra and Michael Wood.

The book has benefited tremendously from the unstinting assistance of Catherine Jordan who chased up numerous scientific papers, often from old or obscure sources.

We also thank Trevor Blake for the drawings transferred from the first book.

Special thanks to Professor Penny Gullan for forthright correction of incorrect terminology as applied to galls and scales and patient identification of numerous photos of species unknown to us. We also thank Penny for providing us with other contacts to assist in identification of unknown creatures. Many other scientists supported the project in various ways. Dr Rolf Oberpreiler provided enthusiastic support, identified numerous weevil specimens, provided important details on their life cycle and brought some scientific papers to our attention. Ted Edwards was always willing to identify moths and caterpillars and pass on interesting details. Don Herbison-Evans also identified caterpillars and moths and provided information about food plants. Other specialists who helped with identification of photos or information include Dr Chris Anderson, Dr Angus Carnegie (myrtle rust), Dr Lynn Cook (eriococcids), Dr Peter Kolesik (cecidomyiid flies), Dr John LaSalle (galls), Dr Gunter Maywald (*Paropsis* beetles), Dr Laurence Mound (thrips), Dr Michaela Purcell (Nema galls), Dr Chris Reid (Chrysomelidae), Dr Adam Slipinski (Coccinelidae), Dr Gary Taylor (lerps and psyllids), Dr Ken Walker (galls), Dr Tom Weir (beetles) and Dr You Ning (Orthoptera).

Michael Wood provided solid support throughout this project and freely shared his experiences gained with many years of expertly growing native plants and controlling pests and diseases.

There are many others who have contributed in various ways such as allowing us to visit their gardens to photograph pests, passing on useful pieces of information or just general support of the project – Kath Carter, Ulli Cliff, Graeme and Denise Crake, Graham and Helen Gray, Liz McDonald, Helen Richards, Lyndal Walters and Jane Wright. Appreciation also to Diane Ward, publisher, New Holland, for her patience and persistence throughout the preparation of this revision; also to the Project Editor, Simon Papps, for his tolerance and understanding.

Finally but by no means least, we express much appreciation to our families who have had to endure all sorts of problems over the long time it took to complete the revision – Barbara, for her incredible patience and solid unstinting support, Gwen who has a wonderful knack of simplifying language without losing the meaning, and George and Joel for accompanying their mother Sandra on travels to South Australia and Western Australia and tolerating her distractions during the finalisation of the book.

PREFACE TO ORIGINAL BOOK

The initial idea for writing this book arose from the success of the *Encyclopaedia of Australian Plants Suitable for Cultivation*. Many readers expressed an appreciation for the approach taken in the pest and disease section of the first volume of that publication and such feedback provided the inspiration for this book. Like Topsy the project grew. It was soon realised that a comprehensive book on the subject was lacking and it is hoped that the present volume fulfills those requirements.

This is not a recipe book advising on how to eradicate plant pests with chemical concoctions. It is, however, a book which aims to help in the identification of pests and diseases which attack native plants and suggests methods of control with the final alternative being the use of safer chemical sprays. It is also a book which emphasises ecological balances. Our view is that a garden full of insects and other small creatures is much more stimulating and interesting than one sterilised by synthetic chemicals.

The emphasis within the text and the arrangement of the text is on the feeding habits of a pest and the symptoms of damage its feeding causes. We make no apologies for this approach in these modern days of broad spectrum insecticides which kill pests irrespective of their feeding habits. It seems apparent to us, as it has been to generations before, that a basic understanding of a pest's feeding habits is of great importance in its control. Many of the modern insecticides may be excellent in commercial crops but because of their toxicity have no application in the home garden.

With regard to sprays we give very few recommendations about the concentration to use. This information is always available on the label of the can or bottle. Always read the label first. If in doubt contact the nearest office of the Department of Agriculture or Primary Industry for advice.

The book covers the major groups of pests encountered in parks, gardens and nurseries. It is not complete and it is doubtful if a comprehensive coverage of pests in all environments of this vast continent would be possible. However it does cover commonly encountered pests from all regions including the tropics.

PREFACE TO REVISIONARY TREATMENT

This book was first published in 1986 under the title of *Pests, Diseases and Ailments of Australian Plants*. It proved to be popular and was reprinted in 1989, 1990 and 1995. Unfortunately its availability lapsed when the original publishers, Lothian Publishing Company, went out of business. The opportunity for this revisionary treatment was taken up by New Holland in 2012.

The original book has become a collectors' item. Numerous readers have supported our approach to identify any creature before taking extreme action against it. This concept is again promoted in this revised edition which is supported by numerous photos to help with recognition. The new book is much larger and more comprehensive, in line with the tremendous expansion of knowledge on the subject that has become available in recent years. Any comprehensive research will uncover new material and numerous new entries on pests, diseases and other problems involving Australian native plants are included in this revision. A substantial number of these problems concern new pests and diseases from overseas. These breaches of biosecurity are a major concern to all Australians since they seem to be occurring with unacceptable regularity.

All public and private gardens support populations of insects and other small invertebrates. Most are completely harmless to us and our plants, and many are directly beneficial. A few are undoubtedly significant pests or potential pests; others are not so bad, causing limited damage and adding diversity and interest to the garden. The serious pests and not-so-bad pests are the main concern in this book, but many of the less problematical and interesting types are also included. We hope that this expanded coverage will encourage gardeners to identify these small animals and consider the role they play in their gardens.

Once again we give very few recommendations regarding which sprays to use for specific problems and their concentrations. The range of chemicals used as pesticides has dramatically increased since the first edition, with many that were originally described found to be unsafe for the user and the environment. Despite the expansion in science around pesticide types, problems such as resistance in mites and many insects groups, especially those with short life spans remain. Impacts on non-target organisms and residual chemicals surviving in the environment and in some cases entering food chains are additional problems. Information for the use of pesticides is included on the labels attached to the spray containers and additional advice can be obtained from the appropriate government authorities.

We, as authors, and experienced gardeners have prepared a book that describes and illustrates, not only the pests, diseases and ailments that can impact Australian plants, in both the garden and natural setting, but also something on the ecology of these species. We hope the book assists readers to recognise the wondrous diversity in their gardens; the role of natural predators in controlling pests; understand how harm from pests can be limited and explore alternative control options that don't involve the wholesale use of non-target chemicals.

CONTENTS

HOW TO USE THIS BOOK

FOR PESTS

1. Determine which parts of the plant are being damaged and what the damage looks like.

2. Has the damage been caused by a chewing pest or a sucking pest?

3. If the pest is available, examine its mouthparts to see if it is a chewing insect or sucking insect.

4. Look at the plants to see if the pests occur singly or in groups.

5. Then go to the *Symptoms and Recognition Features of Pests* in the prelude.

6. Check photos in the troubleshooting section for a shortcut to the problem.

7. After working through the *Symptoms and Recognition Key* go direct to the designated chapter dealing with the particular group of pests and use the symptoms and recognition features included there to narrow the field.

8. After identifying the group to which the pest belongs, check descriptions and feeding habits of the various species.

9. Determine if the pest is still present or has left the scene.

10. Use control techniques if necessary.

11. If identification is difficult send samples to your local Department of Agriculture or Primary Industries office and ask their advice on control.

FOR DISEASES

1. Examine the plant thoroughly looking for a weak root system, wilting, dry appearance, dead patches of foliage or bark, abnormal leaf shedding, pale leaves and the presence of fruiting bodies.

2. Then go to the *Symptoms and Recognition Key* in chapter 19.

3. Don't hesitate to seek help from experts since accurate identification of some diseases is very difficult.

4. After identifying the disease use control techniques if necessary.

FOR NUTRITIONAL DISORDERS

1. Examine the plant thoroughly for symptoms, in particular looking for colour patterns and colour changes in both new and old leaves.

2. Go to symptom key for nutrient deficiency in chapter 20.

3. Check out the details of each element included in chapter 20 and apply fertilisers if necessary.

4. If identification is difficult send samples to your local Department of Agriculture or Primary Industries office.

TROUBLESHOOTING

SYMPTOMS AND RECOGNITION OF PROBLEMS

Each pest or disease leaves clues that can help with its identification. In the case of a pest, its place of feeding on a plant and its mode of feeding can provide useful clues which can lead to its identification. For diseases the point of attack and the symptoms produced give similar vital information. Careful observation of all these clues can be help to determine the identity of the culprit. In this book the major insect pests are dealt with in chapters 4–16. At the start of most chapters is a simple guide based on the symptoms and features that aid with the recognition of each pest group. Reading through the text will hopefully lead to the identification of the culprit. Diseases, including a key guide to their recognition, are dealt with in chapter 19, nutritional problems in chapter 20 and other ailments affecting native plants in chapter 21. The following guide provides the initial breakdown leading to the respective chapters where the problem might occur. The accompanying photos will help to narrow down the problem further.

SYMPTOMS AND RECOGNITION FEATURES OF PESTS

Damage to young shoots

Sudden wilting and collapse of single shoots, prominent hole often present near base......................*shoot wilters Ch. 5*
Webbing filled with faecal pellets..*webbing caterpillars Ch. 8*
Brown bag constructed from leaves with webbing..............................*bag shelter moths Ch. 8*
Crowded colonies of of small plump fleshy insects, winged insects often present.................*aphids Ch. 4*
Small circular, domed or flat waxy structures..*scales Ch. 4*
Fleshy or woody outgrowths on stems, fruit or leaves...*galls Ch. 7*
Black sooty smudging on leaves and stems...*Ch.4, 5, 19*
Sticky secretions, often shiny, on leaves and young shoots..*Ch. 4, 5*
Messy waxy secretions present..*Ch. 4, 5*
Prominent white frothy secretions on stem and leaf axils.......................................*Ch. 5*
White cottony sacs on stems and leaf axils..*mealybugs Ch. 5*
Wilting or distortion of young shoots, colonies of tiny insects feeding on soft growth or
on the undersides of leaves...*Ch. 5*
Leaves rolled and margins joined by silken threads................................*leaf rollers Ch. 8*
Stems swollen or disfigured by a series of bumps....................................*stem galls Ch. 7*
Death of shoot tips...*bugs, borers Ch. 5, 13*
Narrow strips of leaf surface tissue grazed, often slimy...........................*snails, slugs Ch. 12*
Clusters of ants on stems..*collecting honeydew Ch. 4, 5*
Thickened and distorted young leaves.....................................*thrips Ch. 6*
Brown or black watery areas between the veins of fern fronds...........................*leaf nematodes Ch. 6*
Ragged shoots, the leaves often only partly eaten............................*Ch. 8, 9. 10*
Clump of frass and webbing near shoot tip.............................*soft shoot borer Ch. 13*
Surface of potting mix appearing loose or wet.......................*curl grubs, fungus gnats Ch. 14*

Damage to individual leaves

Fleshy or woody growths on leaves..*galls Ch. 7*
Blobs of sticky white waxy material on underside of fig leaves..*fig psyllid Ch. 5*
Sticky secretions accompanied by growth of black sooty mould...*Ch. 4, 5*
Messy waxy secretions on leaf underside...*Ch. 4*
Slender convoluted tunnels between upper and lower leaf surfaces.......................................*leaf miners Ch. 11*
Small circular, domed or flat waxy structures on leaves..*scales Ch. 4*
Papery blisters prominent on upper surface of leaf..*sawflies Ch. 11*
Leaf surfaces very lightly grazed, taking on a dry, mottled or silvery appearance.......................*thrips, mites Ch. 7*
Patches of dry, discoloured tissue on the older eucalypt leaves..*lerps Ch. 5*
Whole leaves eaten, often leaving the petiole stub...................................*cup moth caterpillars Ch. 8*
Whole leaves eaten, often leaving the midrib and main veins...*sawfly grubs Ch. 8*
Colonies of tiny scale-like insects on leaf undersurface, small white flies often present..................*whiteflies Ch. 4*
Groups of small white circular bodies with waxy or floury secretions...*lerps Ch. 5*
Small shell-like or scale-like bodies, often with delicately ribbed patterns...*lerps Ch. 5*
Ragged leaf margins ....*beetles Ch. 9*
Numerous small holes eaten through leaves..*flea beetles Ch. 9*
Narrow dead patches on wattle phyllodes...*mirid bugs Ch. 9*
Large patches eaten from the leaves of epiphytic orchids..*dendrobium beetle Ch. 9*
Large lumps eaten out of leaves or whole leaves chewed to the midrib...*Ch. 8, 9, 10*
Leaves cupped, margins downturned..*whiteflies, soil salinity Ch. 4, 21*
Leaf surface grazed, turning brown, exposing veins..*Ch. 8, 11*
Leaflets rolled or joined together by silken threads..*leaf rollers Ch. 8*
Strips of surface tissue grazed, often slimy...*snails, slugs Ch. 12*
Adjacent leaves webbed together, some leaves often dead...*Ch. 8*

Damage to bark and/or wood

Circular or elliptical tunnels in wood or bark...*borers Ch. 13*
Concentric rings of bark eaten on stems and twigs...*twig girdlers, weevils Ch. 9, 13*
Bark on stems and twigs lifted in a concentric ring exposing cambium ....*twig girdlers Ch. 13*
Bark grazed in narrow strips..*weevils Ch. 9*
Bark lifted in a series of raised strands...*cicadas Ch. 5*
Clusters of miniature cicada-like insects on bark, often with ants..*Ch. 5*
Clumps of brown faecal pellets joined by webbing..*borers Ch. 13*
Slender convoluted tunnels mined in smooth eucalypt bark.......................*scribble moth caterpillars Ch. 13*
Green bark eaten...*snails, slugs Ch. 12*
Brown bark eaten...*deer, goats Ch. 16*
Branches of shrubs and trees torn off..*goats Ch. 16*
Basal bark gnawed, grooves present..*hares, rabbits Ch. 16*
Narrow black trails on trunk and large branches..*termites Ch. 15*

Damage to roots

White fluffy or waxy secretions present among roots...*root coccoids Ch. 14*
Tunnels eaten in sapwood of roots ....*caterillars of hepialid moths Ch. 14*
Root tips eaten ....*Ch. 11, 14*
Roots misshapen, swollen and distorted...*nematodes Ch. 6*

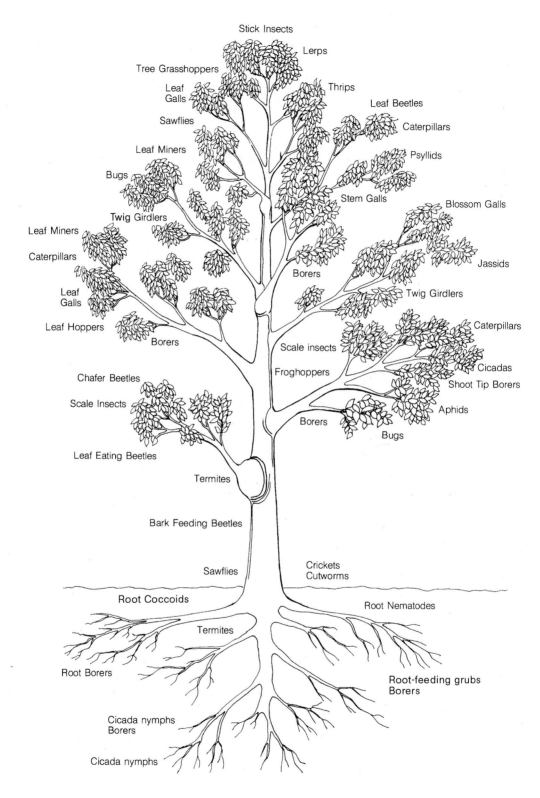

A schematic tree showing the parts affected by various types of pests. [T. BLAKE]

PHOTO SHORT CUTS

It is intended that this selection of photographs will aid with the recognition of pest problems in native plants. These images are linked with the following chapters that deal with specific groups of pests and other problems.

A badly stressed young eucalypt that has been attacked by many types of pest and is struggling to survive. [D. JONES]

This stressed young eucalypt is suffering badly from attacks by galling wasps Ch. 7 and gum tree scale Ch. 4. [D. JONES]

The leafless area around the crown of this Myoporum parvifolium *indicates an insect problem, in this case scale insects Ch. 4.* [D. JONES]

Young caterpillars often cluster together Ch. 8. [D. JONES]

Two pests have been active on these elkhorn fronds – the fern spore caterpillar (large brown areas) Ch. 8 and the Elkhorn fern beetle (small round brown areas) Ch. 9. [D. JONES]

Two pests have been active on this Scaevola, *stem borers Ch. 13 and leaf miners Ch. 11.* [D. Jones]

A dead branch in a healthy canopy may be innocuous or indicate the presence of borers, twig girdlers Ch. 13 or disease Ch. 19. [D. Jones]

These dying leaves indicate attack by an insect when they were young, in this case armoured scale insects Ch. 4. [D. Jones]

The dead stems indicate that something is amiss with this Hakea. [D. Jones]

This dead branch in Persoonia pinifolia *indicates root Ch. 14 or borer problems Ch. 13.* [D. Jones]

Dead shoots in a bottlebrush are often caused by large sap-sucking bugs Ch. 15. [D. Jones]

The unusual purplish colouration in the leaves of this Corymbia *is a result of root or vascular damage Ch. 14, 19.* [D. Jones]

Gum tree scale is making a mess of this young eucalypt Ch. 4. [D. Jones]

Harlequin Bugs are commonly found on the weed known as Marshmallow Ch. 5. [D. Jones]

Cupped leaves can be a sign of sucking insects on the undersurface Ch. 4, 5. [D. JONES]

Feeding ants often indicate the presence of sap-sucking insects such as these jassid nymphs Ch. 5. [T. WOOD]

Inrolled and distorted young leaves like these are often caused by thrips Ch. 6. [S. JONES]

Large patches of stripped eucalypt leaves indicate feeding by large caterpillars or swarming beetles Ch. 8, 9. [D. JONES]

Messy white waxy material indicates the presence of mealybugs on this Calostemma purpurea *Ch. 4.* [D. JONES]

This type of damage on an Acacia *phyllode is caused by a sap-feeding bug, note the puncture hole in the centre of each wound. Ch. 5* [D. JONES]

Damage of this type to to wattle phyllodes is usually caused by mirid bugs Ch. 5. [S. JONES]

Narrow strips eaten in eucalypt leaves indicate weevil feeding Ch. 9. [D. JONES]

Narrow tunnels in leaves are caused by leaf miners Ch. 11. [D. JONES]

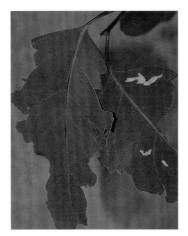

Ragged leaf margins indicates beetle feeding Ch. 9. [D. Jones]

Shredded leaves are an indication of caterpillar feeding Ch. 8. [D. Jones]

Symptoms like these dark veins and pale leaf blades indicate a nutritional problem Ch. 20. [D. Jones]

The grazed areas on these leaves have been caused by snails Ch. 12. [R. Elliot]

The numerous small holes in this leaf, termed shot-holing, are the result of feeding by flea beetles Ch. 9. [D. Jones]

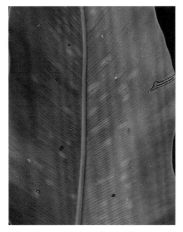

The yellow patches on this Asplenium *frond are caused by the toxic saliva of fern scale Ch. 14.* [D. Jones]

These blobs of wax are covers over a scale insect Ch. 4. [D. Jones]

These eucalypt leaves have been damaged or killed by a severe attack by gum tree scale Ch. 4. [D. Jones]

These leaves have been surface-grazed by leaf skeltonising caterpillars Ch. 8. [D. JONES]

These leaves have been eaten by caterpillars Ch.8. [D. JONES]

These leaves of Leptospermum laevigatum *have been eaten by case moth caterpillars Ch. 8.* [D. JONES]

These waxy deposits on Eucalyptus robusta *are caused by lerps, the black soot is the fungus disease sooty mould Ch. 5.* [D. JONES]

These waxy structures along the midrib of this leaf are soft scales, each blob covers a scale insect Ch. 4. [D. JONES]

This type of feeding on bottlebrushes is caused by sawfly grubs Ch. 11. [D. JONES]

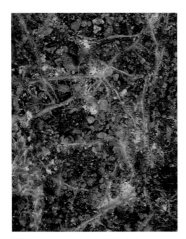

White waxy material around plant roots is a sure indication of the presence of root coccoids Ch. 14. [D. JONES]

These tracks along the bark are caused by termites Ch. 15. [D. JONES]

Some caterpillars hide in simple shelters formed by joining young leaves together Ch. 8. [D. JONES]

Many caterpillars live communally in nests made from webbing and eaten plant parts Ch. 8. [D. JONES]

The brown 'bags' in the crown of this young kurrajong are caused by Bag Moth caterpillars Ch. 8. [D. JONES]

Borer damage has lead to stem breakage in this Acacia sophorae *Ch. 13.* [D. JONES]

Borer damage in stems of Stenocarpus *indicated by dead leaf and piles of frass Ch. 13.* [D. JONES]

Clusters of frass indicate the presence of borers Ch.13. [C. FRENCH]

Damage of this type in a eucalypt trunk indicates persistent borer attack Ch. 13. [D. JONES]

Sudden breakage can indicate the presence of borers Ch. 13 or disease Ch. 19. [D. JONES]

A dry appearance is often the first sign of root problems in members of the Proteaceae Ch.19. [D. JONES]

Dark patches in fern fronds can be caused by nematodes Ch. 6 or root problems Ch.14.
[D. Jones]

Wilting is an indication of root problems, most likely a root-rotting disease Ch. 19.
[D. Jones]

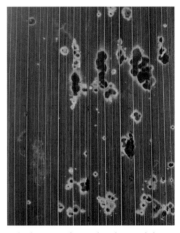

Discolouration and a dry appearance in the leaves of grass trees is usually an indication of root problems Ch.14, 19. [D. Jones]

Exudation of gum in the absence of obvious insect damage is often an indication of disease Ch. 19. [D. Jones]

A leaf-spotting fungus has damaged this palm leaflet, note the prominent yellow halo around each site of attack Ch. 19. [D. Jones]

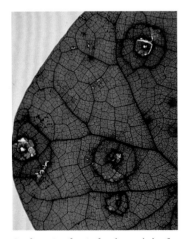

Leaf-spotting fungi often have a halo of pale-coloured tissue around the site of infection Ch. 19. [D. Jones]

Sudden death in a Banksia *is usually caused by root rot Ch. 19.* [D. Jones]

These yellow spots on Agonis flexuosa *are caused by myrtle rust Ch. 19.* [D. Jones]

PEST CONTROL TABLE

PEST	CHEMICAL CONTROL	OTHER CONTROL
Aphids	neem, imidacloprid, malathion, pyrethrum	soap spray, water jets, squashing
Beetles	malathion, pyrethrum	squashing
Borers	not recommended	removing and destroyed affected parts, bluestone paste as deterrent
Caterpillars	contact spray, spinosad, pyrethrum, malathion, neem	*Bacillus thuringiensis*, hand picking
Cockroaches	baits	deterrents made from boric acid, garlic and cayenne pepper, sticky traps
Crickets, grasshoppers	protecting key plants with malathion, pyrethrum or pyrethroids	catching and squashing
Earwigs	malathion, synthetic pyrethroids	beer traps, iron based baits, catching and squashing
Fungus gnats	neem	*Bacillus thuringiensis*
Leaf miners	white oil, neem	destroy effected tips and shoots
Leafhoppers	contact sprays of pyrethrum, malathion	water jets, squashing
Lerps	imidacloprid	hand removal
Mealybugs	white oil, malathion, imidacloprid	squashing, dabbing with cotton buds soaked in methylated spirits
Mites	soap and oil sprays, miticide, neem	water jets, dust of lime and sulphur powder
Nematodes		molasses drench to soil, immersion in hot water
Psyllids	imidacloprid	yellow sticky traps
Sap sucking bugs	pyrethrum based contact spray	knock bugs from trees, collect and destroy, hand picking
Sawflies	malathion or pyrethrum	knock to ground, spraying with water jets, squashing
Scales	malathion, pyrethrum, white oil	soap spray, physical removal, squashing, water jets
Slaters	synthetic pyrethroids	catching, squashing, traps
Snails and slugs	baits based on iron chelate	squashing at night, beer traps
Thrips	malathion, pyrethrum, soap spray, white oil	yellow or blue sticky traps to monitor, soap spray
White Flies	imidacloprid, synthetic pyrethroids	water jets, yellow sticky traps, soap spray

CHAPTER 1

INTRODUCTION

The environment where the majority of Australians live today has been changed so much from its original state that it is difficult to envisage what most areas were like. We now live in modified environments, creating parks and public gardens and tending our own little private plot of lawns, selected ornamental plants, vegetables and fruit crops.

Each of these gardens, private and public, is full of insects and other small invertebrates. Most of the insects are completely harmless to us and our plants, and many are directly beneficial. A few are undoubtedly significant pests or capable of being pests, others are not so bad and add interest to a garden, even if they do cause some plant damage. The serious pests and not-so-bad pests will be the main concern in this book, but many of the less problematical types as well as beneficials are also included.

Even the pests have a special niche in the environment, and as gardeners and horticulturists we tend to forget this natural relationship among living creatures. After all we have changed the environment and introduced foreign plants which we wish to grow, so it is not surprising to find a few creatures feeding on those plants we treasure. Faced with this situation we must be careful not to overreact by spraying to kill all creatures indiscriminately. Such a practice only creates more problems than it solves because new insects may become pestiferous when indiscriminate spraying kills their natural enemies.

The theme of this book is the accurate identification of an invertebrate pest before the safest techniques of control are employed. In other words learn the name and characteristics of that creature before you take steps to kill it, because you might be doing yourself a favour by leaving it alone.

It is possible to achieve a three-way balance in any garden between the pests, their natural enemies and the plants. Sporadic outbreaks of damaging pests will occur but if a tolerant attitude is adopted then in most cases the situation will return to normal. Sometimes assistance is needed to restore the balance. Hopefully this will be after due consideration of all the factors involved and the control applied will be accurate and not indiscriminate. This book aims to help gardeners achieve this aim.

It must be recognised that in commercial situations such as forestry, fruit and vegetable production and plant nurseries, regular chemical application is an integral part of the production process. Commercial growers cannot afford damaged or diseased produce which they cannot sell. Even in such commercial situations where regular spraying is so important the industry is adapting continually. Research into more efficient pest management is being carried out by entomologists and new chemicals and biological agents are regularly being tested for their efficacy and safety in the environment. The emphasis should always be on correct and safe use of pesticides.

PEST ECOLOGY

Australian plants in their natural environment suffer from attacks by a large range of creatures, and it is unusual to see a plant which has not been damaged in some way or another. Most of these attacks are of a minor nature causing superficial damage and slight setbacks to the plants. Sporadic attacks which are of major significance can also occur resulting in stunting or even death of the plant. These sporadic

OPPOSITE: *Two species of Christmas beetle.* [D. JONES]

attacks often occur when the natural ecological balance is disrupted as discussed further below.

By contrast, when grown away from their natural environment, such as overseas or even interstate, plants are often placed in a situation free of their indigenous pests and do not suffer the constant depredations experienced in their native environment. Consequently they often grow faster and with an unblemished appearance. This is often a feature of eucalypts grown overseas. Macadamia nuts in their natural situation grow in rainforests of NSW and Qld. When grown commercially in these areas they suffer from constant attacks by pests. By contrast if they are grown in inland or semi-desert areas and are regularly watered they suffer fewer attacks and their foliage is dark green, lustrous and unblemished.

Caterpillar of the Pasture Day Moth, Apina callisto, *feeding on* Wurmbea dioica. [T. Wood]

Caterpillar of the Black-and-white Tiger Moth, Ardices glatignyi, *feeding on a sun orchid.* [E. Noble]

Severe attacks which cause major setbacks or even kill plants occur in one of two ways:

1. When an insect is introduced into a new environment which is free from its natural enemies. Such pests are mainly those which have been introduced from overseas as have most of the aphids in Australia. Pests from interstate may also come under this category.

2. When the natural balance between a pest and its predators is upset. This natural balance is delicate and subject to small fluctuations, but can be upset by outside influences such as bushfires, clearing of vegetation, or spraying with insecticides. If the balance is tilted in favour of the pest, then its numbers build up very rapidly until it reaches plague proportions. At very high population levels the pest causes severe damage and only a corresponding increase in the numbers of its natural enemies will bring it under control. If this happens quickly then the plants will usually recover but with a minor setback (even after complete defoliation). If the attacks persist at serious levels over several seasons then stunting or even death of the plants can follow. Sometimes natural catastrophes upset the balance as in outbreaks of wattle blight, cup moths, sawflies and phasmids. More frequently, humans upset the balance.

For reasons which are not well understood some insects undergo population explosions at irregular intervals. Their numbers increase enormously, sometimes suddenly, or more often over a period of time. The insects may reach plague levels and devastate large areas or be confined to individual trees. These sporadic increases in the populations of insects cannot be predicted and considerable damage may be caused. Sometimes the outbreaks decline as rapidly as they occurred.

Pests in gardens are subject to control from natural enemies and have similar population fluctuations as they have in the bush. Outbreaks of pests are frequently a problem in new gardens but their incidence lessens as the gardens become established. When new suburbs are created much of the natural bush is cleared for houses, roads, and such, and only a skeleton of the original vegetation remains. Often some plants of the original species will become weakened or die out under the new conditions of increased light and transpiration imposed by the clearing and only those adaptable species will survive. All of these changes affect the balance between insect pests and their predators, and new gardens planted in such estates are often ravaged as a result of this imbalance. As an example, borers are commonly a problem attacking a variety of plants grown in such areas. These pests are usually well entrenched in the weak or dying eucalypts and wattles which have been left, and are almost impossible to adequately control.

As a new suburb settles down and gardens become established a balance is again achieved between the pests and their natural predators and parasites, although this will probably be different to the balance in the original bush. This new balance is subject to similar fluctuations as was the original balance and outbreaks of pests will occur when the balance is upset.

INVERTEBRATES

Insects belong to the group of animals known as invertebrates or animals without backbones. Over 95 per cent of all animals on earth are invertebrates, with over 80 per cent of invertebrates grouped into the single Phylum Arthropoda which includes spiders, mites, millipedes and insects.

Insects are the dominant life form of the earth and the main consumers of plants. In fact there are more species of insects than all other groups of animals and plants put together. Most of the pests that cause problems are insects and hence an understanding of insects' structural features and life cycles is essential in the use of this book.

INSECT STRUCTURE

The body of an insect is composed of three distinct sections: head, thorax and abdomen. The head bears sensing organs such as a pair of antennae, one or more pairs of eyes and mouthparts with their associated sensory palps. The thorax bears three pairs of jointed legs and in winged species has one or two pairs of wings. The abdomen is usually elongated and fleshy. It contains the reproductive structures and is usually divided into a series of segments.

Because there are such huge numbers of insects, many variations occur on this basic structural theme. Caterpillars, for example, have up to five pairs of fleshy false legs called prolegs. These aid in locomotion and ensure a secure grip when feeding. Loopers, which are specialised caterpillars, often only have a single pair of prolegs and adopt a specialised looping motion when on the move. The hind pair of prolegs are often further modified into a structure known as the anal clasper.

Often the body parts of an insect can be difficult to discern because the outer parts are covered with a jointed armour coat known as an exoskeleton. When an insect has grown to its size limit, the outer coat is shed and the insect forms a new coat which hardens leaving room for further growth.

METAMORPHOSIS AND LIFE CYCLES

The word metamorphosis is used for the successive changes of form through which an insect passes in its passage from egg to adult. Most insects undergo some form of metamorphosis in their life cycle. Insects such as beetles and butterflies undergo complete metamorphosis which involves major changes in shape and morphology between the larvae and the adult stages. The development of these insects starts with an egg which hatches to become a larva (in this case a grub or caterpillar) which grows as it feeds, eventually becoming a pupa. Vast changes occur in the pupal stage and eventually an adult insect emerges (in this case a beetle or butterfly). Other insects such as bugs, scales or grasshoppers exhibit incomplete metamorphosis and grow through gradual stages of development known as nymphs. Nymphs shed their old skin between stages of growth (termed moulting). Nymphs are miniature versions of the adults but do not have all of their features. For example, nymphs have wing buds which are very small in the early stages of growth but develop and become fully functional wings when the individual reaches maturity.

It is useful to be able to recognise the different life stages of an insect in order to understand its role as a potential pest and as a consequence implement control measures at the most suitable time. Different life stages often require different food types and therefore an insect may be a pest while in early development, but not as an adult. Consider for example a voracious leaf-eating caterpillar in comparison with the adult butterfly which is completely harmless and a joy to have in the garden. Similarly some beetles do not feed at all, whereas their larval stages can be serious pests. Additionally, insect larvae can change their feeding habits as they grow. Very young leaf-eating caterpillars often feed in groups, whereas older larvae feed singly. Young caterpillars often graze the surface tissues of a leaf whereas larger caterpillars eat whole chunks of leaf tissue. Similarly phasmid nymphs favour young soft leaves, while adults feed on harder mature leaves. Adult scale insects are covered by a waxy structure that is impervious to chemical sprays. By contrast, very young scales, termed crawlers, have no waxy covering and are much more susceptible to control measures.

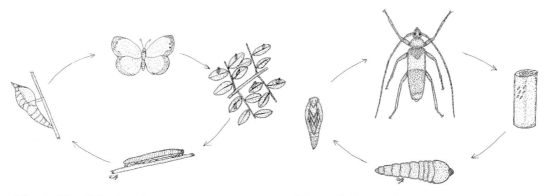

Life cycle of Grass Yellow Butterfly. [D. JONES]   *Life cycle of a Longicorn Beetle.* [D. JONES]

INSECT GROUPS

The following groups of insects are familiar to gardeners because they include most types that are pests of plants. This introductory guide provides some basic information on these groups (each group is an insect Order) and further details can be found in the following chapters.

Blattodea (Cockroaches)

Chewing mouthparts; oval and flattened body shape with a thorax covered by a large plate (the pronotum), which extends partly over the head; two pairs of wings often present, with forewings more hardened than hindwings. Many species release a noxious odour when disturbed or killed.

Coleoptera (Beetles and Weevils)

Chewing mouthparts; complete metamorphosis; two pairs of wings, the hardened forewings (termed elytra) cover the hindwings when at rest; larvae are grubs with or without legs, or wireworms.

Dermaptera (Earwigs)

Chewing mouthparts; incomplete metamorphosis; flattened elongated body with hardened pincer-like structures on the end (termed cerci); two pairs of wings with short hardened forewings, the membranous hindwings folded fan-like beneath the forewings when not in use.

Diptera (Flies and mosquitoes)

Larvae are legless maggots with chewing mouthparts, adults have piercing and/or sucking mouthparts; complete metamorphosis; one pair of wings, the hindwings modified into a knobbed balancing structure called a haltere.

Hemiptera (True Bugs, Aphids, Leafhoppers, Mealybugs, Scales)

Piercing and sucking mouthparts; incomplete metamorphosis; two pairs of wings, the forewings thickened at the base and covering the hindwings when at rest.

Hymenoptera (Bees, Wasps, Ants, Sawflies)

Chewing mouthparts; two pairs of wings, both similar; complete metamorphosis; larvae are grubs.

Isoptera (Termites)

Chewing mouthparts; pale, elongate body; two pairs of wings present in reproductive castes only, social insects living in colonies with a queen, sterile workers and soldiers.

Lepidoptera (Moths, Butterflies)

Larvae are caterpillars with chewing mouthparts, adults have sucking mouthparts; complete metamorphosis; two pairs of wings, both similar.

Orthoptera (Grasshoppers, Locusts, Crickets, Katydids)

Chewing mouthparts; incomplete metamorphosis; two pairs of wings, the forewings thickened and covering the hindwings when at rest; hindlegs often modified for jumping.

Phasmatodea (Stick Insects)

Chewing mouthparts, elongate and cylindrical or flattened body shape that resembles sticks, leaves or grass; incomplete metamorphosis; two pairs of wings with short hardened forewings which form a protective cover over only part of the hindwings.

Thysanoptera (Thrips)

Minute insects with chewing mouthparts; incomplete metamorphosis; two pairs of wings (often fringed), both wings the same.

OTHER PEST ARTHROPODS

The following groups of arthropods include most types that are pests of plants.

Acarina (Mites and Ticks)

Sucking mouthparts, often tiny with individuals having no external segmentation of the abdomen and appearing as a single body mass.

Annelida (Earthworms)

Long cylindrical bodies consisting of many similar segments, no obvious head, lack legs or antennae.

Gastropoda (Snails and Slugs)

Feed with rows of rasping teeth on a tongue like structure known as a radula; soft bodies without body segmentation and often having an external shell made of calcareous material; use a modified fleshy foot to move by gliding along mucus that is produced by glands in the foot.

Diplopoda (Millipedes)

Chewing mouthparts; two pairs of legs per body segment; simple eyes if present, although many have no eyes; often roll into a tight spiral to discourage predators and discharge a foul smelling distasteful substance.

Isopoda (Slaters)

A type of crustacean with chewing mouthparts, heavily armoured flattened body and seven pairs of legs.

Nematoda (Nematodes)

Nematodes are mostly worm-shaped multicellular animals that are commonly found as internal parasites of other animals, but also occur in soil and aquatic environments. Nematodes range in size from fractions of millimetres up to nearly 10 metres long. Some species are problematic plant pests that can impact on agricultural production, commercial forestry as well as home garden plants.

VERTEBRATE PESTS

Some vertebrates can become significant pests to commercial crops or plantations, but usually do minimal or only nuisance damage in domestic gardens. Pest groups are restricted to mammals such as native possums, rats, kangaroos and bandicoots, and introduced species such as rabbits, hares and deer. Birds can be pests, mainly to fruit and vegetable crops, but seldom require control by the home gardener. Most perform an extremely beneficial role in invertebrate control and are valued visitors to home gardens.

Control for vertebrate pests in a home garden usually involves barriers such as netting to protect individual plants or temporary or permanent fencing to restrict access. Chemical baiting is available for introduced pests such as rabbits, however, this form of control is not available for domestic situations.

THE INFLUENCE OF CLIMATE

Invertebrate pest species are influenced by seasonality and environmental conditions, particularly temperature. There is much debate on climate change and the potential effects that may be introduced by the future alteration of Australia's climate. These changes have the potential to influence pest species and how they behave in the garden environment. Many scientific studies are being undertaken to predict the responses of particular pest groups, with results suggesting that the geographical ranges of some pests will increase and that they will spread to new areas. Recent reports have described an alarming trend of pest species spreading away from the equator and towards the north and south poles. The highest rates of movement have been for butterflies at up to 20km per year. There is also the potential for pest outbreaks to increase because of more storm and extreme temperature events and for pests and their natural predators to become out of sync, resulting in greater numbers of pests.

For the home gardener, these changes may bring new pest species and more frequent pest outbreaks. Maintaining a collection of healthy plants within the garden and being vigilant to any pest activity, especially things that you haven't been seen before, will assist in responding to any challenges our changing climate may bring.

CHAPTER 2

SYSTEMS OF PEST CONTROL, CHEMICALS AND SPRAYING EQUIPMENT

Control of pests and diseases is a continual process in crop production or plantation forestry where losses are of economic significance. In most parks and domestic gardens, however, outbreaks are sporadic and pest control occupies only a small proportion of the time spent in maintenance. This time can be kept to a minimum by maintaining healthy plants in a diverse garden environment.

PRINCIPLES OF PEST CONTROL

There are seven basic principles which should be followed to protect your garden plants from being impacted by pest outbreaks.

1. Vigorous healthy plants are able to resist attack better than weakened plants. If pests continually attack plants then this could be an indication of some other ailment, e.g. waterlogged soil weakening the tree. The answer in this case is to improve the drainage of the soil. The use of suitable fertilisers often dramatically increases a plant's resistance to pests, especially if the plants have been deprived of nutrients.

2. Identify accurately the culprit causing the damage. This is of paramount importance because the particular control system chosen will vary with the species of pest.

3. Ensure that the pest is still active and causing problems, e.g. the scale-like covering in lerps is conspicuous but control measures are not warranted unless the lerp insects are still alive underneath the cover. Spraying would only kill other insects, including harmless and beneficial types.

4. If natural control systems are becoming established do not interfere — see Biological Control.

5. If the pests are potentially serious, such as aphids, scale insects, leaf skeletonisers or borers, act as soon as the symptoms are noticed. Such pests can build up in number rapidly and cause considerable damage in a short time; if action is delayed then control may be much more difficult to achieve.

6. Do not try to eliminate all pests, only attempt to reduce them to safe levels.

7. Use the safest control systems first and only use toxic sprays as the last resort.

BIOLOGICAL CONTROL

The majority of insects are not harmful to humans, animals or plants and many are indeed beneficial. Some are deadly enemies of the worst plant pests and their activities are to be encouraged in gardens. If sufficient numbers of such beneficial insects can be established, then outbreaks of damaging pests are greatly reduced.

This system is known as biological control and offers the best hope of controlling the worst pests.

In Australia, there are approximately 25 species of beneficial insects that are sold commercially as biological control agents (often abbreviated to biocontrol) for the control of pests. Additionally CSIRO has released some biological control agents, for example, the parasitic fly *Tricopoda giacomellii* for the control of Green Vegetable Bugs (*Nezara viridula*).

Biological control is based on the balance achieved by nature. Fluctuations occur between a pest and its natural enemies but overall these are balanced out. Thus when a particular insect builds up to plague proportions so its natural enemies also build up and eventually reduce its numbers back to a tolerable level. There is always a lag time until the predator catches up with the pest. Such a balance is possible to achieve in a garden or any natural system and can be aided if three simple guidelines are applied.

1. Limit use of pesticides

Random or indiscriminate spraying with all but very safe pesticides should be abandoned because of the danger of upsetting the balance between pests and their natural enemies. Often a spray applied to control one pest will result in the upsurge of another because the spray has destroyed or reduced the other pest's natural enemies

2. Foster a diverse garden

A garden planted with mixed species is better able to cope with pests than expanses of one kind of plant. Mixed plantings are better able to support beneficial insects than stands of a single species. There are a number of plants that are recommended to encourage beneficial insects.

3. Avoid conditions that favour pests

Remove weeds which are often a harbour for pests and diseases, e.g. marshmallows are a haven for harlequin bugs, while thistles and capeweed provide excellent shelter and food for aphids, slaters and earwigs. Excess litter and debris is a haven for cockroaches, slaters and earwigs.

INTEGRATED PEST CONTROL

Integrated pest control combines the best features of both biological control and pesticide application. Every effort is made to achieve biological control but, where necessary, pesticides are applied strategically to achieve control. Selection and application of pesticides is made after due consideration of the natural balance that exists between pests and their natural enemies.

OTHER CONTROLS

Barriers

Barriers are used to prevent pests from reaching their target plant or from moving up the trunk of a tree to reach the foliage. Barriers are a simple method that the home gardener can use. Examples are included below.

- **Banding** A strip of hessian or cloth tied around the trunk of a tree offers a haven for some pests. The material can be inspected and removed at intervals and the pests destroyed. This technique was commonly employed in the early years of fruit culture in Australia. The strip of material should be tied to the tree in such a way that a flap hangs downwards and encourages the entry of travelling pests.

   Banding has proved very successful for trapping the adult weevils of root borers when they migrate to the parts of the plant which are above ground. It is also extremely effective against white cedar caterpillars, so much so that it need be the only control system adopted for this pest.

- **Floating Covers** Translucent white polyester-based fabric can be used as a temporary barrier to cover plants or crops to protect them from pests at key stages, e.g. as seedlings or when the targeted pest is most active and visible in the garden. The material can be draped over plants, or over frames and secured to the ground. Covers like these can be useful for deterring mobile pests such as aphids, some beetles and cabbage moths.

Sticky Traps

A modification of the simple band is to smother it with a sticky non-drying adhesive material, which effectively traps travelling insects. This system is particularly effective for preventing the passage of ants up a tree. Commercial traps are available for hanging or placing around the garden, or you can make your own using coloured rigid material and a glue such as "Tangle-trap." Very simple traps can be made by cutting yellow card or old yellow plastic buckets into rectangles 10 cm x 15 cm and coating both sides with petroleum jelly or a similar sticky substance. Although different colours attract different pests, yellow traps work for a large range of insects including whiteflies, fruit flies, fungus gnats, thrips and pysllids. A tie-wire can be used to hang the traps in areas where insects are a problem. Insects attracted to the trap will become stuck. Once the sticky substance is covered with dead insects, it can be discarded and replaced.

Deterrents

Some materials are useful because they act as a deterrent to pests and will protect plants if used in their vicinity. Garlic sprays are effective against pests and have the added bonus of deterring cats. Lime is a good deterrent to many ground-dwelling pests. The fungicide thiram is an effective deterrent of rabbits and hares.

Borers can be prevented from laying eggs by painting the trunks and limbs of trees with deterrent pastes. The most commonly used paste is bluestone paint made up of 750g of copper sulphate, 500g of quicklime, and 5 litres of water. The copper sulphate is first mixed into a paste then dissolved in approximately half the amount of water and the lime in the remainder. The two are then mixed together in a wooden, earthenware or copper container (never iron) to form the paint which can be brushed over the trunk. A small amount of linseed oil will help the material spread and stick.

Copper is an effective deterrent of slugs and snails. A copper band or wire wound around a pot will stop them from reaching valuable plants. Similarly pots placed on a copper sheet are fully protected from these pests. Laying fine dry sawdust or crushed eggshells around plants is thought to discourage snails and slugs from travelling across it.

Solutions made from quassia chips are a useful deterrent for pest animals including possums, rats and wallabies. These chips are obtained from two species of plants in the family Simaroubaceae; one species, the quassia tree, *Picrasma excelsa*, is a tall tree native to parts of Central America and the West Indies; the other, *Quassia amara*, known commonly as amargo, is a shrub that is widespread in South America. The chips of both species are used identically. The wood contains a resinous material called quassin which has a very bitter taste. To prepare a deterrent solution soak about 200g of quassia chips in 2 litres of water for at least 24 hours. Strain off the liquid, heat for about 30 minutes then dilute with 10–20 litres of water. This solution can be used as a spray or drench. Adding 5g (or one teaspoon) of soap flakes per litre of spray increases its effectiveness. Spray the plants to be protected for at least five days in a row. Respray after rain. Do not spray on herbs and vegetables that are to be eaten. Quassin is also a strong insecticide. Quassia chips are available from specialist herb suppliers and from some nurseries and chemists.

Attractants, Pheromone lures or traps

Many insects communicate through pheromones, which are powerful chemical scents used to attract members of the opposite sex. Some pheromones have been copied and applied to bait traps to lure insects. Most lure traps only attract male insects and therefore need to be used as an early warning system in conjunction with a specific control.

Food attractants, such as fermenting sugars, lure a wide range of insect species, whereas pheromone attractants are specific to a species. Pheromone attractants have been used with some success against fruit fly infestations and are available for purchase by the home gardener.

MECHANICAL REMOVAL

Hand Picking

Individual large pests such as solitary caterpillars or case moths can be removed by hand and squashed. Pests which cluster together such as aphids, sawflies and processional caterpillars can be removed by snipping off the shoot where they are living or feeding, and squashing or burning it. Pests such as leaf skeletonisers or leaf beetle larvae are often gregarious while young but disperse as they get older. If caught early enough whole colonies can be squashed before they have done much damage. Colonies of eggs can be destroyed before they hatch, e.g. *Paropsis* beetle on eucalypts. Many scale insects are quite obvious and easy to squash by hand or can be removed by using an old toothbrush brushed over the leaf or stem.

Water Jet

A reducing nozzle attached to the end of the garden hose narrows the stream and increases the pressure of the water. Such a jet will not damage shrubs or trees but is very effective at dispersing colonies of pests such as aphids, sawflies, scales, beetles and bugs. Aphids readily drown in water and a powerful jet may stun larger pests such as bugs and sawflies rendering them susceptible to attack by predators. A hand-held spray bottle also works well for individual or delicate plants.

CHEMICAL SPRAYING

Spraying with pesticide should only be used as a last resort when all other methods have failed and the pests are doing intolerable damage to the plants.

Unfortunately there is a general tendency to use sprays as the first control measure but in the garden this approach usually creates more problems than it solves.

Most pesticides are harmful to the environment, affecting birds and other wildlife, fish, spiders and beneficial insects such as bees. Pesticides frequently upset the natural balance between predators, parasites and the pests they control. Thus it is not uncommon for a pesticide to cause an outbreak of a pest different from the one which it has been used to control, e.g. the pesticide kills ladybirds and allows aphids to build up.

Pesticides frequently achieve spectacular results when they are first applied against the specific pest they are designed to control. After a while, however, the results are not so spectacular and higher concentrations may be needed to achieve the same control. After continual spraying it is common for pests to

Foliar application of pesticides demonstrating apropriate protective equipment. [M. Wood]

become resistant to the effects of the spray either by a change in their habits or genetic changes which induce immunity. If there is a genetic change then continued spraying effectively selects those which are resistant, and soon this control measure does not work. Thus a new pesticide must be found to replace the ineffective one and the system becomes a vicious circle with the pests adapting and the environment suffering from pesticide pollution.

Unfortunately in some cases it is only possible to achieve control by using a chemical spray. If spraying is unavoidable then spray properly and observe all of the correct procedures.

Spraying Rules

1. Choose the safest and most effective pesticide for the job.

2. Carefully read the label and follow the instructions on the label, using only at the recommended strength and for the purpose stated.

3. Mix sufficient spray to cover the affected plants only.

4. Spray thoroughly to wet the pests but not so that a great deal of excess spray drips to the ground.

5. Spray on a still, cool, calm day (avoid spraying on windy or hot days at all costs).

6. Preferably spray towards evening to lessen the chance of birds contacting the fresh spray.

7. Spray only the affected plants and if there is excess spray do not apply it to other plants at random. Such random spraying could easily result in the destruction of beneficial insects and an upsurge in the levels of previously insignificant pests.

8. Never spray near fish ponds or pools as fish, frogs and other water life are easily killed by chemicals.

9. If there is excess pesticide dispose of it through the appropriate means. Most local councils can provide information on free drop off days for left over chemicals and empty containers.

Legal Requirements

All chemical products used for pest or disease control (generically termed pesticides) must be registered with the Australian Pesticides and Veterinary Medicines Authority (APVMA). The APVMA determines conditions for the use of pesticides and these conditions are specified on the product label. The use, storage and disposal of pesticides is controlled in the various states by state legislation - for example, in NSW, the Pesticides Act 1999 and Pesticides Regulation 2009. Pesticide products can only lawfully be used for the purposes which they are registered and as specified on their product label. The onus is on the user to meet this requirement.

It is required by law that all pesticides sold must be include information on the label regarding the active ingredients and the amount of these active ingredients contained in the product. The label also contains other information such as recommended strengths to use and compatibility with other sprays. The labels of pesticide products should always be read carefully before a spray is chosen and applied.

Poison Scheduling

The Poison Schedule is the system used to classify substances based on their human health risks. The heading on a product indicates which poison schedule it belongs to. Pesticides can fit into one of four categories.

1. **Schedule 7 (S7)** poisons are substances with high to extremely high toxicity which can cause death or severe injury at low exposures. They require precaution in their manufacture, handling or use and are too hazardous for domestic use. Home garden products cannot be S7 poisons, however some are available for agricultural or horticultural use. They include the following headings on the label: 'DANGEROUS

POISON – KEEP OUT OF REACH OF CHILDREN,' 'Read Safety Directions Before Opening or Using' and 'Can Kill if Swallowed'.

2. **Schedule 6 (S6)** poisons have moderate to high toxicity which may cause death or severe injury if they are ingested, inhaled or come in contact with skin or eyes. They include the heading: 'POISON – KEEP OUT OF REACH OF CHILDREN' and 'Read Safety Directions Before Opening or Using' on the label.

3. **Schedule 5 (S5)** poisons have low toxicity or a low concentration of an active ingredient and are a low to moderate hazard to humans. They are capable of causing only minor adverse effects to human beings in normal use and require caution in handling, storage or use. They have the heading 'CAUTION – KEEP OUT OF REACH OF CHILDREN' 'Read Safety Directions Before Opening or Using' on the label.

4. **Unscheduled substances** are not considered poisons, however they may be capable of causing minor adverse effects to humans and require caution in handling, storage and use. They include the heading 'CAUTION KEEP OUT OF REACH OF CHILDREN' 'Read Safety Directions Before Opening or Using' on the label.

Classes of Pesticide

When pesticides are registered for use in Australia, they are broadly split into two groups:
1) agricultural and commercial products,
2) those suitable for home garden and domestic pest control.

   While both groups may include the same active ingredient, that ingredient is usually at a much lower concentration in home garden products.

Agricultural and commercial pesticides

These products are designed for use by professional operators or primary producers and are often only available from specialist outlets in large volumes. They are not registered for use in home gardens and **should not be used in this way**. They may be highly toxic and can pose a significant threat to the environment and human health. The use of these products is guided by mandatory instructions on how they are used.

Home Garden Pesticides

These products are either exempt from poison scheduling or are Schedule 5 or 6 poisons. They are sold in smaller packets than pesticides designed for agricultural and commercial use and are often of lesser strength. Instructions for their general use are included on their labels.

Action of Sprays

Sprays basically kill pests in one of three ways and an understanding of the mode of action is an important facet of effective control and also ensures that the correct spray is chosen for the job.

Contact Sprays

As the name suggests these kill on contact. They will kill virtually any insect contacted whether it is good or bad. They are most effective when moist and fresh but some sprays are persistent and retain some contact properties for a few weeks after spraying.

Stomach Poisons

These kill after they have been ingested by the insect. They are only effective against *chewing* insects such as caterpillars, locusts and sawflies.

Systemic Sprays

These are absorbed into the sap stream and distributed throughout the plant. They are effective against *sucking* insects such as aphids and mealybugs which feed on the sap. Once absorbed by the sap of the plant they kill only sucking insects (not chewing insects) and hence are safe for wildlife. Unfortunately many of the systemic sprays have toxic properties, are dangerous to handle and their widespread use is not encouraged.

Naming

A trade or brand name refers to the name that a pesticide is sold under. Each commercially available pesticide product also includes the active agents which are the chemicals within a product that provide the active control. This book does not recommend any specific brand or trade names, and only refers to the chemical agents of control. Gardeners should carefully review any commercial products they purchase to identify the active constituents.

Pest Feeding Habits

Pests primarily feed on plants by either chewing the parts or sucking the sap. It is important to determine the method of feeding since this should influence the choice of control measures to be used. The major pests can be split into two groups depending on how they feed.

SUCKING		CHEWING	
Aphids	Leafhoppers	Borers	Locusts and grasshoppers
Bugs	Lerps	Caterpillars and grubs	Millipedes
Cuckoo spit	Mealybugs	Cockroaches	Sawflies
Galls	Mites	Earwigs	Slugs and snails
Planthoppers	Scale insects	Galls	Thrips
Jassids	Thrips	Leaf-eating beetles	

Spray Toxicity

Toxicity is used to describe the ability of a pesticide to cause injury or death to its target pest. The toxic properties of pesticides vary greatly between species and the effects can be influenced by a variety of environmental factors. Toxicity is usually described by the term LD^{50} and refers to the lethal dose required to kill 50 per cent of a group of test animals. A lower figure refers to a more toxic material. Information on the toxicity of a pesticide product can be found on the label or its associated product information.

TYPES OF SPRAYS

Safer Sprays

Some sprays are relatively safe to use and do not have such a drastic effect on the environment. If used correctly they can give as effective a control as the more toxic sprays. Many of the safer sprays listed below can be prepared by the home gardener from ingredients already present around the house. There are various versions of these recipes available; we have included the most commonly used ones below.

Bacterial Spore Suspensions

The bacterium *Bacillus thuringiensis* is a very effective antagonist of caterpillars. Its spores are available in various preparations and sold under commercial trade names. They are sprayed onto the foliage where caterpillars are active and, when ingested, germinate in the caterpillars gut, producing a toxin which kills it. A strain of this bacterium is under trial for its effectiveness against fungus gnats.

The bacterial spores themselves are readily killed by ultraviolet light from the sun and to be effective should be applied in the evenings or during cloudy weather. For continuous protection sprays must be applied every 10 to 14 days.

Bracken

A bracken preparation has been found useful for controlling aphids and many other insect pests. About 25g of dried bracken fronds are soaked in 1 litre of water for 24 hours. The liquid is strained and used in the dilution of 1ml to 1 litre of water.

Chilli Spray

Chilli spray is a favourite of many gardeners for controlling ants, aphids, and many other insects. Some people use it as a deterrent for possums but its success may be dependent on the individual possum. Blend 40–50 small hot chillies (or substitute chilli paste or powder) with one litre of water until thoroughly mixed. Strain the mixture, discarding the solids. Mix with another litre of water and 5g of pure soap flakes dissolved in hot water (or substitute for a few drops of liquid soap). Use the undiluted spray.

Coriander

Oil of the herb coriander has been recorded as an effective control for Two Spotted Mite. It should be applied as a 2 per cent emulsion with water. The addition of a wetting agent, such as soft soap will make the applications more effective.

Eucalyptus Sprays

Extracts of the essential oils of certain *Eucalyptus* species may kill some pests or at least discourage their activities. The effectiveness of the extracts is enhanced in some commercially available formulations by the addition of a pyrethrum compound. A few drops of eucalyptus oil can also enhance the effect of soap sprays.

Garlic Spray

A natural spray can be made from cloves of garlic. It is reported to be effective against a wide range of pests including aphids and caterpillars but the results are not always satisfactory. It acts as both contact spray and a deterrent and is safe for the handler. It is a very effective deterrent for cats. To make garlic spray, crush 85g of garlic cloves and mix with 10ml of paraffin oil. Leave for about 48 hours and then add about 560ml of water and 7g of an oil-based soap. Mix thoroughly, filter out the lumps and store the concentrate in a plastic drum. This will keep forever and can be used as needed. Dilutions range from 1 in 50 to 1 in 100 parts with water.

Horticultural Oils, including White Oil

Horticultural oils are viscous materials that are extremely valuable for controlling a wide range of insect pests (including aphids, caterpillars, mites and scale insects) without having to resort to toxic chemicals. There are a number of products available which are sometimes preparations or mixtures of organic oils. Some are vegetable oils with additional substances such as eucalyptus and/or tea-tree oil as additives. White oil is a viscous material which can be based on petroleum or paraffin oil.

Horticultural oils kill by coating the insect with a layer of oil which cuts off its air supply and the insect suffocates. A thorough coverage is essential for effective control. Two or three sprays at close intervals can also increase its effectiveness, as can the inclusion of a contact insecticide such as pyrethrum or maldison. One drawback with using horticultural oils too regularly is that these materials can also kill beneficial creatures such as ladybird larvae. The oils may stunt the growth of some plants such as ferns and orchids and for these it is best used at about half to three-quarters strength. High temperatures accentuate damage to plants and horticultural oils should not be applied on hot days (above 25°C). White oil should only be applied to leafless deciduous plants during winter months.

Neem (Azadirachtin)

Neem is a naturally occurring product that has low toxicity to humans, other mammals and beneficial insects. It is also biodegradable. It is extracted from the seeds of the Neem Tree, *Azadirachta indica*, a fast-growing species from the Indian subcontinent. Azadirachtin is the name of the chemical compound derived from Neem seeds. Traditionally Neem seeds have been ground into a paste, mixed with water and applied to crops to control a wide range of chewing and sucking pests and some diseases such as black spot, powdery mildew, anthracnose and some rusts. Neem does not kill insects, but rather acts as a repellent and disrupts feeding and growth. It is suitable for use on pests such as two-spotted mites, aphids, curl grubs, caterpillars, citrus leaf miner, white fly and fungus gnats. There are some instances where Neem applications have burnt soft young foliage, so it is best applied in cool weather.

Molasses Spray

Molassess spray has been used successfully for many years in combating leaf-chewing insects. It is also regarded as a successful possum repellent and can be used to deter nematodes. It needs to be sprayed regularly over the leaves of all plants being attacked by caterpillars and other chewing pests to be effective. Dissolve 1 tablespoon of molasses into 1 litre of warm water plus 1 litre of liquid soap. Shake to mix and apply liberally with a spray bottle.

Pyrethrum

Pyrethrum is a natural compound found in daisy flowers of the genera *Pyrethrum* and *Chrysanthemum*. The compound is extracted from the flower head and contains six esters that are collectively known as pyrethrins. These pyrethrins are the active ingredients within Pyrethrum which have insecticidal properties. (Pyrethroids are man-made compounds that have a chemical structure that is similar to the naturally occurring pyrethrins and are dealt with under Types of Less Safe Sprays).

Pyrethrins act by entering the body of the insect through its outer skin and shutting down the nervous system. Pyrethrins do not persist long in the environment as they are degraded by high temperatures and exposure to ultraviolet light. Pyrethrins are broad-spectrum insecticides and will affect a wide range of insects, including those that are beneficial.

Soap Sprays

Soap sprays are commercially available or can be made easily by the home gardener. A soapy-type solution of very low toxicity can be made by mixing water with detergent or soap and applying with a trigger bottle. Soap spray has the same properties as a contact material and also kills by coating the insect and blocking its air supply. It can also impact on the epidermis of the insect by reducing its repelling properties. Soap sprays have some effect on caterpillars but are best for controlling aphids and scales. Two or three sprays at intervals of one to two days may be necessary to clean up persistent attacks. A simple spray can be made by mixing 1–2 tablespoons of liquid soap with 1 litre of boiling water. Allow to cool overnight, put in a spray bottle and apply. Soap spray can be used as a base, with organic gardeners often adding aromatic herbs to improve effectiveness. A few drops of eucalyptus oil can also improve its action.

Spinosad

Spinosad is based on a waste product of the soil bacterium *Saccharopolyspora spinosa*, which is produced during fermentation. It is a natural product, has a broad insect pest spectrum, and is approved for use in organic agriculture. It is highly active and works by both contact and as a stomach poison, although in some species it only affects adult pests. It can provide effective control of caterpillars, flies, some beetles, grasshoppers and thrips, but it has little effect on sucking insects. It is also effectively used as the pesticide component of bait sprays and sticky traps against fruit flies.

Sulphur

Sulphur powder has been used as a dust against sucking insects but it is very difficult to dissolve in water. A product called Wettable sulphur is sometimes available and this can be dissolved in water. It has some properties which act as an insecticide although it is mainly used as a fungicide.

Tea Tree Oil

A small amount of tea tree oil is sometimes added to sprays of horticultural oil. It is also used in conjunction with eucalyptus oil sprays.

Tomato Leaf Spray

Tomatoes are members of the nightshade family and contain toxic compounds called alkaloids in their leaves. A spray for controlling aphids can be made by chopping 1–2 cups of tomato leaves and soaking overnight in two cups of water. Strain the leaves, retaining the liquid. Add another 2 cups of water to the reserved liquid. Apply from a spray bottle to the stems, leaves and flowers of infested plants.

TYPES OF LESS SAFE SPRAYS

Carbamates

Carbamate pesticides act on both target and non-target species by inhibiting the enzyme acetylcholinesterase. Carbamates do not persist in the environment and most are unlikely to contaminate groundwater. They usually remain active for a few hours to a few months in soils and crops, and are rapidly degraded in more alkaline conditions. Carbamates can be toxic to birds and mammals and are highly toxic to bees and aquatic invertebrates. They are suitable for use on chewing pests. They can be very effective against slaters and earwigs when used in areas where these pests congregate.

Examples include: Carbaryl, Bendiocarb.

Imidacloprid

Imidacloprid is a neonicotinoid-type of systemic insecticide which acts as a neurotoxin. Neonicotinoids are synthetic forms of the natural insecticide nicotine. Imidacloprid is currently the most widely used insecticide in the world, however, there are concerns that it has contributed to colony collapse disorder that has led to the widespread decline of honey bees. Imidacloprid occurs in many home garden products that target sucking pests and is also available in tablet form for use at planting time or in pots as a systemic bait taken up through the root system. It is strongly recommended that label information is followed closely and no spraying occurs when non-target species are present or bees are foraging.

Organophosphates

Organophosphorus pesticides inhibit and inactivate the enzyme acetylcholinesterase, affecting nerve activity and leading to overstimulation of muscles. Most organophosphorus chemicals do not persist for long in the environment and are not very mobile in the soils that have high organic content and are more stable under acidic conditions. They are however toxic to target and non-target species, particularly aquatic species, and they may be linked with egg-shell thinning. Most of these chemicals are unsuitable for home garden use, however, there are some products that are available, especially for the treatment of mites.

Examples include: Dicofol (mites), Chlorpyrifos, Profenofos, Diazinon (soil pests) and Dimethoate (sucking pests).

Maldison or Malathion

Maldison and Malathion refer to a manufactured chemical which acts by killing insects on contact. It is effective against most insects including caterpillars, aphids and bugs. It is most effective while fresh but has some persistence on the foliage. Its effectiveness is increased if used in combination with white oil. It is one of the less toxic sprays but should still be used with due care and only if safer sprays have proved ineffective.

Piperonyl Butoxide

Piperonyl butoxide is a pesticide synergist and commonly appears on home garden pesticide labels. Being a synergist means that when it is included with another product, such as pyrethrin, it acts to increase the effectiveness of that product.

Pyrethroids

Pyrethroids are synthetic compounds that act in the same way as naturally occurring Pyrethrins, but tend to have a longer persistence in the environment. Many home garden pesticides contain a pyrethroid as the active ingredient that works as a contact poison. Pyrethroids tend to be suitable for use on a wide range of pests including both sucking and chewing species. They will also kill beneficial insects.

Examples include Permethrin, Bifenthrin, Tau-fluvalinate, Bioallethrin, Bioresmethrin and Esfenvalerate.

Combination Sprays

Some sprays are more effective when used in combination than when used separately. The effectiveness of white oil is increased by adding a contact insecticide such as malathion or pyrethrum. This combination also increases the effectiveness of the insecticide because the white oil aids its spreading and sticking powers. It is emphasised that not all chemicals are compatible and some cannot be mixed together. Information on compatibility is contained on the label of the pesticide.

Wetting Agents

Wetting agents are important materials which increase the effectiveness of sprays. They are usually soapy or detergent-type solutions which overcome the waxy nature of leaf surfaces and help the spray to spread evenly over the whole surface. They must be used at no higher than recommended strength as they can cause damage to foliage at higher rates.

OTHER CHEMICAL CONTROL

Trunk Injections

Control of pests on large trees is generally impractical. Trunk injection has been employed with varying degrees of success and is often recommended by professional arborists on significant individual trees. A hole is drilled to varying widths and depths, depending on the target pest and chemical used, and a solid implant or liquid injection of pesticide is inserted at various depths within the tree trunk. The rationale for using this technique is that the injected or deposited insecticide is taken up in the sapstream and transported to the leaves where it is ingested by chewing or sucking insects. Application of a pesticide through trunk injection is therefore indiscriminate and will kill any insect feeding on leaves at the time of the injection and can therefore enter the broader food chain. Trunk injection is often recommended for use on borers, however, injected insecticides move through outer layers of the trunk and do not reach the heartwood where borers are often most active. Research into specific pesticides for trunk injections is continuing but the authors do not recommend it as a solution for home garden pest control.

Baits

Baits containing stomach poisons are effective against pests such as cutworms, armyworms, slugs, snails, earwigs and millipedes. The baits are generally scattered in the vicinity of the plants to be protected and may be effective for several weeks, although this depends on the weather conditions. Baits can be used strategically if something is known about the habits of the pest to be controlled. Examples are baiting dark, humid areas for millipedes, baiting on humid or rainy nights for slugs and snails and baiting weedy areas for cutworms.

Commercially prepared baits are effective against the pests mentioned above. Alternatively, an effective bait can be prepared by mixing 5g of Paris green or metaldehyde with 120g of bran. Immediately before use, sufficient water is added to form a crumbly mash and the bait can then be placed in small heaps. This bait is best applied in late afternoon or early evening as it becomes less attractive to caterpillars when it dries. Another useful bait can be made by mixing one part by volume pyrethrum powder to two parts flour.

Baits are poisonous and should not be placed where they can be eaten by pets or children.

Dusts

Insecticides contained in powder can be applied to foliage or soil as dusts, however, care must be taken during application to avoid inhalation. Commercially prepared home garden dusts contain active ingredients such as synthetic pyrethroids or are sulphur- or copper-based.

Finely ground sulphur can be applied as a dust against aphids. A dust composed of equal parts by weight of fine sulphur and hydrated lime has been used with some success against mites. Leaf-eating beetles can be controlled by dusting with a mixture of one part by weight of pyrethrum powder to two parts by weight of talc. Finely ground tobacco can be effective against white fly and slugs. Foliage damage may result from dusts if used at temperatures above 30°C.

CHAPTER 3

BENEFICIAL INSECTS AND OTHER USEFUL ORGANISMS

An established garden has an ecology that has developed over the years, with a wide range of vertebrate and invertebrate creatures sharing the environment. The vast majority of these creatures do not feed on plants and should not be regarded as being pestiferous just because of their presence. In fact the proportion of pesky critters in a garden is usually quite low, although it can change from season to season. That leaves a high proportion of creatures that are either benign or are useful in some way. The latter group includes predators and parasites that eat other creatures, including the pests that damage plants. While these creatures can be useful at removing plant pests and should be encouraged, it also should be noted that predators feed opportunistically and they will dine on any creature, good or bad, that crosses their path.

As in natural bushland, the more diverse a garden, the more likely the plants, pests and beneficial species will be in balance. It will therefore be less likely that major pest outbreaks will occur as the plants will be healthy and natural predators will be present to keep pest numbers to a minimum. Gardeners are encouraged to grow a wide range of plants, including groundcovers, midstorey and canopy species and consider providing habitat for some vertebrate species such as birds, reptiles and frogs (discussed below). Even if the garden is small, beneficial species can be encouraged through careful planting or with commercial seed mixes that include a collection of plant species designed to provide resources for beneficial insects. Most importantly, it is useful for gardeners to be able to recognise these beneficial species.

BENEFICIAL INSECTS

Ants

Every garden supports ants of one kind or another. There are probably more than 3,000 species of ants occurring naturally in Australia. Several exotic species have also become naturalised (see chapter 15). Ants, which are social insects in the family Formicidae, are very important in the ecology of gardens and the bush. Significant roles include soil aeration, mixing layers of soil together, distributing nutrients, dispersing seeds and carcase disposal. They also provide an important food source for other animals including lizards, birds, spiders and other insects. Some ants are diurnal, others nocturnal. Many ants, especially some of the larger species, are active predators, killing and consuming small creatures, including pests that feed on plants. As an example, some ants of the genus *Pheidole* are active predators of the citrus gall wasp. The Green Tree Ant, *Oecophylla smaragdina*, although often detested by

Ant carrying a leafhopper. [T. Wood]

Bullant with cicada nymph. [T. Wood]

Antlion. [T. Wood]

tropical gardeners, exerts a significant degree of control over insects on the trees it inhabits, reducing the impact of pests on crops such as mangos and citrus. Ants, however, can also cause problems, especially when they invade houses or are involved with sap-sucking pests (see chapters 4 and 5). A number of serious ant pests that have become naturalised in Australia are discussed in chapter 15.

Antlions

Antlions are unusual insects that are placed in the family Myrmeliontidae, order Neuroptera. The adults have a long slender abdomen, knobbed antennae and two pairs of gauzy wings similar to lacewings. They are weak fliers, flying mainly on warm summer nights in search of a mate. Because of their clumsy flight with all wings seeming to act independently they are sometimes called helicopters by children. Their eggs are laid in the ground. Antlion larvae are flattish oval or spindle-shaped brown insects that grow to about 1cm long. Each has three pairs of legs and large prominent hooked jaws for grabbing prey and tubular mouthparts for sucking body fluids. They live singly at the bottom of a small cone-shaped pit which is excavated in sandy soil, often in dry sites. The larvae feed on ants and other small insects which blunder into the pit. When the insect tries to escape up the loose sides of the pit the larva flicks sand to impede its progress, grabbing it with its large jaws.

Antlion pits. [S. Jones]

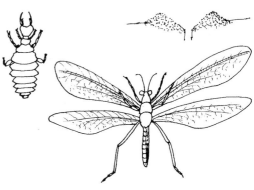

Antlion and larva. [T. Blake]

Hanging fly with a caterpillar. [T. Wood]

Hanging Flies or Scorpion Flies

These insects are placed in the order Mecoptera, family Bittacidae. They are not true flies because they have two pairs of wings. They hang in the foliage of plants and feed on passing insects which they catch with their powerful back legs. They not only feed on crawling insects such as caterpillars but can also catch flying insects that pass within range.

Lacewings

There are about 400 native species of these attractive insects, with most occurring in the tropics. They are placed in the order Neuroptera, with species in the families Chrysopidae and Hemerobiidae being of special significance to gardeners. Adults and larvae feed on insects. The adults are green to brown or black insects with gauzy wings that cover the body in a tent-like arrangement when at rest. They also have prominent eyes and slender antennae. Their distinctive eggs, each held upright on a thin stalk to avoid predation, are laid in rows, often along stems or on the underside of a leaf. The larvae, which grow to about 0.8cm long, range from short and fat to narrow and slender depending on the species. They have large prominent hooked jaws for grabbing prey and tubular mouthparts for sucking body fluids. The larvae can often be seen wandering on plants, usually staying in the vicinity of pest insect colonies. They attack a range of small insects including thrips, scales, aphids, mealybugs and white fly; they also take insect eggs and mites. Chrysopidae larvae have small spines on their bodies on which they impale remains of their prey. This acts as camouflage while they are feeding. The larvae of specialised scale-eating lacewings have been observed to prise the covering off scales with their powerful jaws and feed on the insect beneath. Some lacewings can be purchased from biocontrol firms, usually being distributed as eggs. The Green Lacewing, *Mallada signata*, is a widely distributed and common native species whose larvae feed on many different pests. The lacewing *Oligochrysa lutea* has a slender green body and broad gauzy wings. Its larvae feed on mealybugs, particularly the citrus mealybug. A large brownish species with dark wings, *Nymphes myrmeleonides*, is common in temperate and subtropical gardens. Its adults can be more than 4cm long.

Green Lacewing. [T. Wood]

Brown lacewing. [T. Wood]

Gold lacewing. [T. Wood]

Lacewing, Nymphes myrmeleoides. [T. Wood]

Lacewing eggs. [T. Wood]

Gold lacewing. [T. Wood]

Mantidflies feed on small insects.
[T. Wood]

Owlfly. [T. Wood]

Larva of a ladybird. [T. Wood]

Mantidflies

These distinctive insects, placed in the family Mantispidae, order Neuroptera, resemble small to miniature praying mantids. Their front legs are strongly modified for grasping prey and their abdomen is covered with two pairs of gauzy wings. They are cumbersome fliers, mostly at night, and are attracted to lights. The adults feed on small insects which they mostly catch while at rest. The larvae of some species are predators of small insects, others are parasitoids of wasp, bee and scarab beetle larvae. The larvae of one specialist group feed on spider eggs.

Owlflies

Owlflies are distinctive insects which are somewhat similar to antlions and lacewings. They are placed in the family Ascalaphidae, order Neuroptera. There are about 40 species in Australia. The adults have a long slender abdomen, two pairs of gauzy wings and long antennae, each with a prominent apical knob (sometimes two-coloured). They are strong fliers during summer and feed on a range of insects which they catch on the wing. Some species fly during the day, many are active at night. They can release a strong smell when disturbed or handled. Stalkless eggs are laid in batches on grass stems. The larvae, which look just like those of antlions (complete with large wicked jaws), hide in bark and litter, and ambush passing insects.

PREDATORY BEETLES

Beetles are a large group of insects and it is not surprising to find specialised species which are predators of other insects. These beetles are usually active runners with relatively long legs and ferocious mandibles. Frequently their larvae are also predators.

LADYBIRDS OR LADY BEETLES (FAMILY COCCINELLIDAE)

There are some 57 genera and 260 named species of these popular insects native to Australia. More species await formal recognition. Readily recognised by their neat domed shape (0.2–1cm long), the majority of ladybirds are beneficial, feeding on small soft-bodied insects such as aphids, mealybugs, mites, leafhoppers, psyllids and scales; they also feed on the eggs of some insects. One species also feeds on leaf fungi, including powdery mildew. Note however, that ladybirds in the genus *Epilachna*, including the 28-spotted ladybird, are not beneficial but feed on plant tissue.

Predatory ladybird adults and their larvae are voracious feeders and it has been estimated that a single fully grown larva can eat 40–100 aphids in a day. Ladybirds and their larvae are mostly active during the day and often remain close to their food source. The larvae are unusual creatures that can have prominent mandibles; some are covered with white mealy powder and cottony threads. Adult ladybirds are easily handled and can be readily moved onto colonies of pests. Some species are available commercially from biocontrol firms. Because of their importance in ensuring a natural balance in the garden, several examples are included here.

Black Ladybird *Lindorus lophanthae.* A small black or brownish-black ladybird about 0.3cm long with short grey hairs over the body. The larvae are greyish. Both stages feed on aphids and mealybugs in the autumn-winter months. Young larvae eat scale eggs and crawlers. This species was introduced into the USA as a biocontrol agent and is now well established in California.

The Common Spotted Ladybird,
Harmonia conformis. [T. Wood]

Larva of the Common Spotted Ladybird.
[T. Wood]

Pupa of the Common Spotted Ladybird.
[T. Wood]

Common Spotted Ladybird *Harmonia conformis.* A familiar species which is yellow or orange with about 23 large black blotches on the wing-covers. The larvae, which are black with two yellow transverse bands and long legs, are ferocious feeders. Adults (0.8cm) and larvae feed on aphids, scale insects and mites.

Fungus-eating Ladybird *Illeis galbula.* A bright yellow ladybird (0.6cm) with black cross bands on the wing-covers and a central black line where they join. The larvae, which can grow up to 1cm long, are whitish with rows of black dots on the back. Adults and larvae feed on leaf fungi. They are valuable because they feed on powdery mildew and may contribute to the control of small infections of this disease. Some biologists, however, believe that they may also be responsible for spreading this fungus. When disturbed the adults drop quickly to the ground.

The Fungus-eating Ladybird,
Illeis galbula. [D. Jones]

Gumtree scale Ladybird *Rhyzobius ventralis.* A small round dull black ladybird about 0.4cm long. The adults appear furry because most parts of the body are covered with short fuzzy hairs. Adults and larvae feed on all stages of the gumtree scale (*Eriococcus coriaceus*). The ladybird was introduced into New Zealand to help control this scale insect.

Mealybug Ladybird *Cryptolaemus montrouzieri.*

Larvae of this species, which are very prominent, grow to about 0.5cm long. They are white with long, waxy marginal filaments and the body appears as if dusted with flour. The adults, which are about 0.4cm long, are dull-black or greenish black with an orange head and thorax. They appear furry because most parts of the body are densely covered with short fuzzy hairs. Both stages feed actively on mealybugs and some scales, including cottony cushion scale. This ladybird is available from biocontrol firms.

Larvae of the Mealybug Ladybird,
Cryptolaemus montrouzieri. [T. Wood]

Larvae of the Mealybug Ladybird,
Cryptolaemus montrouzieri, *feeding on mealybugs on fruit clusters of* Hydriastele.
[D. Jones]

Cast skins of One-spot Ladybird pupae.
[D. Jones]

One-spot Ladybirds, Orcus bilunulatus, *on* Casuarina cunninghamiana. [D. Jones]

One-spot Ladybird *Orcus bilunulatus.* A distinctive dark bluish-black ladybird, 0.5–0.7cm long with a single orange spot on one wing-cover. Both the adults and larvae feed on soft-bodied sucking insects.

Red Chilocorus Ladybird *Chilocorus circumdatus.* An Asian species that somehow became established in Queensland where it was found feeding on citrus snow scale. It also feeds on red scale, oriental scale, oleander scale and white louse scale. The adults are about 0.5cm long and range in colour from reddish brown to bright orange. They are dome-shaped and have a fine black border around the wing-covers. The larvae have prominent black spots and numerous soft black spines. This ladybird is available from biocontrol firms.

A ladybird larva, Chilocorus *sp.* [T. Wood]

Spidermite Ladybird *Stethorus nigripes.* A very tiny black native ladybird about 0.1cm long. Both the adults and larvae feed on mites. In 1974 this species was unsuccessfully tested in the USA as a possible biocontrol agent.

Steelblue Ladybird *Halmus chalybeus.* A conspicuous species, about 0.4cm long with the males deep blue with a metallic sheen and the females green. The larvae are slender and yellow or greyish. Both stages feed on aphids, scale insects and mites during the warm months of the year.

The Steelblue Ladybird, Halmus chalybeus, *on* Pandorea. [D. Jones]

Striped Ladybird *Micraspis frenata.* A striking species which is 0.4–0.5cm long and bright orange with about five longitudinal black stripes on the wing-covers. The adults feed on nectar, grass pollen and aphids. Female beetles have been observed to eat up to 60 aphids in a day. This ladybird, which is common in eastern Australia, occurs in gardens and is also found in crop plants and weeds.

Three-banded Ladybird *Harmonia octomaculata.* An orange ladybird 0.6–0.8cm long with eight to eleven black spots on the wing-covers. The spots are aligned more or less in three transverse rows. This ladybird feeds on aphids and whiteflies.

Tortoise-shelled Ladybird *Harmonia testudinaria.* A yellow ladybird which is 0.6cm long and boldly marked with a network of black lines. The larvae are brownish with cream markings. It occurs mainly in tropical areas.

Transverse Ladybird *Coccinella transversalis.* A colourful orange or yellowish species about 0.5cm long with four long black blotches on the wing-covers and a dark line where the wings join. The adults and larvae eat aphids, often feeding together. This ladybird is available from biocontrol firms.

The Transverse Ladybird, Coccinella transversalis. [T. Wood]

Variable Ladybird *Coelophora inaequalis.* A very common species in which the 0.5cm-long adults show great variation in colour (yellow to orange) and markings on the wing-covers that range from small dots to larger blotches and even longitudinal stripes. The larvae are brownish black with yellow markings. This species feeds on aphids.

Vedalia Ladybird *Rodolia cardinalis.* A small, dull red ladybird about 0.4cm long with five black spots on the wing-covers. The body is densely covered with short fuzzy hairs. Both adults and the reddish larvae are common predators of Cottony Cushion Scale, *Icerya purchasi.* This ladybird, which was introduced into the USA in the late 1800s, was responsible for the resurrection of the Californian citrus industry after it had been devastated by Cottony Cushion Scale. It is still well established in California and Florida and is available from biocontrol firms. More recently this ladybird has been introduced to the Galapagos Islands to control infestations of Cottony Cushion Scale attacking a range of plants native to the islands.

White-collared Ladybird *Hippodamia variegata.* A red ladybird with black spots on the wing-covers (two spots are larger than the others) and a white transverse band behind the head. It feeds on aphids and thrips.

These soldier beetles, Chauliognathus tricolor, *feed on soft-bodied pests.* [T. Wood]

Yellow-shouldered Ladybird *Apolinus lividigaster.* A tiny black beetle (0.3cm) that has two prominent yellow shoulder patches. It also has a covering of short fuzzy hairs over most of its body. The larvae are brown with soft spiky marginal projections. The adults and larvae eat scale insects.

Soldier Beetles (family Cantharidae)

Some soldier beetles are useful predators of soft-bodied pests. Both the adults and larvae of *Chauliognathus pulchellus* and *C. lugubris* feed on caterpillars and other soft-bodied pests including gum scale. The adults are slender greenish-brown beetles about 1.5cm long, with yellow or red on the thorax. They cluster in large numbers on the flowers of eucalypts and other Myrtaceae over summer.

Ground Beetles (family Carabidae)

These are medium to large very active beetles with slender legs, longish antennae and rows of tiny pits on the wing-covers. They give off an unpleasant small when handled. Some species have a metallic greenish or violet sheen. They hunt at night and hide under litter and in cracks during the day. Many of these beetles are active predators, running their prey to the ground and literally tearing them to pieces. They are particularly fond of soft grubs such as

A pest-eating soldier beetle, Chauliognathus lugubris. [T. Wood]

Carab beetles. [T. Blake]

armyworms and cutworms. The larvae, which are slender grubs with well developed legs and strong jaws, are also predators. The green carab beetle, *Calosoma schayeri*, is a widespread and common species that is active in late spring and summer. The adults and larvae of the widely distributed bombadier beetle, *Pheropsophus verticalis*, feed on other insects. When disturbed, this beetle can release puffs of hot acrid gas from its rear end, accompanied by a distinct pop.

A carab ground beetle. [D. Jones]

Bombadier beetles run down other insects. [D. Jones]

A clerid beetle. [T. Wood]

Clerid beetle larva. [D. Jones]

Clerid Beetles (family Cleridae)

Many of these slender, hairy beetles are predators which feed on other insects. Some others feed on pollen. Individuals can bite savagely when handled. The predators, which are active during the day, can often be seen running on tree trunks in very hot weather. The larvae are also predatory. Some clerid beetle larvae are highly specialised and feed on borer larvae, crawling through borer tunnels and eating the grubs.

Click Beetles (family Elateridae)

These are common slender beetles that right themselves when upside down by a using specialised jumping mechanism, producing an audible click in the process. The adults feed on nectar and pollen. The larvae, called wireworms, are slender with a hard outer skin. The larvae of some species feed on caterpillars and grubs, others eat plant roots.

Click beetle. [D. Jones]

Some wireworms feed on insects, others eat plant roots. [D. Jones]

Soft-winged Flower Beetles (family Melyridae)

These small but colourful beetles, which are often found on flowers, feed on pollen and small insects. One species, the Red-and-blue Beetle, *Dicranolaius bellulus*, feeds on young caterpillars and aphids. One study showed that an adult beetle can eat about 40 aphids in a day.

The Red-and-blue Beetle feeds on caterpillars and aphids. [T. Wood]

Tiger Beetles (family Cicindelidae)

These are fast-running, often brightly-coloured predator beetles with long, thin legs and a downturned head. The adults feed on a range of insects captured while running or flying. Their larvae hide near the end of a tunnel in the ground and ambush insects that walk past.

Rove Beetles (family Staphylinidae)

These beetles, which superficially resemble earwigs, can be recognised by their long, slender body and very short wing-covers which leave most of the abdomen exposed. Numerous species occur in Australia, ranging in size from less than 0.1cm long to about 2cm. Many species are predators of soft-bodied insects including thrips and fungus gnats, others feed on fungi or pollen. Some species produce a smelly liquid when disturbed and the larger species can bite strongly. One predatory species is available from biocontrol firms.

PREDATORY BUGS

Several groups of bugs in the insect order Hemiptera can be regarded as beneficial to gardeners.

Assassin Bugs (family Reduviidae)

These are ugly, dull-brown or blackish predatory bugs which ambush other insects and suck out their body juices through a large, curved, needle-like beak or rostrum that extends down from the head. These true bugs are particularly fond of large, soft-bodied grubs and caterpillars but will attack any insect, good, bad or otherwise (bees are another common target). The adults, which range from 1–3cm long, and all nymphal stages are active feeders. Many species move slowly and are not easily disturbed, but some, especially those which live among litter, move quickly and run down their prey. Assassin bugs frequently bite people if handled, inflicting a painful burning sting. Adults of a common species, *Pristhesancus plagipennis* (often called the 'bee killer'), that is found in eastern state gardens, are about 3cm long. They are pale brown but the strikingly different nymphs are black with a bright red abdomen. The Red-and-black Assassin Bug, *Coranus trabeatus*, ferociously attacks a wide range of insects. A fast moving species, *Ectomocoris decoratus*, is bluish-black with an orange blotch on the back. It commonly lives under bark and in litter. Some Australian assassin bugs have their hind legs intricately clothed with large brushes of hair.

Red assassin bug feeding on a hoverfly. [T. Wood]

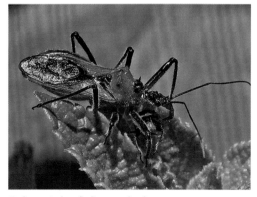

Red assassin bug feeding on a beetle. [T. Wood]

Mirid Bugs (family Miridae)

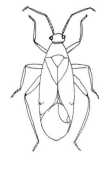

Mirid bug. [T. Wood]   *Mirid bug.* [T. Blake]

Mirid bugs are mostly small and quite active insects. Although many mirids are plant pests, others also feed on small soft-bodied insects and insect eggs. For example, the adults, which are only about 0.4cm long, and nymphs of the brown smudge-bug, *Deraeocoris signatus,* feed on mites, aphids and the eggs of moths and butterflies. The Apple Dimpling Bug, *Campylomma livida*, also feeds on mites and moth and butterfly eggs, including those of species in the genus *Heliothis*, but it can also cause damage to apple fruit and cotton leaves.

Damsel Bugs (family Nabidae)

Damsel bugs are slender, slow to active bugs with strong front legs that grasp prey. They feed on the eggs of moths and butterflies, and on small, soft-bodied insects such as aphids, caterpillar and grubs. They may also attack leafhoppers and mites. A common predatory species, *Nabis kinbergii,* feeds on a wide range of caterpillars and is an important controlling agent of *Heliothis* caterpillars. It is a fast-moving, slender, dull grey to pale brown bug about 0.8cm long with long antennae and prominent eyes.

Shield Bugs (family Pentatomidae)

Predatory shield bug feeding on beetle larvae. [D. Jones]

Shield bugs feed on a range of insects. [T. Wood]

These are moderately large, shield-shaped bugs that feed on caterpillars. The Glossy Shield Bug, *Cermatulus nasalis,* is a common species that often targets the boldly-striped soft caterpillars of the Wanderer Butterfly. It is a shiny brown bug which is 1.5–2cm long with a black patch on each wing-cover. The bright red and black nymphs look more like small spiders than bugs. Adults and nymphs of the Spined Shield Bug, *Oechalia schellenbergii*, vigorously attack the grubs of tortoise beetles (*Paropsis* species), spearing them with their proboscis and sucking their juices. Incredibly they can walk about while the grub is impaled on the proboscis. The mottled brown adults, which are about 1.5cm long, have a pair of prominent thorn-like spines on the thorax. Their nymphs are also red and black.

Pirate Bugs (family Anthocoridae)

These are tiny to small bugs which are sometimes called flower bugs. Both the adults and larvae of the native bug *Orius armatus* suck the juices from species of thrips and other small soft-bodied pests. This species has been found to be effective at controlling Western Flower Thrips, *Frankliniella occidentalis*, during trials on cut flower crops in Western Australia. This bug, the adults of which can fly, also eats aphids, spider mites and insect eggs. It is available from biocontrol firms.

PREDATORY CATERPILLARS (ORDER LEPIDOPTERA)

The caterpillars of a few species of moths are known to be predacious on other insects, including pests. Perhaps the best known of these is the scale-eating caterpillar, *Catoblemma dubia*, in the family Noctuidae. This rather voracious species relishes pink wax scale, black scale and others, often decorating its body with their waxy coverings. The larvae of the phycitid moth, *Stathmopoda melanochra*, feed on the common and persistent gumtree scale, *Eriococcus coriaceus*. The activities of this predator have been encouraged in eucalypt plantations.

DRAGONFLIES AND DAMSELFLIES (ORDER ODONATA)

There are about 320 native species of these large, familiar flying insects that can be recognised by their narrow body, two pairs of gauzy wings and bulging eyes. They can fly very fast, are highly manoueverable in the air and catch their prey on the wing using their basket-like leg arrangement to grasp the prey, eating it with their powerful jaws. Their prey consists of a range of other flying insects, including gnats, mosquitoes, flies, bugs, bees, moths and butterflies.

Dragonflies feed on insects that they catch on the wing. [T. Wood]

Dragonfly. [T. Wood]

EARWIGS (ORDER DERMAPTERA)

Australia has about 100 species of native earwig, most of which are innocuous or even beneficial to gardeners. Some native earwigs feed on plant parts, some feed on algae and lichens, some feed on decaying plant and animal matter and others eat insect eggs and soft-bodied insects such as aphids. Some species also eat plant organs. The common native brown earwig, *Labidura riparia*, of subtropical and inland areas lives in soil and eats wireworms and cutworms.

A common species of native earwig. [D. Jones]

An unusual native earwig. [D. Jones]

PREDATORY/PARASITIC FLIES (ORDER DIPTERA)

The larvae of bee flies such as Comptosia lateralis *feed on the eggs and larvae of other insects.* [T. Wood]

Flies, which belong to the insect order Diptera, have a single pair of wings, the other pair being replaced by small structures known as halteres. Fly larvae, usually called maggots, are legless and often also blind. Flies generally have a bad repution as nuisance insects, however many species are useful predators and parasites. In some species both the adults and larvae are beneficial.

Bee Flies (Family Bombyliidae)

Adult bee flies feed on nectar and pollen but their larvae feed on the eggs and larvae of other insects. The adults can often be recognised by the unusual manner in which their wings sweep back when at rest.

Hoverflies (Family Syrphidae)

The adults of these insects hover, almost motionless, over plants and flowers, with the wings vibrating so fast they appear as an indistinct blur. Frequently they will dart rapidly away and resume their hovering elsewhere. Some strongly banded colourful species mimic bees or wasps. Hoverflies often gather around flowers, feeding on pollen and nectar; also acting as pollinators. The larvae of many species are predacious. One common species has green larvae with a pale stripe down the centreline of the back. The larvae, which can grow to about 1cm long, are blind but readily find their prey. Eggs, laid among colonies of aphids, thrips, scales, lerps or mites, hatch into legless larvae that resemble a small maggot or slug. These spear the soft-bodied pest with their pointed mandibles and suck it dry, sometimes holding the body aloft in the process. Planting free-flowering annuals is a useful way to attract hoverflies into a garden.

Hoverflies feed on nectar. [T. Wood]

Hoverfly larva. [D. Jones]

A hoverfly. [T. Wood]

Larva of a hoverfly feeding on lerps.
[D. Jones]

Syrphid fly larvae feeding on Cottony Cushion Scale. [D. Beardsell]

Robber fly feeding on a beetle. [T. Wood]   *Robber flies feed on insects that they catch while flying.* [T. Wood]

Robber Flies (Family Asilidae)

These are ugly-looking predators which fly rapidly, capturing other insects on the wing. Their prey is injected with neurotoxins and proteins that liquefy the body contents. These voracious flies are often seen sitting on a convenient perch, sucking the juices out of their victim. Robber flies, which are most active during warm weather, are common garden residents. They often produce a buzzing noise when flying. The larvae of some species are also useful predators.

Tachinid Flies (Family Tachinidae)

These are small to large, active flies covered in prominent bristles. The larger ones are stout insects with prominent gauzy wings (wingspan 2–3cm across). The adult flies feed on nectar and their maggots are internal parasites of other invertebrates, including caterpillars, beetle larvae, sawflies, bug nymphs and spiders. There are about 500 species of tachinid fly in Australia. The females either lay their eggs directly onto the skin of the host or they deposit live larvae onto the host. Another group of tachinid flies lay eggs onto the host's food plant and the host either eats the eggs which hatch internally or hatched maggots attach themselves to the prey as the insect is moving about on the plant. Whatever the method, the maggots tunnel into the internal tissues of the host, feeding while the host remains alive, eventually killing it.

Tachinid flies, such as this species of Rutilia, *parasitise curl grubs.* [T. Wood]

Some tachinid flies are common parasites of large caterpillars and frequently the white egg(s) can be seen, often placed near the caterpillar's head. Other species target beetles and bugs. An important group of tachinid flies in the genus *Rutilia* parasitise the late-stage larvae of scarab beetles that are known commonly as white grubs or curl grubs. One tachinid fly, *Locustivora pachytyli*, is a useful parasite of the Australian Plague Locust. Another species, *Myothria fergusoni*, attacks matchstick grasshoppers in the subfamily Morabinae. Two tachinid flies are successful parasites of the Light Brown Apple Moth. An exotic species, *Trichopoda giacomellii*, has successfully controlled the Green Vegetable Bug in Australia. Similarly a native tachinid fly, *Cryptochaetum iceryae*, has proved successful in control of the pestiferous native Cottony Cushion Scale in some other countries.

Long-legged flies eat small soft-bodied insects such as aphids. [T. Wood]

Long-legged Flies (Family Dolichopodidae)

These small brightly coloured (often green or metallic brown) flies have long slender legs. The larvae, which are often found in soil, among litter or under bark, have no legs. The adult flies eat small, soft-bodied insects including aphids. The larvae of some species are also predators.

PRAYING MANTIDS (ORDER MANTODEA)

These familiar insects are often seen sitting still on plants or flowers in an apparent praying attitude while awaiting an insect or some other food item to come in range. They are masters of camouflage with the added ability to remain absolutely still for long periods. When the prey item is sufficiently close, the mantis pounces, grabbing the prey in its specialised front legs which are adorned with sharp grasping spines. These hold the prey while it is being devoured with sharp mandibles. There are about 200 species of praying mantids native to Australia and they range in size from relatively small to large insects. Some are green, others grey to brown. A specialised group of small grey to black mantids live on bark and tree trunks. The young of some species mimic ants.

Adult praying mantis. [T. Wood]

Praying mantis. [T. Wood]

Green praying mantis. [T. Wood]

Egg case of brown praying mantis. [T. Wood]

Juvenile praying mantis. [T. Wood]

PARASITOID WASPS (ORDER HYMENOPTERA)

A colourful ichneumon wasp. [T. Wood]

Wasps are a huge and common group of insects ranging from tiny species which are hardly visible to very large and colourful species such as the hornets. More than 12,000 species of wasps are native to Australia. Many of the larger wasps, including paper wasps and hornets, are predators of cicadas, caterpillars, grubs and spiders, frequently stinging and gathering these still-living invertebrates in nests as a food source for their larvae. Numerous species of small wasps are parasitoids that lay their eggs inside or on the skin of a

Caterpillar with a group of wasp eggs laid on its back. [T. Wood]

Ichneumon wasp. [T. Wood]

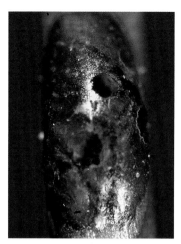

Parasitised scale. [J. Fanning]

range of insects, including caterpillars, borer larvae, thrips, scales, whitefly and mites. The eggs hatch into wasp larvae which steadily consume the body of the host over time until mature, eventually causing its death. The parasitoids then pupate in or near the body of the host, sometimes producing a mass of white woolly cocoons. Parasitoid wasps, which include braconid, chalcid and ichneumon wasps, are of tremendous importance in the control of many pests. Some parasitoids are available from biocontrol firms. These include Trichogramma pretiosum, an egg parasitoid which is used to control caterpillars including Heliothis and some loopers; also Trichogramma carverae, an egg parasitoid of leafrollers; a species of Metaphycus

Ichneumon wasp parasitising beetle larvae. [T. Wood]

which is an egg parasitoid of some scale insects including soft brown scale and black scale; adults of the red scale, oriental scale and oleander scale are killed by the larvae of the parasitoid wasps Aphytis melinus and A. linganensis; also Eretmocerus warrae and Encarsia formosa, both of which are parasitoids of greenhouse whitefly; Aphidius colemani is a parasitoid of green peach aphid.

OTHER WASPS

Paper wasps are social insects that build nests in garden plants and under eaves and other sheltered places around buildings. The nests are made out of grey papery material with special cells on the underside in which the wasp larvae are raised. The larvae are fed on caterpillars collected by the adult wasps. Paper wasps are useful in a garden but can sting painfully if the nest is disturbed. Vespid wasps also feed caterpillars to their larvae, which are raised in subterranean nests.

Paper wasps and nest. [T. Wood]

Vespid wasp. [T. Wood]

A different type of paper wasp nest. [T. Wood]

OTHER BENEFICIAL ANIMALS

Birds

Birds are the major predators of most plant pests and their activities in a park or garden are to be encouraged. Small birds that travel in groups or flocks, such as Silvereyes and thornbills, can have a significant impact on colonies of pests such as scales, aphids and lerps in a short period of feeding. Pardalotes are specialist lerp eaters. Many honeyeaters feed on wide range of insects as well as the nectar from flowers. Cuckooshrikes and thrushes feed on many caterpillars and grubs, including sawfly larvae. Currawongs also eat a range of large insects, often searching under bark for food. Ground foraging birds, such as magpies and ibises specialise on root-feeding caterpillars such as armyworms and curl grubs. Magpies and shrike-thrushes consume numerous caterpillars and grubs, including cutworms. Black cockatoos eat the larvae of wood-boring beetles and moths, in the process sometimes causing more damage than the borers themselves. Birds can be attracted to a garden by growing suitable plants and providing permanent water points.

Black-faced Cuckooshrike with beetle. [M. Sutcliffe]

The Spotted Pardalote feeds on lerps and scale insects. [M. Sutcliffe]

The Yellow Robin helps control pests in gardens. [M. Sutcliffe]

Brown Thornbill eyeing off large grub. [M. Sutcliffe]

Brown Thornbill with large grub. [M. Sutcliffe]

Golden Whistler with large fly. [M. Sutcliffe]

Red-capped Robin with caterpillar. [M. Sutcliffe]

Grey Shrike-thrush eyeing off moth. [M. Sutcliffe]

Grey Butcherbird with curl grub. [M. Sutcliffe]

Centipedes

These are distinctive, multi-legged creatures which are mainly nocturnal. They are fast moving and feed on a range of insects and other small animals including spiders and earthworms. Ground-dwelling caterpillars such as cutworms are also part of their diet. The first pair of a centipede's legs are modified into claws which are used for holding prey and injecting poison. Centipedes normally shelter under rocks, logs or in mulch. They should not be handled as their bites are quite painful.

Centipede. [D. Jones]

Frogs

Frogs add extra interest to a garden as well as being beneficial by consuming numerous pests such as earwigs, slaters, slugs and cutworms. There are many different species of frog but their survival in a garden is largely dependent on the presence of permanent water as well as sheltered sites where they can live. Ensure that frog ponds do not include goldfish or mosquito fish, which will predate on tadpoles and frogs' eggs.

Frogs, such as this Green Tree Frog, feed on a wide range of insects. [D. Jones]

Jacky Lizard waiting to ambush passing insects. [D. Jones]   *Jacky Lizard with a large grasshopper.* [D. Jones]

Lizards

Lizards are predators that are active in hot weather when most insect pests are about. They may live on the ground or in trees and are voracious consumers of insects and other small creatures, both by day and night. The commonly seen blue-tongued lizards, *Tiliqua scincoides* and *Tiliqua occipitalis*, love to feast on introduced garden snails. Small dragons and skinks will eat a wide variety of diurnal insects, while geckos will hunt nocturnally active insects. Lizards are a very interesting and important part of the garden fauna and can be attracted by including cover such as rocks and logs. Species of dragon lizards are particularly fond of rockeries, while geckos like leaf litter, logs, and trees with loose bark under which they can hide. If rocks are not available, artificial cover in the form of roof tiles and PVC pipe can be provided. Lizards like sunny areas to bask in as well as plenty of ground cover to hide from predators such as birds.

Land Planarians (Class Turbellaria)

These worm-like invertebrates hide in moist humid areas during the day and emerge on moist or wet nights to feed. They are often very colourful, with bright hues of yellow, blue and red being known. Land planarians are very sticky to the touch and, like slugs and snails, move on a bed of slime. The slime is so sticky that prey items can become trapped on the slime trails. The planarians sometimes also use the slime to drown their victims. They feed on earthworms, slugs, slaters and millipedes, sucking out their body contents through their single orifice. They also feed on dead animals. There are about 300 species of land planarians in Australia, about half of which have been named. Some exotic species which have been accidentally introduced have become naturalised.

Caterpillar infested with predatory mites. [T. Wood]

Predatory Mites

A number of plant-feeding mites are major pests of plants, causing severe economic damage. Antagonistic to them are ferocious predator mites which can, by their feeding, quickly reduce the numbers of the pest to a tolerable level. The introduced predator mite, *Typhlodromus occidentalis,* is active against two-spotted mites. Two-spotted mites are also eaten by three other predator mites, *Amblyseius womerslyi*, *Neoseiulus californicus* and *Phytoseiulus persimilis*. The latter species also eats Two-spotted Mites. Two other predatory mites feed on thrips. These are *Typhlodromips montdorensis* and *Neoseiulus cucumeris*. The fungus gnat mite or hypoaspis mite, *Stratiolaelaps scimitus*, feeds on the larvae of fungus gnats, spider mites, flower thrips and mealybug crawlers. All of these predatory mites are available from commercial biocontrol firms. The adults and nymphs of velvet mites are also predators on other mites and soft-bodied insects such as caterpillars. They are relatively large for a mite (3–4mm long), usually bright red or brown, and appear velvety from a covering of short hairs. They are often seen when soil litter is disturbed.

Entomopathogenic Nematodes

There are some nematodes (eelworms) that invade the bodies of insects, eventually causing their death. They do not actually kill the insect themselves; that task is carried out by a bacterium that lives symbiotically within the eelworm. On its release from the eelworm, this bacterium enters the insect's bloodstream and rapidly kills the insect. One species of entomopathogenic nematode is available for purchase from biocontrol firms. It has been used against fungus gnat larvae, African black beetles and other scarabs; also vine weevils.

Spiders

Although a number of spiders must be treated with caution, most species are completely harmless to humans and provide a valuable service by reducing insect numbers. Hunting spiders, such as the huntsman spiders and wolf spiders, feed on a range of prey that they either ambush or run down. The tiny jumping spiders and crab spiders attack many small insects including some pests. Web spinners, which are very common garden residents, trap small to large flying insects.

Jumping spider and prey. [T. Wood]

Leaf-curling spider. [T. Wood]

Spider and oversize prey. [T. WOOD]

Spider awaiting prey on a bearded orchid. [T. WOOD]

Green crab spider. [T. WOOD]

Huntsman spider. [T. WOOD]

Huntsman spider with prey. [T. WOOD]

Orb-weaving spider. [T. WOOD]

OTHER CONTROLLING ORGANISMS

Bacteria

Several species of bacteria are antagonistic to insects and some have been tested or are available for use against pests. One of these, *Bacillus thuringiensis* (popularly known as Bt), is used as a biological pesticide. It acts specifically against the caterpillars of moths and butterflies and hence is safe for other forms of life. When, in the course of its normal feeding, the caterpillar ingests a large number of live spores of the bacterium (40,000 to 50,000 is a basic figure), it is killed by toxic protein crystals which the *Bacillus* produces during spore production. These proteins paralyse the digestive system and stop feeding. Commercial sprays containing spores of Bt are available. Because they are degraded by sunlight, the sprays are best applied in late afternoon or early evening. A specific strain, *Bacillus thuringiensis israelensis* (known as Bti), is used in some parts of Australia to help control mosquitoes (it may also have potential against fungus gnats when applied as a soil drench). The spores of another bacterium (*Bacillus popilliae*) are used in the USA to control larvae of the Japanese black beetle. A symbiotic bacterium that lives in some nematodes kills insects after invasion by its host (see entry for Entomopathogenic Nematodes above).

Fungi

Certain fungi are known to be pathogenic to insects (termed entomophagous or entomopathogenic). These fungi are often quite specific, not causing any plant damage and only infecting particular groups of insects. Probably the best known examples are found in the various species of *Cordyceps* which are commonly known as vegetable caterpillars. Ground-dwelling grubs (including many pest species) ingest spores of this fungus while feeding. The fungus germinates and invades the body of the host causing mummification. Eventually a characteristic fruiting head is produced to disperse the spores. Most pathogenic fungi need moisture to develop and under suitable conditions can penetrate an insect's skin. Once inside, the fungus spreads rapidly through the body, killing the insect and eventually producing fungal spores that can be spread by wind or water. Often the affected insect becomes covered with fungal threads. Some pathogenic fungi can spread quickly through colonies of insects. A small number are

Caterpillar killed by parasitic fungus. [D. JONES]

being investigated as possible control agents to use against pest insects (referred to as mycoinsecticides). One fungus, *Verticillium lecanii*, is used in European greenhouses to control sucking pests such as aphids, thrips and whiteflies. Another fungus, *Neozygites floridana*, has been shown to control spider mites. A commonly occurring soil fungus, *Beauvaria bassiana*, has proved to be effective against a wide range of pests, including whiteflies, aphids, grasshoppers, codling moth, mirid bugs and the larvae of many beetles. Another widespread soil fungus, *Nomuraea rileyi*, is being tested in some countries for its effectiveness against caterpillars of the moth family Noctuidae, including cutworms, loopers and *Heliothis*. The red-headed fungus *Fusarium coccophilum* is known to envelope and destroy whole colonies of sucking insects such as aphids and scale. The fungus *Metarhizium anisopliae* is known to infect sucking bugs, including mirid bugs and the green vegetable bug.

Emperor Gum Moth caterpillar killed by parasitic virus. [D. JONES]

Viruses

Some viruses are known to be antagonistic to insects if ingested. Caterpillars of the Emperor Gum Moth are frequently killed by a virus. They become slow in their movements, lose condition and die. Their body contents liquefy after death, often leaking from the body, and the emaciated remains are commonly seen hanging from eucalypt stems. The leaked fluids contain virus particles. A few viruses have been used as pest control agents (termed virus pesticides). Because they are easily destroyed by sunlight they must be applied in late afternoon or early evening. Baculoviruses, which have extremely tiny particles, are pathogenic to many types of insect. Many have a very limited host range and are safe in the environment. One specific baculovirus is used in Europe as a biological insecticide to control a damaging cutworm that feeds on a wide range of crops. In the USA sprays containing virus spores have been found to be effective in controlling cabbage looper and two species of pine sawfly. The USDA Forest Service sprays a specific baculovirus from airccraft to control a serious pest known as the gypsy moth, *Lymantria dispar*. Others have been used to control codling moth in apple crops, armyworms in beetroot, budworms in tobacco and bollworm in cotton.

CHAPTER 4

INSECTS THAT SUCK SAP – 1

(APHIDS, SCALE INSECTS, MEALYBUGS, WHITEFLIES)

Insects that suck sap have modified mouthparts allowing them to pierce plant tissue and drink the sap. The hair-like mouthparts, which are like a hollow needle (called stylets), are pushed into the plant tissue. When the stylets pierce the veins of the plant, the sap, which is under pressure (turgor pressure), actually flows through the insect's mouthparts into its digestive system. Thus the insect actually drinks the sap rather than having to exert suction in a sucking process. Because of the plant's turgor pressure the sap passes fairly quickly through the insect and the digestive system removes any necessary nutrients. Excess fluid is shed as droplets from the rear end. Insects with mouthparts specialised for this type of feeding include aphids, bugs, cicadas, cuckoo spits, eriococcids, froghoppers, jassids, leafhoppers, lerp insects, mealybugs, psyllids, root coccoids, thrips, soft scales and whitefly.

Aphids, scale insects, mealybugs and whiteflies are placed in the suborder Sternorrhyncha of the order Hemiptera. A relatively recent reclassification following molecular studies in 1995 reduced the old group of 'Homoptera' to two suborders: Sternorrhyncha (aphids, mealybugs, psyllids, scale insects, whiteflies) and Auchenorrhyncha (cicadas, leafhoppers, treehoppers, planthoppers). A third suborder (Hemiptera) contains the true bugs.

Honeydew and Sooty Mould: the excess sugary fluid passed out by sap sucking or sap drinking insects accumulates on the plants in the areas where they are feeding. Known as honeydew, this sticky material provides food for bees, wasps and ants, and is also the perfect growing medium for the fungus known as sooty mould. This fungus grows rapidly on the sugary exudates, eventually covering affected areas with a black sooty growth that is not only unsightly, but can also impede photosynthesis and impact on pollination and the appearance and sale of fruit. The presence of sooty mould is a good indication that sucking insects are active and it is often surprising to see how much honeydew and sooty mould can be generated from what appears to be a relatively small infestation of sucking pests. For more details on sooty mould see chapter 19.

The Role of Ants: some groups of sucking pests, especially colonising types such as aphids, froghoppers, jassids, leafhoppers, lerp insects, mealybugs and soft scales, attract the attention of ants. The ants collect the sugary honeydew as food, protect the sucking insects from predators and may even move them to new areas on the plant or construct 'byres' to shelter them. The ants closely attend

Ants construct a covering (byre) over colonies of sap-sucking insects. [D. JONES]

healthy colonies of the insects and may even induce the release of honeydew by tapping the insect with their antennae. They also vigorously defend the colonies of sucking insects when threatened. The presence of ants, especially when they occur in small active groups, is a sure indication that sucking insects are present. Ants that attend sucking insects range from small species to quite large types. The widespread meat ant, *Iridomyrmex purpureus*, of southern areas is commonly found in shrubs and trees in association with many types of sucking insect.

SYMPTOMS AND RECOGNITION FEATURES

Messy patches of waxy secretions and sooty mould around leaf bases, leaf-sheaths, bracts, etc............... *mealybugs*

Tough white cottony sac-like structures on stems ....*mealybugs*

Shiny, messy patches of waxy secretions and sooty mould under leaves, leaf margins often curled downwards ....*whiteflies*

Shiny secretions on young tissue, distortion of shoots and curling of leaves; colonies of small fleshy insects with obvious legs .... *aphids*

Shiny secretions on young stems and the underside of leaves, immobile, round or scale-like creatures with a marginal fringe of waxy filaments, small white-winged adults often present .... *whiteflies*

White fluffy or waxy secretions present among roots..*root coccoids (chapter 14)*

Small soft-bodied insect with prominent eyes, antennae and two tubes protruding from near the end of the back, winged adults often present in colonies..*aphids*

Shiny secretions on stems and older leaves, immobile insects covered with a lid-like structure.........................*scales*

Immobile insects covered with a soft waxy structure...*soft scales*

Immobile insects covered with a hard structure which has a prominent darker spot towards one end........................ ....*armoured scales*

Immobile scale-like insects with a prominent white cottony sac.............*cottony cushion scale*, *pulvinaria scales*

Immobile insects covered with a leathery structure resembling rice bubbles or ticks.....*gum tree scale, tick scale*

Small, soft white or grey slowly mobile insects covered with waxy filaments and mealy powder.............*mealybugs*

Immobile, round or scale-like creatures with a marginal fringe of waxy filaments, small white-winged adults often present ....*whiteflies*

APHIDS

Aphids, also called plant lice, greenfly or blackfly, are placed in the order Hemiptera, suborder Sternorrhyncha, superfamily Aphidoidea. There are about 4,400 species of aphid recognised world wide, placed into about 10 families. Aphids are small soft-bodied insects 0.1–0.2cm long that have prominent dark eyes, antennae and long legs. They also have a pair of tubular structures (cornicles) at the rear end. Aphids commonly congregate in colonies on young growth, particularly fast-growing plant shoots, but also on emerging inflorescences and developing buds. A few species are also known to feed on roots (chapter 14). Most aphids live in colonies but some species occur as sparsely scattered individuals. Colour, which is often indicative of the species, includes grey, black, green, yellow, orange and pink. There are 13 specialised species of aphid that are native to Australia and another 150 or so species that have been introduced (nearly 80 from Europe, four from USA and the rest from Asia). Some aphids are specific to a particular group of plants, whereas others have a wide host range. Some significant pest species of aphids are often found colonising common weeds, including thistles and swan plant. The introduced aphids feed mainly on a range of commercial agricultural and horticultural crops, although some can also damage exotic and native garden plants. Plant collections grown in greenhouses, shadehouses and ferneries are also susceptible.

Life cycle: aphids generally live in mixed feeding colonies of adults and nymphs. The adults can be winged or wingless, depending on the species, time of year and stage of life cycle. Species from temperate climates have males and females that mate normally and produce eggs which can survive climatic extremes (such as winter). These eggs hatch in spring to produce wingless females that produce live young by an asexual process known as parthenogenesis (pregnancy without mating). As many as five young can be produced by an adult female each day. The young aphids, which can already be pregnant when born, feed and produce more asexual progeny, facilitating a rapid build up in numbers. Male aphids are absent and play no part in this type of reproduction. Species from tropical areas often do not produce eggs. At certain times of the year adult females develop wings to aid dispersal. Some species of aphid have the remarkable ability to change their sex depending on the prevailing temperature and the type of plant on which they are feeding.

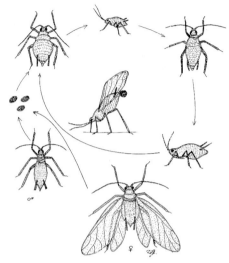

Life cycle of aphids. [D. Jones]

Aphids excrete considerable amounts of sweet sugary honeydew that is relished by ants. Ants closely attend aphid colonies, eating and carrying away the sugary exudates and protecting the colonies from predators. They may even relocate the aphids to new areas on which to feed. In some instances the ants may even construct a protective sheath of debris (known as a byre) to cover feeding aphid colonies.

Plant damage: wilting of soft new growth is a typical symptom of aphid attack. Other symptoms of aphid damage include distortion of shoots and buds, leaves that expand unevenly or are deformed, yellow or brown patches in leaves, and premature fall of leaves or buds. Some species of aphid cause the leaves to curl and the insects then feed within this protected area, others cause the formation of galls. A few species inject toxic saliva that causes browning and death of plant tissue. Attacks on developing inflorescences in native lilies and orchids can result in deformed spikes and premature shedding of buds. Developing fern crosiers are very susceptible to aphid attack, often resulting in incompletely developed or distorted fronds. Extensive attacks by aphids can be severely debilitating to plants because of the large volumes of sap the insects remove (they waste nearly as much as they use). In addition some aphids are vectors of plant virus diseases which can cause all sorts of growth problems (for more details on plant viruses see chapter 19). Sooty Mould invariably follows persistent aphid attacks, literally covering branches and leaves with a coating of black.

Natural control: aphids thrive in suitable climatic conditions and are especially noticeable in spring and autumn. Numbers increase quickly in favourable conditions but can decline equally rapidly following frost, heavy rain and hot windy weather. They are also attacked by a wide range of predators. Many types of ladybird, both adults and larvae, devour aphids with relish and probably provide the most effective natural control of aphid populations. Aphids are also avidly eaten by the larvae of lacewings and syrphid flies which suck the victims dry. The larvae of predatory wasps eat aphids from the inside leaving dried aphid bodies (mummies) as a memento of their work. Small birds such as thornbills, Silvereyes and pardalotes find them tasty but rarely eliminate them from a plant. Lizards, spiders, predatory beetles, bugs and mites also feed on aphids. Although many predators include aphids on their menu, garden infestations are not always controlled by these activities. Entomogenous fungi also kill aphids. Infestations in greenhouses, shadehouses and ferneries can be especially difficult to control by natural means.

Other control: aphids should be controlled as soon as noticed. Small infestations can be squashed between the fingers or hosed with jets of water. Soap solutions applied under pressure may also provide some control. Introducing predators such as ladybirds and lacewings early in the season can effectively control their build-up (predators are available from commercial biocontrol companies). Large or persistent infestations may need to be controlled by spraying with a contact insecticide such as malathion, imidacloprid or a pyrethrum derivative. Excellent control in glasshouses has been achieved by using pest strips that release insecticidal vapours into the atmosphere, but such methods also kill useful insects including predators.

Examples

No attempt has been made to group these insects into families. Formal identification of these pests should be carried out by experts.

Black Citrus Aphid, *Toxoptera aurantii*

Description: adults, which can be winged or wingless, are oval in shape and black to brownish black, sometimes shiny. Nymphs are dark red-brown. Eggs are not produced.
Climatic regions: temperate, subtropical and tropical regions, coastal and inland.

Black Citrus Aphids. [T. Wood]

Host plants: attacks more than 190 genera in 80 families. Crops include avocado, citrus, coffee, custard apple, lychee and mangos. Native garden plants include species of *Billardiera, Dodonaea, Ficus, Harpullia, Hibiscus, Leptospermum, Persoonia, Pittosporum* and *Platysace.* Black Citrus Aphid also attacks young shoots and flower buds of macadamias causing bud-drop, wilting and dieback. It commonly becomes established in collections of ferns and orchids.
Feeding habits: colonies of mixed stages feed on young shoots and can cause distortion and leaf curling.
Notes: another similar species, *T. citricidis,* also confusingly known as the Black Citrus Aphid, is a much more serious pest of citrus, including the native species. It can aggregate into very large dense colonies on young shoots and flower buds. It also attacks members of the family Rutaceae, including *Acronychia, Euodia, Flindersia, Geijera* and *Philotheca.*
Control: as per introduction for aphids.

Cotton Aphid, *Aphis gossypii*

Description: adults, 0.1–0.2cm long, which are usually winged, are oval to pear-shaped and green to creamy yellow or white mottled with dark green. Nymphs are paler.
Climatic regions: temperate, tropics and subtropics.
Host plants: this pest feeds on hundreds of different plants in many plant families. Commercial crops include bananas, citrus, cotton, cucurbits, capsicums and asparagus. Ornamentals include species of *Arthropodium, Abutilon, Dianella, Dichopogon, Eremophila, Grevillea, Goodenia, Gossypium, Hibiscus, Lavatera, Myoporum* and *Thelionema*; also grasses. Attacks occur on nursery plants and garden plants.
Feeding habits: colonies of mixed stages feed on shoot tips and the underside of leaves. Affected leaves curl, become yellow and fall prematurely.

Notes: this pest is widely distributed throughout the tropics. Infestations can cause severe damage. This aphid is a vector of some plant virus diseases.

Control: as per introduction for aphids.

Cowpea Aphid or Black Legume Aphid, *Aphis craccivora*

Description: adults, 0.1–0.2cm long, which can be winged or wingless, are oval to pear-shaped, greyish black to black and shiny. Very young nymphs are whitish then become grey, but dull as if dusted with wax. Winged adults appear as the colonies become dense.

Climatic regions: tropics and subtropics.

Host plants: feeds on *Clerodendrum, Glycine, Grevillea, Hibiscus* and many different legumes. Attacks occur in nurseries and gardens. Also damages commercial peanut crops.

Feeding habits: colonies of mixed stages feed on soft growth and developing buds. Affected growth becomes distorted. Buds fall prematurely.

Cotton Aphids on Thelionema. [D. Jones]

Notes: this pest, which can congregate in very dense colonies, is distributed mainly in the tropics. Infestations can cause severe damage. This aphid is a vector of some important virus diseases that infect legumes.

Control: as per introduction for aphids.

Fig Aphid, *Greenidea ficicola*

Description: adults, which can be winged or wingless, are slender to pear-shaped and yellowish brown to dark brown with long abdominal tubes. Winged adults have a dark abdomen. Nymphs are paler.

Climatic regions: tropics and subtropics.

Host plants: a wide range of *Ficus* species, including the commercial fig, *F. carica*, and commonly grown ornamentals such as *F. benjamina*. Attacks occur on nursery plants and garden plants.

Feeding habits: colonies of mixed stages feed on young shoots and leaves, resulting in sticky white exudates.

Notes: this pest, which is widely distributed throughout the tropics, is significant where figs are planted as ornamentals. Infestations are often sporadic and controlled by natural enemies, but damaging attacks can occur in nurseries.

Control: as per introduction for aphids.

Green Peach Aphid, *Myzus persicae*

Description: adults, about 0.2cm long, which can be winged or wingless, are bright green to yellowish green with a black head and thorax. Sometimes pinkish-winged adults occur in a colony. Nymphs are green to yellow-green.

Climatic regions: temperate and subtropical regions, coastal and inland.

Host plants: this aphid infests hundreds of types of plants in numerous plant families, including a wide range of commercial crops (including *Macadamia*), garden plants and weeds. Numerous native plants are attacked, including species and cultivars of *Ammobium, Anigozanthos, Atriplex, Blechnum, Brachyscome, Grevillea, Hymenosporum, Pteris, Rhodanthe* and *Xerochrysum*. In some seasons severe attacks occur on the new growth of *Hymenosporum flavum* in southern Victoria. Natural populations of native orchids (particularly species of *Acianthus, Caladenia, Diuris* and *Pterostylis*) and lilies (*Dichopogon* for example) are attacked, introducing the possibility of viruses spreading into wild populations. This aphid can be a persistent pest of orchids and ferns in greenhouses.

Feeding habits: colonies of mixed stages feed on tender leaves, young shoots and flowers causing growth distortion, leaf curling and flower malformation. Exudates of honeydew give affected tissue a wet, shiny appearance and eventually lead to the growth of sooty mould.

Notes: this is a widespread (worldwide) and persistent pest that can build up in numbers extremely quickly. In cold climates it overwinters as eggs. The common weed Sow Thistle, *Sonchus oleraceus*, is a favoured host. Green Peach Aphid is a significant vector of many plant virus diseases.

Control: as per introduction for aphids. This pest has developed resistance to many insecticides.

Colony of Green Peach Aphids. [T. Wood]

Green Peach Aphids feeding on an orchid. [J. Fanning]

Green Peach Aphids feeding on Hymenosporum flavum. [D. Beardsell]

Green Peach Aphids or Yellow Orchid Aphids feeding on Acianthus exsertus *in the wild.* [B. Hall]

Myrtle Aphid or Guava Aphid, *Greenidea psidii*

Description: adults, which can be winged or wingless, are slender to oval in shape and dark brown with long paler outcurved abdominal tubes.

Climatic regions: tropics and subtropics.

Host plants: a wide range of Myrtaceae species, including *Callistemon, Eucalyptus, Eugenia, Melaleuca, Metrosideros, Rhodomyrtus, Syzygium* and *Tristania*. Also feeds on *Ficus* species and the commercial guava, *Psidium guajava*. Attacks occur on nursery plants and garden plants.

Feeding habits: colonies of mixed stages feed on young shoots and the underside of leaves causing distortion and leaf curling.

Notes: this pest, which is common in many Asian countries, is spreading. Damaging attacks can occur on nursery plants and commercial guava crops.

Control: as per introduction for aphids.

Pine Aphid, *Cinara tujafilina*

Description: adults, which are pear-shaped and can grow 0.3–0.4cm long, are grey-brown, often with darker lines (sometimes wholly reddish). Winged adults occur late in the season.

Climatic regions: temperate and subtropical regions, mainly inland.

Host plants: feeds on conifers including several species of Native Pine (*Callitris columellaris, C. endlicheri, C. preissii* and *C. rhomboidea*).

Feeding habits: colonies of adults and nymphs feed on young shoots causing browning and dieback. They also produce copious quantities of sticky honeydew which attracts wasps, bees and ants. In severe infestations the ground and any objects under the trees can become sticky from the exudate.

Notes: a sporadic pest which can disfigure native pines.

Control: as per introduction for aphids. Usually controlled by natural enemies.

Root Aphids or Root Coccoids are treated in chapter 14

Other Aphids

- **Yellow Orchid Aphid**, *Macrosiphum luteum*, is an exotic species that has become established in orchid collections in Australia. It thrives in glasshouses and is an important vector of virus diseases. The nymphs are bright yellow with black eyes, legs and antennae. The adults can be wholly yellow or with dark markings on the back. Adults are winged and can disperse by flight. Recently similar aphids have been found on native orchids growing in the wild.

- **Maidenhair Fern Aphid**, *Idiopterus nephrelepidis*, attacks developing fronds of Maidenhair Ferns (*Adiantum* species and cultivars) causing them to blacken and die. It has been found in some plants grown in south-eastern Australia. In some other countries this aphid has also been recorded attacking bird's-nest ferns (*Asplenium* species) and Boston fern (*Nephrolepis* species and cultivars).

- **Two native podocarps** (*Podocarpus elatus* and *Sundacarpus amara*) are attacked by an unknown grey aphid that congregates on the underside of the leaves causing malformation and leaf death. Commonly known as the 'podocarp aphid', this species is most prevalent in the tropics.

Yellow Orchid Aphids on new shoot of Dendrobium speciosum.
[D. JONES]

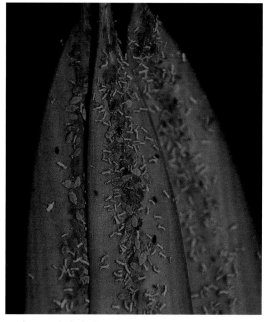

'Podocarp aphids' feeding on leaves of Sundacarpus amarus.
[D. JONES]

SCALE INSECTS AND MEALYBUGS

Native soft scale attended by Green Tree ants on Thespesia populnea. [D. JONES]

The scale insects, also known as coccoids, are highly adapted plant parasites placed in the order Hemiptera, suborder Sternorrhyncha, superfamily Coccoidea. Mealybugs also occur in the Coccoidea but are placed in the family Pseudococcidae. About 8,000 species of scale insect have been recognised worldwide. Scale insects are highly destructive plant pests that conceal themselves beneath a covering which can be waxy, leathery or cottony. They feed by sucking sap. A wide range of plants – exotic or native and including commercial crops such as cut flowers, vegetables, fruit and forestry trees – are impacted by various scale insects. The number of species of scale insects in Australia is uncertain but it would seem to be well in excess of 400 species. Those attacking commercial crops are well known, but there are also a number of species found on native plants that have not yet been studied or even named scientifically. As well as native scale insects there are many serious pestiferous scales that have been introduced from overseas.

Ants commonly attend scales to collect the waste products (honeydew) and generally encourage their activities. Not only do they protect scales from direct predation but they can also move individuals to new feeding areas. In some areas, ants frequently construct a byre (a protective covering of litter) over the colonies of scales which they attend.

Growth forms: several types of scale insects have been separated into families. Two groups can be readily recognised within commonly encountered scale insects – the soft scales (family Coccidae) and the armoured or hard scales (family Diaspididae). Soft scales secrete a waxy covering that is actually part of the insect's body. By contrast the scale that covers the body of an armoured scale is not part of the insect's body, although the scale insect shelters and feeds beneath this structure. Cast skins of nymphal stages (termed exuviae) moulded into the scale can be seen as a patch or dot of a different colour to the rest of the covering.

Life cycle: the actual insect under the scale is soft-bodied and attached to the plant by its needle-like mouthparts (stylets) through

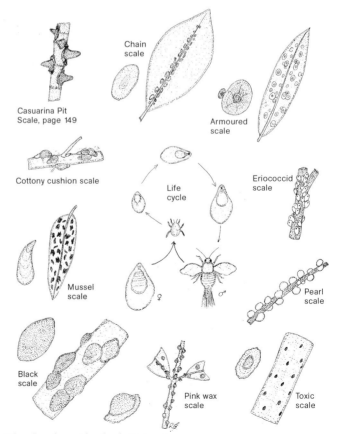

Life cycle and examples of scale. [D. JONES]

which it sucks sap. The adult females of most species of scale insect actually lose their legs, becoming immobile and remaining under the scale until they die (an exception is the Cottony Cushion Scale, which remains mobile). The adult male (if present) is a small, mobile, winged insect which is short-lived and does not feed. The coverings (scales) of female scale insects are usually larger than those of the male and sometimes of a different shape. Male insects are unknown for some species of scale insect, and the females then reproduce asexually by a process known as parthenogenesis. Eggs are either laid under the female scale, hatching normally, or they hatch within the female's body and the nymphs are born alive (termed ovovivipary). The young nymphs, known as crawlers, move actively about on the plant to seek a suitable feeding site. They can also be blown about by air currents and wind, or moved by ants, or on the feet of birds. Once attached to the plant they form a scale and then in most species become immobile. Juvenile scale insects moult up to seven times as they develop. The number of days for each developmental stage depends on environmental factors such as temperature, humidity and rainfall. Colonies of scales usually consist of insects in mixed stages of development.

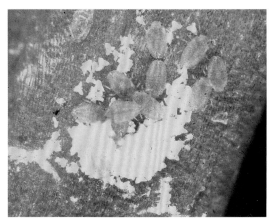

Crawlers of Soft Brown Scale. [J. FANNING]

Adult female Soft Brown Scale and crawlers. [J. FANNING]

Nymph of Soft Brown Scale. [J. FANNING]

Adult Soft Brown Scale on Cryptostylis. [J. FANNING]

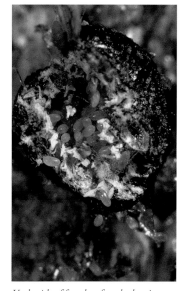

Underside of female soft scale showing eggs and crawler. [J. FANNING]

Plant damage: the damage caused by a single scale insect is minimal, however scale insects can reproduce rapidly in favourable conditions and large populations cause severe plant damage. Subtle impacts such as growth reduction or loss of vigour are not immediately obvious, however, growth distortion, leaf yellowing followed by premature defoliation, stem dieback, flower and fruit shedding are much more apparent. Scale insects generally cluster in colonies and plant damage can be serious. Some species have toxic saliva and cause damage out of all proportion to their size and numbers. Sooty mould usually grows on the sugary exudates (honeydew) secreted by most of these insects.

Natural control: being a group of such diverse insects, the populations of scales are controlled naturally by a wide range of predators and parasites. Small birds such as pardalotes, thornbills and Silvereyes feed on them. Possums such as the Sugar Glider are attracted to the colonies by the sweet exudates and in feeding destroy many of the insects. Predators include assassin bugs, ladybirds, lacewings, hoverflies and specialised scale-eating caterpillars. Thrips, mites and some entomopathogenic fungi may also play a role. Tiny wasps (brachonid, chalcid and ichneumon wasps) often provide effective control. Some of these are available from commercial biocontrol firms.

Bad infestation of Gum Tree Scale on Eucalyptus pyriformis. [D. JONES]

Soft scale damage on Myoporum parvifolium. [D. JONES]

Damage by Waratah Scale to Alloxylon flammeum. [D. JONES]

Damage to Alyxia buxifolia *caused by Chinese Wax Scale.* [D. JONES]

Other control: small or localised infestations can be removed by scraping or scrubbing. Swabbing or rubbing with isopropyl alcohol is also effective (note that other alcohols, including methylated spirits, can penetrate and injure plant tissue). Infested stems and branches can be removed and burnt. Larger infestations should be sprayed with applications of white oil or white oil combination sprays at about two-week intervals until the insects have all been killed. Sprays are most effective when applied at the crawler stage. Complete spray coverage is necessary. White oil may damage young soft growth, or stunt soft species such as ferns. In such cases it should be sprayed at about half to three quarters the recommended strength. Damage is also accentuated by hot weather and white oil should not be applied when the ambient temperature is above 25°C. Some horticultural oils which have added ingredients such as eucalyptus or tea tree oil may provide better control. Soap solutions applied under pressure can also be a useful control for scales. Combination sprays of white oil and pyrethrum or malathion may prove more effective (see combination sprays Chapter 2). Infestations of some scales, such as the wax scales, are persistent and very difficult to kill. Heavily infested trees should also be fertilised and well watered to help restore their vigour.

NOTE: To aid with identification the scale insects included here are grouped into families. Formal identification of these pests should be carried out by experts.

SOFT SCALES (FAMILY COCCIDAE)

This is a large group of scale insects consisting of more than 1,100 species distributed in most regions of the world and feeding on more than 160 genera of plants in over 200 families. They are most prominent on woody shrubs and trees, but also occur on herbaceous plants. In this group of scales the waxy covering (or scale) is attached to the body of the insect. The mouthparts (stylets) pierce the phloem and excess fluid is excreted as honeydew which attracts ants and provides a growing medium for sooty mould. Eggs are laid in a waxy sac or under the body of the female. Female soft scales have two or three immature instars and the males four.

Black Scale or Brown Olive Scale, *Saissetia oleae*

Description: a leathery oval scale 2–4mm long, with a prominent raised 'H' marking on the dorsal surface. Juveniles, produced in summer, are light brown and somewhat papery, adults are brown to black waxy domes. The females often produce eggs without mating, but sometimes males are present. Crawlers settle on the stems of young shoots and along the midrib on the underside of leaves.

Climatic regions: subtropical and temperate regions with high humidity.

Host plants: an economically important pest of olives, citrus, kiwifruit, grapes and *Cymbidium* orchids; also numerous ornamental plants including *Hibiscus*, *Hydrangea*, cycads, ferns and palms; affected native plants include *Avicennia marina*, *Myoporum* spp. *Olearia* spp., *Syzygium* spp., *Lagunaria patersonii*, *Lepidozamia peroffskyana*, *Pittosporum undulatum* and members of the family Rutaceae (*Boronia*, *Citrus*, *Correa*, *Crowea*, *Eriostemon*, *Phebalium* and *Philotheca*).

Feeding habits: a colonising scale that secretes copious quantities of honeydew resulting in a strong growth of sooty mould. Colonies have a very dirty appearance and are usually attended by ants.

Notes: an introduced scale, probably from South Africa, which is a vigorous and serious pest. It dislikes hot dry weather, preferring humid conditions.

Control: as per introduction for scales. Best control is achieved at the crawler stage.

Cottony Scales or Chain Scales, fourteen *Pulvinaria* species

Description: a group of colonising soft scales which develop a covering ranging from circular to elongate or mussel-shaped depending on the species. Immature scales have an almost transparent covering. Adult females become conspicuous with their white cottony sacs into which they lay their eggs. The nymphs of these scales retain their legs and are mobile throughout all their moults until they become adults. They can move about on the stems and leaves, changing their feeding positions.

Climatic regions: temperate, subtropics and tropics, coastal and inland.

Host plants: the hop bush scale or emu bush scale, *Pulvinaria dodonaeae*, attacks species of *Acacia*, *Dodonaea*, *Eremophila* and *Myoporum*; attacks by it on natural populations of *Eremophila margarethae*, from semi-arid areas of Western Australia, have resulted in significant mortality (some 10 per cent of the population or about 90 plants killed). This scale has also been found on *Eremophila forrestii*, *E. frasei* subsp. *galeata*, *E. mitchellii* and *E. spectabilis* subsp. *brevis*.

Hop Bush Scale, Pulvinaria dodoneae, *on* Eremophila.
[S. Jones]

Pulvinaria flavicans attacks species of *Acacia, Bossiaea, Dillwynia, Oxylobium* and *Templetonia*.

Pulvinaria psidi attacks *Alstonia scholaris, Boronia serrulata, Callistemon* sp., *Dodonaea triquetra*. and *Ficus* sp.

Pulvinaria urbicola is a widespread species that has the potential to become a serious pest in Australia. It has been found in Queensland and the Northern Territory on *Adiantum* sp., and several exotic garden plants.

The Saltbush Scale, *P. maskellii*, and Ice Plant Scale, *P. mesembryanthemi*, are treated separately.

Feeding habits: one common name of these scales arises because of the habit of the nymphal stages lining themselves up end to end while feeding, usually along the midrib of leaves. These scales are often attended by ants. Sooty mould is usually bad following attacks of these scales.

Notes: these scales include several native species, some of which can be difficult to control.

Control: as per introduction for scales. Predators include ladybirds and lacewings.

Chinese Wax Scale, *Ceroplastes sinensis*

Description: a whitish or grey-brown waxy scale 0.3–0.5cm long with a more-or-less domed shape and six sunken dark spots (often brown with a white waxy insert) around the margins. Instars have a starry appearance due to conspicuous marginal projections of white waxy material. Second stage instars are wholly white, third stage instars are pink. Adult males are unknown. In temperate areas this scale, which overwinters as juvenile and adult females, produces a single generation per year.

Climatic regions: subtropics and temperate regions, mainly coastal.

Host plants: an economically important pest of citrus, feijoas and the mangrove *Avicennia marina*; also ornamental plants including species of *Achronychia, Banksia, Bursaria, Callistemon, Dodonaea, Melaleuca, Pittosporum, Solanum* and *Syzygium*.

Feeding habits: the young instars feed on the leaves (often on the midribs) before moving onto the stems where they develop into adults, often congregating in clusters and in rows. They exude large quantities of honeydew.

Notes: an introduced scale from Central America or South America. It can be a persistent garden pest and is capable of rapid increase and spread. Although of waxy appearance, the scale covering is hard and difficult to squash.

Control: as per introduction for scales. Spraying is most effective against the instars.

Chinese Wax Scale on Dodonaea. *[D. Jones]*

Chinese Wax Scale on Banksia spinulosa. [D. JONES]

Flat Brown Scale or Tessellated Scale, *Eucalymnatus tessellatus*

Description: females form an oval or pear-shaped (often asymmetrical) very flat light brown to reddish-brown or dark brown wrinkled scale 0.2–0.5cm long. This species reproduces asexually, the females giving birth to live young (the eggs hatch within the body of the female). Males are unknown. Young crawlers are purplish-brown. One or two generations are produced during a year (more in greenhouses). The surface of young scales is noticeably wrinkled, especially on the margins.

Climatic regions: tropics, subtropics and temperate regions, mainly coastal.

Host plants: a significant pest of palms (many species attacked), also ferns and cycads; other native plants include *Callistemon polandii, Eucalyptus* spp*., Syzygium* spp*., Pittosporum* spp. and many rainforest plants.

Feeding habits: the scales are found in scattered groups on both leaf surfaces. Premature yellowing occurs on the tissues where they feed.

Notes: this very flat scale, which originates from South America, is closely appressed to the surface of the host.

Control: as per introduction for scales.

Globular Wattle Scale or Tick Scale, *Cryptes baccatus*

Description: adults form a highly domed or globular, pearl-like, leathery scale 0.5–1cm long. Colour ranges from slaty blue to greyish or white with older adults becoming brown. Adult females lay eggs under the scale and the crawlers move to suitable feeding sites (usually aggregating close together).

Climatic regions: subtropics and temperate regions, inland and coastal.

Host plants: restricted to wattles but attacks many species. Phyllodinous species include *Acacia acinacea, floribunda, A. linifolia, A. longifolia, A. melanoxylon* and *A. viscidula;* ferny-leaved wattles include *A. decurrens* and *A. mollissima.* On occasions this pest causes serious damage to stands of *Acacia aneura* and *A. pendula* in inland areas.

Feeding habits: a vigorous scale which congregates in clusters along branches. Colonies usually contain insects of mixed ages. Sooty mould discolors the bark.

Notes: feeding colonies resemble clusters of bloated ticks. This species usually congregates in groups along stems and branches. It can be a serious native pest, weakening or killing even healthy wattles. It often appears on young wattles in newly established gardens.

Control: as per introduction for scales. Effective control can be difficult. Steps should be taken when the scales are first noticed as they increase in numbers very rapidly.

Ice Plant Scale, *Pulvinaria mesembryanthemi*

Description: a round to oval brown scale that grows to about 0.3cm long. Adult females lay their eggs into a large conspicuous white cottony sac. The crawlers move to suitable sites where they settle, feed, moult and develop a covering (scale).

Climatic regions: tropics, subtropics and temperate regions, mainly coastal.

Host plants: common on species of *Carpobrotus* and some other members of the family Aiozoaceae; also *Atriplex vesicaria.*

Feeding habits: occurs as individuals and small groups on leaves, stems and flower buds.

Notes: this species, introduced from South Africa, is often prominent on ice plants growing on coastal dunes and headlands; also cultivated plants including many exotic succulents. Fairy-wrens have been observed feeding assiduously on this scale.

Control: usually a minor pest which is controlled by natural enemies.

Flat Brown Scale on palm. [D. Jones]

Globular Wattle Scale close up. [T. Wood]

Globular Wattle Scale on Acacia cognata. [R. Elliot]

Ice Plant Scale, Pulvinaria mesembryanthemi, *on* Carpobrotus. [T. Wood]

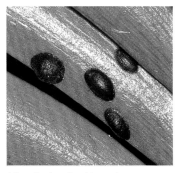

Nigra Scale on Lepidozamia. [D. JONES]

Nigra Scale, *Parasaissetia nigra*

Description: females form a flattish leathery, oval, dark brown to black, shiny, smooth scale 0.3–0.5cm long (colour and size can vary with the host plant). Larvae settle on young shoots and along the midrib of leaves. This species reproduces asexually, the females laying eggs without fertilisation by males (which are unknown). The eggs, which are pink, hatch in three to four days. Crawlers are yellowish pink. This species produces two to six generations per year.

Climatic regions: tropics, subtropics and temperate regions, mainly coastal.

Host plants: recorded as attacking plants from more than 75 plant families. A wide range of native plants are hosts, including species of *Allocasuarina, Callistemon, Casuarina, Correa, Eugenia, Ficus, Hibiscus, Macadamia,* and *Syzygium*; also gingers. Indoor and glasshouse ferns, palms and orchids are also attacked.

Feeding habits: a colonising scale that secretes masses of honeydew resulting in sooty mould growth. Young scales frequently lodge on the adult covering.

Notes: an introduced pest from Africa that is now established in many countries. Host-specific strains or geographical races of this scale may occur. It dislikes hot weather and low humidity.

Control: as per introduction for scales. These scales are easily dislodged and if only a few individuals are present it is a wise precaution to remove them before they spread.

Pink Wax Scale and sooty mould on Syzygium floribundum. [D. JONES]

Pink Wax Scale or Red Wax Scale, *Ceroplastes rubens*

Description: a conspicuous waxy, white, pink or reddish scale 0.3–0.5cm long and domed to a similar height with four small wing-like lobes on the margins and an apical depression. Adults look like hard blobs of white, pink or reddish wax. Adult females lay red eggs under the scale and the crawlers move to suitable feeding sites. This species reproduces sexually with two generations per year (usually summer).

Climatic regions: tropics, subtropics and warm temperate regions, mostly humid coastal districts.

Host plants: a commercial pest of citrus that also attacks a wide range of native and exotic garden plants including *Acronychia, Callistemon, Calophyllum, Dodonaea, Elaeocarpus, Eugenia, Garcinia, Melaleuca, Pittosporum* and *Syzygium* species. Attacks are common on the umbrella tree, *Schefflera actinophylla*. Ferns, orchids, gingers and palms are also hosts.

Feeding habits: a colonising scale found on leaves (mainly the upper surface), stems and branches. It often forms in lines along the midrib. It reproduces very freely and spreads rapidly, quickly colonising new growth and spreading to new plants. Leaves and stems become very dirty with sooty mould which grows on the copious exudates. Ants are attracted to the colonies.

Notes: a pantropical scale that is a major garden pest in many areas.

Control: as per introduction for scales. Spraying is most effective against the tiny crawlers. Also controlled by predators and parasites.

Saltbush Scale, *Pulvinaria maskellii*

Description: an oval white scale which grows to about 0.5cm long. Adult females lay their eggs into a conspicuous white cottony sac. The crawlers move to suitable sites where they settle, feed, moult and develop a covering (scale).

Climatic regions: temperate regions, widespread inland (Western Australia, South Australia, Victoria, New South Wales).

Host plants: attacks plants in at least 17 genera and 11 families including various species of Chenopodiaceae, especially *Atriplex nummularia* and its cultivars that have been developed for rangeland planting; also *Rhagodia hastata*; in 1973 this scale is reported as killing up to 17 per cent of the branches of *Atriplex vesicaria*; also *Callitris preissii, Dodonaea* sp., *Gastrolobium bilobum, Hakea ilicifolia, Jacksonia* sp., *Nuytsia floribunda, Oxylobium lanceolatum, Santalum* sp. and *Viminaria denudata*. The inland wattle, *Acacia coriacea*, is frequently badly attacked by this scale.

Feeding habits: dense colonies on leaves, in leaf axils and on younger branches. In severe infestations the plants appear as if they have burst into masses of white flowers.

Notes: a native scale of sporadic occurrence. It has come to significance recently following serious attacks on clonally propagated populations of cultivated saltbush which have resulted in stem dieback and plant death. One bad outbreak killed about 40 acres of saltbush near Carinda in northern New South Wales. Colonies of this scale are attended by ants.

Control: usually a minor pest which is controlled by natural enemies, but becoming increasingly significant in recent years, especially on *Atriplex* cultivars.

Soft Brown Scale, *Coccus hesperidium*

Description: a soft, oval, fairly flat scale; adults 0.3–0.5cm long, greenish, brown or mottled yellow-brown with a series of faint lines across the scale; crawlers greenish-yellow. This species mainly reproduces asexually, the females giving birth to live young (the eggs hatch within the body of the female), but sexual reproduction occasionally also occurs. Three to five generations are produced during a year (continuous in the tropics).

Climatic regions: tropics to temperate regions, coastal and inland.

Host plants: a wide range of trees and shrubs including citrus, *Ficus* spp., *Schefflera* spp., palms, ferns and orchids. A significant pest of plants grown in greenhouses and glasshouses.

Feeding habits: this species often does not spread widely over a plant but commonly clusters in small dense colonies on isolated leaves, branches and twigs. It produces an abundance of honeydew which results in a heavy sooty mould infestation and vigorous ant activity. The honeydew frequently drips onto lower areas.

Notes: a cosmoplitan species which is a serious pest in some areas. Its presence is revealed by the dirty appearance of the colonies.

Control: as per introduction for scales. Usually controlled by parasitic wasps and predators such as ladybirds and scale-eating caterpillars.

Soft Brown Scale on Asplenium cuneatum. [D. Jones]

Soft Brown Scale on Correa. [D. Jones]

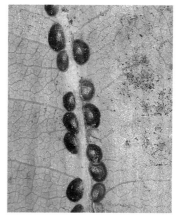

Soft Brown Scale on Pittosporum. [D. Jones]

White Wax Scale, *Gascardia destructor*

Description: soft-shelled scale that forms a large, conspicuous, white, waxy blob up to 1cm long and 0.6cm high. Nymphs are only partially covered with wax. Adult females are white. Males are unknown in Australia.

Climatic regions: subtropics and temperate regions, mainly coastal.

Host plants: attacks a wide variety of native plants including *Aronychia* spp., *Boronia* spp., *Bursaria spinosa*, *Cryptocarya triplinervis*, *Dodonaea* spp., *Elaeodendron* spp., *Eriostemon* spp., *Philotheca* spp. and *Syzygium* spp. Also infests ferns and is a common pest of citrus.

Feeding habits: eggs are laid in October. Crawlers, which emerge in December-January, frequently colonise leaf midribs and veins on the upper surface and as they develop move onto the stems. Adults, which are completely hidden within the waxy blob, nearly always live on the stems. Copious quantities of honeydew are secreted, resulting in extensive sooty mould growth and ant activity.

Notes: an introduced pest which has declined in importance following the release of wasp parasites introduced from South Africa. Severe outbreaks can occur on stressed plants.

Control: as per introduction for scales. Adults can be removed by hand picking. Birds often clean up infestations. Spraying the nymphs is more effective than spraying the adults.

White Wax Scale on Dodonaea.
[D. Jones]

Other Soft Scales

- **Fig Wax Scale or Chinese Tortoise Scale**, *Ceroplastes rusci*, attacks a range of plants in tropical and subtropical parts of the world, including Australia. It has a domed pale pink waxy scale with red lines and a flat indented skirt-like margin, the whole structure resembling the shell of a tortoise. It is a pest of bananas, citrus, figs and a range of native plants.

Fig Wax Scale on a rainforest plant.
[D. Jones]

ARMOURED SCALES (FAMILY DIASPIDIDAE)

The largest group of scale insects consisting of more than 2,370 species distributed in most regions of the world and feeding on nearly 1,400 genera of plants in 180 families. They favour long-lived hosts such as woody shrubs and trees, but many significant pests occur on soft herbaceous plants. In this group of scales the waxy covering (or scale) is not attached to the body of the insect. It consists of wax with a darker area visible through the scale which is actually the cast skins (termed exuviae) of juvenile moults. The mouthparts (stylets) of armoured scales pierce the plant cells and feed on the contents, killing the cells in the process. Affected areas around the scale become yellow and die. Honeydew is not produced by armoured scales. Female armoured scales have two immature instars and the males four.

Armoured scale on palm. [D. Jones]

Boisduval Scale or Orchid Scale, *Diaspis boisduvalii*

Description: females form a more or less circular creamy white to light brown scale measuring 0.15–0.25cm across with a dark centre; males form a slender white scale about 0.2cm long with parallel sides and three longitudinal ridges. Adult females lay eggs under the scale and the pale yellow crawlers move to suitable feeding sites. This species reproduces sexually with several generations per year.

Climatic regions: tropics, subtropics and temperate regions, mainly coastal.

Host plants: many species of palm, bromeliads, cacti, cycads, ferns, epiphytic orchids and foliage plants.

Feeding habits: the female scales are relatively inconspicuous but the clusters of narrow male scales and associated massed white detritus they produce are much more prominent. The cottony detritus can be easily confused with the mess made when mealybugs feed. Tissue around the feeding of adult scales yellows (visible from both surfaces). Badly affected organs wilt, twist and die prematurely.

Notes: this introduced scale, native to tropical America, is one of the most serious pests of greenhouse plants, particularly orchids. It can also infest many garden plants, especially in the tropics. Toxic chemicals in the saliva kill plant tissue resulting in an expanding yellow halo around each scale. Severe infestations can quickly kill weakened or neglected plants and can seriously debilitate even healthy plants. The scale is mainly spread by the distribution of infested plants. Crawlers are dispersed by wind and air currents (including currents from fans).

Control: as per introduction for scales. This species is often very difficult to control without the application of strong chemicals. Small pockets of this scale living in the shelter of sheaths and bracts often survive spraying (even thorough spraying) and then quickly multiply and spread to form new infestations. Infestations often expand rapidly and quick control action is necessary.

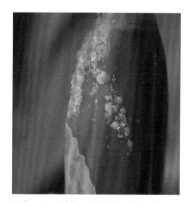
Infestation of female Boisduval Scale.
[D. JONES]

Infestation of male Boisduval Scale.
[D. JONES]

California Red Scale or Red Scale, *Aonidiella aurantii*

Description: females form a hard, circular conical scale 0.15–0.2cm across which can be grey with a reddish centre or wholly reddish, often with paler margins (the red body of the female insect is visible through the scale). Male scales are paler and more elongate. Adult females reproduce asexually, giving birth to live young (the eggs hatch inside the female's body). Crawlers and adult males are yellowish.

Climatic regions: subtropics and temperate regions, coastal and inland.

Host plants: known to feed on plants in more than 75 plant families, including commercial crops such as citrus (it is the most important pest of citrus worldwide), mango, apple, plums and pears. Native plants include species of *Abutilon*, *Acacia*, *Alstonia*, *Araucaria*, *Callistemon*, *Casuarina* (*equisteiifolia*), *Doryanthes*, *Ficus*, *Hibiscus*, *Pandorea* and *Schefflera*. It also infests palms, cycads and ferns.

Red Scale on leaf undersurface of Syzygium smithii. [R. ELLIOT]

Feeding habits: this species congregates in colonies of mixed stages on all above-ground parts of the plant. It can cause leaf and fruit yellowing and premature shedding and stem dieback.

Notes: an introduced armoured scale that apparently originated in China. Once established it can spread rapidly.

Control: as per introduction for scales. This can be a very difficult pest to control.

Circular Black Scale or Florida Red Scale, *Chrysomphalus aonidum*

Description: females form a circular, flattish, purple-black scale about 0.2cm across with a conical pale to reddish point; male scales are slightly more elongate. Adult females lay eggs under the scale and the crawlers move to suitable feeding sites. This species reproduces sexually with three to six generations per year (continuous in the tropics). The crawlers are dispersed readily on wind currents.

Climatic regions: tropics, subtropics and temperate regions, coastal and inland.

Circular Black Scale on Dendrobium kingianum. [D. Jones]

Host plants: a significant pest of citrus, bananas, mangos and papaya; can also be damaging on palms, ferns, cycads and epiphytic orchids; garden plants also affected.

Feeding habits: colonises most parts of a plant, especially young shoots, stems, leaves (both surfaces) and fruit. The tissue around the scale yellows and dies. Infested leaves and fruit drop prematurely.

Notes: an introduced pest from Asia which can quickly increase in numbers and be very destructive. It is killed by heavy frost and heavy rain. Infestations are worst in dry conditions.

Control: as per introduction for scales. Natural control is by parasitic wasps (both native and introduced species).

Coconut Scale, Fern Scale or Aspidistra Scale, *Pinnaspis aspidistrae*

Description: females form a semi-transparent oyster-shaped or mussel-shaped scale 0.1–0.25cm long, with a pale yellow to light brown region at the narrow end; males form a slender white powdery scale about 0.1cm long with parallel sides and three longitudinal ridges. Adult females lay eggs under the scale and the crawlers move to suitable feeding sites. This species reproduces sexually with several generations per year. The crawlers can be dispersed by animals and air currents.

Climatic regions: tropics, subtropics and temperate regions, mainly coastal.

Host plants: mainly palms (many species), cycads, ferns (many spp.), epiphytic orchids and some foliage plants (dracaenas, cordylines, African violets). Often a severe pest of bird's-nest fern (*Asplenium australasicum*), elkhorn (*Platycerium bifurcatum*), maidenhairs (*Adiantum* spp.) and tassel ferns (*Huperzia* spp.).

Male (white) and female (brown) Coconut Scale on Asplenium australasicum. [D. Jones]

Feeding habits: infestations of this scale resemble shredded coconut sprinkled over the leaves and stems of the plant. These are the coverings of the male scales, the females being duller and less conspicuous. On ferns the pest congregates in the leaf axils and among the sporangia on the underside of the leaves. Tissue around the feeding of an adult scale dies and forms a yellow halo which is visible from both surfaces. Wilt is also common. Badly affected fronds die prematurely.

Notes: a common pest in the tropics and also frequent in greenhouses and glasshouses in cooler climates. Severe infestations can kill weakened or neglected plants and can have a severe impact even on healthy plants. The scale is mainly spread by the distribution of infested plants. It is often confused with Boisduval Scale, which has rounder females and longer male scales.

Control: as per introduction for scales. Can be very difficult to control and is worse on neglected plants.

Mussel Scales, *Lepidosaphes* spp.

Description: females are slender, often curved scales with a hard grey to brown or purplish mussel-shaped covering (often with a paler border) 0.15–0.35cm long and 0.1cm wide. Males, which are rare, are smaller and of a different shape. Adult females lay eggs under the scale covering without the intervention of a male.

Climatic regions: tropical to temperate regions, coastal and inland.

Host plants: two native species, *Lepidosaphes macadamiae* from Queensland and *L. macella* from New South Wales (both known as Macadamia Mussel Scale), attack species and cultivars of *Macadamia*. Heavy infestations can result in reduced growth, premature leaf fall and nut shedding. *Lepidosaphes machilii* is an exotic species that attacks *Cymbidium* orchids. *Lepidosaphes gloveri* occurs in Queensland feeding on species of rainforest plants. Some mussel scales are host-specific and are restricted to a single species of plant.

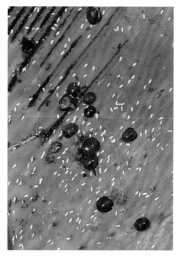

Soft Brown Scale and a mussel scale on Asplenium australasicum. [D. Jones]

Feeding habits: those species which form dense colonies on the bark of stems and twigs can cause bark cracking and stem dieback. Others that feed on leaves and developing fruit cause premature shedding.

Notes: native species of this genus are poorly known and several species are undescribed.

Toxic Scale or Latania Scale, *Hemiberlesia lataniae*

Description: females form a circular conical scale 0.15–0.2cm across which is greyish, brownish or dirty white, often with concentric rings and a yellowish to brown spot situated more or less centrally. Male scales are smaller and oval in shape. This species either reproduces sexually or asexually (males are unknown in some races) with the eggs laid under the female scale.

Climatic regions: tropics, subtropics and warm temperate regions, mainly coastal.

Host plants: recorded worldwide from species in more than 250 plant genera, including many commercial crops; natives include species of *Acacia, Ficus, Grevillea, Hakea, Isopogon, Lomatia, Melia, Passiflora* and *Macadamia*; also orchids, cordylines, ferns, palms, gingers and cycads.

Feeding habits: an inconspicuous scale that can be found on all parts of the plant, usually in colonies. Adult scales can become partially embedded in the bark.

Notes: a cosmopolitan scale that causes damage out of all proportion to its size and numbers, because it injects very toxic saliva while feeding. Entire dead branches are not an infrequent feature of this pest's activities and plant deaths have also been recorded.

Control: as per introduction for scales. A very difficult pest to control.

Tropical Palm Scale, *Hemiberlesia palmae*

Description: females form a circular or oval very convex scale 0.175–0.23cm across which is tan to dark brown with a darker brown spot situated more or less centrally. Males are unknown. This species reproduces asexually.

Climatic regions: tropics and subtropics, mainly coastal.

Host plants: feeds on a wide range of plants, including many commercial crops such as bananas; natives include species of *Alpinia, Barringtonia, Cordyline, Hibiscus, Syzygium* and *Macadamia*; also many species of palms (including coconut), orchids and ferns.

Feeding habits: forms colonies on the underside of leaves and fronds.

Notes: this species can become established as a persistent pest of glasshouse plants.

Control: as per introduction for scales.

Parlatoria Scale or Palm Scale, *Parlatoria proteus*

Description: hard-shelled scale with a domed shell 0.2–0.4cm long, round to ellipsoid, whitish, greyish or brown with a yellowish or darker tip, sometimes with a couple of dark, concentric rings near the top. Crawlers are dark brown.

Climatic regions: tropics and subtropics, mainly coastal.

Host plants: a severe pest of palms; also attack cycads, orchids, cacti, *Ficus* spp. and *Schefflera* spp.

Feeding habits: this species congregates in colonies of mixed stages in sheltered areas including leaf sheaths, folded leaves, leaflets and inflorescence bracts. In severe infestations it can also be found on less protected sites including stems and petioles. Severe attacks in palms cause yellowing and premature leaf fall. Sooty mould grows on the exudates.

Notes: this is a severely debilitating pest. Control measures should be initiated as soon as practical. Weakened palms are particularly susceptible. The females of some populations look like miniature fried eggs.

Control: as per introduction for scales. Spraying is often needed.

Parlatoria Scale on Acacia triptera. [D. JONES]

Male (elongated) and female (round) Parlatoria Scales. [D. JONES]

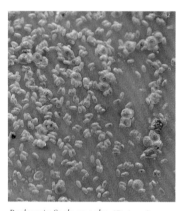

Parlatoria Scale on palm. [D. JONES]

Waratah Scale or Macadamia White Scale, *Pseudaulacaspis brimblecombei*

Description: females form a wedge-shaped or mussel-shaped scale 0.2–0.3cm long which is white with an orange-yellow tip (the insect). Male scales are smaller and brown. Adult females lay eggs under the scale and the crawlers move to suitable feeding sites.

Climatic regions: subtropics and temperate regions, mainly coastal.

Host plants: species of Proteaceae, particularly Waratah (*Telopea speciosissima*) and *Macadamia* species and cultivars; also severe on *Alloxylon flammeum*.

Damage by Waratah Scale to Alloxylon flammeum. [D. JONES]

Male (elongated) and female (mussel-shaped) Waratah Scales on Alloxylon flammeum. [D. JONES]

Macadamia White Scale. [D. JONES]

Feeding habits: this species occurs in colonies of mixed stages and feeds on the underside of leaves, sometimes along the midrib. Infested leaves curl, become yellow to brown and fall prematurely.

Notes: a native pest that can be troublesome on stressed plants. The name *Pseudaulacaspis eugeniae* is sometimes applied to this species.

Control: as per introduction for scales.

Other Armoured Scales

The Cherry Ballart Scale, *Poliaspis exocarpi*, attacks species of *Exocarpus* and *Santalum*; also many pea-flowered shrubs including species of *Daviesia*, *Dillwynia*, *Oxylobium* and *Pultenaea*. The female scales are narrow and nearly straight-sided, white with a yellow tip. Male scales are slightly wider.

Armoured scale on Exocarpos cupressiformis. [T. Wood]

GIANT SCALES (FAMILY MONOPHLEBIDAE)

A small family of specialised scale insects that feed on woody shrubs and trees. The juvenile stages are often covered with wax or hairs (sometimes densely so), whereas the adults are often hairless. The adult females, which commonly grow to more than 1cm long, retain their legs. Female giant scales have four immature instars and the males five.

Cottony Cushion Scale, *Icerya purchasi*

Description: females have a plump soft orange, reddish, yellow or brown body about 0.5cm long, hairy and with black legs. The females, which are always hermaphrodite, develop a conspicuous fluted white egg-sac which is up to 1cm long and covered with cottony threads. At this stage the insect itself appears to be standing on its head. The eggs within the sac are red. The tiny dark red males are winged and the crawlers dark red. The females can either reproduce asexually (without mating) resulting in hermaphrodite progeny or after normal mating which results in a mixture of males and hermaphrodites.

Climatic regions: temperate and subtropical regions, inland and coastal.

Host plants: a wide range of plants including species of *Acacia*, *Eucalyptus*, *Grevillea*, *Hakea*, *Kunzea*, *Melaleuca* and *Pittosporum*,

Cottony Cushion Scale on wattle. [T. Wood]

and *Exocarpos cupressiformis*. Species of the succulent genus *Carprobrotus* are a favourite host. Also exotic succulents and cacti.

Feeding habits: the crawlers aggregate in clusters and rows along leaf midribs and veins (often on the leaf underside), and on twigs and smaller stems. The adults seem to prefer the bark of larger branches and even the trunk. Honeydew is secreted copiously, attracting ants and causing the growth of sooty mould.

Notes: this scale, which is a native insect, was accidentally introduced into California during the late 1880s and devastated citrus orchards. It was finally brought under control when the Australian Verdalia Ladybird, *Rodolia cardinalis*, and a parasitic tachinid fly, *Cryptochaetum iceryae*, were introduced. Successive outbreaks of Cottony Cushion Scale occurred in other countries and this species is now distributed worldwide. Control is usually gained by the introduction of these natural control agents.

Control: outbreaks occur sporadically but control measures are rarely necessary since the pest is usually cleaned up within a few months by parasites and predators such as ladybirds. Fairy-wrens also feed actively on this scale. Colonies are easily squashed between the fingers or disrupted by jets of water.

Giant Mealybugs or Snowball Mealybugs, several species of *Monophlebulus*

Description: several species involved. Adults, which grow 2–2.5cm long, can move slowly. The females are wingless. Some species are completely covered with white waxy threads, others are nearly naked and often have orange and bluish markings. The young of some species are densely covered with waxy threads (resembling 'snowballs'). Reproduction is from eggs.

Climatic regions: relatively common in the tropics and subtropics, much less common in temperate areas.

Host plants: mainly members of the family Myrtaceae, particularly species of *Eucalyptus*, *Corymbia*, *Callistemon* and *Melaleuca*; also *Acacia* species.

Feeding habits: juveniles congregate on young shoots, the adults, which are mostly solitary, are often found on hardened growth.

Notes: there are about nine species of giant mealybugs native to Australia.

Control: these unusual insects, which cause minimal damage, add interest to a garden.

Giant mealybug. [T. WOOD]

Giant mealybug. [T. WOOD]

Giant mealybugs attended by ants. [S. JONES]

FELT SCALES (FAMILY ERIOCOCCIDAE)

Another large family of scale insects with about 550 species recognised worldwide. They feed on plants in more than 45 families. More than 140 species are recorded from plants in the family Myrtaceae. Female eriococcids either live beneath a waxy covering or within a gall they induce (see chapter 7). Female felt scales have two or three immature instars and the males four. There are many species of native eriococcids that have been poorly studied, some examples are included here.

Native eriococcid feeding on Casuarina cunninghamiana. [D. JONES]

Native eriococcid and sooty mould on Westringia rosmarinifolia. [D. JONES]

Native eriococcid and sooty mould on Alyxia buxifolia. [D. JONES]

Gum Tree Scale, Eriococcid Scale or Rice Bubble Scale, *Eriococcus coriaceus*

Description: a leathery, rounded scale closely packed along branches. Adult females form a rice-bubble like brownish to reddish scale 0.2–0.4cm long with an opening at one end. Male scales, which are smaller and white, occur in groups close to the female scales. Adult males are dark brown and with wings. This species reproduces sexually with two to five generations per year depending on the temperature. Mated females lay eggs under the scale and after hatching the crawlers emerge through the opening and move to a feeding site. Crawlers can be dispersed on the wind or as passengers on birds.

Climatic regions: temperate regions and subtropics, inland and coastal.

Host plants: this scale attacks many species of *Eucalyptus*. Attacks can be especially severe on eucalypts with blue-green foliage such as *E. globulus* and *E. nitens*. Weak trees are more susceptible to attack as are also some ornamental eucalypts such as *E. caesia, E. crenulata, E. morrisbyi, E. perriniana* and *E. pulverulenta*.

Feeding habits: scales cluster in dense masses along the stems and branches and exude a reddish sticky gum on which sooty mould grows. Colonies are usually attended by ants which defend their territory vigorously. Squashed scales leak a reddish staining juice.

Notes: a very serious pest which can weaken or even kill vigorous trees. In recent times this species has become a serious pest of plantation eucalypts grown for timber. Persistent attacks weaken growth and affect apical dominance, resulting in trees with an unusual branching habit. Severe attacks result in stem dieback and plant death.

Control: as per introduction for scales. Control should be initiated when first noticed because once established these scales can be very difficult to eradicate. They are controlled naturally by a range of predators including lacewings, ladybirds, scale-eating caterpillars and entomophagous fungi, but these

Ants attending a dense cluster of Gum Tree Scale. [T. WOOD]

Gum Tree Scale attended by ants. [T. WOOD]

Gum Tree Scale on Eucalyptus globulus. [D. JONES]

Ants attending a small infestation of Gum Tree Scale. [D. JONES]

Gum Tree Scale on eucalypt stems. [D. JONES]

Severe infestation of Gum Tree Scale on Eucalyptus pyriformis. [D. JONES]

antagonists seem to be absent in the early stages of infestation, allowing the scales to rapidly build up numbers. Small infestations can be cleaned up by squashing between the fingers (very messy) or removing and burning the infested branches. Some birds such as pardalotes, Silvereyes, thornbills and Noisy Miners feed on concentrations of this scale.

Macadamia Felted Coccid, *Eriococcus ironsidei*

Description: this species appears on infested tissue as tiny whitish flecks. It is a small, dirty white to pale yellow-brown scale 0.1–0.15cm long that lacks waxy threads. Adult females, which are immobile and wingless, are covered with a felt-like hairy sac. They also have marginal bristles. Adult males have wings but do not feed. Eggs are laid within the female sac and after hatching the tiny crawlers move about or can be dispersed by wind, birds or on vehicles.

Climatic regions: tropics and subtropics.

Host plants: *Macadamia integrifolia, M. tetraphylla* and their cultivars.

Feeding habits: this pest congregates in colonies on all parts of the tree. It causes twisting and other distortion of new growth, leaves and flower racemes, leaf yellowing, stunting, stem dieback. It also reduces yield.

Notes: this is a serious pest of commercial *Macadamia* plantations and can also be a problem in home gardens. It is very difficult to eradicate once established in an area.

Control: as per introduction for scales. Spraying is usually necessary.

Teatree Scale or Manuka Blight, *Eriococcus orariensis*

Description: females form an oval pale red, bright red or brownish scale 0.1–0.13cm long with projecting waxy threads and a pair of small lobes at the rear end; after mating the female encloses her whole body in a white felted sac into which she lays eggs. The males form a smaller cottony sac. This species reproduces sexually with two or three generations per year. The crawlers are dispersed by wind and on birds and animals.

Climatic regions: subtropical and temperate regions, mainly near-coastal.

Host plants: *Leptospermum* species and cultivars (especially *L. scoparium* and its hybrids); also some *Kunzea* spp.

Feeding habits: scales cluster in colonies along the stems and branches, often partly hidden within cracks or crevices on the stems or under strips of bark. A sticky gum is exuded on which sooty mould grows.

Notes: a damaging native pest. Severe attacks, which weaken plants considerably, are invariably accompanied by sooty mould growth which gives the whole plant a scorched appearance. Persistent attacks can result in plant death. This scale became established in New Zealand in the middle of the 20th century, resulting in the death of large patches of the the native shrub Manuka, *Leptospermum scoparium*. Known as Manuka Blight, its spread was eventually curtailed by an entomopathogenic fungus, *Myriangium thwaitesii*. It is now largely contained within New Zealand.

Control: as per introduction for scales. Can be a very difficult scale to control.

Other Felt Scales

Kunzea felt scale, sometimes called the snowball mealybug on Kunzea *stem.*
[D. Jones]

- **Kunzea Felt Scale**, *Callococcus acaciae*, feeds on the stems of *Kunzea*. The red- to burgundy-coloured globular female scale is covered white waxy threads that form a conspicuous curly snowball-like mass.

LAC SCALES (FAMILY KERIIDAE)

There are about 90 species of lac scales distributed around the world. Shellac resin is obtained from a secretion of a female lac insect, *Kerria lacca*, that occurs in Thailand. The Australian species have been poorly studied and it appears that less than 10 species have been described. Two serious pestiferous species present on Christmas Island would have devastating impacts on Australian native plants if they became established on the mainland.

Red Mulga Lac Scale, *Austrotachardia acaciae*

Description: hard-shelled scale that forms a waxy, bright orange, somewhat globular structure, although adjacent scales often coalesce together in masses.

Climatic regions: temperate regions, mainly inland.

Host plants: mainly found on Mulga, *Acacia aneura*.

Feeding habits: little is known of its life cycle. The scales form conspicuous patches of colour along stems and branches. Honeydew is secreted resulting in ant activity and sooty mould growth.

Notes: a native scale which has limited impact on gardens. The sweet exudate, particularly from fresh bright scales, is eaten by Aboriginal people.

Control: as per introduction for scales.

Red Mulga Lac Scale on branches of Mulga. [C. FRENCH]

Other Lac Scales

- **Melaleuca Lac Scale**, *Austrotachardia melaleucae*, forms hard red-brown to dark brown globose coverings on the branches of species of *Melaleuca* and *Allocasuarina* in semi-arid regions.

- **Native Pine Lac Scale**, *Austrotachardia* sp., forms crowded masses of white scales on some species of *Callitris* in inland areas. Each female scale, which is more or less circular in shape, has several irregular bumps around a sunken dark central area.

- **Lobate Lac Scale**, *Patatachardina pseudolobata*, is a widely distributed, devastating scale that is known to infest more than 300 species of plant. It is not known from mainland Australia but does occur on Christmas Island. The scales have a domed, brittle, dark red-brown to black covering with two pairs of lobe-like extensions on the margins. The females reproduce parthenogenetically. The crawlers are dark red.

- **Yellow Lac Scale**, *Tachardina aurantiaca*, also occurs on Christmas Island where it is having serious impacts on the composition of native rainforests. It is known to infest about 29 species of plant native to the island and has a mutual association with the Yellow Crazy Ant *Anoplolepis gracilipes*. The scales are bright yellow with a brownish central area. The crawlers are orange.

Native pine lac scale on Callitris *stem.* [D. JONES]

Lac scale on Kunzea. [S. JONES]

MEALYBUGS (FAMILY PSEUDOCOCCIDAE)

Stem mealybugs attended by ants. [D. Jones]

Mealybugs, of which there are more than 2,000 species, are part of the superfamily Coccoidea, in the family Pseudococcidae. Mealybugs are soft, plump insects that exude a covering of waxy powder or waxy threads which act as a protective device, sheltering the insect and reducing water loss, as well as being water-repellent. They congregate in colonies and suck the sap from a range of plant tissue, including leaves, buds, flowers and fruit. Some species even feed on bark. They are messy feeders with white waxy detritus littering the site where they are active. Mealybugs are widely distributed around the world, attacking a wide range of crops and ornamental plants. About 200 species occur in Australia, including some important species that have become established as pests in other countries.

Life cycle: mealybugs are very active in spring and autumn when conditions are warm and humid, but not too hot, cold or dry. After mating the adult females lay eggs in a wax-covered sac or cocoon (some species lay naked eggs which hatch very quickly, giving the impression of live birth). Young mealybugs, in common with scale insects, are known as 'crawlers'. The juveniles are often yellow but soon exude a waxy or mealy covering. They move to a feeding site where they suck sap and grow, moulting as they develop. Juveniles and adult females are mobile but are sluggish and tend to be sedentary. Adult females are wingless and resemble a large crawler, whereas the adult males are winged. It does not feed at all, just mates and dies.

Detritus from a mealybug infestation on a palm. [D. Jones]

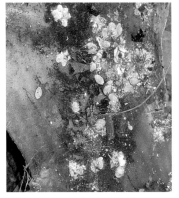

Infestation of an unknown mealybug; note the discarded wax and sooty mould. [D. Jones]

Plant damage: attacks by mealybugs cause a range of symptoms, including discoloured tissue, misshapen leaves, distorted shoots, bud drop and misshapen fruit. Although they attack a wide range of plants, the most destructive and commonly encountered species are found on ferns, orchids, palms, bulbs and large-leafed woody-liliaceous species such as cordylines. Attacks on palms and cordylines centre around the leaf sheaths in the developing crown and can be very persistent and severe. Emerging leaves become misshapen and stunted, and prolonged attacks weaken and even kill plants. Succulent- and cactus-growers also have to contend with these same pestiferous species. Some other species of mealybugs also attack woody garden plants including eucalypts, grevilleas and wattles.

Mealybugs favour dry conditions and prefer to feed under cover, congregating in leaf axils, leaf sheaths, leaf undersides, curled or folded young leaves and in cracks and crevices. Their depredations are frequently severe on plants that have been weakened by external factors, such as insufficient water and poor nutrition. They are often a continual nuisance on potted plants, especially those used for indoor decoration. Congregations of mealybugs are frequently attended by ants which protect the bugs, collect their sugary excretions and may even move the mealybugs about or construct 'byres' to shelter them. Exudates from mealybugs are sugary and favour the development of sooty mould. Note that the colour of their exudates can be a useful guide to the identity of the species involved.

Natural control: infestations of mealybugs fluctuate with the weather, declining when the temperatures become too high or too cold and when humidity is low. Mealybugs are attacked by a range of predators and parasites. Small birds including thornbills, Silvereyes and pardalotes feed on them avidly, as do hunting spiders, parasitic wasps, damsel bugs, adults and larvae of ladybird beetles, and the larvae of hoverflies and lacewings. Some fungi which grow on their exudates can become parasitic, devastating the colonies.

Other control: exposing the pests by removing their cover aids in predation; for example removing the sheathing bases of dead palm fronds. Hosing with jets of water also disrupts the colonies, but mealybugs are hard to wet because of their waxy coating. Because mealybugs favour warm, humid conditions, control of infestations on plants grown indoors and in greenhouses can be difficult. The introduction of parasitic wasps, available from commercial biocontrol firms, is a safe way to dramatically reduce numbers. Small infestations can be cleared by squashing or dabbing the insects with cotton buds soaked in methylated spirit. Severe infestations are much more difficult as mealybugs develop resistance to insecticides fairly quickly. Spraying with a mixture of white oil and malathion (see combination sprays, chapter 2) can be effective. Imidacloprid can also be very useful in controlling this pest.

EXAMPLES
(formal identification of these pests should be carried out by experts)

Bulb Mealybug, *Vryburgia amaryllidis*

Description: adults, which grow 0.2–0.4cm long, have an elongated, elliptical, red to purplish body covered with white waxy powder. The margins lack waxy threads and two pairs of broad waxy filaments protrude from the anal region. The short legs are often hidden. Reproduction is from eggs which are pink.
Climatic regions: tropics, subtropics and temperate areas.
Host plants: succulents and bulbs, including native species of *Calostemma* and *Crinum*.
Feeding habits: feeds in crevices around the base of the plant, the lower leaf sheaths, and on the roots. affected plants lack vigour and feeding areas become blotched and messy.
Notes: an exotic pest that has been recorded from Queensland and South Australia.
Control: very difficult, spraying is usually necessary.

Casuarina Mealybug, *Pseudoripersia turgipes*

Description: adults immobile, 0.3–0.4cm long, grey to pale brown, elliptical, concave and scale-like with faint body segments. The margins lack waxy threads. The females are wingless. Reproduction is from eggs.
Climatic regions: subtropics and temperate areas.
Host plants: species of *Allocasuarina* (*A. distyla*, *A. leuhmannii*) and *Casuarina* (*C. cristata*, *C. cunninghamiana*, *C. glauca*).
Feeding habits: juveniles and adults congregate on young shoots, empty adult shells persist on stems. Masses of sooty mould discolor the growth and indicate the presence of the mealybugs.
Notes: this species superficially resembles a scale insect.
Control: damage is usually minor and control is rarely needed, although nursery seedlings are sometimes attacked.

Citrus Mealybug, *Planococcus citri*

Description: adult females are elliptical or oval, 0.3–0.4cm long, with numerous short marginal filaments, covered with conspicuous white powdery wax. Anal filaments are prominent but less than one quarter of the body length. Exudate is dark yellowish. Adult males are winged. Large numbers of tiny eggs are laid inside a sac. Live young may also be born over summer.
Climatic regions: tropical to temperate regions.

Host plants: feeds on citrus and many other species of the family Rutaceae including *Acronychia, Boronia, Crowea, Eriostemon* and *Philotheca;* also *Macadamia*, ferns and orchids.

Feeding habits: colonies of mixed stages feed on young shoots and leaves; a different form of the pest also attacks plant roots.

Notes: a moderately common garden pest.

Control: as per introduction for mealybugs.

Cottonwool Mealybug, *Erium globosum*

Cottonwool Mealybugs, Erium globosum, *on Acacia; note the presence of* Cryptolaemus *larva.* [D. Jones]

Description: adults covered with an irregularly shaped white cottony sac 0.5–0.6cm long. Adults immobile, 0.3–0.4cm long, reddish with white waxy meal, fleshy. The females are wingless. Reproduction is from eggs laid within the sac. Crawlers move to new feeding sites.

Climatic regions: temperate regions.

Host plants: species of *Acacia*, garden-grown plants and natural plants in the wild.

Feeding habits: this mealybug forms colonies along stems, the cottony sacs often covering the bases of phyllodes. In severe infestations the cottony sacs overlap and run together. New sacs are stark white but become stained and creamy-white with age.

Notes: severe attacks cause growth distortion, leaf and bud drop and stem dieback. This mealybug is attacked by the larvae of the Mealybug Ladybird, *Cryptolaemus montrouzieri.*

Control: this pest can cause significant damage in newly planted gardens. Spraying is difficult as the insect is protected by the water-repellent cottony sac.

Grevillea Mealybug, *Australicoccus grevilleae*

Severe damage caused by the Grevillea Mealybug, Australicoccus grevilleae. [D. Jones]

Description: adults covered with a white cottony sac 0.4–0.5cm long. Adults immobile, 0.3–0.4cm long, almost globular, dark purplish-black with white waxy meal, fleshy, purple juice when squashed. The females are wingless. Reproduction is from eggs laid within the sac. Crawlers move to new feeding sites.

Climatic regions: subtropics and temperate regions.

Host plants: species and cultivars of *Grevillea*, both garden-grown plants and natural plants in the wild.

Feeding habits: this species congregates on buds, leaf axils and young shoots. In bad infestations the cottony sacs overlap and run together. Masses of sooty mould discolor the growth and indicate the presence of the mealybugs.

Notes: the white cottony sacs of this species are conspicuous among the foliage. Severe attacks cause growth distortion, leaf and bud drop and stem dieback. This mealybug is attacked by the larvae of the Mealybug Ladybird, *Cryptolaemus montrouzieri.*

Control: this pest often persists on established *Grevillea* plants without obvious major damage, but can be a significant pest of stressed plants and in newly planted gardens. Spraying is difficult as the insect is protected by the water-repellent cottony sac.

Long-tailed Mealybug, *Pseudococcus longispinus*

Description: adults are elliptical to oval, 0.3–0.6cm long, with conspicuous segments and short marginal wax filaments. The surface is covered with a white waxy or mealy powder. The anal filaments are as long as the body and the exudate is clear. Females are wingless and can move slowly. The short-lived males are winged. Yellowish eggs hatch quickly after being laid and the yellowish crawlers disperse to feeding sites.

Climatic regions: tropics and temperate areas.

Host plants: recorded feeding on plants from more than 85 plant families. This includes a wide range of indoor and greenhouse plants such as bulbs, bromeliads, cacti, cycads, ferns, palms, succulents, gingers and orchids. Also found on outdoor palms, many garden plants and commercial crops, including citrus, mangos, coffee and macadamias.

Infestation of Long-tailed Mealybugs.
[J. FANNING]

Feeding habits: colonies of mixed stages feed on young shoots, leaves, inflorescences, buds, flowers and fruit, usually congregating in crevices or other hidden places. Feeding sites develop discoloured or distorted tissue and the area is littered with masses of white, waxy secretions and covered with sooty mould.

Notes: a serious and common pest that prefers warm humid conditions, but often feeds in dry sheltered sites. It can build up in numbers very quickly and is almost impossible to eradicate from plant collections.

Control: as per introduction for mealybugs. The larvae of parasitic wasps *Anagyrus fusciventris* and *Leptomastix dactylopil* parasitise the Long-tailed Mealybug. These commercially available wasps are active from late spring to autumn.

Long-tailed Mealybug adult and nymphs.
[J. FANNING]

Wattle Mealybug, *Melanococcus albizziae*

Description: adult females are wingless and remain fixed in place. They are oval to elliptical, segmented, about 0.4–0.5cm long, dull purple to blue-black, partially covered with a white waxy secretion which forms a brood sac beneath the female's body. The margins of the body have short waxy filaments. Live young, which are born into the brood sac, disperse as reddish crawlers, often on air currents.

Climatic regions: tropics and temperate regions, coastal and inland.

Host plants: many species of *Acacia,* both bipinnate and phyllodinous spp.; also *Albizzia, Paraserianthes* and *Grevillea banksii*.

Feeding habits: these bugs range from individuals or small groups to crowded congregations. They feed on young shoots and developing stems, producing copious waxy secretions and honeydew. Large congregations result in stem dieback and successive infestations can kill trees.

Wattle Mealybug on Acacia dealbata. [T. WOOD]

Notes: a sporadic native pest that is sometimes very damaging in newly planted gardens.

Control: usually by natural enemies.

Other Mealybugs

Wattle Mealybug on Acacia rubida.
[T. Wood]

- **Cotton Mealybug**, *Phenacoccus solenopsis*, a native of North America, was identified on cotton crops in various parts of Queensland in 2009-2012. It is known to attack more than 150 plant species and will probably spread onto native species of *Gossypium* and other native members of the Malvaceae.

- **Lantana Mealybug**, *Phenacoccus parvus*, also a recent invader, may also pose problems for native members of the Verbenaceae such as species of *Callicarpa* and *Clerodendron*.

- **Golden Mealybug**, *Nipaecoccus aurilanatus*, feeds on conifers of the genus *Araucaria*. It has become a serious pest in the USA and in 1994 caused significant damage in Queensland to a planting of clonally propagated hoop pine, *A. cunninghamii*, that was being grown for seed production. This infestation, which eventually was controlled by natural predators and parasites, caused defoliation, stem dieback, impeded cone development and killed up to 10 per cent of young trees.

THRIPS

Thrips drink sap which oozes from lacerations in plant tissue caused by their chewing (not sucking) habits. For more information see Chapter 6.

WHITEFLIES OR SNOWFLIES (FAMILY ALEYRODIDAE)

Whiteflies or snowflies, of which there are more than 1,550 species, are small sucking insects belonging to the order Hemiptera, suborder Sternorrhyncha and superfamily Aleyrodoidea. The adults, which have four wings, are covered with fine white powder which gives rise to the common names. They feed in colonies and rise in clouds from foliage when disturbed. As well as a few introduced species which are considered to be significant pests, about 25 native species, some highly specialised, feed on a range of native plants.

Life cycle: after mating the adult females lay eggs on the underside of leaves, often choosing younger leaves towards the top of a shoot. The eggs hatch to produce crawlers that move to a site on the leaf underside where they feed and moult to produce scale-like nymphs. Often the nymphs have a border of waxy filaments. The nymphs continue to grow and moult, eventually becoming adults which can live for two to six weeks depending on the season.

Plant damage: whiteflies, which occur in large numbers in some years and are absent in others, are mainly a pest of vegetables, but also attack a range of plants grown in glasshouses and greenhouses, especially ferns and orchids. Some species are found in natural habitats, especially rainforest, and can also damage garden plants. Whiteflies are mostly active during warm weather. They mainly feed on the underside of leaves but can also congregate on developing buds. Affected plant parts often change colour. *Alyogyne* and *Hibiscus* leaves often take on bluish or purple tones; other plants become yellow or white and papery; attacks on soft growth are followed by wilting. Affected buds may abort. Sooty mould grows on their copious exudates. Some species of whitefly can transmit plant viruses.

Natural control: whitefly numbers are usually controlled by tiny wasps which parasitise the young stages. The introduced wasp *Encarsia formosa* is an effective parasite of pestiferous whiteflies, especially in the warmer summer months. The female wasps lay eggs in late stage nymphs and pupae, the parasitised whiteflies turning brown or black. This valuable parasite is available from commercial biocontrol firms.

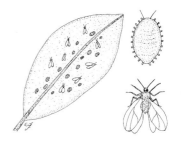

Whitefly and larva. [D. Jones]

Other control: small infestations can be cleared up by dabbing with methylated spirit. Strong jets of water greatly disrupt colonies. Whiteflies are attracted to the colour yellow and sticky traps can be used to measure numbers and reduce population levels. Severe infestations are very difficult to control chemically and a sequence of sprays may be necessary. Imidacloprid and pyrethrins are commonly recommended for control. Because these pests develop resistance quickly it is best to rotate pesticides and avoid spraying with broad-spectrum insecticides.

EXAMPLES

(formal identification of these pests should be carried out by experts)

Ash Whitefly, *Siphoninus phillyreae*

Description: the adults are small and moth-like, about 0.15cm long, with red eyes, yellowish body and white wings. Cream to greyish eggs are laid in circles on the underside of leaves. The crawlers are light brown. Final stage nymphs are light brown with one or two prominent longitudinal crown-like tufts of white waxy material and rod-like marginal filaments, each tipped with a globule of clear glassy wax.
Climatic regions: subtropics and temperate regions.
Host plants: recorded as a pest on citrus, apples, pears and other fruit crops; also many common garden ornamentals (both evergreen and deciduous), including natives.
Feeding habits: small to large colonies of mixed stages suck the sap from the underside of leaves leading to wilting and premature leaf-fall. Heavy feeding is associated with sooty mould. When disturbed the adults tend to swarm around the host leaf where they are feeding or over the whole plant. They also swarm when deciduous trees are shedding leaves in autumn/winter, often moving onto evergreen trees.
Notes: this species, native to Europe and North Africa, was first recorded in Adelaide in 1998 where it has badly damaged trees in streets and gardens. It has now spread to Queensland, New South Wales, Australian Capital Territory, Victoria and South Australia.
Control: as per introduction for whitefly.

Banksia Whitefly, *Aleurocanthus banksiae*

Description: the adults are small and moth-like, about 0.15cm long, the wings covered with floury powder. The nymphs are glossy, flattened and scale-like.
Climatic regions: subtropics and temperate regions.
Host plants: *Banksia* species.
Feeding habits: colonies of mixed stages suck the sap from young shoots and leaves.
Notes: a native pest of sporadic occurrence.
Control: usually controlled by natural enemies.

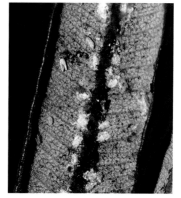

Banksia Whitefly on Banksia integrifolia.
[D. Beardsell]

Coconut Whitefly, *Aleurodicus destructor*

Description: the adults and larvae closely resemble Spiralling Whitefly (see below). White waxy threads are exuded by the nymphs to form a tangled mess that becomes a site for sooty mould.

Climatic regions: tropics.

Host plants: attacks a range of palms, including Coconut Palm (*Cocos nucifera*), and some fruit crops; natives include species of *Acacia, Banksia, Ficus, Ptychosperma* and *Syzygium.*

Feeding habits: colonies feed on the leaf undersides causing puckering and the leaf margins to curve downwards forming a shelter for the pests.

Coconut Whitefly on Syzygium alatoramulum. [D. Jones]

Whitefly, possibly Aleurodiscus *sp.* [J. Fanning]

Notes: this species is native to tropical parts of Australia and adjacent countries. Severely infested plants develop a thick covering of sooty mould; sometimes the whole plant is blackened.

Control: a sporadic pest that is usually controlled by natural enemies.

Common Whitefly or Greenhouse Whitefly, *Trialeurodes vaporariorum*

Description: the adults are about 0.1cm long, pale yellow with white waxy wings that are folded and overlap when at rest. The eggs, which are often laid in a circle, have thin filaments around the margin. They are yellow-green, turning grey to black before hatching. The crawlers are light green with red eyes. Nymphs are pale yellow to translucent and scale-like with fine waxy marginal filaments.

Climatic regions: tropics, subtropics and temperate regions.

Host plants: attacks a wide range of vegetables and other crop plants, garden annuals and weeds; natives include species of *Alyogyne, Bauera, Boronia, Brachyscome, Correa, Crowea, Eriostemon, Goodenia, Grevillea, Hibiscus, Phebalium, Pimelea, Philotheca, Prostanthera, Scaevola* and *Xerochrysum.* Also glasshouse-grown ferns and orchids.

Feeding habits: colonies of mixed stages suck the sap from the underside of leaves. Adults usually occur near the tops of shoots, nymphs lower down. Sooty mould grows on the copious exudates.

Notes: a serious and sometimes persistent cosmopolitan pest that favours warm conditions. It can be a significant and persistent nuisance in greenhouses and glasshouses. As well as causing physical damage and leading to sooty mould growth, it can also transmit plant viruses.

Control: as per introduction for whitefly.

Hakea Whitefly, *Synaleurodicus hakeae*

Description: the adults are about 0.15cm long, yellow, moth-like, the wings covered with floury powder. The nymphs are scale-like, flat to slightly convex and pale yellow with numerous marginal filaments.

Climatic regions: temperate regions.

Host plants: *Hakea* species.

Feeding habits: colonies of mixed stages suck the sap from young shoots and leaves.

Notes: a native species of sporadic occurrence in south-west Western Australia.

Control: usually controlled by natural enemies.

Silverleaf Whitefly, *Bemisia tabaci*

Description: the eggs are tear-shaped with no marginal filaments. The adults are small and moth-like, about 0.15cm long, with red eyes, yellow body and white wings. When resting there is a noticeable gap between the wings. White to brown eggs are laid haphazardly on the underside of leaves. Crawlers are greenish. Nymphs are pale yellow. Moulting nymphs cast silvery skins which remain on the leaves.
Climatic regions: tropics and subtropics.
Host plants: this species has been recorded from more than 500 different species of host plants including cotton, tomatoes and sweet potato.
Feeding habits: colonies of mixed stages suck the sap from young leaves.
Notes: native to the tropical parts of some countries including Australia. The native strain is fairly innocuous but the worst variant is a pesticide-resistant strain introduced into Australia from overseas. Where present this strain displaces the native strain. As well as damaging cotton crops (and others) this whitefly has become a serious pest of greenhouses and glasshouses in many countries.
Control: as per introduction for whitefly. The parasitic wasp *Eretmocerus hayati* has been released in Queensland to control this pest in cotton crops.

Spiralling Whitefly, *Aleurodicus dispersus*

Description: the moth-like adults are about 0.2cm long with white wings (the forewings are sometimes spotted). White, fluffy, waxy material, deposited in a more or less spiral pattern, is associated with the yellow to pale brown eggs. Crawlers move to a feeding site and become sedentary larvae that are oval and pale green with waxy tufts.
Climatic regions: tropics.
Host plants: attacks a wide range of plants including fruit crops, vegetables and garden plants; natives include species of *Acacia*, *Cordyline*, *Crotalaria*, *Gossypium*, *Hibiscus*, *Spathoglottis*, *Terminalia*, *Murraya* and *Scindapsus*.
Feeding habits: colonies of mixed stages suck the sap from young shoots and leaves reducing vigour and causing leaf drop. The eggs, along with conspicuous waxy material, are usually deposited on the underside of leaves.
Notes: this is an introduced species originally from South America and the Caribbean region. It has become established in north-eastern Queensland (north of Mackay) and around Darwin in the Northern Territory. Most attacks occur during the dry season.
Control: as per introduction for whitefly.

Other Whiteflies

- A number of species of native whitefly are of sporadic appearance on native plants.

- Several species in different genera feed on species of *Acacia*. Other native whiteflies are known to feed on species of *Leucopogon* and *Monotoca elliptica*.

- **Dryandra Whitefly**, *Aleurotradielus dryandrae*, with yellow adults and black scale-like nymphs, is known to feed on a range of Proteaceous genera in eastern and western parts of Australia, including *Banksia*, *Dryandra*, *Grevillea* and *Hakea*.

- **Saltbush Whitefly**, *Lipaleyrodes atriplex*, feeds on saltbush in inland areas.

- An unknown species, probably exotic, feeds on epiphytic orchids and ferns in glasshouses. It has round brown to black nymphs with a conspicuous marginal fringe of waxy filaments.

PHOTO EXAMPLES

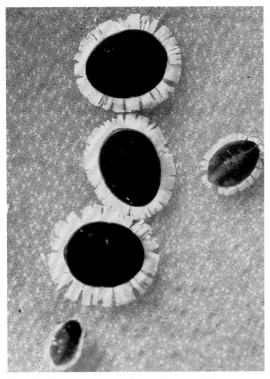

Nymphs of an unknown whitefly that feeds on orchids.
[J. FANNING]

Whitefly larvae on Dendrobium *leaf.* [J. FANNING]

Native whitefly on Leucopogon parviflorus. [D. JONES]

Whitefly infestation on the underside of Syzygium *leaves.* [D. JONES]

Whitefly nymphs on a rainforest tree.
[D. JONES]

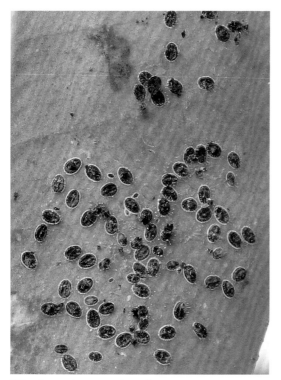

Whitefly nymphs on a rainforest plant. [D. Jones]

Whitefly eggs and nymphs on a daisy. [D. Jones]

CHAPTER 5

INSECTS THAT SUCK SAP – 2

(BUGS, CICADAS, CUCKOO SPIT, FROGHOPPERS, JASSIDS, LEAFHOPPERS, LERPS, PLANTHOPPERS, PSYLLIDS, TREEHOPPERS)

SYMPTOMS AND RECOGNITION FEATURES

Shiny sap on branchlets, small black or metallic, cicada-like insects which move around the branch when approached, hopping as a last resort .... *jassids*

Shiny sap on branchlets, tapering calcareous tubes attached to twigs and branchlets........................... *froghoppers*

Sticky, shiny sap on branchlets and leaves, colonies of small, mobile, pale-coloured insects which hop if disturbed
.... *leafhoppers*

Prominent white frothy secretions on stems and in leaf axils.. *cuckoo spit*

Sticky white sap or white waxy threads covering young shoots... *psyllids*

Blobs of sticky white waxy material on the underside of leaves... *psyllids*

Sudden wilting and collapse of single shoots, prominent hole often present at the base of the shoot...................
.... *large bugs (orange, shield, tip, holy cross)*

Localised wilting, colonies of small active, dull-coloured insects.... *small bugs (Rutherglen, coon, stainer, mirid, lace)*

Localised wilting, colonies of medium-sized, slow-moving, brightly-coloured insects .... *harlequin bugs*, jewel bugs

Patches of dry discoloured tissue on the older leaves.. *lerps*

Distortion and malformation of growing tips and/or flower buds... *psyllids*

Solitary green, grey or brown small insects, which can hop or fly, wings meet tent-wise .... *leafhoppers*

Solitary large insect on branches or trunk, with prominent eyes and gauzy wings.............................. *cicadas*

Domed circular or shell-shaped structures on leaves... *lerps*

Tiny hairy insects that jump suddenly when disturbed.. *leafhoppers*

SAP-SUCKING BUGS

True bugs of the insect order Hemiptera, suborder Hemiptera, are sucking insects with powerful needle-like mouthparts. Many of the commonly encountered species are shaped like a shield but the size and shape of bugs varies greatly. Some species are predators, attacking other insects and spiders (see also chapter 3); other bugs feed on plants, usually sucking the sap from soft growth. A number of native species are pests of plants and a few overseas species have also become naturalised in Australia. Many species of bug congregate in colonies that consist of adult and immature stages, others are solitary insects. Some solitary species inject toxic saliva and cause damage out of proportion to their feeding activity. When handled, most bugs emit a nauseous odour and a few can squirt a caustic smelly fluid which can damage sensitive tissues such as the eyes and nasal membranes.

Life cycle: eggs, which are often laid in crowded groups, hatch into small nymphs. The nymphs, which gather in groups, are usually of quite different shape and colouration to the adults, often resembling small spiders or ticks. They moult as they grow, eventually developing into an adult which is usually winged. Some adult bugs stand guard over their eggs until they hatch.

Plant damage: the symptoms of bug attack vary with the feeding habit of the particular group of bugs involved. Wilting of shoots is primarily caused by large solitary bugs and sometimes their feeding holes can be seen low down in the afflicted shoots. Leaf yellowing and dehydration of leaf tissue is more commonly associated with the colonising types of bug. Bugs that inject toxic saliva usually cause brown patches in the affected tissue.

Natural control: many predators feed on bugs including birds, hunting spiders, scorpions, assassin bugs, tachinid flies and chalcid wasps. In tropical regions parasitic fungi often destroy bugs during the wet season.

Other control: small infestations of bugs cause minor plant damage and control is not generally warranted. Large numbers, especially when the bugs gather in feeding swarms, should be controlled before they cause significant damage. It should be noted however, that some native bugs congregate in large non-feeding groups at certain stages in their life cycle. Although highly visible at this stage, they do not damage plants. Complete control of pestiferous bugs is difficult, but populations can be reduced by shaking the colonies into a collecting device such as an upturned umbrella or sheet of plastic. The collected insects can be killed by burning, immersion in hot water or water topped with kerosene. Bugs are sluggish in the early morning or during cool weather and are then easier to catch. If spraying is necessary a pyrethrum-based contact spray is usually successful.

NOTE: To aid with identification the bugs included here are grouped into families.

SHOOT WILTERS OR COREID BUGS (FAMILY COREIDAE)

There are about 60 native species of these sturdy bugs, each 2–3cm long, many of which suck the sap from developing plant shoots. The majority of species feed on eucalypts, causing wilting and dieback. Often a feeding hole left by the bug near the base of the afflicted shoot reveals the cause of the damage. The adults of some species produce a foul-smelling fluid when disturbed.

Death of a eucalypt shoot after feeding by shoot wilting bug – note puncture hole.
[D. Jones]

Coreid bug emerging from moult.
[T. Wood]

Coreid bug. [T. Wood]

Eucalyptus Tip Bug, *Amorbus alternatus*

Description: a stout, shield-shaped bug, 1.8–2.2cm long with a short spiny process on each shoulder. Adults are dull and blackish with a yellowish inverted V-shape on the wing-covers. The blue-and-orange nymphs, which taper rapidly to a narrow head, contrast markedly in colour with the adults.

Climatic regions: tropics and subtropics, coastal and inland areas.

Host plants: species of *Corymbia* and *Eucalyptus*.

Feeding habits: adults and nymphs suck the sap from young shoots causing wilting and dieback.

Often the feeding hole is visible at the base of the wilted shoot. These insects can be solitary or feed in small groups.

Notes: a widespread native bug that causes minor damage to garden eucalypts. Adults are slow-moving, whereas the nymphs move quickly.

Control: control is not usually warranted. Hand collection or disruption by smacking with sticks is usually sufficient.

Eucalypt Tip Bug, Amorbus alternatus. [T. Wood]

Coreid bug nymphs, Amorbus *sp.* [T. Wood]

Holy Cross Bug and other Crusader Bugs, *Mictis profana*

Description: large dark-coloured bugs 2–3cm long with a distinctive white or yellow St Andrew's cross on the back. Nymphs are brown or black with two small orange spots on the abdomen.

Climatic regions: tropics, subtropics and temperate areas, coastal and inland.

Host plants: species of *Acacia, Cassia, Corymbia, Eucalyptus, Hardenbergia, Hibiscus* and *Senna*; also a wide range of other plants; a serious pest of citrus.

Feeding habits: these are usually solitary insects that suck the sap of young shoots, causing wilting and dieback. They can also be found in larger infestations of adults and nymphs, becoming particularly active during hot weather.

Notes: the holy cross bug and other crusader bugs are large, conspicuous insects which are particularly prominent in tropical and subtropical regions. They eject a foul smelling caustic liquid when disturbed and on no account should be handled. This liquid can damage sensitive areas such as the eyes.

Control: as per introduction for bugs.

Holy Cross Bug. [T. Wood]

Tip damage by nymphs of Holy Cross Bug to Hardenbergia violacea. [S. Jones]

Crusader bug nymph. [T. Wood]

Callistemon Tip Bug, *Pomponatius typica*

Description: a slender, slow-moving, brown bug, 1.8–2cm long, with a pointed abdomen. Immature stages are similar to the adults, but smaller.

Climatic regions: tropics and subtropics, mainly coastal.

Host plants: *Callistemon* species and cultivars and some broad-leaves species of *Melaleuca*; sometimes severe on *Callistemon viminalis*.

Feeding habits: a solitary insect that sucks the sap from young growths causing wilting and premature death. Evidence of attack shows up as brown shoots amid clusters of normal healthy shoots. Close examination of afflicted shoots will show the feeding hole produced by the bug, often near the base of the shoot.

Notes: this pest resembles a dead leaf and can be extremely difficult to spot on the damaged growths. Its attacks can be annoying but are rarely serious and over time probably encourage branching.

Control: spraying is not usually warranted. Hand collection should be sufficient.

Callistemon Tip Bug damage. [D. Jones]

SHIELD BUGS (FAMILY PENTATOMIDAE)

A very large group containing nearly 400 species of native bugs. They are mostly shaped like a shield, broadest just behind the head and tapered to the wings. They can also be recognised by their antennae which have five segments. Many produce a smelly fluid when disturbed. Most are plant feeders although some are useful predators.

Mating shield bugs. [T. Wood]

Adult shield bug. [T. Wood]

Shield bugs feeding on sheoak seeds. [D. Jones]

Acacia Shield Bug, *Alcaeus varicornis*

Description: very flat, shield-shaped bugs 2–2.5cm long. The adults are greyish with dark brown areas and irregular mottling. The nymphs are dark grey to black with cream to white markings

Climatic regions: tropics, subtropics and warm temperate regions, mainly coastal.

Host plants: species of *Acacia*; also *Exocarpus cupressiformis*.

Feeding habits: adults and nymphs feed on young phyllodes and developing pods. Affected phyllodes develop brown blotches where feeding has occurred.

Notes: the adults are difficult to see when at rest on the bark. They also hide in crevices during the day.

Control: usually controlled by natural enemies.

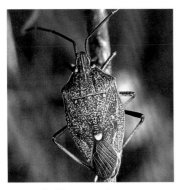

Brown Shield Bug Poecilometis patruelis. [T. Wood]

Nymph of the Brown Shield Bug Poecilometis patruelis. [T. Wood]

Brown Shield Bugs, several *Poecilometis* species

Description: stout, shield-shaped bugs 1.5–2.5cm long. Adults range from yellowish brown to dark brown, usually with darker spots and mottles. The tips of the hindwings protrude in a diamond-shape through a gap at the end of the forewings. The nymphs are oval to nearly circular wingless bugs, usually dark grey to black with cream to white lines and markings

Climatic regions: tropics, subtropics and temperate regions, coastal and inland.

Host plants: species of *Angophora*, *Banksia*, *Callitris*, *Corymbia* and *Eucalyptus*; also citrus and some rainforest trees.

Feeding habits: adults and nymphs feed on young shoots causing wilting, dieback and growth distortion. Adults tend to be solitary while nymphs congregate in crowded groups (the nymphs are often seen in groups on the bark of smooth-barked eucalypts).

Notes: adults and late-stage nymphs eject a foul smelling yellow liquid that can cause irritation to sensitive skin and membranes. One common species, *P. patruelis*, feeds on a wide range of eucalypts.

Control: usually controlled by natural enemies.

Green Shield Bug, *Ocirrhoe unimaculata*

Green Shield Bug, Ocirrhoe unimaculata. [T. Wood]

Description: a shield-shaped, brightly-coloured bug about 1cm long. Adults have a distinctive wedge-shaped structure on the back and a small patch of clear wing tissue. Juveniles are smaller green nymphs.

Climatic regions: subtropics and temperate regions, mainly coastal.

Host plants: this bug feeds on a wide range of plants including *Acacia sophorae*, *Chamaelaurium uncinatum*, *Correa alba*, *Leucopogon parviflorus* and *Myoporum insulare*.

Feeding habits: a solitary insect that is common in coastal districts. It feeds on buds and immature fruit.

Notes: this species is rarely a problem and adds interest to a garden.

Control: as per introduction for bugs.

Green Vegetable Bug, *Nezara viridula*

Nymph of the Green Vegetable Bug, Nezara viridula. [T. Wood]

Description: shield-shaped bugs, 1–1.5cm long. Adults are usually bright green but can be yellowish; they become brownish-green in winter. The nymphs are black or green with red, white, orange and green spots.

Climatic regions: tropics, subtropics and temperate regions, coastal and inland.

Host plants: a wide range of native and exotic garden plants; an important pest of vegetables; also citrus and passionfruit.

Feeding habits: adults and nymphs feed on young shoots causing wilting and growth distortion. They also damage buds, developing fruit and seeds. The adults tend to be solitary, whereas the nymphs congregate in groups.

Notes: an introduced species that is believed to be native to Ethiopia. It is now widely distributed in Australia. It is mainly present from spring to autumn in temperate regions but breeds all year in the tropics. All stages emit a strong unpleasant smell if disturbed.

Control: about four species of wasps that parasitise the eggs have been released to control this pest; also an exotic tachinid fly, *Trichopoda giacomellii*, has been successful as a biological control agent.

Young nymph of the Green Vegetable Bug, Nezara viridula . [T. Wood]

Spiny Orange Bug, *Biprorulus bibax*

Description: a stout, shield-shaped bug up to 2cm long. Adults are green to pale green with a prominent spine on each shoulder. The nymphs are brightly marked with orange and black and at first glance could be mistaken for a bush cockroach.

Climatic regions: tropics, subtropics and temperate regions, mainly coastal.

Host plants: species of *Citrus* including the native species *C. glauca* (syn. *Eremocitrus glauca*) and those formerly included in *Microcitrus*.

Feeding habits: adults and nymphs feed on young shoots and fruit causing wilting, dieback and distortion. Adults tend to be solitary while nymphs congregate in small groups.

Notes: later stage nymphs can eject a foul smelling yellow liquid that can cause irritation to sensitive skin and membranes. This bug is of limited significance in native gardens but can be a significant pest of commercial citrus orchards.

Control: usually controlled by natural enemies.

JEWEL BUGS
(FAMILY SCUTELLERIDAE)

These are colourful bugs, often with a metallic sheen. There are about 30 native species and they can be recognised by the wing-covers extending to the tip of the abdomen and completely enclosing the wings. Most are plant feeders although some are useful predators.

Jewel bug nymphs. [T. Wood]

Cotton Harlequin Bug, *Tectocoris diopthalmus*

Description: a stout, brilliantly-coloured bug which grows up to 1.5cm long. The adults are strikingly dissimilar, the males being metallic-blue with red blotches, whereas the females are a dull-orange with blue blotches. The nymphs range from blue to yellow with red or black markings.

Climatic regions: tropics and subtropics, mainly coastal.

Host plants: mainly native species of *Abutilon, Brachychiton* and *Hibiscus*. In coastal districts these insects are very fond of the Cottonwood (*Hibiscus tiliaceus*).

Feeding habits: this is a gregarious insect which gathers together in groups of mixed stages of development with no two individuals being of exactly the same colour pattern. They feed on young shoots, leaves, flower buds and seeds, causing wilting and dieback.

Notes: this brightly-coloured although variable bug is a common inhabitant of tropical gardens but is rarely a troublesome pest. The females guard their eggs until they hatch.

Control: as per introduction for bugs. Spraying is rarely necessary.

Cotton Harlequin Bug nymphs sheltering or feeding on Brachychiton *seeds.*
[D. Jones]

Green Jewel Bug, *Lampromicra senator*

Description: adults are colourful bugs, 1–1.4cm long with a metallic sheen. They are commonly bright green with a transverse orange band on the back and orange-and-green legs.

Climatic regions: tropics and subtropics.

Host plants: rainforest plants, especially *Ficus* species and some ferns.

Feeding habits: these insects feed on soft growth causing wilting and death. Attacks can be severe on developing fern fronds. This can be a persistent pest in ferneries, greenhouses and shadehouses.

Notes: these colourful insects gather together in non-feeding clusters in hot dry weather. They are often found resting in damp sites such as on shrubs and trees in moist gullies and along stream banks.

Control: as per introduction for bugs. Spraying is sometimes necessary to clean up persistent attacks.

Metallic Jewel Bug, *Scutiphora pedicellata*

Description: a shield-shaped, brightly-coloured bug about 1cm long. Adults are deep metallic blue mottled with black and with two bright red blotches on the thorax.

Climatic regions: tropics and subtropics, mainly coastal.

Host plants: this bug feeds on a variety of plants including *Ficus*, *Melaleuca* and *Pomaderris* species.

Feeding habits: a gregarious insect which in some seasons may swarm over the trees on which it feeds. They are particularly fond of *Ficus* fruit. Their feeding may be followed by sap exudation.

Metallic Jewel Bug feeding on Pomaderris *fruit.* [D. JONES]

Nymphs of Metallic Jewel Bug on Pomaderris. [D. JONES]

Notes: this brightly-coloured bug is a decorative addition to the garden, although sometimes it reaches nuisance levels and must be controlled. Another species, *S. pedicellata*, which is shiny metallic green with black spots and orange patches, is common in subtropical and temperate areas. It is commonly seen sucking juices from young leaves, buds and developing fruit of a range of garden plants.

Control: as per introduction for bugs.

Mallotus Harlequin Bug, *Cantao parentum*

Winter cluster of Mallotus Harlequin Bug, Cantao parentum. [D. JONES]

Description: distinctive orange bugs about 2cm long with prominent black spots (two spots on the thorax, eight spots on the abdomen).

Climatic regions: tropics and subtropics, mainly coastal and near coastal.

Host plants: species of *Acalypha*, *Croton*, *Macaranga* and *Mallotus*; sometimes also on species of *Araucaria*.

Feeding habits: these insects feed on young growth from spring to autumn, causing wilting and dieback.

Notes: these colourful insects gather together in highly visible non-feeding clusters during winter. The females guard their eggs and young nymphs against predators.

Control: as per introduction for bugs. Control is rarely necessary.

STAINER BUGS (FAMILY PYRRHOCORIDAE)

An important group of plant-feeding bugs on a world scale, however the family is less common in Australia and only two species are known to be pests. The bugs are often colourful and gather together in groups, feeding during the day.

Stainer Bug, *Dysdercus sidae*

Description: slender, long-legged bug about 1.2cm long. Adults are brownish to orange with a prominent dark spot on each wing-cover.

Climatic regions: subtropical and temperate regions, mainly inland.

Host plants: a very wide range of native plants; also cotton.

Feeding habits: an active bug which congregates on young shoots, buds and the underside of leaves. Its feeding causes wilting, tissue death and premature leaf-fall.

Notes: a native bug that can be a nuisance in some years. Its numbers are subject to considerable fluctuations.

Control: as per introduction for bugs.

Harlequin Bug, *Dindymus versicolor*

Description: adults up to 1cm long, red and black with a green or yellowish abdomen. Nymphs are wingless and even more brightly coloured than the adults.

Climatic regions: temperate areas, coastal and inland.

Host plants: species of *Abutilon, Alyogyne, Brachychiton, Callistemon, Grevillea, Hibiscus* and *Thomasia*, especially *T. macrocarpa*.

Feeding habits: the bugs either cluster in colonies consisting of adults and nymphs in various stages of development, or spread out to feed individually or in mated pairs. They feed on young shoots causing wilting and dieback, and on the underside of leaves resulting in dead patches and premature leaf-fall.

Notes: this pest is a familiar sight in waste areas, often feeding on weedy species of *Malva* (Mallows). Outbreaks of this pest are sporadic but it can severely damage or even kill young plants in new gardens. If handled these bugs emit a noxious odour.

Control: as per introduction for bugs. Spraying is rarely necessary. Cleaning infested areas reduces numbers as these pests shelter in accumulations of rubbish, weeds, bark and mulch.

Harlequin Bugs on Hibiscus heterophyllus. [D. JONES]

Harlequin Bugs on Correa. [S. JONES]

STINK BUGS (FAMILY TESSARATOMIDAE)

An important group of plant-feeding bugs on a world scale, however the family is less common in Australia with only one species considered to be a pest.

Bronze Orange Bug, *Musgraveia sulciventris*

Description: a stout, broad bug up to 2.5cm long. Adults are bronze to nearly black with a prominent black patch on the abdomen. The nymphs are much flatter than the adults and develop through a range of colours including green, salmon, orange and pink.

Climatic regions: tropics, subtropics and temperate areas, mainly coastal.

Host plants: native species of *Boronia, Citrus, Phebalium, Philotheca* and many other species of the family Rutaceae; also a pest of cultivated citrus.

Feeding habits: adults and nymphs congregate on young shoots causing wilting and dieback.

Notes: a repulsive native bug that ejects an evil smelling caustic fluid at the slightest disturbance. The smell of this fluid lingers for a long time and contact with the eyes and nasal passages should be avoided as it is caustic and can damage tender membranes. Although of limited importance in native gardens, bronze orange bug is a significant pest of commercial citrus groves. Despite their repugnant nature and smell these insects are avidly eaten by Spangled Drongos. When these birds are feeding on an infestation of this pest the whole area reeks from the exudation of the bugs.

Control: as per introduction for bugs. Usually controlled naturally by assassin bugs and birds.

SWARMING BUGS (FAMILY LYGAEIDAE)

A large group of smallish bugs that feed on sap with many species feeding on immature and ripe seeds. Members of this family are common in the tropics and are familiar to many gardeners, especially as these bugs gather in large swarms in late autumn and early winter, often smothering foliage.

Example of a Lygaeid bug. [T. Wood]

Coon Bug or Cottonseed Bug, *Oxycarenus luctuosus*

Description: an active slender bug about 0.5cm long. Adults, which have membranous wings, are black and white, the nymphs bright red.

Climatic regions: tropics and subtropics, coastal and inland.

Host plants: mainly species of the family Malvaceae, including *Abutilon, Gossypium, Hibiscus* and *Lavatera plebeia*; also some species of *Lasiopetalum*.

Feeding habits: the bugs feed on soft growth causing wilting; also developing fruit and seeds. Hardened plant tissue becomes dry and papery. This pest congregates in large swarms consisting of adults and nymphs. Although they do not feed while swarming, the combined weight of a large swarm can cause significant physical damage to plants.

Notes: the numbers of this native bug fluctuate greatly from season to season.

Control: as per introduction for bugs. Control measures will be necessary in some seasons to prevent rapid build up of numbers.

Swarming Seed Bug, *Graptostethus servus*

Description: an active, slender bug that swarms in the dry season The adults, which are about 0.9cm long, are orange with black markings and a distinctive orange cross on the back. The nymphs are orange and black.

Climatic regions: tropics and subtropics, coastal and inland.

Host plants: native and exotic legumes and grasses.

Feeding habits: the bugs feed on developing fruit and seeds. This pest forms large swarms of adults and nymphs after the wet season. Although they do not feed while swarming, the combined weight of a large swarm can cause significant physical damage to plants.

Notes: a seasonal pest.

Control: as per introduction for bugs.

Rutherglen Bug, *Nysius vinitor*

Description: an active, slender bug about 0.5cm long. Adults are a dull grey to brown with a couple of darker blotches. Nymphs are smaller and of similar colouration.

Climatic regions: tropics, subtropics and temperate regions, coastal and inland.

Host plants: feeds on a wide range of exotic and native plants but this pest seems to be particularly fond of daisies such as species of *Brachyscome, Helichrysum, Rhodanthe* and *Xerochrysum.*

Feeding habits: this pest appears during spring and summer and in some seasons congregates in thousands, the swarms consisting of winged adults and nymphs. They feed by sucking sap and the combined effect of a swarm severely debilitates the affected plant. Soft growth wilts and mature tissue takes on a dry appearance. While on the plant these insects not only feed but also move about actively. When disturbed the adults rise in clouds.

Notes: this native pest and related Grey Cluster Bug, *N. clevelandensis*, are widely distributed through eastern Australia. In some seasons they cause serious damage and may need to be controlled.

Control: as per introduction for bugs. These pests overwinter on weeds, especially in wet years. They should be controlled when first noticed as they can build up in numbers quickly. Outbreaks usually last for about two months.

MIRID BUGS (FAMILY MIRIDAE)

A very large group of slender bugs with numerous species occuring in Australia. Many feed on plants but a few are important predators of other insects (see chapter 3).

A typical mirid bug. [T. Wood]

Adult and nymph mirid bugs and phyllode damage to Acacia longifolia. [S. Jones]

Mirid bugs clustered on palm leaf. [D. Jones]

Fern Mirid Bug, *Felisacus glabratus*

Description: a slender active bug that grows to about 1.5cm long. Adults can be brown but are more commonly shiny green.

Climatic regions: tropics and subtropics, mainly coastal.

Host plants: various species of fern, particularly *Asplenium.*

Feeding habits: a solitary insect that attacks the developing croziers, sucking the sap from young tissue and causing papery patches. Damaged fronds can die prematurely.

Notes: generally a minor pest that appears sporadically. Occurs in many countries.

Control: not usually warranted. Hand picking is successful.

Green Mirid Bug. [T. Wood]

Green Mirid Bug, *Creontiades dilutus*

Description: a slender active bug that grows to about 0.7cm long. Adults are pale green, often with red markings. They have long antennae and clear gauzy wings folded flat on the back.

Climatic regions: tropics, subtropics and temperate regions, coastal and inland.

Host plants: a wide range of native and exotic garden plants, especially those with soft growth; often damages the young growth of *Alyogyne, Hibiscus* and *Swainsona*.

Feeding habits: can be solitary but more usually in groups.

Notes: a fast-moving native insect that appears sporadically in large numbers.

Control: as per introduction for bugs with spraying needed in some seasons.

Myrtle Mirid Bug or Leaf-blotching Mirid Bug, *Eucerocoris suspectus*

Description: slender, long-legged bug 0.6–1cm long. Adults are orange with black legs, long black antennae and gauzy wings. Nymphs are elliptical in shape and orange with bands on the legs and antennae.

Climatic regions: tropics and subtropics, mainly coastal.

Host plants: broad-leaved species of *Melaleuca*, including *M. cajuputi, M. leucadendra, M. quinquenervia* and *M. viridiflora*; also some species of *Callistemon,* including *C. polandii* and *C. viminalis.*

Feeding habits: adults and nymphs suck the sap from young leaves and shoots. Their highly toxic saliva causes large patches of tissue to die around the areas where their mouthparts have pierced. This results in brown sunken patches on the leaves that become papery; they can also severely distort new growth. Surprisingly, despite the very obvious results of this pest's activities, very few of them can be located on an affected plant.

Notes: a serious and debilitating pest which causes damage out of all proportion to its size and abundance. Attacks mainly occur during December to March and are worse during very wet seasons. This pest, which can cause severe damage to nursery plants, is being tested as a biological control agent for weedy populations of *Melaleuca quinquenervia* naturalised in southern parts of the USA.

Control: as per introduction for bugs. Spraying small plants is often necessary.

Acacia Spotting Bug, *Eucerocoris tumidiceps*

Description: a dull brown slender bug about 0.5cm long with antennae about as long as its body. Nymphs are smaller versions of the adult.

Climatic regions: subtropics and temperate region, mainly coastal.

Host plants: phyllodinous wattles.

Feeding habits: adults and nymphs suck sap from the phyllodes. Their saliva must be toxic for patches of tissue collapse and die around where they feed. Badly affected phyllodes may fall prematurely.

Notes: these bugs are very difficult to see and are usually not present in large numbers.

Control: as per introduction for bugs. Spraying may be necessary.

LACE BUGS (FAMILY TINGIDAE)

A small group of native bugs with a distinctive pattern of reticulate veins on the forewings.

Olive Lace Bug, *Froggattia olivinia*

Description: slender bug about 0.3cm long; adults pale to dark brown with lighter blotches, the hindwings transparent and lacy; nymphs are covered with stiff spines.
Climatic regions: subtropics and temperate regions, coastal and inland.
Host plants: species of the family Oleaceae such as *Notelaea longifolia;* also a serious pest of cultivated olives.
Feeding habits: the bugs congregate on the undersides of the leaves which become covered with tarry excrement and fall prematurely.
Notes: a native bug with messy habits but of limited significance for native gardens. It does however cause significant damage to commercial olive plantations, passing through several generations in a season. It also has impacts on populations of feral olives.
Control: as per introduction for bugs. Spraying is rarely necessary in native gardens.

SOAPBERRY BUGS (FAMILY RHOPALIDAE)

These bugs, commonly known as scentless plant bugs, feed on the fruit and seeds of some rainforest plants, especially members of the Sapindaceae. They are fast runners and do not emit a smell when disturbed.

Red-eyed Bug, *Leptocoris mitellatus*

Description: slender bug, 1.2–1.5cm long; adults dull red with a bright red abdomen, shiny red eyes, black legs and antennae, the hindwings are dark red-brown, transparent and lacy; nymphs are red and shortly hairy.
Climatic regions: subtropical and temperate regions, coastal and inland.
Host plants: *Alyogyne, Atalaya, Atriplex, Callicarpa, Cupaniopsis, Dodonaea*; also passionfruit, peaches, tomatoes and other vegetables.
Feeding habits: the bugs congregate in colonies of mixed stages, often on the undersides of leaves.
Notes: a native bug that is attracted to lights and sometimes shelters in large numbers in buildings, creating mess and other problems. This bug can be a nuisance to fruit and vegetable growers.
Control: as per introduction for bugs. Spraying is rarely necessary in native gardens.

Clusters of Red-eyed bugs, Leptocoris mitellata. [T. Wood]

PSYLLIDS AND LERPS

Psyllids, also known as jumping plant lice, are placed in the family Psyllidae of the suborder Sternorrhyncha. Although tiny, being only 0.2–0.8cm long, they are among the most damaging of plant pests, sucking sap from the leaves of a wide range of native plants. The adults resemble miniature cicadas, holding their wings in an inverted V-shape over their bodies. The nymphs are often flattened on the leaf surface. These insects can be especially damaging to eucalypts in natural populations and cultivated plants in parks, gardens and plantations, causing premature leaf-fall, weakened growth, dieback and sometimes even tree death. Many psyllids have a restricted host range (often only a single plant species is attacked), others are much less specialised. Australia has about 330 species of psyllid.

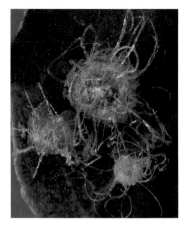

Lerps of the genus Glycaspis – *the insect is the orange blob.* [J. FANNING]

Two groups of psyllids can be readily recognised. One group, known as the 'free-living psyllids', move about freely, attacking the soft tissue of new shoots. Some free-livers form galls or pits (see chapter 7), others live within leaves that have curled as a result of their feeding activities. The second group, known as 'lerp insects', build a soft or hard covering under which they shelter and feed (the structure itself is the lerp – derived from an aboriginal word, either 'lerep' or 'larp'). Some lerp insect species are well known for the waxy shell-shaped covering they construct. Shelters can be intricate and beautifully patterned and are unique in shape and structure for each species. They are mostly waxy white but one species constructs shelters that are bright scarlet. Often lerp insects are found feeding together in colonies of mixed species. Lerp insects, which sometimes explode in numbers to form huge infestations, were gathered by Aboriginal people in many parts of Australia as a food source.

Lerps of the genus Creis or Hyalinaspis *on young leaves of Red Box.* [D. JONES]

Life cycle: the adults, which are winged, are capable of flying (some species are strong fliers, others weak, many can jump) but are often dispersed by wind. Feeding colonies consist of a mixture of adults and nymphs in various stages of development. Adults mate and the females lay stalked eggs in clusters, rows or circles on the leaves of the host plant. The newly hatched nymphs move to a suitable site to feed and develop. The lerp insects construct the lerp shelter from their waste products which harden on contact with the air (the waste from free-living psyllids stays as a liquid). Most psyllids occur from spring to autumn, some species overwinter as nymphs.

Plant damage: psyllids and lerp insects attack new shoots and leaves, causing a range of symptoms. Free-living psyllids cause leaf-curling and shoot-distortion. The feeding of some species causes the development of small pimple-like galls on the affected leaves, others develop sunken pits. The feeding activities of many lerp insects results in small to large discoloured dry or papery patches developing on the affected leaves. This damage, which can seem out of proportion to the number of feeding insects, probably results from the injection of toxic saliva. Affected areas on the leaves, which are yellowish at first, soon become pink, purplish or brown and contrast with the normal green of unaffected tissue. In severely infested trees the whole crown takes on a purplish to bronze appearance. Afflicted leaves are shed prematurely and severe infestations may partially defoliate trees. If attacks persist over successive seasons the tree becomes weakened and dieback of the branches can occur. Honeydew is secreted by the feeding insects.

Serious damage to Eucalyptus blakelyi *by the White Lace Lerp,* Cardiaspina albitextura. [D. Jones]

Severe lerp damage to leaves of Eucalyptus botryoides. [D. Jones]

Distorted growth on Acacia terminalis *caused by psyllids.* [D. Jones]

It has been shown that periodic attacks by lerp insects on forestry eucalypts reduce vigour and slow growth. Severe attacks can cause defoliation and result in the production of epicormic shoots which are themselves susceptible to attack by other pests such as stem borers. In some seasons huge areas of various species of forestry eucalypts have been defoliated by lerp insects. Persistent attacks reduce wood quality and value. Cultivated plants of the mahogany gums, *Eucalyptus botryoides* and *E. robusta*, are very prone to attacks by lerps, rendering the trees unsightly.

Natural control: psyllids and lerp insects are eaten by a range of predators including many small birds (pardalotes and Willie Wagtails are very fond of them – also lorikeets and Swift Parrots), hoverflies, hunting spiders, scorpions, assassin bugs, ladybirds and lacewings. Eggs and nymphs are parasitised by a range of tiny wasps.

Interaction with Bell Miners: the Australian native bird known as the Bell Miner or 'bellbird', *Manorina melanophrys*, feeds on the sugary coating of the lerp insect, but often does not eat the psyllid hiding beneath the lerp. These birds, which have increased dramatically in numbers in psyllid-infested forests, also

Dieback in Eucalyptus botryoides *caused by a combination of lerps and a high population of Bell Miners.* [D. Jones]

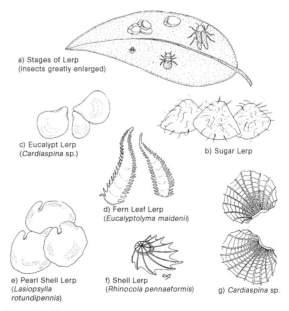

a) Stages of Lerp (insects greatly enlarged)

c) Eucalypt Lerp (*Cardiaspina* sp.)

b) Sugar Lerp

d) Fern Leaf Lerp (*Eucalyptolyma maidenii*)

e) Pearl Shell Lerp (*Lasiopsylla rotundipennis*)

f) Shell Lerp (*Rhinocola pennaeformis*)

g) *Cardiaspina* sp.

Examples of lerps. [D. Jones]

aggressively defend their territory, driving off other birds that might feed on the psyllids. The action of the Bell Miners, together with a range of other complex factors, including land clearing, forest fragmentation, weed invasion, fire and soil water changes, have led to large areas of forest, including some World Heritage sites, suffering decline and dieback in eastern Australia. Affected forests are distributed from southern Queensland to Victoria and, although the Bell Miners play a significant role in the decline and dieback, it is difficult to isolate their impact from the complex of other issues involved.

Other control: adult psyllids are attracted to yellow colours and can be trapped on yellow sticky traps. Psyllid and lerp insect infestations in parks and gardens fluctuate considerably each season. Although often not a major problem, lerp insects in particular can increase dramatically in some years, disappearing just as quickly. Control of severe lerp insect infestations in gardens is generally not practical and the populations are hopefully soon controlled by natural enemies. Persistent infestations on relatively small trees can be reduced or controlled by spraying with imidacloprid. For large trees control is generally impractical; trunk injections of systemic materials can be used, but are not recommended for environmental reasons (see chapter 2).

EXAMPLES

Blue Gum Psyllid, *Ctenarytaina eucalypti*

Description: a free-living psyllid that jumps when disturbed. Adults 0.3cm long, brown with milky to pale yellow wings, cream to orange patches on thorax. Very young nymphs (which look like aphids) are pale yellow, older nymphs are brown. The nymphs become covered with white, fluffy, waxy secretions.
Climatic regions: temperate areas.
Host plants: *Eucalyptus* species with blue juvenile foliage, particularly *E. globulus*, *E. leucoxylon* and *E. nitens*.
Feeding habits: colonies of mixed stages feed on seedlings and the juvenile growth of adult plants. Young shoots become distorted and deformed. Affected areas become covered with white waxy secretions.
Notes: a sporadic pest that appears in significant numbers during some seasons, but is often controlled by parasitic wasps. Blue Gum Psyllid has become established in commercial eucalypt plantations in many countries and also affects ornamental eucalypts grown in England.
Control: as per introduction for psyllids and lerps.

Christmas Bush Psyllid, *Cerotrioza* species

Description: a free-living psyllid that jumps when disturbed. Adults 0.3cm long, yellowish with darker concentric bands and distinctive dark markings on the inner wing margins. The nymphs are yellow with red eyes.
Climatic regions: warm temperate regions.
Host plants: New South Wales Christmas Bush, *Ceratopetalum gummiferum*, and Coachwood, *Ceratopetalum apetalum*.
Feeding habits: the young growth is attacked causing the leaf margins to roll inwards and the whole leaf to partially curve like a new moon. Adults and nymphs live and feed within the protective leaf margins. Severe infestations cause growth distortion, stunting, dieback, blackening from sooty mould and loss of flower (calyx) production.
Notes: a significant pest that causes commercial damage to cut flower plantations. Attacks have occurred on young nursery plants and mature plants grown for cut flower production. Attacks are worst on vigorous, well-fed plants.
Control: as per introduction for psyllids and lerps. Control is difficult because the psyllids are protected from contact sprays by the inrolled leaf margins.

Acacia Sucker, *Acizzia uncatoides*

Description: a free-living psyllid. Adults resemble pale yellow to light brown cicadas and are 0.2–0.3cm long with clear wings. Nymphs are small yellow to pale orange insects about 0.2cm long, with red eyes.

Climatic regions: tropics, subtropics and temperate regions, chiefly coastal.

Host plants: Known to feed on about 50 species of *Acacia* (including species native to Africa and Hawaii); also several species of *Albizia*.

Feeding habits: the insects congregate on young shoots, sometimes forming dense colonies. Waxy secretions and honeydew (followed by sooty mould) build up on infested shoots. Stem dieback and breakage can occur following severe infestations.

Notes: a native pest that has become established in many countries overseas where it causes serious damage. A biological control program has been initiated in Hawaii because of the threat posed to the endemic *Acacia koa*.

Control: as per introduction for psyllids and lerps.

Black Wattle Bunching Psyllid, *Acizzia acaciaedecurrentis*

Description: a free-living psyllid. Adults are brown to black, 0.3–0.4cm long, with paler patches. Nymphs are small, yellow to orange, oval, mealybug-like insects about 0.3cm long, often hidden among curled leaves or waxy secretions.

Climatic regions: temperate regions.

Host plants: species of *Acacia*, including *A. decurrens*, *A. longifolia*, *A. mearnsii* and *A. pycnantha*.

Feeding habits: young growth is attacked resulting in severely deformed leaves and shoots. The dwarfed or compacted shoots bend strongly and the leaves are bunched together and curl, providing ideal shelter for the nymphs. Often white waxy secretions are also present on infested shoots. Sooty mould is also a common feature.

Notes: a significant pest that is usually prominent on young seedlings and saplings. Its attacks often result in dwarf stunted plants that become unthrifty and die. This pest can be locally common, affecting every young tree in a locality.

Control: as per introduction for psyllids and lerps. Control is difficult because the psyllids are protected from contact sprays by the bunched shoots.

Black Wattle Bunching Psyllid damage on Acacia mearnsii. [D. Jones]

Cottonwood Psyllid or Hibiscus Psyllid, *Mesohomotoma hibisci*

Description: a free-living psyllid that jumps when disturbed. Adults 0.5cm long, head and thorax orange, abdomen green, wings clear with dark veins. Nymphs exude masses of white waxy threads.

Climatic regions: tropics and subtropics, mainly coastal.

Host plants: species of the family Malvaceae, including *Abutilon*, *Hibiscus* and *Thespesia*; this insect is particularly fond of Cottonwood, *Hibiscus tiliaceus*.

Feeding habits: colonies of mixed stages feed on young shoots and leaves which can wilt and fall prematurely. Affected areas develop yellow patches and become covered with white waxy threads.

Notes: a sporadic pest which appears in significant numbers during some seasons.

Control: as per introduction for psyllids and lerps.

Eucalyptus and Bloodwood Lerps, species of *Cardiaspina, Glycaspis, Spondyliaspis* and *Eucalyptolyma*

Description: winged adults (0.3–0.5cm long) mate and females lay clusters of eggs on young eucalypt leaves which hatch into nymphs. The nymphs move to suitable sites, feeding and constructing a lerp under which they feed and shelter. The lerp is enlarged as the nymph grows. The shape and other features of the lerp are characteristic of each species. Lerps in the genus *Cardiospina* produce intricate lacy or lattice-patterned fan-shaped or basket-like lerps, species of *Spondyliaspis* produce shell-like lerps (they resemble scallop shells), whereas species of *Glycaspis* either produce soft white conical lerps with a sugary texture, or round lerps that resemble armoured scales. These lerps often also have waxy threads that extend from the margins. Lerp insects of the genus *Eucalyptolyma* produce elongate lerps with ragged or lobed margins.

Cardiaspina *lerps on* Eucalyptus botryoides. [D. BEARDSELL]

Climatic regions: tropical to temperate regions, inland and coastal.

Host plants: various *Eucalyptus* and *Corymbia* species.

Feeding habits: the lerp insects feed on both surfaces of young leaves. Toxic substances in the saliva of *Cardiaspina* species cause localised patches of leaf tissue to die, initially turning reddish to purplish, later bronze to brown. In severe attacks the lerp insects cluster together, killing leaves and causing premature leaf-fall. Leaves affected by species of *Glycaspis*, *Eucalyptolyma* and *Spondyliaspis* do not develop discoloured or dead patches but become messy from sugary exudates, waxy secretions and sooty mould.

Notes: eucalypt lerp insects, especially species of *Cardiaspina*, are a widespread, very common and significant group of pests that occasionally undergo tremendous population explosions. Healthy eucalypts can cope with severe attacks over a couple of seasons, but repeated defoliation severely weakens trees and may result in stem dieback. A number of lerp insects have become established as serious pests in overseas gardens and eucalypt plantations causing premature leaf fall, dieback, stunting and tree death.

Eucalypt lerp species and hosts: lerp insects of the genus *Cardiaspina* attack a wide variety of eucalypts. In northern New South Wales the flooded gum, *Eucalyptus grandis,* is attacked by two species: Finger Lerp, *C. maniformis,* feeding on the upper surface of the leaf and Brown Basket Lerp, *C. fiscella*, on the lower surface.

White Lace Lerps, Cardiaspina albitextura, *on* Eucalyptus blakelyi. [D. JONES]

Lerps on Eucalyptus melliodora. [D. JONES]

Rectangular Lerps, Glycaspis siliciflava, *on* Eucalyptus robusta. [D. JONES]

Ants attending lerps of the genus Glycaspis *on a* Eucalyptus *leaf.* [T. Wood]

Stem lerps of the genus Phyllolyma *on Red Box, with nymph visible at the bottom.* [D. Jones]

Lerp of Spondyliaspis plicatuloides. [T. Wood]

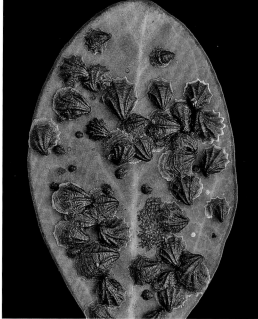

Shell lerps, Spondyliaspis bancroftii. [T. Wood]

- *Cardiospina fiscella* is also responsible for severe attacks on two species of mahogany gums, *E. botryoides* and *E. robusta*, in extensive areas of New South Wales and eastern Victoria.

- Grey Box Psyllid: an outbreak of an as yet unidentified species of *Cardiospina* has recently affected 70,000 hectares in western Sydney, severely impacting areas of threatened Cumberland Plain Woodland and in particular Grey Box, *E. moluccana*.

- Ironbark Lerp, *Cardiospina artifex*, has been recorded defoliating Grey Ironbark, *Eucalyptus paniculata*, in central New South Wales and Flooded Gum, *E. grandis*, in south-eastern Queensland.

- The red gums, *Eucalyptus blakelyi*, *E. camaldulensis* and *E. tereticornis*, are at times severely attacked by White Lace Lerp, *Cardiospina albitextura*, and Red Gum Basket Lerp, *C. retator*.

- Host-specific lerp insects include: Mountain Ash Psyllid, *Cardiospina bilobata*, which has caused extensive defoliation of Mountain Ash, *E. regnans*, regrowth in Victoria; Pink Gum Lace Lerp, *C. densitexta*, attacking Pink Gum, *E. fasciculosa*, in south-eastern South Australia; and Manna Gum Lerp, *C. squamula*, on Manna Gum, *E. viminalis*.

- Yellow Box, *E. melliodora*, is attacked by two different lerp insects: Yellow Box Lace Lerp, *Cardiospina tenuitela*, in the Hunter Valley, NSW; and Yellow Box Lerp, *Lasiophylla rotundipennis*, in the ACT and western NSW. The latter species not only attacks leaves but also forms colonies on stems and young branches.

- Species of *Glycaspis* attack a range of eucalypts but often cause only minor damage. Red Gum Lerp, *Glycaspis brimblecombei*, which produces hard, round, scale-like lerps (regarded as a relatively minor pest in southern Australia), is a serious pest of plantations in USA, Italy and South Africa.

- The Rectangular Lerp, *Glycaspis siliciflava*, commonly attacks *Eucalyptus robusta* in coastal districts. It forms a distinctive rectangular-shaped white covering over the yellowish insect. Other species of *Glycaspis* form soft fibrous lerps.

- Eucalypts and bloodwoods are also attacked by a range of lerp insects other than species of *Cardiaspina* and *Glycaspis*. The lerp of Spotted Gum Psyllid, *Eucalyptolyma maidenii*, which attacks the Spotted Gum, *Corymbia maculata*, and Lemon-scented Gum, *C. citriodora*, is elongated with lobed margins and resembles the leaf of a fern.

- The lerp of *Eucalyptolyma irregularis* is irregularly shaped. This latter species sometimes appears in huge numbers on garden-grown eucalypts and bloodwoods in subtropical regions.

- The familiar manna deposit found on the ground beneath peppermint forests in south-eastern Australia is the accumulated scales of the Sugar Lerp, *Spondyliaspis eucalypti*. This material, which is sweet and sugary, can be eaten or mixed with water to make a drink.

- Two species of *Spondyliaspis* have scallop-shell-shaped lerps: *S. plicatuloides* is common in south-eastern Australia, often on Snow Gum, *E. pauciflora*; *S. bancroftii* occurs in Queensland and New South Wales.

- Lerps on eucalypt stems are commonly of the genus *Phylloloma*.

Control: as per introduction for psyllids and lerps. Control of lerps is often difficult to achieve or impractical; impossible on large trees.

Eucalypt Gall Psyllids, species of *Schedotrioza* and some *Glycaspis* – see Chapter 7.

Fig Psyllid, *Mycopsylla fici*

Description: a lerp insect that lives in colonies under communal shelters. Dense clusters of brownish eggs laid commonly on the leaf margins hatch simultaneously and the nymphs feed together in a group. A sticky communal lerp is formed over the feeding group consisting of wax and honeydew secreted by the flattish greenish-yellow nymphs. Winged adults, brownish and about 1cm long, fly to new areas to feed and mate.
Climatic regions: tropics, subtropics and temperate regions.

Sparse crown of large Ficus macrophylla *suffering from severe infestation of Fig Psyllids.* [D. Jones]

Early symptoms of Fig Psyllid damage to Ficus macrophylla. [D. Jones]

Leaves of Ficus macrophylla *affected by Fig Psyllid.* [D. Jones]

Psyllid infestation of Ficus australis. [D. Jones]

Host plants: a serious pest of Moreton Bay Fig, *Ficus macrophylla*, less serious on *F. rubiginosa* and *F. obliqua*, which seem to be better able to resist attacks.

Feeding habits: feeding groups of nymphs occur on young shoots and the underside of leaves. The lerps, formed as a result of their feeding, are thick sticky blobs of white waxy material. Affected leaves develop yellow patches that become necrotic before the leaves fall prematurely. Fallen leaves are very sticky to touch and become a nuisance to pedestrians and vehicles.

Notes: a sporadic pest that periodically reaches plague proportions. Infested trees lose leaves and develop a sparse canopy. Severe attacks can lead to complete defoliation, resulting in sunburnt and cracked bark on exposed limbs. Persistent attacks weaken trees, rendering them susceptible to attack by other pests (such as borers) and wood-decaying diseases. Cases of death are known. Healthy, regularly watered trees are much less susceptible to psyllid damage than weakened trees.

Control: as per introduction for psyllids and lerps. Control is impossible on large trees.

Syzygium Leaf Gall Psyllid or Pimple Psyllid, *Trioza eugeniae* – see Chapter 7.

Other Psyllids and Lerps

- Two species of free-living psyllids attack the Kurrajong, *Brachychiton populneus*, and cause severe damage in some seasons. *Aconopsylla sterculiae* congregates in mixed colonies of adults and nymphs on young twigs and new shoots, secreting large amounts of honeydew, whereas the star psyllid, *Tyora sterculiae*, attacks the leaves. The nymphs of the latter species have a star-shaped tuft of white waxy filaments at the end of the body.

- Several species of free-living psyllids (mostly in the genus *Acizzia*) feed on wattles, *Acacia* spp., and cause sporadic problems. Many are host-specific, often only attacking a single species of wattle. They are only noticeable when present in large numbers, usually attacking young shoots and covering the affected tissue with abundant honeydew. In subtropical areas the flower buds and young shoots of *Acacia fimbriata* are often distorted by a species of psyllid.

- The Lemon Gum Psyllid, *Cryptoneossa triangula*, is a free-living species that attacks the young shoots of Lemon-scented Gum, *Corymbia citriodora*, covering the growth with waxy secretions and causing growth distortion.

- Bunching psyllids attack the young shoots of some native plants, causing the shoots to become bunched and deformed. Small yellow to orange nymphs hide among the affected growth. These insects have toxic saliva and can cause serious growth malformation. Spraying may be necessary for control.

- A destructive group of psyllids attacks the flower buds of many plants causing distortion and interrupting flowering. In subtropical areas the flowers of some species and hybrids of racemose *Grevillea* are attacked, causing distortion and abortion of the buds. White waxy threads are a clue to the presence of the pest.

CICADAS

Cicadas are not only interesting insects, but are a delight to children because of their appealing appearance and the noises they emit. They possess the best-known sound apparatus among insects and their orchestras enliven warm summer days and evenings. Many cicadas are large and colourful and in some countries they are treated as pets with the adults kept in cages. There are about 220 species of cicada native to Australia, including the common and very noisy Green Grocer Cicada, *Cyclochila australasiae*, which has adapted extremely well to city life.

Cicadas are readily recognised by their characteristic shape, bulbous eyes set on either side of the head and large transparent wings marked by a network of darker veins. The wings are held flat or in an inverted V-shape above the body. Cicada adults commonly range in size from 2–8cm long with the occasional giant species reaching 12–15cm in length. Mostly seen in summer they are renowned for the calls of the males, which range from continuous buzzes to strident sibilations. Cicadas fluctuate greatly in numbers, emerging in abundance in some years, less so in others.

Life cycle: adults, which usually only live for a few weeks, mate and the females lay eggs in notches and slits she cuts into bark. On hatching the nymphs drop to the ground and live underground where well-developed front legs enable them to burrow through the soil. The nymphs shed their skin as they grow. The period spent underground varies with the species (commonly two to seven years, but as long as 15–17 years in some American species). When mature they burrow to the soil surface (usually in the evening), emerge and climb the nearest obstacle (usually a bush or tree). Here the adult emerges from its skin for the last time, leaving the shed skin (exoskeleton) clinging to the support.

Plant damage: cicadas are sucking insects that feed on sap. The nymphs suck the sap from roots and the adults from the stems of a wide range of native and introduced plants. They feed by inserting their large mouthparts into the veins of a plant. Some species feed on soft or developing shoots causing wilting, others feed on mature trunks and branches, with no apparent detrimental effects. Large quantities of honeydew are expelled during feeding – it literally rains down when large numbers of cicadas are feeding in a tree. Adults may also damage the bark of branches and twigs during egg-laying. Despite these activities, overall plant damage from cicadas is relatively minor.

Natural control: numerous species of bird find cicadas a tasty part of their diet, often only eating the soft abdomen and discarding the rest often while the poor insect is still alive. Possums include them in

Early stages of adult cicada emerging. [T. Wood]

Adult cicada nearly fully emerged from pupa. [T. Wood]

Recently emerged adult cicada. [T. Wood]

their diet as also do bats, hunting spiders and lizards. They are also predated by mantids, tree crickets and large ants. Predatory wasps, known as 'Cicada killers' also use them as food for their young. Specialised beetles parasitise the nymphs.

Other control: cicadas are not regarded as major garden pests and in fact add interest and sound to a garden. Chemical control is unnecessary and if numbers are a nuisance they are readily dispersed by hosing, sticks and other missiles.

EXAMPLES

Bladder Cicada, *Cystosoma saundersii*

Description: adults about 5 cm long, a delicate pale green with opaque green wings and a conspicuously inflated abdomen.

Climatic regions: Queensland tropics and subtropics, northern New South Wales.

Host plants: *Eucalyptus* species and some native rainforest plants.

Feeding habits: solitary insect. Adults suck sap from young shoots, nymphs from tree roots.

Notes: an interesting weak-flying cicada that drops to the ground when disturbed and 'plays dead'. It calls in the evening and at night, remaining still and quiet during the day to avoid predation.

Bladder Cicada. [D. Jones]   *Green Grocer Cicada.* [T. Wood]

Control: natural enemies.

Green Monday Cicada or Green Grocer Cicada, *Cyclochila australasiae*

Description: adults 6–8 cm long, a delicate pale-green with clear wings.

Climatic regions: temperate areas.

Host plants: many *Eucalyptus* species and a wide range of other plants, both native and exotic.

Feeding habits: solitary or in loose colonies. Sucks sap from shoots and branches. Nymphs suck sap from tree roots.

Notes: a very familiar large, noisy insect (possibly the noisiest insect in the world) which appears in large numbers in some years.

Control: natural enemies.

FROGHOPPERS, SPITTLEBUGS AND CUCKOO SPIT

The free-living adults of these insects can jump or hop strongly when disturbed. They are small – mostly 0.4–0.6cm long – and resemble miniature cicadas. Some species have broad cicada-like heads, others have narrow, tapered or horn-like heads. The adults, which are often solitary or occur in small groups, are readily mobile and not easily observed. The nymphs, which often congregate together, are also difficult to observe but their presence is often announced by actively feeding groups of ants. Both the adults and their nymphs feed by sucking plant sap.

The nymphs of froghoppers construct hard calcareous tubes attached to the branchlets on which they feed. These tubes are open at the wide end and taper to a narrow bottom. The nymphs feed within the tubes (one nymph per tube), gaining protection from attack by predators. They live immersed in liquid secretions which prevents desiccation. The tubes, which are extended as the insect grows, remain long after the insects have matured into adults.

Larval tubes of froghoppers. [D. JONES]

The nymphs of spittlebugs cover themselves with a white, frothy secretion derived from the plant sap. This material, known as cuckoo spit or frog spit, provides insulation, prevents desiccation and also hides the nymphs from predators. The material also has a taste that deters predators. It is made when the nymph blows air through soapy waste fluid exuded from its posterior glands.

Plant damage: froghoppers are mainly found on small stems and branchlets of eucalypts and wattles. Occasionally attacks by froghoppers can be severe enough to stunt growth but mostly these are minor pests. Sooty mould grows on their exudates. Ants often attend froghoppers and collect the sugary secretions.

Spittlebugs attack a range of succulent plants including saltbushes, *Goodenia* and *Scaevola* species. They also feed on the young growth of some woody plants including sheoaks, wattles, baeckeas, grevilleas and many rainforest plants. The native spittlebugs are rarely a problem in a garden and can add interest. Some exotic species however, which are not found in Australia, occur in huge infestations and cause significant damage to garden plants and crops such as sugar cane.

Natural control: the adults of froghoppers and spittle bugs are eaten by small birds such as thornbills and pardalotes, as well as hunting spiders, lizards, assassin bugs, robber flies, ladybirds and lacewings. The nymphs are parasitised by small wasps.

Other control: froghoppers are rarely troublesome and can easily be squashed while in the tube stage. Badly affected stems or branches can be cut off and burnt. Spittlebugs can be removed by squashing or dispersing with a jet of water. The appearance of both froghopper and spittlebugs in a garden is mostly temporary, damage is usually minor and chemical control is unnecessary.

EXAMPLES

Common Froghopper, *Chaetophyes compacta*

Description: small cicada-like insect, 0.6–0.8cm long. Females are greenish-brown while the smaller males are dark brown to black, sometimes with a green head.
Climatic regions: tropics and temperate regions of eastern states.
Host plants: mainly *Acacia* and *Eucalyptus* species.
Feeding habits: solitary or in small colonies on young shoots. Larval tubes often occur in clusters. Adults hop readily when disturbed.
Notes: a widespread and common species.
Control: usually controlled by natural enemies.

Spine-tailed Froghopper, *Machaerota finitima*

Description: an unusual small brown insect, 0.6–0.8cm long with lacy wings and a prominent curved, spine-like appendage on the tip of the abdomen.
Climatic regions: tropics and subtropics.
Host plants: species of *Acacia, Callistemon, Casuarina, Eucalyptus* and *Melaleuca*; also some rainforest plants.
Feeding Habits: usually solitary but sometimes in small groups.
Notes: an interesting species of minor significance.
Control: usually controlled by natural enemies.

Common Spittlebug, *Philagra parva*

Description: small, brown, solitary, cicada-like insect about 0.8cm long with a curved horn on the head. The soft grey to brown nymphs cover themselves with white froth.
Climatic regions: tropics and temperate regions.
Host plants: a wide range of plants, including many rainforest plants and species of *Acacia, Eucalyptus, Casuarina* and *Allocasuarina*.
Feeding habits: nymphs, covered by frothy spittle, often lodge in leaf axils on young shoots. Dark marks on the stem indicate feeding sites.
Notes: a very common insect that is of usually minor significance. Infestations are often sporadic and short-lived but sometimes numerous plants in an area can be attacked. Occasionally heavy infestations can cause premature leaf fall.
Control: by hosing. Natural enemies usually keep numbers down.

Cuckoo Spit, Dark Spittlebug, *Bathylus albicinctus*

Description: small, beetle-like insect about 0.5cm long. Adults are brown to black with white markings. Nymphs feed beneath a covering of white froth.
Climatic regions: tropics and temperate regions.
Host plants: herbaceous plants including grasses, lilies and palms; also young growth of eucalypts and *Callitris*.
Feeding habits: nymphs, covered by frothy spittle, often lodge in leaf axils on young shoots.
Notes: a common insect that is usually of minor significance.
Control: by hosing. Natural enemies usually keep numbers down.

Cuckoo Spit on Acacia. [D. Jones]

LEAFHOPPERS, TREEHOPPERS, PLANTHOPPERS AND JASSIDS

These common names are applied loosely to a very large group of sap-sucking insects that feed on a wide range of plants. More than 20,000 species have formally named in the world and estimates suggest that another 80,000 or so species are yet to be described. As a group they are particularly well represented in the tropics. Leafhoppers embrace a wide range of shapes, forms and colours. Variation in size and colour is sometimes displayed even within a species. Many leafhoppers are of drab colouration, commonly in shades of green and brown, but some species are very brightly coloured. Many leafhoppers resemble miniature cicadas with wings meeting tentwise over the body. Others have broad, lacy wings that meet in a high or low V-shape.

Most leafhoppers are highly specialised and feed on a single plant species or within a small group of related plant species. Some have a wider host range and a few naturalised exotic species cause damage to some important crop plants, including cotton, lucerne, potatoes, sugar, tomatoes and grapes. Most native species feed on eucalypts, wattles and sheoaks, mostly causing minor damage overall, but with sporadic severe infestations. Many poorly known species feed on rainforest plants.

Life cycle: adults mate and the female lays eggs in slits made in the bark of branches and twigs. The eggs hatch and the young nymphs feed on plant sap, passing through a series of moults before developing as a winged adult. The nymphs of some species have prominent white waxy projections on their hindquarters which appear as tails. Some leafhoppers are solitary, others feed together with nymphs in mixed groups. When disturbed the adults jump or fly. The nymphs cannot fly and some species cannot hop or jump either (others can jump strongly). These nymphs characteristically move to the far side of a stem or leaf when disturbed.

Common Jassid nymphs of different ages feeding together. [T. Wood]

Waxy threads from leafhoppers feeding on Melaleuca bracteata. [D. Jones]

Leafhopper nymphs. [T. Wood]

Leafhopper nymphs exposed when 'byre' is removed. [D. Jones]

An example of a leafhopper. [T. Wood]

Adult Leafhopper, Brunotartessus fulvus. [T. Wood]

Plant damage: damage usually consists of distorted growth, wilting and dieback. The saliva of some species is toxic and causes dead papery patches in affected plant tissues, also resulting in premature leaf-fall. A few species of leafhoppers are known to be vectors of plant viruses and phytoplasma, adding greatly to their significance as pests. Leafhoppers feed on a wide range of plants, particularly eucalypts, wattles and sheoaks. Ants often attend feeding colonies of leafhoppers and collect the sweet secretions. Sooty mould frequently grows on the exudates and is associated with severe infestations.

Natural control: leafhoppers are preyed upon by a wide range of animals including small birds such as thornbills, tree climbing skinks, hunting spiders, scorpions, assassin bugs, ladybirds, robber flies and lacewings. Eggs and nymphs are parasitised by chalcid wasps and predatory mites. Entomopathogenic fungi also kill leafhoppers.

Other control: leafhoppers are usually a minor garden pest and only a problem in some years when they can gather in their hundreds. Large clusters are readily dispersed by jets of water and many can be killed by squashing or hitting. In cases of severe infestation, as frequently occurs with Passion Vine Hopper, chemical spraying must be used. Contact sprays such as pyrethrum and malathion are effective.

EXAMPLES

Common Brown Leafhopper, *Orosius orientalis*

Description: small brown cicada-like insect with paler mottling on the wings. Nymphs are brown.
Climatic regions: a native species from temperate regions.
Host plants: a wide range of native plants, garden plants, crops and weeds.
Feeding habits: colonies of mixed stages feed on young shoots. This leafhopper is a vector of plant virus diseases including Tobacco Yellow Dwarf Virus which affects beans and tobacco. It can also carry phytoplasma diseases
Notes: this species is a serious pest of vegetables, tobacco and grapes.
Control: as per introduction for leafhoppers. Spraying is often necessary in crops but rarely in garden plants.

Common Jassid, *Eurymela fenestrata*

Description: small cicada-like insect, 1.2–1.5cm long. Adults have a red, orange or yellow head and thorax and dark brown to black wings (purplish in the sun) with white or yellowish markings. Very young nymphs are black with faint red markings, later nymphs red to orange with black markings.

Ants attending nymphs of the Common Jassid. [T. Wood]

Two adults and a nymph of the Common Jassid. [T. Wood]

Climatic regions: subtropics and temperate regions.

Host plants: *Eucalyptus* species.

Feeding habits: colonies of mixed stages feed on young shoots. Ants are attracted to the colonies and feed on honeydew, which also grows sooty mould. Individual adults or small groups are frequently seen on larger branches or sheltering on the bark of the trunk.

Notes: a familiar garden insect which is sometimes a serious pest of forestry trees.

Control: as per introduction for leafhoppers. Usually controlled by natural enemies.

Green Leafhopper, *Siphanta acuta*

Description: adults 0.8–1cm long, the wings folded at a high angle in a triangular shape, bright-green with small yellowish dots over the surface and red dots on the edges. Nymphs brownish or yellowish.

Climatic regions: tropics and temperate regions of many countries.

Host plants: mostly eucalypts and wattles, but also *Macadamia* and a wide range of garden plants.

Feeding habits: adults are generally solitary. Immature stages congregate on young shoots.

Notes: this is a familiar garden insect, the adults resembling a narrow, green triangle. It has become a serious pest of gardens and commercial crops in Hawaii. Numbers fluctuate seasonally.

Control: usually unnecessary.

Green Leafhopper nymph. [J. Fanning]

Adult Green Leafhopper. [T. Wood]

Green Wattle Hoppers on Acacia dealbata. [T. Wood]

Green Wattle Hopper, *Sextius virescens*

Description: small cicada-like insect, 0.5–0.8cm long. Adults are pale green with two short but prominent, brown-tipped, horn-like spines on the upper thorax (just behind the head). Nymphs are greenish with an extended anal tube which they waggle when disturbed.

Climatic regions: temperate regions.

Host plants: feather-leaved wattles, particularly *Acacia dealbata* and *A. decurrens.*

Feeding habits: dense colonies of mixed stages occur on young shoots, especially vigorous young plants. A heavy exudate of dark honeydew attracts ants and supports sooty mould.

Notes: a common species which in some years is a significant pest. Ovipositor slits in the bark exude gum freely and can be unsightly. The adults hop when disturbed, making a click noise as they do.

Control: as per introduction for leafhoppers. Usually controlled by natural enemies.

Passion Vine Hopper, *Scolypopa australis*

Description: adults about 0.7cm long and 1cm across, triangular with clear wings bordered by brown and black bands. The wings are generally held at a low angle and the adults jump readily when disturbed, making a loud click. Immature stages are small whitish insects with a prominent anal tuft of white, waxy, hair-like filaments which are held erect. They also jump readily.

Climatic regions: subtropics and temperate regions, inland and coastal.

Host plants: various crops and garden plants including *Grevillea* spp., *Hymenosporum flavum*, *Passiflora* spp., *Syzygium paniculatum* and *S. smithii*. They also severely attack the croziers and uncoiled fronds of many ferns including *Asplenium australasicum*, *Cyathea australis*, *C. cooperi*, *Dicksonia antarctica* and *Polystichum proliferum*.

Feeding habits: the immature stages feed in colonies on young growth, whereas the adults tend to be solitary or mingle with the nymphs. Their feeding causes dry papery patches of tissue. Sooty mould grows on their exudates.

Notes: a common and widespread species which is a serious pest in some seasons. It may also be the vector of a phytoplasma disease. Both adults and nymphs jump actively when disturbed. The nymphs are popularly known as 'hairy rockets' or 'fluffy bums'. This species has become a serious pest of kiwifruit in New Zealand.

Control: as per introduction for leafhoppers. Hosing disrupts the colonies. Spraying may be necessary to control serious infestations.

Adult Passion Vine Hopper. [T. Wood]

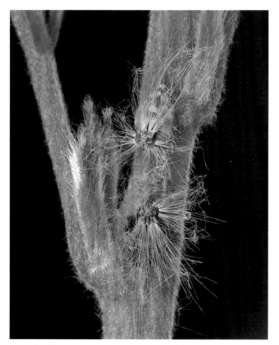

Nymphs of the Passion Vine Hopper (popularly known as 'Hairy Rockets'). [T. Wood]

Passion Vine Hoppers, adults and nymphs. [R. Elliot]

Saltbush Leafhopper on Atriplex nummularia. [S. JONES]

Damage to Old Man Saltbush caused by Saltbush Leafhoppers. [S. JONES]

Saltbush Leafhopper, possibly *Austroasca merredinensis*

Description: small, grey leafhopper, 0.8–1cm long with transparent lacy wings that spread at a wide angle. Nymphs are miniature versions of the adults.
Climatic regions: temperate regions, mainly inland.
Host plants: species of *Atriplex,* particularly *A. nummularia.*
Feeding habits: the adults feed singly or in groups with nymphs on the stems and leaves. Infestations are sometimes severe, causing leaf distortion and premature leaf-fall. The stems and leaves also become covered with sooty mould which grows in the copious exudates of the leafhopper.
Notes: The greyish insects blend in well with the leaves but are more prominent on the stems.
Control: controlled by natural enemies.

Spiny Leafhopper or Horned Treehopper, *Sertorius australis*

Description: small, brown, cicada-like insect, 0.6–0.8cm long with transparent lacy wings. Adults are hump-backed, with a short, hard, sharp spine on either side of the head.
Climatic regions: subtropics and temperate regions.
Host plants: *Acacia, Banksia, Eucalyptus, Grevillea, Hakea* and *Telopea.*
Feeding habits: adults mostly solitary, immature stages in colonies on young shoots.
Notes: an interesting insect. The head spines are noticeable if the adults are picked up between the fingers.
Control: controlled by natural enemies.

Other Leafhoppers

- A closely related and similar-looking greenish species to the Green Leafhopper is *Colgar peracutum* from tropical and subtropical regions. It feeds on *Melaleuca*, eucalypts and rainforest plants.

- *Sephena cinerea* also has triangular adults but these are of bluish-grey hues, not green. This species is common in the tropics and sometimes gathers in dense colonies.

- *Eurymelops bicolor* is a very colourful jassid, 1.2–1.4cm long with orange and black wings.

CHAPTER 6

OTHER SAP-FEEDING PESTS THRIPS, MITES, NEMATODES

The three specialised groups of sap-feeding creatures detailed here are separated from those in the previous two chapters because they lap or drink the sap that leaks from plant tissue damaged by their feeding activities.

SYMPTOMS AND RECOGNITION FEATURES

Silvery, mottled or bronzed leaves with a dry appearance...*mites*, *thrips*
Tiny (hand lens) slender insects often in colonies, adults if present have fringed wings......................*thrips*
Tiny (hand lens) rounded animals moving about on fine webs..*mites*
Small black 'island-like' areas between leaf veins..*nematodes*
Abnormally branched roots...*nematodes*

THRIPS

Thrips, which belong to the insect order Thysanoptera, are minute to tiny (less than 0.1cm to 0.5cm long) slender insects that are very difficult to see as individuals, but the combined effects of their feeding activities on plants can become very obvious. They attack most above-ground organs, particularly leaves, buds and flowers. Thrips congregate in colonies and feed by rasping the surface of the plant tissue with their asymmetric mouthparts, lapping up the sap which leaks out. As a consequence of their feeding the damaged tissue desiccates and takes on a dry appearance. Thrips can fly, but poorly, and are mostly dispersed by wind. Modern transport systems have distributed many species around the world and some of the more significant pest species are now cosmopolitan. Thrips build up rapidly to plague proportions in years with a warm winter, and swarm in immense numbers. Some thrips feed on a relatively narrow range of plant species while others have a very broad host range.

It is estimated that there are about 700 native and 60 introduced species of thrips in Australia. About 35 species of thrips are serious pests that damage commercial crops in Australia, with many of these serious pests having been inadvertently introduced from overseas. Not all thrips attack plants. Some feed on fungi, others predate small invertebrates such as mites and others play a major role in plant pollination, especially of subtropical rainforest plants and some cycads.

Thrips damage: thrips that feed on leaves congregate on the lower surface in mixed colonies of nymphs and adults. Feeding sites are usually littered with small dark blobs of excreta. Thrips

Thrips damage on fern. [D. Jones]

Thrips damage on Acacia *growth.* [D. Jones]

Damage caused by leaf-feeding thrips. [D. Jones]

feeding often results in a characteristic change in colour of the leaf surface, usually a silvering or bronzing. Many thrips feed on very soft young leaves while they are still tightly packed in developing shoots, these emerging and developing abnormally, often distorted. Damaged buds often twist and flowers become misshapen, do not open properly, develop colour streaks or dry patches, and finish quickly. Feeding by thrips can also impede pollination and the development of seeds. A few species of thrips form galls (see chapter 7). Thrips are also associated with disease. More than 20 plant-infecting tospoviruses are known to be carried by thrips, many of these organisms causing significant disease in commercial crops.

Life cycle: thrips can be winged or wingless. Winged adults have slender knobbed wings edged with a delicate fringe of hairs. Eggs, which can be produced either after mating or without mating (parthenogenesis), are laid on the host plant and hatch into tiny larvae known as nymphs. These are smaller and paler than the adults, ranging from white to pale yellow or orange. Nymphs pass through two instars, moulting at each stage before becoming adults.

Natural control: thrips are greatly influenced by climatic conditions and can be destroyed by extremes of climate, especially heavy rain. Predators include small birds, frogs, bugs, spiders, hoverflies, mites and fungi. Some tiny wasps also parasitise thrips.

Other control: thrips are attracted to bright colours, especially yellow, and colourful sticky cards can be used to monitor their presence. Thrips are difficult to control with chemicals because there is a high level of resistance to many horticultural sprays. They are also difficult to target because they are often hidden within plant structures such as flower buds, leaf buds, bracts and rolled leaves. The eggs are often laid directly into plant tissue. A contact spray such as malathion or a pyrethrum-based spray may be effective. Predatory mites are available for use in greenhouses with species such as *Typhlodromips montdorensis* and *Neoseiulus cucumeris* being preferred control for thrips in protected environments. These mites work best at minimising thrips build-up when applied early in the growing season at the first sign of thrips.

EXAMPLES

Blossom Thrips or Plague Thrips, *Thrips imaginis*

Description: the adults are very slender grey insects, about 0.12cm long with narrow, fringed wings. Nymphs are smaller wingless versions of the adult.

Climatic regions: tropics, subtropics and temperate regions.

Host plants: buds and flowers of many garden plants with thin-textured petals, such as species of *Alyogyne, Baeckea, Eucalyptus, Hibbertia, Hibiscus, Lechenaultia,* and *Leptospermum.* The flowers of kangaroo paws are also commonly damaged by these thrips. The flowers of the New South Wales Christmas Bush, *Ceratopetalum gummiferum,* an important commercial species, are vulnerable to attack. The feeding activities of the thrips on the calyces causes them to become brown and fall without developing into the attractive red colours for which the species is renowned.

Feeding habits: the thrips congregate in the flowers causing premature browning and shortening the flowering period.

Notes: blossom thrips numbers fluctuate greatly and in warm to hot dry periods can quickly build up to plague levels. These thrips also reduce fruit set and seed formation.

Control: as per introduction for thrips. Spraying with malathion or pyrethrum may be successful if the thrips are not hidden. Heavy rain kills large numbers and hosing garden plants may have a similar impact.

Glasshouse Thrips, *Heliothrips haemorrhoidalis*

Description: the adults are about 0.15cm long, very slender with narrow, fringed wings. Eggs are laid just under the surface of a leaf. The nymphs are smaller wingless versions of the adult.

Climatic regions: tropics, subtropics and temperate regions.

Host plants: these thrips thrive in glasshouses and greenhouses and in hot wet summers can also cause problems with garden plants. Susceptible glasshouse-grown plants include brachyscomes, ferns, peperomias, scaevolas and orchids. In the garden they feed on a wide range of soft-leaved plants including *Brachyscome, Cissus, Dampiera, Goodenia, Passiflora, Scaevola* and *Xerochrysum.*

Feeding habits: the thrips congregate in colonies on the leaves. They attack maturing leaves causing a silvery, mottled appearance followed by yellowing and premature shedding.

Notes: glasshouse thrips favour warm moist conditions and their numbers are drastically reduced by hot dry weather. They can be distributed in a glasshouse by air movement from ventilators and fans.

Control: as per introduction for thrips. Spraying with malathion or pyrethrum may be successful if the thrips are not hidden. Blue sticky traps can be used to monitor thrips populations in glasshouses. Predatory mites and lacewing larvae may be available from biocontrol firms; also a small wasp is an effective parasite of this species. Spraying with soap spray and horticultural mineral oils may also provide control.

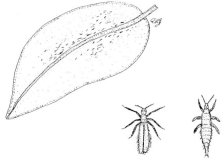

Glasshouse Thrips. [D. JONES]

ORCHID THRIPS

Several species of thrips damage orchids and can become a significant problem in orchid collections. Blossom Thrips and Glasshouse Thrips (both detailed above) commonly damage orchid flowers, and Onion Thrips, *Thrips imaginis,* and Citrus Thrips, *Pezothrips kellyanus,* have been reported as occasional pests in collections. Most of these thrips feed on flowers, causing distortion and browning. Recently another species, *Helionothrips errans,* a serious pest of glasshouse plants in the USA and Europe, has been found feeding on cultivated orchids in New South Wales. It has also been discovered in Perth. This species, with yellow nymphs and distinctive dark adults which are nearly black, feeds on orchid leaves causing significant damage. Affected areas develop a silvery appearance, eventually becoming brown. This pest, which is likely to be widely distributed by plant trading, poses a significant threat to orchid collections in Australia. It may also impact on native orchids in the wild.

Western Flower Thrips, *Frankliniella occidentalis*

Description: adults are very small (0.1–0.2cm), thin and yellow to pale brown in colour. The females, larger than the males, can reproduce parthenogenetically. The males are always pale yellow.

Climatic regions: subtropics and temperate regions.

Host Plants: flowers and foliage of many exotic garden plants and crops; also found on native species such as *Anigozanthos humilis, Boronia herophylla, Carprotus edulus, Hibbertia hypercoides* and *Petrophile linearis*.

Feeding habits: these thrips attack the flowers and foliage of plants causing early flower-fall, premature senescence, surface scarring, stunting and deformity.

Notes: this species, first discovered in Australia in 1993, originates from south-western USA. It is now widespread around the world and a significant pest of crops. It also spreads Tomato Spotted Wilt Virus and is of significant concern to Australia's cut flower industry. The life cycle of this species in heated glasshouses is almost continuous with 12–15 or more generations per year.

Control: as per introduction for thrips. Spraying with malathion or pyrethrum may be successful if the thrips are not hidden, however the species is known to be resistant to many pesticides. This species can be monitored with yellow sticky cards. The predatory mites *Typhlodromips montdorensis* and *Neoseiulus cucumeris* are available as biological control agents and can be used early in the growing season to prevent thrips building up. Another mite, *Hypoaspis miles*, normally used for fungus gnat control, can also be effective.

OTHER THRIPS

- Specialised thrips in the genus *Dichromothrips* feed on the flowers of native terrestrial orchids in Australia and New Zealand, causing distortion and premature browning of the petals and sepals. *Dichromothrips spiranthidis* has been found on species of *Microtis, Prasophyllum* and *Spiranthes*; *D. australiae* on *Diplodium* (*Pterostylis*) *atrans*.

- The Kauri Thrips, *Oxythrips agathidis*, feeds on the young soft growth of kauri seedlings (*Agathis* species) in Queensland and is reported to cause damage to seedlings in forestry nurseries.

Damage to Prasophyllum *flowers by* Dichromothrips spiranthidis. [D. JONES]

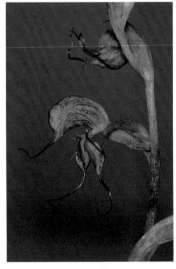

Damage to Pterostylis *flowers by* Dichromothrips. [D. JONES]

MITES, SPIDER MITES AND TWO SPOTTED MITE

Mites are tiny animals that are placed in the animal order Acarinae. There are thousands of species of mites known and many thousands that are yet to be recognised. Mites live in most types of habitat and feed on a wide range of materials. Many are parasites on other animals, some are beneficial because they are active predators of small creatures (some species have been used as biocontrol agents), and large numbers break down dead organic matter, including plant litter. An important group also feed on plants and estimates suggest that worldwide more than 6,000 species of mites are plant-feeders. Most of these play an important role in the natural ecology and are not a problem to gardeners, but a few species have become serious pests of crops and ornamental plants.

Mites are not insects but are closely related to spiders and scorpions. The adults of many species have eight legs, although the first larval nymphs often only have six legs (a specialised group, the erinose mites have only four legs). Mites also lack antennae. Mites reproduce by eggs (some can also reproduce asexually) and they pass through a series of nymphs before becoming an adult. Adult mites are often so small that they are difficult to detect, even with the aid of a magnifying glass.

Plant Damage: mites attack a huge range of plants and are frequently a serious pest of plants grown in enclosed structures such as greenhouses and glasshouses. They feed by piercing the leaf tissue with stylets and lapping up the oozing sap. Numerous plant cells can be destroyed in a very short time when the mites are feeding actively. Afflicted leaves at first take on a dry appearance and then become silvery, bronze or mottled with pale yellow spots and patches. In severe infestations the leaves become bronze, brown or yellow and fall prematurely. In temperate regions mites are a major problem during the summer when conditions are generally dry and humidity low. Their attacks become even more serious during long dry spells and droughts when plants become stressed by the lack of water. In subtropical and tropical gardens they are more commonly a pest during late winter and spring, less commonly a pest during the wet season.

Mite-damaged leaves of Correa bauerlenii. [D. JONES]

Mite damage to Notelea. [D. JONES]

Mites favour dry conditions in gardens and are often noticed as a pest on plants growing in protected situations such as in the lee of a wall or under eaves where rain doesn't penetrate. For the same reason they are often a serious pest of plants used for indoor decoration, particularly those that are neglected. Mites are also a problem in bush-houses and glasshouses, attacking ferns, orchids and palms. Again their attacks are worse on plants in dry corners or plants that are neglected, underwatered or in need of repotting. Mites are not commonly a nuisance on healthy garden shrubs and trees, but sick or weak shrubs and trees are prone to attack.

Groupings: several groups of plant feeding mites can be recognised (treated here at family rank).

- **Tetranychidae (Spider Mites):** this family, most if not all members of which feed on plants, contains some of the most serious and persistent pests. These mites, which occur in a range of colours, have a round to oval body 0.04–0.08cm long. They commonly feed on the underside of leaves and reproduce very rapidly during warm, dry weather, quickly building up into noticeable infestations. Many species spin fine, delicate webs on which they move about – a feature which helps to reduce the effectiveness of sprays used against them.

- **Eriophyidae (Gall Mites, Erinose Mites):** a very large group of specialised mites, all of which feed on plants. They have a tubular body and only two pairs of legs. For more details on these mites see chapter 7. The growth malformation known as witches' broom (see chapter 19) may also be caused by these mites.

- **Tenuipalpidae (False Spider Mites, Bunch Mites):** these are very small mites (0.02–0.04cm long) with a flattened reddish or brownish body. They are so small that they are very difficult to discern, even under a hand lens. These mites, which are mostly found in the tropics, are sometimes called bunch mites because their feeding often results in bunched and deformed shoots. These attacks can persist on a plant for many seasons.

- **Tarsonemidae (Broad Mites, Thread-footed Mites):** most of these mites feed on fungi but a few feed on plants, with a couple of species that inject toxins while feeding causing serious plant damage.

- **Penthaleidae (Earth Mites):** a group of mites that are active in winter. One introduced species poses a serious threat to some native ground orchids.

Natural control: mite numbers are greatly influenced by climatic conditions. They are also attacked by ladybirds and their larvae (small black ladybirds of the genus *Stethorus* feed actively on mites), many different predatory mites, tiny parasitic wasps, lacewings, hoverflies, damsel bugs, tiny spiders and predatory thrips in the genus *Scolothrips*. Predatory mites can be purchased from biological control companies.

Other control: control measures should be initiated as soon as the mites are noticed because they build up in numbers very quickly. Malformed growths and affected or fallen leaves should be removed and burnt to reduce build-up of populations. Jetting the underside of leaves with water helps reduce the mite populations. A dust consisting of a mixture of equal parts by weight of fine sulphur and hydrated lime has been used with some success, but only in cool weather. Soap and oil sprays (not in hot weather) can also be used. A predator mite *Typhlodromus occidentalis* has been released onto some commercial crops with considerable success in reducing the pest mite populations. Other predator mites, *Amblyseius womersleyi* and *Phytoseiulus persimmilis*, have been found to control two-spotted mite in subtropical areas. Spraying with a miticides (or acaracide) may be necessary to clean up bad infestations. However, because of their short generation times, mites can quickly build up genetic resistance to chemical sprays. These sprays are also indiscriminate, killing beneficial animals as well as the pests. To reduce the build-up of genetic resistance it is a wise policy to alternate different miticides if several applications are needed.

EXAMPLES

Broad Mite or Glasshouse Mite, *Polyphagotarsonemus latus*

Description: a tiny eight-legged mite (family Tarsonemidae) about 0.02–0.03cm long (visible under a hand lens) which is shiny translucent white, pale yellow or green. The eggs have numerous tiny, white, wart-like bumps over the surface. The development from egg to adult takes 4–10 days and there are 20–30 generations in a year. Adults live 5–13 days, producing up to 75 eggs in this time.

Climatic regions: mainly the tropics, subtropics and warm temperate regions, chiefly coastal; less common in temperate regions.

Host plants: garden plants include species and cultivars of *Alyogyne, Brachyscome, Cissus, Ficus, Goodenia, Hibiscus, Scaevola* and *Schefflera*; indoor and glasshouse plants include ferns, orchids, *Cissus* and *Tetrastigma*; also numerous vegetables and crop plants.

Feeding habits: the mites congregate on the underside of soft new leaves on the growing tips of young shoots. They inject toxins with their saliva that inhibit leaf development and cause distortion. Other symptoms include thickened brittle leaves, bronzing, downcurved leaf margins (cupping), curling and puckering. Affected shoot-tips often die back. The mites also feed on the sepals and calyces of flower buds, resulting in bud-drop and distortion of the flowers. Even small populations of broad mites can cause significant plant damage.

Notes: this mite is a serious pest in tropical regions and is also a persistent pest of plants grown in glasshouses and greenhouses. It is worst in warm, moist conditions. The males can transport the females around on their bodies and mites attached to the legs of whiteflies can also be transported to new hosts.

Control: as per introduction for mites. The predatory mite *Neoseiulus cucumeris* is available from biocontrol firms for use in glasshouses.

Bunch Mites or Brevipalpus Mites, several *Brevipalpus* species

Description: tiny, flattened, eight-legged mites (family Tenuipalpidae), 0.02–0.04cm long, dull brown to reddish and lacking any blotches.

Climatic regions: tropics, subtropics and temperate regions, mainly coastal.

Host plants: attacks occur on a wide range of native and exotic plants.

Feeding habits: these mites inject toxic saliva while feeding. Circular patches of damaged tissue appear where they have fed. Their feeding also results in compaction and bunching of new shoots, leaf drop and bud malformation.

Notes: these are serious and destructive pests. Some species are important vectors of plant viruses. Several species of *Brevipalpus* are native to Australia and feed on native plants. Plants affected include *Atriplex nummularia, Grevillea barklyana, G. longifolia, Hakea petiolaris, H. plurinervia, Myoporum* spp. and *Solanum* spp. Some exotic mite species have also become naturalised as pests: *Brevipalpus californicus* attacks a wide range of plants and is of significance because it can transmit the orchid fleck virus; *B. phoenicis*, the Passionvine Mite, is a pest of citrus and passionfruit and is also the vector for a passionfruit virus; and *B. lewisii* attacks a wide range of garden plants.

Control: as per introduction for mites.

Red-legged Earth Mite, *Halotydeus destructor*

Description: the adult mites (family Penthalidae), which are about 0.1cm long, have a black velvety body and eight red legs. Nymphs are smaller versions of the adult, also with eight legs. The development from egg to adult takes 4–14 days and there are 20–30 generations in a year.

Climatic regions: temperate regions, chiefly coastal, near coastal and adjacent areas.

Host plants: a wide range of plants, particularly delicate species or those with tender shoots. It is known to damage and kill native orchids, particularly species of *Chiloglottis, Corysanthes* and *Pterostylis*; in

Flower of Pterostylis wapstrarum *damaged by Red-legged Earth Mite.* [P. NORRIS]

Flower of Wurmbea dioica *damaged by Red-legged Earth Mite.* [P. NORRIS]

Red-legged Earth Mites on bud of Pterostylis wapstrarum. [P. NORRIS]

Red-legged Earth Mites attacking buds of Pterostylis wapstrarum. [P. NORRIS]

Rosettes of Pterostylis wapstrarum *damaged by Red-legged Earth Mite.* [P. NORRIS]

Tasmania this mite has had serious impacts on the populations of two threatened species, *Pterostylis wapstrarum* and *P. ziegeleri*; it also attacks species of *Caltha, Centella, Eremophila, Nymphoides, Wurmbea* and *Villarsia.*

Feeding habits: the mites suck the sap from the surface of the leaf or flower which then dries out, becomes silvery and turns white. These symptoms can be confused with frost damage. Young shoots may be distorted. When disturbed the mites drop to the ground and hide in the soil. The mites also move freely over the soil to feed on adjacent plants.

Notes: this species, which is native to South Africa, was introduced to Australia in the early 20th century. It is now widespread in south-eastern Australia (including Tasmania) and the south-west of Western Australia. This pest is mainly active in winter and attacks pastures, crops, weeds and a wide range of native plants. Although mainly a pest of crops, this mite is known to impact on garden plants and natural populations. The mites survive summer heat and dryness as dormant (diapause) eggs.

Control: as per introduction for mites.

Two Spotted Mite or Red Spider Mite, *Tetranychus urticae*

Description: tiny eight-legged mites (family Tetranychidae), 0.03–0.05cm long, dull-yellow to greenish (newly hatched larvae have six legs). In summer the development from egg to adult can be as short as four days and there are several generations in a year. The females have two conspicuous dark blotches on the back (males have less conspicuous blotches). Each female lays 70–100 eggs in her lifetime. Winter non-feeding populations of adult females are orange to red all over. They overwinter in the soil and crevices in bark.

Climatic regions: tropical to temperate regions, coastal and inland.

Host plants: a wide range of garden ornamentals including many natives; plants such as ferns and orchids grown in glasshouses are also susceptible; also vegetables, fruit trees, nursery and crop plants.

Feeding habits: adults and immature stages congregate on the underside of leaves and on new growth. They spin fine webs as they feed, the webbing becoming denser as the feeding progresses. As the colonies grow the mites move to the upper surface of the leaves. Mites and their webs can also often be seen on young shoots of infested plants. Affected leaves appear dry, become silvery, yellow or mottled and fall prematurely.

Notes: a serious and destructive pest. Its spread and impacts are worse in warm to hot dry weather

Two Spotted Mite and egg. [J. Fanning]

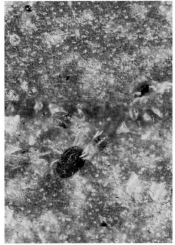

Two Spotted Mite. [J. Fanning]

Two Spotted Mites. [J. Fanning]

Two Spotted Mites and eggs. [J. Fanning]

Two Spotted Mites feeding on Pandorea *shoot, note webbing.* [D. Beardsell]

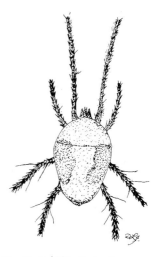

Two Spotted Mite. [D. Jones]

with low humidity. This pest can also be very severe on plants covered in dust and those growing in sheltered dry sites, such as under eaves. In areas with a cool winter the adult female mites become red all over and hibernate as a non-feeding stage. This gives rise to their alternate common name of Red Spider Mite. Mites in warm climates and glasshouses do not hibernate but feed and reproduce all year. **Control:** as per introduction for mites. Regular hosing or misting of foliage reduces the impact of this pest. The predatory mites *Phytoseiulus persimilis* and *Typhlodromus occidentalis* are effective predators of Two Spotted Mite. They are used in integrated pest management control systems and are available from biocontrol firms.

OTHER MITES

- The Cyclamen Mite, *Steneotarsonemus pallidus* (family Tarsonemidae), is sometimes found on glasshouse plants. It feeds on new shoots as they emerge, usually feeding between the bracts or very small developing leaves. It is easily confused with Broad Mite but can be distinguished by its smooth eggs.

- Several species of *Bryobia* mite (family Tetranychidae) feed on plants in gardens and glasshouses. Adults, which can be brown or reddish, are quite large (up to 1cm long) and more visible than most mites. They congregate on the underside of leaves but do not spin webs.

NEMATODES OR EELWORMS

Nematodes or eelworms are microscopic to tiny roundworms (a few reach 0.5cm long) that are ubiquitous in most natural systems on earth, including soils, freshwater and marine environments. There are thousands, if not millions, of species of nematodes worldwide, all placed in the phylum Nematoda. Not all nematodes are plant pests – a large number are involved in breaking down organic matter and recycling nutrients. Some eelworms are beneficial in gardens, killing other nematodes and pests such as cutworms. Numerous species however, are parasitic on plants and can cause considerable damage to plants in nurseries, gardens, horticultural and agricultural crops. Some nematodes can spread plant viruses. Parasitic nematodes move through films of water on the surface of plant organs, some species feeding on surface tissue, others entering internal tissues through stomata or damaged sites. They feed on cell juices which they suck up after injecting saliva. Toxic chemicals in the saliva cause cell death and growth distortion. A group of species attack subterranean plant parts (root systems, bulbs and rhizomes) while others are more commonly found on the above-ground parts (stems, leaves and flowers).

Soil nematodes occur in a wide range of soil types but are usually uncommon to absent from very dry soils and waterlogged soils. They are tolerant of a wide range of soil pH (acidity or alkalinity) and salinity. Nematodes prefer infertile soils rather than highly fertile soils and are most prevalent in soils where the structure has been destroyed and the organic content depleted.

Symptoms of nematode attack include stunting, growth malformation, leaf yellowing, distortion and various deficiency symptoms. Affected plants often wilt readily when placed under stress. Nematodes can also cause swollen stems, lesions and galls. Leaf nematodes cause small black, brown or yellow patches on the leaves in areas restricted by the surrounding veins.

Natural control: a group of soil-living fungi such as the fungus *Clonostachys rosea* are parasitic on soil nematodes.

Other control: correct identification of the culprit is a very important step in the control process and must be carried out by experts. Soil populations of plant parasitic nematodes can be reduced by rotating with crops of resistant plants such as grasses, or growing plants that release chemicals into the soil that are antagonistic to nematodes. These include African marigold (species and hybrids of *Tagetes*), Indian Mustard (*Brassica juncea*) and species of *Crotalaria*. These plants are most effective when dug in just before maturity. Soil organic matter also discourages nematodes and should be kept at high levels by digging in green crops, compost, manure and mulching. Potted plants can be freed of leaf nematodes by immersion in very warm water (40°C) for 15 minutes. Sugar or molasses applied to the soil at 500 grams per square metre will kill all nematodes, good and bad. Control of root-feeding nematodes in nurseries is only possible by pretreating potting materials with steam-air mixtures (also kills beneficial nematodes). Specific chemicals (nematicides) are available for commercial growers but not for home garden use.

ROOT NEMATODES

Root Burrowing Nematode, *Radopholus similis*

Description: microscopic eelworms that burrow in the soft outer tissues of roots. The whole life cycle is completed within the root tissue, the juveniles either migrating in the root tissue or leaving to reinfect another root. Congregations of the nematodes cause localised areas of heavy damage.

Climatic regions: tropics and subtropics, coastal and inland.

Host plants: many native species including *Lagunaria patersonii, Lophostemon confertus, Musa* species, and *Syncarpia glomulifera*; also avocado, bananas, citrus, coffee and many vegetables.

Feeding habits: the eelworms burrow in the outer root tissues. Lesions formed by the nematodes develop into cankers which restrict the movement of water and nutrients. Root death is common. Badly damaged plants can topple over or blow over in wind. Some trees

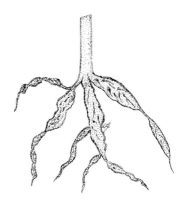

Root Burrowing Nematode. [D. Jones]

lose all of the small feeder roots resulting in leaf yellowing, stunting and dieback.

Notes: a significant introduced pest which debilitates and kills plants. The nematode burrows can form entry points for pathogenic fungi such as Honey Fungus, *Armillariella luteobubalina*. The impacts of this nematode are worse in coarse soils devoid of organic matter.

Control: as per introduction for nematodes.

Root Knot Nematode, several species of *Meloidogyne*, especially *M. incognita*

Description: microscopic eelworms that burrow into roots and cause irregular swellings and galls on the main and lateral roots. This feeding results in the formation of enlarged (giant) cells and the surrounding tissue becomes a gall in which the nematodes live. The galls restrict the movement of water and nutrients

Climatic regions: tropics, subtropics and temperate regions, coastal and inland.

Host plants: a wide range of exotic and native plants including species of *Alyogyne, Abutilon, Eremophila, Grevillea, Hibiscus, Myoporum* (particularly *M. parvifolium*), *Prostanthera, Solanum, Swainsona* (particularly *S. formosa*) and *Xerochrysum bracteatum*. Also found in more than 20 species of crops and vegetables.

Feeding habits: the eelworms congregate within the cells of the roots causing growth distortion.

Notes: there are about 60 species of *Meloidogyne* distributed worldwide and they attack the roots of thousands of species. They are a widespread serious pest of horticulture and also impact on gardens. They cause severe stunting and plant death due to disruption of water and nutrient uptake and transport. Affected plants have knotted roots. The whole root system can be severely stunted and open to entry by fungal diseases.

Control: as per introduction for nematodes. A strain non-toxic to humans (strain 251) of a widespread and common fungus, *Purpureocillium lilacinum,* is being used in some countries as a biocontrol agent for Root Knot Nematode.

Root Lesion Nematodes, several species of *Pratylenchus*

Description: microscopic eelworms that cause stunting and abnormal branching in roots. Affected tissue initially appears as water-soaked areas which turn into brown or black lesions. Fine roots blacken and die.

Climatic regions: tropics, subtropics and temperate regions, coastal and inland.

Host plants: some species of the family Myrtaceae, including *Eucalyptus globulus*; also *Grevillea alpina* and *Hardenbergia violacea*; also a wide range of agricultural and horticultural crops, including bulbs, fruit and vegetables.

Feeding habits: eelworms congregate near the root-tips, killing them and causing abnormal branching. They also spread to other parts of the root system.

Notes: several species of *Pratylenchus* occur in Australia but their impacts on native plants are largely unknown. They are certainly widespread and have the capacity to kill healthy plants.

Control: as per introduction for nematodes.

ABOVE-GROUND NEMATODES

Leaf Nematodes, several species of *Aphelenchoides*

Description: microscopic eelworms attacking the interveinal areas of leaves, causing brown discolouration.

Climatic regions: tropics, subtropics and temperate regions, chiefly coastal.

Host plants: a wide range of native plants in the family Asteraceae (e.g. *Brachyscome*, *Xerochrysum*) and Goodeniaceae (*Goodenia*, *Scaevola*), as well as kangaroo paws (*Anigozanthos* species and hybrids); most Australian native ferns seem to be susceptible to leaf nematodes, some more so than others (species of *Asplenium* are particularly susceptible).

Feeding habits: the eelworms move over the leaf surface in a film of water and enter through the stomata. They feed in the leaf cells causing them to collapse. Initial spread is limited by the veins, hence the damaged tissue remains as small, discoloured islands between the veins. Eventually the eelworms move in surface water to new areas and whole leaves can die. Some plants with soft foliage develop watery spots in the leaves which spread and turn brown.

Notes: there are several species of *Aphelenchoides* which attack plants. They are serious pests which can easily kill healthy plants and are an increasingly significant pest in the nursery industry, particularly

Leaf nematode damage to Asplenium *fern frond.* [C. GOUDEY]

Leaf nematode damage on Asplenium *fern.* [C. GOUDEY]

for foliage plant growers. *Aphelenchoides fragariae* attacks a wide range of plants (more than 250 species worldwide), including figs, lilies, bulbs, ferns and many foliage plants, as well as crops including strawberries.

Control: severely affected leaves should be removed and/or whole plants burnt. Potted plants can be freed of leaf nematodes by immersion in hot water at a temperature of 40°C for 15 minutes. Water plants via drippers and avoid wetting the fronds of susceptible ferns, such as species of *Asplenium*.

Stem Nematode or Bulb Nematode, numerous species of *Ditylenchus*

Description: microscopic eelworms attacking the stems and branches of trees and shrubs causing blackened swollen areas, distorted leaves and dieback. This nematode also attacks bulbs which become distorted and develop lesions, malformed leaves and flowers.

Climatic regions: subtropics and temperate regions.

Host plants: a wide range of ornamental and nursery plants including many natives such as species of *Acacia, Cassia, Clematis, Hardenbergia, Pandorea, Scaevola* and *Senna*; also a serious pest of a wide range of bulbs.

Feeding habits: congregates in areas on the stem causing blackening. Also forms galls on some species. Affected bulbs become soft and spongy with brown rings.

Notes: there are more than 80 species of *Ditylenchus* which attack plants. *Ditylenchus dipsaci* is one of the worst, attacking more than 1,200 species of plants worldwide. It also has morphologically indistinguishable races which attack different plants under different conditions. These nematodes, which are at their worst under cold moist conditions, can complete their life cycle in less than 30 days. They can also survive in organic matter for up to 10 years. The impact of these nematodes on native plants is poorly known. They can be transported on seed but the eelworms are killed if the seeds are treated with hot water prior to sowing.

Control: as per introduction for nematodes.

OTHER NEMATODES

- Specialised nematodes that form symbiotic galls on the stems of plants are dealt with in chapter 7.

CHAPTER 7

GALLS

Galls are curious abnormal growths that can be found on many types of plant. While often prominent on stems and leaves, galls can occur on most plant parts, including the roots, bark, vegetative buds, flower buds, flowers (sometimes on the petals themselves), fruit and seed. They are caused by a wide range of external agents which induce a reaction in the plant that results in the growth malformation visible externally on the plant as a gall. Commonly the gall encloses the organism that induced it, with or without an external orifice for mating or the elimination of waste such as honeydew. However, more open types of structure can also be classed as a gall. Galls range in size from pinhead bumps to grossly swollen structures several centimetres across. They also range in complexity from simple structures enclosing a single organism to complex multibranched structures containing colonies of organisms. Many galls are short-lived, lasting just a few months; others can last several years. Shapes not only include obvious round, swollen or elongated structures, but less prominent types such as sunken pits,

Bladder-like galls on the fruit of Syzygium smithii. [D. Jones]

Unknown stem gall. [C. French]

Stem galls on Olax phyllanthi. [C. French]

Unknown male and female galls on Casuarina cunninghamiana. [D. Jones]

Blister galls on Eucalyptus *leaf.* [T. Wood]

Blister galls on Eucalyptus robusta. [D. Jones]

Bud gall on Amyema miquelii. [C. French]

Eucalypt leaf completely changed by massed galling into a hard structure. [D. Jones]

Flower galls on Dianella caerulea *induced by an unknown agent.* [D. Jones]

Flower galls on Eremophila longifolia. [D. Jones]

Large stem gall on Eucalyptus pauciflora. [D. Jones]

Woody stem gall on Leucopogon verticillatus. [S. Jones]

blisters, pouches, felted masses, thickened areas and distorted plant organs including enlarged buds and thickened leaves with inrolled or folded margins. Galls can be smooth, but are more often bumpy, irregularly roughened or even hairy. They are commonly of different colouration to the normal plant tissue and many take on pink or reddish tones as they age. Single plants of some species, particularly eucalypts and wattles, can carry several distinctive types of gall on different parts of the plant.

Gall formation is a very complex process and the type and shape of a gall is often unique, depending on the interaction between the causative agent and the particular plant species. Gall-inducing agents include most major insect groups, mites, nematodes, bacteria and even fungi. Galling insects, which are by far the commonest gall-inducing group, include aphids (not known on native plants), flies, psyllids, coccoids, thrips, beetles, wasps and moths. Galling agents probably interact in various ways with the internal cells of the plant, inducing them to change their normal growth patterns and resulting in the formation of a gall. For instance, chemicals injected by a female insect at the time of egg-laying may well initiate gall formation. Subsequent gall development is maintained as long as the galling insect is still alive and producing more chemicals in its saliva. In scale insects, gall induction begins after the insertion of the thread-like feeding organ (stylet) of first stage instars and nymphs. The stimulus appears to be unknown chemicals in the saliva. The stimulus provided by the causative agent probably also results in an increased level of plant hormones and other materials flowing to the site of irritation. Bacteria and fungi probably induce gall formation by the release of chemicals as they invade the plant tissue.

Galls provide shelter and food, allowing the causative agent to complete its life cycle. Insect galls commonly have one or more internal chambers in which the larvae feed and grow. Some galls have a layer of nutritious tissue lining the inner walls on which the larvae feed. Pupation often occurs within the gall and the adults emerge through a hole they cut in the surface. Some insects such as specialised thrips live and breed solely inside the gall. Most galls last for only a few months but some of the larger galls are present for one or more years and woody galls can persist after the causative agent is dead. The smaller galls grow rapidly to peak size and last a few months or less. The longer-lived galls increase in size as they develop. Some female scale insects can live inside the gall for five years, whereas adult males only live one or two days. Galls are also strongly seasonal in their appearance, with most of the causative agents only being able to attack the plant organ at a suitable stage of its development. Predation and parasitism of the gall-inducers also takes place within the gall. Usually these interactions are complex and highly specialised. Other insects (secondary invaders) may also take advantage of the gall by feeding on its tissue or simply living within its walls.

The interaction between plants and galling agents is a tremendously interesting subject and an ideal nature study exercise for both children and adults. Galls of some form or shape can be found in almost any garden. Some galls have a shape so characteristic and repeatable that it can be used in their identification. Many galling agents exist in very close relationship with their plant host (termed host-specific). Sometimes the galls caused by male and female insects of the same species are vastly different in size and shape, indicating different stimuli produced by the larvae. In a similar way different species of insects, even within the same natural group (such as wasps, flies etc), induce galls of a different size or shape on the same plant. Such complex relationships are not easily understood and provide an insight into some of the remarkable relationships that exist in nature. There are a huge array of galls' causal agents which are yet to be identified. Some examples are included in the following photos.

GENERAL CONTROL OF GALLS

Most attacks by gall-inducing agents are of a minor nature but some can seriously impede growth and impact on the appearance of a plant. Minor gall infestations cause limited problems and are probably best left alone and valued for the interest they add to the garden. If seen as a problem, small infestations of galls should be removed and destroyed when first noticed, preferably before the adults have emerged. This helps to curtail a build-up of the pest. Severe infestations of galls however, are particularly difficult to control. Chemicals are generally ineffective because the larvae are enclosed within the gall and their efficacy cannot be gauged. Once emergence holes are visible in the galls it is too late to initiate control procedures. Trees and shrubs that are persistently and badly attacked are probably best removed. Practices such as watering during dry periods (especially droughts), mulching with organic materials and the application of suitable fertilisers or manures promote plant health and reduce the impact of galls.

SYMPTOMS AND RECOGNITION FEATURES

Note – galls are so complex that this can only be regarded as an introductory guide.

Tiny pimple-like swellings on leaves, often reddish..*wasps*

Round green or red galls on eucalypt leaves...*wasps*

Irregular conjoined swellings on eucalypt stems...*wasps*

Tiny hard green galls on leaves and phyllodes...*eriophyid mites*

Clusters of reddish, hard, horn-like galls on eucalypt stems..*wasps*

Bubble-like galls with an apical opening, usually on leaves.....................*psyllids, cecidiomyiid flies*

Pimple-like galls, blisters and sunken pits on *Syzygium* leaves ....*pimple psyllid*

Brown scaly gall on *Allocasuarina* resembling sheoak fruit................................*Cylindrococcus*

Large woody gall on *Callitris* stems resembling a cone .... ***Callitris gall fly***
Hard winged capsule-like gall with two or four apical horns .... ***Apiomorpha***
Slender hard red to brown horn-like galls, curved or straight .... ***Apiomorpha***
Hard round to acorn-like galls with waxy orifice .... ***Apiomorpha***
Hard elongated torpedo or cigar-shaped galls with waxy orifice .... ***Apiomorpha***
Bell-shaped galls on eucalypt leaves, the underside white with numerous pores .... ***Apiomorpha***
Dense witches' broom-like cluster of galls in eucalypts .... ***witches' broom scale***
Inflated green to red gall on leaf or stem tip with numerous chambers .... ***nema galls***
Young leaves with strongly inrolled margins resembling a tube .... ***thrips***
Round green galls on *Geijera* leaves .... ***thrips***
Hard, irregular-shaped brown galls with a powdery surface .... ***fungus galls***
Swollen hard stem galls on eucalypts with numerous chambers, surface roughened .... ***gregarious gall weevils***
Irregular conjoined swellings on roots .... ***nematodes***
White hairy galls on leaves of *Banksia coccinea* .... ***Banksia gall fly***
Hard green misshapen galls on leaves of *Banksia menziesii* .... ***erinose mite***
Tiny pimple-like blister galls on young leaves *Agathis* .... ***Kauri giant scale***
Irregular round galls on wattle inflorescences .... ***wattle gall flower flies***
Roughened blister-like galls on leaves of *Corymbia* .... ***bloodwood blister mite***

GALL-INDUCING SCALE INSECTS

The scale insects, superfamily Coccoidea (order Hemiptera, suborder Sternorrhyncha), are mainly dealt with in chapter 4. Some specialised groups which induce galls in plants are detailed here.

FELT SCALES (FAMILY ERIOCOCCIDAE)

Felt scales are common in Australia with at least 85 species inducing galls on Australian plants in the family Myrtaceae. Some amazing galls induced by specialised scale insects are found on eucalypts and sheoaks. These insects induce characteristic galls that range from small pits to large complex structures. Female eriococcids have two immature instars, the males four. The largest gall in the world is induced by a species of native eriococcid scale in the genus *Apiomorpha*. The adult female eriococcid insects are fleshy, wingless and with short stubby legs, whereas the males are winged. Intriguingly, the male and female insects of many gall-inducing eriococcids induce galls of very different shapes. Galls induced by female insects can be quite large and sometimes woody, whereas those of the male insects are smaller and often cylindrical, tubular or horn-shaped. Both sexes frequently make galls in localised groups, sometimes clustered together, often with the male galls growing on the side of a female gall, or on a nearby stem or leaf. The shape of the gall induced by an eriococcid is a useful means of

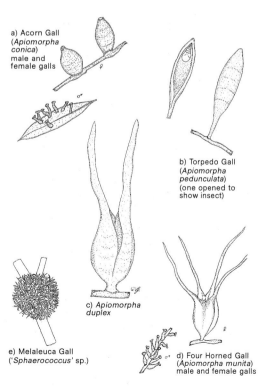

a) Acorn Gall (*Apiomorpha conica*) male and female galls

b) Torpedo Gall (*Apiomorpha pedunculata*) (one opened to show insect)

c) *Apiomorpha duplex*

d) Four Horned Gall (*Apiomorpha munita*) male and female galls

e) Melaleuca Gall ('*Sphaerococcus*' sp.)

Some types of eriococcid galls. [D. JONES]

identification, being constant within the species irrespective of the host plant.

When female eriococcid galls are split open a squat fat insect can be found inside, nestling in a tubular or cylindrical cavity. The female insect remains encased within the gall for all of her life, whereas the adult male leaves the gall to mate. Mating occurs through an opening found at the apical end of the female gall. The offspring escape through the same opening.

Natural control: gall-inducing eriococcids are controlled by specialised parasitic wasps which insert their ovipositor through the gall and lay an egg on the insect inside. Birds also crack open the galls to eat the larva. Some individual trees seem more prone as a host to gall-inducing eriococcids than others.

Other control: eriococcid gall infestations are usually of a minor nature and control is rarely warranted. Early removal and destruction of the galls will reduce infestation. Chemical control of these galls is unsatisfactory because the insect is protected within the gall.

EXAMPLES

Apiomorpha Galls, many species of *Apiomorpha*

Description: about 40 species of native eriococcids are currently known in the genus *Apiomorpha*, all associated with species of *Eucalyptus*. Recent studies have shown the presence of numerous species that are yet to be named, including several cryptic species that have been recognised in laboratory studies. The galls produced by male and female *Apiomorpha* insects are remarkably different and also constant within a species. Both types are hard and become woody with age. *Apiomorpha* galls can occur as single individuals, localised groups or crowded in odd-shaped clusters on stems and leaves. The galls induced by female insects can be relatively large and occur in a wide variety of shapes; often rounded, ellipsoid, cylindrical or spindle-shaped, sometimes even adorned with spreading horn-like structures. A single plump, teardrop-shaped mealy insect, 0.3–3.4cm long, cream, white, yellow or pinkish, is found in a chamber inside the female gall. The galls induced by the male insects are smaller than the female galls and often cylindrical or tubular. They are usually found on stems or leaves close to the female galls or

Male Apiomorpha munita *galls on* Eucalyptus polyanthemos.
[T. Wood]

Cluster of male Apiomorpha *galls, possibly* A. munita *(some tiny scale insects are just visible)*. [D. Jones]

often attached to them. The female insects can live for up to five years, whereas the males only last up to about one year.

Climatic regions: tropical, subtropical and temperate regions, coastal and inland.

Host plants: *Eucalyptus* species.

Feeding habits: larvae suck sap inside the gall. Each gall contains a single insect.

Notes: the coccid insect controls the shape of the gall (not the host) and the distinctive galls of more than one species of *Apiomorpha* can be found on the same tree. The female *Apiomorpha* insects secrete honeydew from the apical opening. They also secrete a waxy substance that protects their body from contamination by the honeydew. The honeydew is collected by ants and the waxy secretions can often be seen at the apical opening of the gall.

Control: as per introduction for eriococcid galls.

Male and female galls of Apiomorpha munita *attended by meat ants.* [T. WOOD]

Female galls of Apiomorpha pharetrata *on stringybark.* [D. JONES]

Male and female galls of Apiomorpha munita. [D. JONES]

Large female galls and smaller male galls of Apiomorpha pileata. [D. JONES]

Single female gall of Apiomorpha thorntoni. [D. Jones]

Male galls of Apiomorpha strombylosa *on* Eucalyptus polyanthemos. [T. Wood]

Apiomorpha species

- *Apiomorpha attenuata* induces green to greyish spindle-shaped to cigar-shaped female galls on the stems of *Eucalyptus camaldulensis*.

- *Apiomorpha conica* induces smooth green to grey ovoid or ellipsoid female galls with a circular apical orifice attached to stems (sometimes also larger branches), whereas the small, horn-like galls of the males are clustered on leaves.

- *Apiomorpha duplex* (Twin-tailed Gall) induces one of the largest galls in the world. It forms a four-sided chamber with a pair of horns up to 20cm or more long. The whole structure resembles a *Hakea* fruit.

- *Apiomorpha gullenae* induces cigar-shaped galls up to 1.3cm long on the leaves of ironbarks and *E. populnea* in southern Queensland. Male galls are unknown.

- *Apiomorpha malleeacola* is a specialised species that induces green to grey ellipsoid to ovoid female galls on the gumnuts of some mallee eucalypts.

- *Apiomorpha minor* induces smooth, bright green, ellipsoid female galls attached to stems, whereas the small, horn-like galls of the males are solitary or in small groups on leaves.

- *Apiomorpha munita* (Four-tailed gall) is a common species that induces a four-horned female gall. The gall itself is more or less four-sided, with each ridge ending in a slender spreading flattened horn. These galls are sometimes found in large clusters. The slender, tubular male galls, which are often bright pink or purple, cluster on nearby twigs or on the sides of the female galls. *Apiomorpha munita* is quite variable in its morphology and chromosome numbers. Studies have shown the presence of three recognisable variants which have been treated as morphological subspecies (subps. *munita*, subsp. *tereticornuta* and subsp. *malleensis*). Within each subspecies are chromosomal forms that appear to be restricted to a limited host range. *Eucalyptus ovata* is a host of subsp. *tereticornuta*.

- *Apiomorpha nookarra* induces an ovoid green or reddish female gall about 1.2x0.6cm on the stems of *Eucalyptus racemosa*. Unlike most other species this gall has two internal chambers, arranged one above the other. The male gall is a tubular structure about 0.6cm long.

- *Apiomorpha pedunculata* induces hard, cylindrical, torpedo-like female galls 5–8cm long that jut out from eucalypt stems and branches.

- *Apiomorpha pharetrata* induces clusters of ovoid or tubular ribbed galls on leaves and stems.

- *Apiomorpha pileata* induces 'acorn galls' so named from their green, ovoid, acorn-like female galls which have a slit-like pit at the top covered with a waxy secretion.

- *Apiomorpha pomaphora* is a specialised species from the Irwin Botanical District in Western Australia, inducing galls on *Eucalyptus eudesmoides* and *E. gittinsii*. The ovoid to ellipsoid, green to grey, female stem galls, 2–3cm long, have an apical cap that becomes deflected to one side with age. Male galls, which are found on the leaves, are elongated with a flared apex.

- *Apiomorpha spinifer* induces unusual saucer-like or bell-shaped female galls that are green on the upper surface and white beneath with numerous pores. Each is attached to a eucalypt leaf by a thick stalk. Male galls are small cylindrical structures.

- *Apiomorpha spinifer* induces green pear-shaped female galls on the stems of eucalypts.

- *Apiomorpha strombylosa* induces irregular swellings covered with knobbly protruberances.

- *Apiomorpha thorntoni* induces reddish-brown, deeply ridged female galls on the leaves and petioles of some stringybarks in south-eastern Australia. The males can be found as a cluster of tubes on the side of a female gall.

- *Apiomorpha urnalis* (Urn Gall) induces a spindle-shaped female gall parallel to a stem of Yellow Box, *Eucalyptus melliodora*. The apex of the gall often swells like the lip of an urn.

- *Apiomorpha variabilis* induces pear-shaped female galls to 4.5cm long on the stems of *Eucalyptus globulus*. This gall has two internal chambers. The female insect is 2.5–5cm long.

Sheoak Cone-mimicking Galls, *Cylindrococcus* species

Description: there are three species of these eriococcids endemic in Australia, two named and one undescribed from Western Australia. They induce remarkable galls on some species of sheoaks. The male and female galls have a similar appearance but are of different size and shape.

Climatic regions: temperate regions, coastal and inland.

Host plants: species of *Allocasuarina*.

Feeding habits: larvae feed on sap inside the gall. Each gall contains a solitary insect.

Notes: the galls can occur singly or in groups, sometimes clusters. Often a single tree within a group of sheoaks will carry numerous galls but few or none occur on adjacent trees.

Control: as per introduction for eriococcid galls.

Male and female cone-mimicking galls on Allocasuarina littoralis. [D. Jones]

Cylindrococcus species

- *Cylindrococcus spiniferus* has female galls that exhibit a remarkable mimicry of the sheoak fruiting structure. They are conical to spindle-shaped, green to brown structures, 1.5–3cm long, covered with three to six rows of overlapping bracts, each with a narrow drawn-out apex. A single opening occurs at the apex of the gall. A plump pinkish insect about 0.4cm long occurs inside the female gall. The male insects induce shorter, narrower, tapered galls. They are often found together with the female galls or sometimes occur singly towards the shoot apex. This species commonly induces galls on *Allocasuariana littoralis* but other hosts include *A. monilifera*, *A. muelleriana*, *A. nana*, *A. paludosa* and *A. paradoxa*.

- *Cylindrococcus casuarinae* induces slender to broad, smooth galls, each gall subtended by a basal cluster of papery bract-like structures. Sometimes a single slender gall will be found on the tip of a shoot. Male galls are short and slender. A common host is *Allocasuarina verticillata*; others include *A. huegeliana* and *A. luehmannii.*

Bloodwood Apples or Bush Coconuts, *Cystococcus* species

Description: there are three species of these native eriococcids, two named and one undescribed. The

females induce large woody galls on the stems of inland species of bloodwoods. Both sexes live within a gall, feeding on the nutritious lining of a large internal chamber.

Climatic regions: temperate inland regions.

Host plants: species of *Corymbia,* especially the Desert Bloodwood, *C. opaca.*

Feeding habits: larvae feed on sap inside the gall. Each gall contains a solitary female and several males.

Notes: The galls often occur in groups, sometimes crowded close together. The female insect and the moist lining of the internal chamber are edible and can be eaten as bush tucker.

These galls are commonly known as Bloodwood Apples. [D. Jones]

Control: control is not necessary.

Cystococcus species

- *Cystococcus pomiformis* induces hardy, grey, woody, globose galls that can be as large as a tennis ball. The surface is mostly smooth to slightly roughened. The gall has a large chamber that contains a single yellowish female insect which can grow to about 0.4cm long. The gall has an apical opening which is usually sealed by the tip of the female's abdomen. Early in her life the female produces numerous male nymphs which feed on the nutritious lining of the gall until mature. When they leave the gall as adult males about 0.8cm long, they carry some first-instar females clinging to their abdomen. These females then disperse on the host as crawlers after the male lands, eventually inducing new galls on young stems.

- *Cystococcus echiniformis* induces large, hard, dark galls that have a rough to knobby surface.

OTHER GALL-INDUCING ERIOCOCCIDS

- **Melaleuca Cluster Galls** ('*Sphaerococcus*' species). Several gall-inducing eriococcids, originally described in the mealybug genus *Sphaerococcus*, are the subject of detailed research and currently await transfer to one or more new genera. They induce complex galls on paperbark species of *Melaleuca* in the tropics and southern Australia. Gall types include clusters of overlapping bracts that resemble buds or

Flowerbud-mimicking 'Sphaerococcus' gall on Melaleuca viridiflora. [D. Jones]

'Sphaerococcus' gall on Melaleuca leucadendra. [D. Jones]

flowers, tight groups of short, projecting, twig-like structures with swollen heads, globose fruit-like structures and circular blisters on the bark.

- **Teatree flower bud/fruit galls** (*Eremococcus* species). Four species in this genus induce galls on species of *Leptospermum*. The globose to pear-shaped galls look similar to the flower buds or fruit of the host plant. One species, *E. pirogallis*, infests the foliage of species of *Leptospermum* in areas around Sydney and the Blue Mountains. In some seasons large areas of *Leptospermum* are affected, the leaves covered with galls that pass through shades of green to pink, red and brown as they mature.

- **Teatree Stem Gall**. Females of the native eriococcid *Callococcus leptospermi* induce woody galls on the stems and twigs of some species of *Leptospermum*. The galls can be single or fused together to form an irregular-shaped woody structure that extends along a stem. The males do not induce galls but first instar male nymphs feed inside the female gall. This eriococcid is being tested as a possible biocontrol agent to be used against naturalised populations of *L. laevigatum* in South Africa.

Heavy gall infestation on eucalypt leaves induced by males of a species of Opisthoscelis. [D. JONES]

- The males and females of species of *Opisthoscelis* and *Tanyscelis* induce distinctly different galls on many species of *Eucalyptus*. Shapes include small craters, pits, pimples, bubbles and woody projections. Sometimes these galls are abundant on a single tree, impairing normal growth and development.

PIT SCALES (FAMILY ASTEROLECANIIDAE)

This is a group of scale insects that form pits or induce unusual galls on plants. They can also cause growth deformation. Three species of the genus *Frenchia* cause galls on some plants in Australia.

Natural control: probably by small parasitoid wasps. Some individual trees seem more prone as a host to gall-inducing pit scales than others.

Other control: these gall infestations are usually of a minor nature and control is rarely warranted. Early removal and destruction of the galls will reduce infestation. Chemical control of these galls is unsatisfactory because the insect is protected within the gall.

Casuarina Pit Scale, *Frenchia casuarinae*

Description: females induce galls on twigs and stems of the host. Small galls containing a single insect occur on twigs, larger galls containing two or three insects occur on thicker stems. The gall consists of a hollow basal chamber topped with a 0.4cm-long, stick-like, black, hollow tube which has a terminal orifice. The enclosed female scale insect is pink and jelly-like. First instars settle on young branches and cause swellings around the point of attachment. Males are unknown.
Climatic regions: subtropics and temperate regions.
Host plants: species of *Allocasuarina* and *Casuarina*.

Feeding habits: a long-lived scale that usually develops as individuals or in small groups. Single swellings occur on small twigs, multiple swellings on larger stems, eventually developing into a woody gall. Occasional severe attacks can weaken trees and cause stem breakage (at the point of the gall) and dieback.

Notes: a specialised native scale which is usually sporadic in its occurrence and causes limited damage.

Control: as per introduction for pit scales. Galls can be removed by hand. Severely damaged branches should be removed.

Other Gall-inducing Pit Scales

- Females of the Banksia Pit Scale, *Frenchia banksiae*, induce unusual hairy galls on both surfaces of the leaves of *Banksia serrata*. The galls contain tufts of hairs which are actually modified leaf hairs of the host. The males do not form galls.

- An unnamed species of *Frenchia* induces similar hairy galls on *Banksia integrifolia*.

Unusual furry galls of Frenchia *sp. on* Banksia integrifolia. [D. JONES]

GIANT SCALES (FAMILY MARGARODIDAE)

These scale insects gain their common name because of the large size of the females. Although there are more than 400 species worldwide only three species in three genera are known to induce galls. Two of these occur in Australia. One species causes serious defoliation and dieback of kauri pine plantations.

Kauri Giant Scale, *Coniferococcus agathidis*

Description: the feeding of first instar nymphs induces tiny pimple-like blister galls on very young leaves. Second instars are legless. Mature females, about 0.3cm long, can leave the gall or mate through an opening. Each female, which can move slowly, lays up to 200 eggs in a sac which is placed on the host in sheltered sites among older leaves or among leaf scales. There are several generations each year with the scale peak coinciding with the spring flush of growth on the kauris.

Climatic regions: tropics and subtropics.

Host plants: species of *Agathis*, especially *A. robusta* and *A. palmerstonii*.

Feeding habits: the scales suck sap inside the gall.

Notes: a native scale that has become a serious pest of plantations of kauri pines. Large numbers of nymphs can distort new growth, resulting in premature defoliation. Successive attacks over several years result in stem dieback and tree death. Attacks are so persistent and debilitating in some areas, such as the Mary River Valley in south-eastern Queensland, that plantation growth has been abandoned. Damage to natural stands of kauri has also been observed on Fraser Island in south-eastern Queensland and on the Atherton Tableland in the tropical north.

Control: no practical control.

Other Gall-inducing Giant Scales

- The Hoop Pine Giant Scale, *Araucaricoccus queenslandicus*, attacks plantations and natural stands of Hoop Pine, *Araucaria cunninghamii*, in south-eastern Queensland, causing relatively minor damage.

GALL-INDUCING ARMOURED SCALES (FAMILY DIASPIDIDAE)

Armoured scales are dealt with in detail in Chapter 4. A few species induce galls. In Australia scales of the genus *Maskellia* induce unusual galls on some species of *Eucalyptus*. The male and female galls are of different size and shape and can resemble those induced by some species of *Apiomorpha* in the Eriococcidae (see above). Several undescribed species of *Maskellia* induce galls on a range of eucalypts in southern Australia. Clusters of female galls can be very dense on young shoots causing misshapen growth, reduced leaf size and coalesced swellings. The accompanying photo shows a conglomerate of male and female galls induced by an unknown *Maskellia* species on a eucalypt near Gosnells in Western Australia.

Galls induced by an armoured scale in the genus Maskellia – *the green swellings are female and red ones male.* [C. FRENCH]

Witches' Broom Scale, *Maskellia globosa*

Description: female galls develop on stems of the outer canopy forming irregular swellings and witches' broom-like clusters of stems. The conical, pimple-like male galls develop on leaves. The female galls can occur in abundance, resulting in reduced leaf size and stem dieback. Epicormic shoots are often produced lower down on affected stems. Severe infestations can lead to the tree death.
Climatic regions: temperate regions, mainly coastal.
Host plants: *Eucalyptus blakelyi, E. gomphocephala* and *E. macrocarpa.*
Feeding habits: the scales feed within the confines of the gall.
Notes: The damage caused by this scale is often prominent in relict patches of vegetation including those by roadsides. It mainly impacts on large trees and is a major cause in the decline of Grey Box, *E. macrocarpa.*
Control: control is impractical.

Witches' broom gall. [S. JONES]

Diaspidid gall on Ironbark. [S. JONES]

GALL-INDUCING FLIES (FAMILY CECIDOMYIIDAE)

It is perhaps surprising to consider that flies can be associated with the formation of galls in plants, however there are numerous small to minute dipterans that can induce this response. Fly galls can be formed on the leaves, stems and buds of plants. In eucalypts, the galls range in shape from tiny pimples to large spongy or woody structures. The galls can be pale green or take on strong reddish tones. The fly larvae (maggots) feed within the gall, emerging as adults. Some galls contain only a single maggot, others several maggots.

Members of the fly family Cecidomyiidae are a common cause of galls in plants and for this reason are known as gall gnats or gall midges. Not all members of this family are associated with gall formation; others feed on plant organs such as flowers without inducing galls, some predate mites or scale insects. Adult gall midges are small flies – usually 0.1–0.3cm long – with long antennae and distinctively patterned wings that are often hairy. The larvae are legless maggots. Thousands of species of gall midges are distributed throughout the world, including some that are serious pests of crop plants. The family is well represented in Australia but is largely unstudied and many species remain poorly understood. It is known that more than 100 species of gall midges cause galls on eucalypts, often in highly specific associations, with gall shapes being specific (and thus recognisable) between the plant host and gall insect. Another important group of gall midges forms galls on species of *Acacia*, impeding fruit and seed formation.

Natural control: gall-inducing flies are mainly controlled by specialised parasitic wasps which insert their ovipositor through the gall and lay an egg on the maggot inside.

Other control: gall-inducing fly infestations are usually of a minor nature and control is rarely warranted. Early removal and destruction of galls will reduce infestation. Chemical control is unsatisfactory because the insect is protected within the gall.

Galled buds or inflorescences of Acacia mearnsii *possibly induced by a gall fly.* [D. Jones]

EXAMPLES

Banksia Gall Fly, *Dasineura banksiae*

Description: eggs laid on soft new growth in late spring-summer hatch to form small yellow maggots which burrow into leaf tissue. These maggots induce globular to irregularly rounded white hairy galls 0.5–0.6cm across on the leaves of *Banksia coccinea*. Commonly the leaf underside is affected but galls can also occur on the upper surface. Up to five maggots can occur in each gall. The margins of infected leaves often curl making the galls prominent. The galls, which become brown with age, can occur in significant numbers to make the leaves unsightly.
Climatic regions: temperate regions, mainly coastal.
Host plants: *Banksia coccinea* and perhaps other species.
Feeding habits: fly maggots feed within the gall.
Notes: a native gall fly that has become a significant pest of commercial cut flower farms. The prominent galls reduce the value of stems used for cut flowers.
Control: as per introduction for galls. Chemical control measures are being tested.

Galls on Banksia coccinea *induced by the Banksia Gall Fly.* [S. Jones]

Teatree Gall Flies, two *Dasineura* species

Description: galls are formed on vegetative and floral buds. The cone-like galls consist of overlapping papery bract-like structures. The tiny yellow to orange larvae live within the gall and pupate in silky cocoons between the outer bracts. The adult brownish-orange flies are about 0.2cm long.
Climatic regions: temperate regions, mainly coastal.
Host plants: *Leptospermum laevigatum.*
Feeding habits: fly maggots feed within the gall.

Gall on Leptospermum laevigatum *induced by the Teatree Gall Fly,* Dasineura tomentosa. [D. Jones]

Notes: two species of native gall fly are involved. *Dasineura strobila* produces hairless, brownish, elongated galls about 1cm long. By contrast *Dasineura tomentosa* produces distinctive globose galls that are covered with silky white hairs. *Dasineura strobila* has been introduced to South Africa to help control naturalised populations of *L. laevigatum*. The other species is being tested for this purpose.
Control: as per introduction for galls.

Paperbark Gall Fly, *Lophodiplosis trifida*

Description: induces swollen galls in young shoots of *Melaleuca quinquenervia*, impacting on the stem growth. Older galls become woody with numerous exit holes made by the departing adults.
Climatic regions: tropics and subtropics, mainly coastal.
Host plants: *Melaleuca quinquenervia*.
Feeding habits: fly maggots feed within the gall.
Notes: this gall fly has been successfully introduced into Florida where it is assisting in the biological control of naturalised populations of *Melaleuca quinquenervia*.
Control: as per introduction for galls.

Saltbush Gall Fly, *Lophodiplosis trifida*

Description: induces spongy balloon-like galls 1.5–2cm across on the flowers of *Atriplex vesicaria*. There is a single hard-walled chamber inside the gall containing a single orange maggot.
Climatic regions: inland temperate regions.
Host plants: *Atriplex vesicaria*.
Feeding habits: fly maggots feed within the gall.
Notes: this gall fly has sporadic impacts on seed set in this saltbush, which is an important foraging plant in semi-arid rangelands.
Control: as per introduction for galls.

Wattle Flower Gall Flies, several *Dasineura* species

Description: female flies lay eggs in the open flowers and the larvae enter the wattle ovaries to feed, causing the formation of a cluster of galls that replaces a flowerhead. Each gall arises from a single ovary and the whole structure is sometimes called a 'floret gall'. Individual gall shapes include globes, balls, bag-like structure and elongate tubes. Some galls are hairy. Initially green, the infested flower heads eventually become brown and woody. Gall shapes are consistent within a species of gall midge.
Climatic regions: subtropics and temperate regions, coastal and inland.
Host plants: numerous species of *Acacia*.
Feeding habits: fly maggots attack the flower-heads disrupting fruit and seed formation.
Notes: although they impede seed formation, these pests are usually of minor significance. Sometimes however, whole populations of wattles in an area will be affected. These insects have the potential to be used as biological control agents to limit seed formation in naturalised populations of *Acacia* in South Africa.
Control: as per introduction for galls.

Dasineura species

- *Dasineura acaciaelongifoliae* induces globose, tightly packed galls on the flowers of *Acacia implexa, A. longifolia, A. maidenii, A. sophorae, A. stricta* and *A. oxycedrus*.

- *Dasineura glauca,* known as the Grey Fluted Galler, induces grey, sparsely hairy, tubular galls on the flowers of *Acacia pendula* and *A. omalophylla*. The galls are 1–1.6cm long and 0.3–1.3cm wide. Each gall has up to 12 chambers and each chamber is occupied by a single larva. This species is abundant, in some seasons completely preventing seed formation on some trees and sometimes impeding seed formation over large areas.

Galls induced by a Wattle Flower Gall Fly on a flowerhead of Acacia baileyana.
[D. JONES]

Tubular galls on Acacia pendula *induced by a Flower Gall Fly,* Dasineura glauca.
[D. JONES]

Galls induced by a Wattle Flower Gall Fly on flowerheads of Acacia mearnsii.
[D. JONES]

- *Dasineura glomerata* induces irregularly-lobed, ball-like galls on the flowers of *Acacia deanei, A. elata, A. hakeoides, A. melanoxylon, A. provincialis, A. pycnantha, A. retinodes* and *A. schinoides*.

- *Dasineura pilifera* induces hairy, inflated, bag-like galls on the flowers of *Acacia baileyana, A. dealbata* and *A. decurrens*.

- *Dasineura rubiformis* induces soft round to pear-shaped galls the flowers of *Acacia deanei* and *A. mearnsii*

- *Dasineura sulcata* induces small galls hidden among the remnants of the flowers of *Acacia saligna*.

Other examples

Wattle Pinnule Gall Fly, *Austroacaciadiplosis botrycephalae*

Description: spherical galls, 0.3–0.4cm across, initially green, becoming red, develop on the upper surface of pinnules. Each gall contains a single larva. Mature galls split at the apex to release the larvae which pupates in soil.

Climatic regions: temperate regions.

Host plants: *Acacia baileyana, A. deanei, A. dealbata* and *A. decurrens*.

Feeding habits: the larvae feed within the galls.

Notes: this gall fly has the potential to be used as biological control agent against naturalised populations of *Acacia* in South Africa.

Control: as per introduction for galls. A very minor pest.

Galls induced by the Wattle Pinnule Gall Fly on Acacia dealbata. [T. WOOD]

Galls induced by a Wattle Flower Gall Fly on flowerheads of Acacia granitica. [D. JONES]

Galls induced by a Wattle Flower Gall Fly on flowerheads of Acacia pendula. [D. JONES]

Galls induced by Wattle Flower Gall Fly on Acacia falciformis. [D. JONES]

Callitris Gall Fly, *Mesodiplosis callitridis*

Description: smooth, woody, yellowish brown or bluish brown galls up to 0.8cm long and 0.5cm wide, formed on the branchlet tips of cypress pines. Each gall contains a single maggot. When disturbed the maggot spins rapidly on its longitudinal axis and rolls away. The galls split into three valves when old, mimicking a ripe *Callitris* cone.

Climatic regions: subtropics and temperate regions, mainly inland.

Host plants: *Callitris* species, especially *C. glaucophylla, C. gracilis* and *C. endlicheri.*

Feeding habits: the fly maggot feeds within the gall.

Notes: a very minor pest that exhibits a fascinating case of mimicry with the gall closely resembling the fruiting cone of the host.

Control: as per introduction for galls.

Eugenia Gall Fly, *Dasineura eugeniae*

Description: this insect attacks developing fruit to form hard, misshapen galls with an irregular surface. Each gall contains several maggots.

Climatic regions: tropical and subtropical regions, mainly coastal.

Host plants: species of *Syzygium* including *S. cormiflorum, S. francisii, S. moorei* and *S. oleosum.*

Feeding habits: the fly maggots feed within the gall.

Notes: the galls resemble a misshapen fruit that does not ripen. Whole crops of fruit can be affected in some seasons.

Control: as per introduction for galls. Control is impractical on large trees.

Other Gall Flies

- Several species of gall flies induce galls in many different species of eucalypts. The Eucalypt Gall Fly, *Harmomyia omalanthi*, induces tiny pink to red pimple-like galls on the midribs of some species of *Eucalyptus*.

Galls on Eucalyptus macorrhyncha *leaf induced by gall flies.* [D. JONES]

Galls on Eucalyptus polyanthemos *leaf induced by gall flies.* [T. WOOD]

SYMBIOTIC GALL-INDUCING NEMATODES/FLIES

A unique gall-inducing relationship exists between maggots of some small flies in the genus *Fergusonina*, family Fergusoninidae, and nematodes of the genus *Fergusobia*. Larval nematodes are incorporated within the eggs of the female fly at the time of laying. The larval nematodes initiate gall formation before the eggs of the fly hatch. Both organisms then work in a remarkable symbiotic association, feeding inside the gall that has been induced as a result of their activities. Symbiotic galls of this type are found over much of Australia, most commonly on species of *Eucalyptus*. Other genera known to be hosts (all in the family Myrtaceae) include *Angophora*, *Corymbia*, *Melaleuca*, *Metrosideros* and *Syzygium*. Infested tissue includes leaf buds, stem tips and flower buds. The galls range from small pimple-like or pea-sized structures to much larger flat leaf galls. Infestations are mostly of a minor nature but sometimes huge infestations occur that can result in the loss of branches from the sheer weight of the galls. Infestations of the flower buds can also have significant impacts on honey and seed production. Small wasps may also breed in these galls. Large old galls show numerous exit holes left by departing insects.

Life cycle: adult flies mate and the female lays eggs on young plant tissue. The larval nematodes begin feeding before the eggs hatch, causing a rapid proliferation of plant tissues. On hatching, the fly larvae form chambers in which they live in harmony with the nematodes. Small galls usually have only one chamber, large galls can have hundreds. The nematodes breed within the cavity and when a female fly larvae matures and is about to pupate, two fertilised female nematodes enter its body. Larval nematodes penetrate the oviduct of the fly and accompany any eggs that are laid.

Natural control: gall-inducing flies are mainly controlled by specialised parasitic wasps and moths. Their larvae tunnel through the gall, eating any fly larvae they find.

Other control: infestations by these symbionts are often minor and control is rarely warranted. Early removal and destruction of galls will reduce infestation. Chemical control is unsatisfactory because the insect is protected within the gall.

Nema Galls or Fergusoninid Galls, *Fergusonina/Fergusobia* species

Description: a range of gall sizes and shapes developed on vegetative buds, flower buds and shoot tips. Flower bud galls are not uncommon in eucalypts, the galled buds being several times larger than normal buds. Galls on eucalypt leaves and shoot tips are irregularly bumpy and often reddish. Shoot galls on *Melaleuca* are variable in shape but usually hairy.
Climatic regions: tropics, subtropics and temperate regions, coastal and inland.
Host plants: various genera in the family Myrtaceae, especially species of *Corymbia*, *Eucalyptus* and *Melaleuca*.
Feeding habits: both the fly larvae and the nematodes feed inside the gall that has been induced as the result of their activities. The nematodes increase in numbers while in the gall and invade the female fly before it emerges to lay eggs, thus completing the life cycle.
Notes: there are many species of *Fergusonina* that induce galls on species of Myrtaceae and all have associated nematodes living in the galls. Preliminary studies indicate that some of these combinations might be unique and specific to a plant host. One species of *Fergusonina*, which attacks the leaves and developing flower buds of *Melaleuca quinquenervia*, is being investigated as a possible biological control agent to be used against naturalised populations of this species in the southern states of the USA.
Control: as per introduction for galls.

Fergusonina species

- *Fergusonina brimblecombei* induces galls in the flower buds of *Eucalyptus crebra*, *E. melanophloia*, *E. odorata* and *E. hemiphloia*.

- *Fergusonina centeri* induces irregular conical galls on the terminal shoot buds of *Melaleuca leucadendra*.

- *Fergusonina eucalypti* induces galls in the flower buds of *Corymbia maculata*.

- *Fergusonina evansi* induces galls on the leaves of *Eucalyptus melliodora*.

- *Fergusonina greavesi* induces galls in the stem tip buds of *Eucalyptus polyanthemos*.

- *Fergusonina lockharti* induces galls in the stem tip buds of *Eucalyptus rudis*.

- *Fergusonina makinsoni* induces lettuce-like galls in the vegetative buds of *Melaleuca dealbata*.

- *Fergusonina newmani* induces galls in the leaf buds of *Eucalyptus gomphocephala*.

- *Fergusonina nicholsoni* induces galls in the flower buds of *Eucalyptus macrorhyncha*, upsetting flowering, honey and seed production.

- *Fergusonina thornhilli* induces galls in the terminal leaf buds of *Eucalyptus dalrympleana*.

- *Fergusonina tillyardi* induces galls in the flower buds of *Eucalyptus blakelyi*, *E. camaldulensis* and *E. tereticornis*.

- *Fergusonina turneri* induces vegetative and floral galls on natural populations of *Melaleuca quinquenervia*. This species is being investigated as a possible agent for the biological control of naturalised populations of *Melaleuca quinquenervia* in Florida.

- *Fergusonina williamensis* induces galls in the terminal leaf buds of *Eucalyptus baxteri*.

- *Fergusonina* sp. – an undescribed species that induces galls in the flower buds of *Corymbia ptychocarpa*.

Nema galls on Melaleuca leucadendra.
[D. JONES]

Other examples

Other Gall-inducing Nematodes

- The Root Knot Nematode, *Meloidogyne incognita*, which forms galls on the roots of many plants, is dealt with in chapter 14.

Nema galls on Eucalyptus robusta.
[D. JONES]

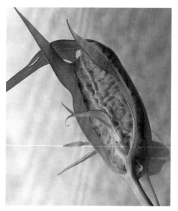

Nema gall on tip of eucalypt shoot.
[D. JONES]

GALL-INDUCING PSYLLIDS

Psyllids are common sucking insects of plants that are dealt with in detail in chapter 5. A few specialised psyllids are galling agents with at least one being a very tenacious and severe pest of *Syzygium* species and cultivars. Juvenile psyllids are flightless nymphs whereas the adults are tiny winged insects. Galls produced by some psyllids are quite substantial and out of all proportion to the size of the insect. Large numbers on leaves interfere with photosynthesis and cause premature leaf-drop. Commonly the galls are round to ovoid, but elongate horn-shaped galls and sunken pits are also known. Some species have galls with an apical orifice that is kept plugged by waxy secretions. Others, which lack an orifice, split open at maturity. The galls can occur singly or in groups with the walls fused together. Young green galls become yellowish, purple or red at maturity.

Natural control: psyllids in the nymph stages are attacked by tiny parasitic wasps; the adults are preyed upon by ladybirds, lacewings, ants and beetles.

Other Control: the galls are best removed and destroyed at an early stage of development to prevent build-up of numbers. Chemical control is ineffective because the insects are protected within the galls.

EXAMPLES

Eucalypt Gall Psyllids or Bubble Galls, *Schedotrioza* and *Glycaspis* species

Description: two groups of free-living psyllids – about 12 species of *Schedotrioza* and 180 species *Glycaspis* – many of which form inflated, spongy, bubble-like or sac-like galls on the upper surface of eucalypt leaves. Adults emerge in spring, coinciding with flushes of new growth on the host. Tiny eggs are laid in groups or rows under the epidermis of the leaf. The nymphs burrow into the leaf, resulting in the formation of galls. Usually only one nymph occurs in each gall, but sometimes adjacent galls become fused together as they grow, occasionally forming multi-chambered structures.

Psyllid galls of the genus Glycaspis *on leaf of* Eucalyptus delegatensis. [T. Wood]

Climatic regions: tropics, subtropics and temperate regions.

Host plants: species of *Eucalyptus*. Some gall-inducing psyllids are specific to a single species of *Eucalyptus*, others feed within a range of related species.

Feeding habits: the nymphs feed inside the gall (one per chamber). Mature galls mostly become red but in one species the galls age yellow.

Notes: most of these psyllids cause limited damage to the host tree. One species however, *S. distorta*, which causes leaf twisting and distortion when present in large numbers, has caused significant damage to *Eucalyptus leucoxylon* and ironbarks in South Australian gardens.

Control: as per introduction for gall-inducing psyllids.

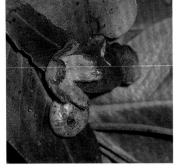

Flower-like psyllid galls of the genus Schedotrioza *on* Eucalyptus saligna. [D. Jones]

Psyllid galls of the genus Schedotrioza *on* Eucalyptus rossii. [T. Wood]

Syzygium Leaf Gall Psyllid or Pimple Psyllid, *Trioza eugeniae*

Description: a free-living psyllid that forms small, pimple-like galls, blisters or sunken pits. Winged adults about 0.2–0.3cm long, are brown with a white band around the abdomen. Adult females lay yellow eggs on the margins of new leaves. Nymphs are yellowish with red eyes. The feeding activities of the nymphs result in the formation of galls or pits, often with the accumulation of wax.

Climatic regions: tropics, subtropics and temperate regions, mainly coastal.

Host plants: species of *Syzygium,* particularly *S. australe, S. floribundum, S. moorei, S. paniculatum* and *S. smithii.*

Feeding habits: psyllids feed within the galls or pits (one nymph per site of attack) causing discoloration and distortion of the leaves and shoots. Very young leaves often develop incorrectly or abort. The galls and surrounding tissue often become bright red or purplish, develop incompletely and contrast with normal growth.

Notes: a widespread pest that is tenacious and persistent once established. It not only renders trees unsightly but can also cause dieback of branches. This pest is particularly severe on two commonly grown species, *S. australe* and *S. paniculatum*, but also affects others. Several cultivars purportedly resistant to psyllid attack have been marketed in recent years. However, some of these cultivars have eventually succumbed to infestation. Some species, such as *S. leuhmannii*, appear to be resistant to this pest.

Control: spraying new growth prior to infestation with imidacloprid or malathon is usually necessary to control this pest. Spraying infested growth will be unsuccessful since the psyllid is protected from contact by the walls of the gall. Prune off and destroy infested growth and spray any new growth that appears. Psyllid numbers are reduced by heavy frost and hot weather. Cultivars supposedly resistant to sustained attack are available to home gardeners.

Pit galls and pimple galls on Syzygium paniculatum *caused by the Pimple Psyllid.* [D. Jones]

Deformed stems and leaves of Syzygium paniculatum *caused by the Pimple Psyllid.* [D. Jones]

Pimple Psyllid galls on Syzygium australe *'Orange Twist'.* [R. Elliot]

Other Gall-inducing Psyllids

• The leaves of some rainforest plants are galled by poorly studied species of psyllid.

GALL-INDUCING THRIPS

Thrips are tiny insects that are dealt with in chapter 6. A few specialised thrips are associated with galls and a unique group of Australian native thrips induce the formation of galls on the stems, leaves and phyllodes of some native plants. These thrips are particularly active in inland regions and usually involve a range of gall types, including inrolled leaf margins, pits, pouched galls and cylinders or tubes. As well as inducing galls, many species of thrips construct simple shelters by joining adjacent leaves or flat phyllodes together with glue secreted from the anus. Narrow phyllodes can be joined together by threads in a webbing-type arrangement.

If a gall is opened and found to contain numerous tiny insects then these are undoubtedly thrips as, unlike most gall-formers, these insects continue reproduction inside the gall. Some thrips-induced galls can be long-lived. Over 1,000 individuals have been recovered from a single gall on *Acacia pendula*.

Natural control: gall-inducing thrips are mostly controlled by parasitic wasps and predatory mites. They are also attacked by other groups of thrips.

Other control: cut off and destroy affected growth. Spraying may be necessary to control severe infestations.

LEAF DISTORTION THRIPS

Several species of thrips in more than one genus induce growth distortion in plants. Commonly these thrips attack young developing shoots causing a range of symptoms including leaf curling, folding or rolling of the leaf margins and crumpling of the leaf surfaces. The thrips gain protection and feed on surfaces within the distorted leaves. These growth distortions are called 'leaf roll galls' by some authorities.

Distortion and inrolling of young Callistemon *leaves caused by leaf distortion thrips.* [D. Jones]

Leaf roll galls on Callistemon *growth caused by leaf distortion thrips.* [D. Jones]

EXAMPLES

- Callistemon Distortion Thrips, *Teuchothrips disjunctus*, affects species and cultivars of *Callistemon*, particularly *C. citrinus* and *C. viminalis*. Affected leaves become folded, twisted, reduced in size and often turn reddish. This is a persistent and disfiguring pest which makes badly affected plants appear unsightly. Similar damage to *Callistemon salignus* is caused by *Teucothrips minor*.

- Pittosporum Distortion Thrips, *Teucothrips ater*, causes prominent distortion on developing shoots of *Pittosporum undulatum*, *P. angustifolium* and *P. revolutum*.

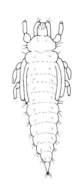

Leaf distortion thrips. [D. JONES]

Damage caused by leaf distortion thrips to Callistemon citrinus *(left) and* Ficus benjamina *(right).* [D. JONES]

Black thrips on distorted leaves of Pittosporum angustifolium. [S. JONES]

Leaf roll galls on Pittosporum angustifolium *induced by Pittosporum Distortion Thrips.* [S. JONES]

Damage to Pittosporum undulatum *caused by Pittosporum Distortion Thrips.* [R. ELLIOT]

- Paperbark Distortion Thrips, *Teucothrips melaleucae*, causes growth distortion in *Melaleuca leucadendra*.

- Ficus Distortion Thrips, *Gynaikothrips australis*, causes twisted and rolled leaves on shoots of *Ficus macrophylla*. Similar symptoms have been observed on *Ficus benjamina, F. obliqua, F. rubiginosa* but the culprit is unconfirmed.

- Olearia Distortion Thrips, *Klambothrips oleariae*, induces leaf distortion in young shoots of some species of *Olearia*.

Control: these thrips are very difficult to control because they remain hidden within the leaves. Badly affected shoots should be cut off and burnt to reduce population levels.

Fig Gall Thrips, *Gynaikothrips* species

Description: some thrips induce galls in the young leaves of native figs. The galls are created by leaf distortion or the infolding of the margins. The galls may only be active for a couple of months but appear at intervals on flushes of new growth over the warm months.
Climatic regions: tropics and subtropics, mainly coastal.
Host plants: some species of *Ficus*.
Feeding habits: the thrips live and breed inside the gall.
Notes: these thrips can impair the appearance of pot-grown figs used for indoor display.

Control: as per introduction for gall-inducing thrips. Spraying when young may be necessary to control these pests.

EXAMPLES

- The leaves of *Ficus microcarpa* are attacked by *Gynaikothrips ficorum*. It feeds on the upper surface of young leaves causing folding and distortion.

- The leaves of *Ficus macrophylla*, *F. obliqua* and *F. rubiginosa* are attacked by *Gynaikothrips australis*. It feeds on the lower surface causing the margins to fold inwards.

Wattle Gall Thrips, *Kladothrips* species

Description: about 30 species of native thrips induce galls in the young phyllodes of native wattles, many in host-specific relationships. Often a single female induces the gall and a colony builds up within the enclosed structure. The galls range from tubular structures to hollow spheres and pouches. Some galls have projecting spikes or prominent ridges. One significant group of thrips live within shelters constructed by gluing or sewing adjacent phyllodes together.

Climatic regions: tropics, subtropics and temperate regions, coastal and inland

Host plants: numerous species of *Acacia*.

Feeding habits: the thrips live and breed inside the gall.

Notes: wattle gall thrips are a unique group within the native gall-inducing thrips. Many species form a specific insect-plant relationship.

Control: as per introduction for gall-inducing thrips.

EXAMPLES

- Mulga trees, *Acacia aneura,* commonly have round to ovoid, green to brownish galls on the phyllodes caused by *Kladothrips arotrum*.

- *Acacia harpophylla* develops curved galls that start off as an outgrowth on a phyllode margin, eventually becoming a hollow sphere.

- *Acacia pendula* has hollow, smooth, spherical to somewhat irregular bubble-like galls formed on the phyllodes by *Kladothrips rugosus*. These galls, commonly known as boree galls, appear in large numbers over extensive areas in some years.

- *Acacia melanoxylon* often has sausage-shaped hollow galls on the phyllodes caused by *Kladothrips rodway*. Very young galls appear on the phyllodes as small ridges.

Other Gall-inducing Thrips

- Leaves of the Wilga, *Geijera parviflora*, frequently carry unusual galls formed by tightly inrolled leaf margins. These galls are induced by the feeding of a complex of about ten species of thrips in the genera *Sacothrips*, *Choleothrips* and *Moultonides*.

- Distinctive swollen pale galls on the leaf margins of *Aurantiaca rhombifolia* are caused by *Neocecidothrips curviseta*.

- Unusual bulbous galls on young stems of *Casuarina cristata* and *C. obesa* are caused by *Lotatubothrips crozieri* and *L. kranzae*. These galls can be very long-lived.

- Distinctive irregularly roughened galls on the young leaves of *Piper novae-hollandiae* are caused by species of *Liothrips*, the adults of which are black. The greyish galls, which develop when the margins of infested leaves roll inwards, contrast with the dark green of normal leaves.

GALL-INDUCING MOTHS

The larvae of some species of small moths produce galls in native plants. These are not well known and none can be considered as pests of significance. The galls, which are mainly formed in the stems of the host plant, are woody with a large central chamber where the caterpillar lives and feeds. The moths emerge through a hole cut in the gall. Plants known to be hosts of gall-inducing moths include *Banksia integrifolia*, mistletoes, chenopods, a *Canthium* species, and some species of *Eremophila*.

Natural control: largely unknown but probably controlled by parasitoid wasps.

Other control: if seen as a problem, cut off and destroy the affected growth.

GALL-INDUCING WASPS

Wasps that are responsible for the formation of plant galls are common in Australia. Most gall-inducing natives are chalcidoid wasps which belong to the superfamily Chalcidoidea; four species of *Banksia* wasp gallers are ichneumons in the family Braconidae, superfamily Ichneumonoidea. Gall-inducing wasps are tiny – usually 1–0.3cm long – dark-coloured wasps, sometimes with a metallic sheen. Many often have a solitary lifestyle and in some species the females do not need to mate. Eggs are laid under the bark or leaf epidermis of the host plant and the larvae feed within the tissue of the gall that their activities induce. Typically the gall has an inner layer of nutritious cells on which the larvae feed, surrounded by a hard layer of protective cells. The galls range from small single-celled structures with one larva per gall to complex structures containing up to 100 cells and larvae. Pupation occurs in the gall and the adult wasp cuts a prominent hole on emergence.

Natural control: the most common parasitoids of gall-inducing wasps are other tiny wasps that lay their eggs directly into the gall using long, slender ovipositors. Their larvae feed on the insects in the galls, in some cases burrowing from one cell to another.

Other control: most gall wasps are minor pests and control is rarely necessary. Sometimes individual trees may carry huge infestations on their leaves, marring their appearance, resulting in premature leaf-fall and hindering growth and causing dieback. Trees that are persistently infested may be best removed. If seen as a problem the galls should be removed from young trees and destroyed when first noticed to prevent build up of numbers. This is an impossible task on large trees. Chemical control is usually ineffective and its efficacy is masked by the structure of the gall.

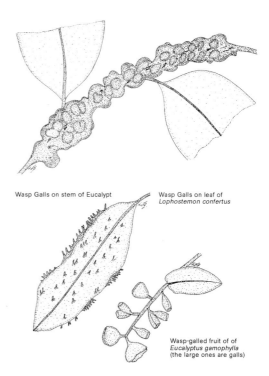

Wasp Galls on stem of Eucalypt

Wasp Galls on leaf of
Lophostemon confertus

Wasp-galled fruit of of
Eucalyptus gamophylla
(the large ones are galls)

Examples of galls caused by wasps. [D. Jones]

EXAMPLES

Citrus Gall Wasp, *Bruchophagus fellis*

Description: adult females lay several eggs under the bark of a new shoot in spring. The legless, grub-like larvae feed in the stem tissue inducing the formation of an elongated, irregularly-swollen, woody gall. Adult wasps emerge in spring, leaving prominent small holes in the gall, to begin a new life cycle (only one generation per year).

Climatic regions: tropics, subtropics and temperate regions, coastal and inland.

Host plants: species and cultivars of *Citrus*, including native species. Plants of *Acronychia littoralis* develop very similar woody galls on the stems and leaf midribs (see photo) but it is uncertain if they are caused by the Citrus Gall Wasp.

Possible Citrus Gall Wasp damage on Acronychia littoralis. [R. Elliot]

Feeding habits: the larvae feed within the gall.

Notes: this native pest is now widely distributed over most of the citrus-growing areas in Australia, most likely transported on infested nursery plants.

Control: remove and burn or solarise all galls before wasp emergence (September). This important pest is parasitised by two native wasps, *Megastigmus brevivalvus* and *M. trisulcus*, which lay eggs onto larva inside the gall. Up to 90 per cent of the larvae inside a gall can be parasitised. These wasps are available from some biocontrol firms.

Geraldton Wax Gall Wasp, *Oncastichus goughi*

Description: adult female wasps, which are about 0.2cm long and black with red eyes, lay eggs under the bark of new shoots. The larvae feed in the wood and induce a swollen ellipsoid gall that becomes yellowish or reddish with age, sometimes causing branch deformation. Small holes indicate where adult wasps have emerged.

Climatic regions: subtropics and temperate regions.

Host plants: cultivars of *Chamaelaucium uncinatum*.

Feeding habits: the larvae feed within the galls.

Notes: although a minor native pest of garden plants, this species is of commercial significance for cut flower growers since galled stems are considered unsuitable for sale. It has also become established in California and Israel

Control: as per introduction for gall-inducing wasps. Regular spraying is necessary in commercial plantations.

Eucalypt Gall Wasps, superfamily Chalcidoidea

Description: galls caused by chalcidoid wasps are common on eucalypts in Australia and in fact it is believed that every eucalypt species (in the genera *Angophora*, *Corymbia* and *Eucalyptus*) is host to at least one species of gall-inducing wasp, if not several. Mostly these relationships appear to be specific between the host plant and the wasp. Wasp galls range from small protruding pimple-like structures on leaves, each containing a single insect, to complex elongated woody structures on stems that may contain up to 100 individuals. The galls often become yellowish, pink or red as they age.

Climatic regions: tropics, subtropics and temperate regions, coastal and inland.

Host plants: species of *Angophora*, *Corymbia* and *Eucalyptus*.

Feeding habits: the larvae feed within the galls.

Notes: within a species some trees appear to be more susceptible than others. Attacks range from minor to debilitating. Badly affected trees are unsightly.

Control: as per introduction for gall wasps. Most gall wasps are minor pests and control is rarely necessary. Sometimes individual trees may carry huge infestations on their leaves and stems, marring their appearance, resulting in premature leaf-fall, hindering growth and causing dieback. Trees that are persistently infested may be best removed. If seen as a problem the galls should be removed from young trees and destroyed when first noticed to prevent build up of numbers. This is an impossible task on large trees.

EXAMPLES

Several species of native gall wasp have become established in many countries overseas, impacting significantly on commercial eucalypt plantations and ornamental plantings, resulting in major economic losses and distress. These damage caused is worse in overseas environments because of the absence of natural enemies.

- *Selitrichodes globulus* causes the formation of massed disfiguring purple-black stem galls on *E. globulus* in California.

- *Ophelimus maskelli* causes serious blanketing infestations of pimple galls on the leaves of several eucalypts in the Mediterranean region.

- *Lectocybe invasa*, although apparently unknown in Australia, causes irregularly swollen, bumpy galls on the stems and leaf midribs of several eucalypts in many countries. It has serious impacts on seedlings and young plants, including impairment of apical dominance and even plant death.

- *Epichrysocharis burwelli* seriously galls the leaves of *Corymbia citriodora*, impacting on essential oil production and also reducing the appeal of nursery plants.

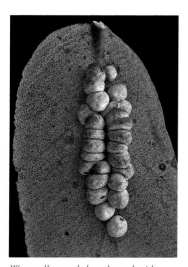

Wasp galls crowded on the underside midrib of a eucalypt leaf. [T. WOOD]

Chalcidoid wasp galls on Eucalyptus calophylla. [C. FRENCH]

Colourful wasp galls on eucalypt leaves. [D. JONES]

Distortion of Tallowood stems caused by wasp galls. [D. JONES]

Large wasp galls on leaves of Eucalyptus robusta. [D. JONES]

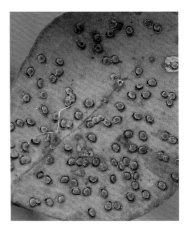

Wasp galls on a Eucalyptus polyanthemos *leaf.* [D. JONES]

Wasp galls on the stem of a eucalypt. [D. JONES]

Pimple Galls induced by a wasp on underside of eucalypt leaf. [D. JONES]

Wasp gall on the stem of a eucalypt. [D. JONES]

Wattle Gall Wasps, *Trichilogaster* species

Description: several species of chalcidoid wasps induce galls on phyllodinous wattles. The galls, which can be formed on phyllodes, vegetative buds and flower buds, range from globose structures to irregular-shaped galls containing many larvae.

Climatic regions: subtropics and temperate regions, mainly coastal.

Host plants: species of *Acacia*.

Feeding habits: the larvae feed within the gall.

Note: galls induced by these wasps are often large.

Control: as per introduction for gall wasps. Most wasp gall-formers on wattles are minor pests and control is rarely necessary.

EXAMPLES

More than a dozen species of *Acacia* have become established as invasive weeds in South Africa. Two species of native chalcidoid wasps have been released as biological control agents. They cause heavy infestations of weighty galls that not only impede seed formation but also cause limb breakage.

- *Trichilogaster acaciaelongifoliae* attacks the young flower buds of *A. longifolia*, *A. sophorae* and *A. floribunda*, causing the formation of large, smooth, misshapen galls.

- *Trichilogaster signiventris* attacks young flower buds and vegetative buds near the stem tips of *A.*

pycnantha causing the formation of large globose galls covered with pimply projections.

- *Trichilogaster maidenii* induces smooth galls on young flower buds of *A. longifolia*, *A. implexa*, *A. floribunda* and *A. maidenii*.

- *Trichilogaster pendulae* induces smooth galls on flower buds of *A. pendula*.

- Several species of wasp induce galls on Mulga, *A. aneura* (see also thrips). One significant species, *Trichilogaster esculenta*, induces marble-sized galls on the flower buds. These galls, which are covered with small pimply projections, are known as mulga apples. The gall and the enclosed female insect are edible and eaten as bush tucker, either raw or after cooking in hot coals. The other wasp galls are apparently not edible.

Galls induced by a Wattle Gall Wasp, Trichilogaster acaciaelongifoliae, *on the flowers of* Acacia sophorae. [D. Jones]

Galls induced by a wattle gall wasp, Trichilogaster signiventris, *on the flowers of* Acacia pycnantha. [T. Wood]

Wasp gall on Acacia aneura, *known as a mulga apple.* [C. French]

OTHER EXAMPLES

Other Gall-inducing Wasps

- Clusters of unusual red or purplish horn-like galls on eucalypt stems are caused by chalcidoid wasps of the family Pteromalidae, probably species in the genus *Terobiella*. A single wasp larva occurs in each gall.

- Round to spindle-shaped stem galls 3–15cm long on *Banksia marginata* are caused by larvae of the braconid wasp *Mesostoa kerri*.

- Leaves of Brush Box, *Lophostemon confertus*, can be rendered unsightly by colonies of hard, pointed wasp galls which are quite prickly.

- Hard, round, pink to green galls among the seed pods of *Bossiaea eriocarpa* in Western Australia are caused by a small black wasp in the genus *Tanaostigmodes*. The wasp lays an egg in the flower or developing seed pod. Each gall contains a single wasp larva living in a central chamber.

- Small bubble-like galls on the flower petals of the kurrajong *Brachychiton populneus* are caused by the wasp *Megastigmus sulcicollis,* while rounded lumpy galls on the young branches are the result of feeding by larvae of the wasp *Decatomothorax gallicola.*

- Unusual bud-like galls on *Allocasuarina* stems are caused by *Lisseurytoma* species.

Galls on Acacia *induced by a wattle gall wasp – note protruding phyllode.* [T. Wood]

Horn-like galls probably induced by a chalcidoid wasp. [D. Jones]

Horn-like galls probably induced by a chalcidoid wasp. [D. Jones]

Wasp galls on young kurrajong stems. [D. Jones]

GALL-INDUCING BEETLES

The larvae of some specialised beetles induce galls on stems and growing tips. The damage is mostly very limited and these insects are generally of minor significance to plant growers. The larvae of gall-inducing beetles are small, fleshy, whitish grubs which may or may not be legless. The larvae are fairly obvious when the galls are broken open and provide a useful way to identify the type of causative insect.

Natural control: tiny parasitoid wasps lay their eggs in the galls. Birds may also crack open the galls to eat the larvae.

Other control: remove and burn the galls if they are causing problems.

EXAMPLES

Gregarious Gall Weevil, *Strongylorhinus* species

Description: female weevils lay their eggs in a series of 'ringbarking' holes chewed in the bark at a particular point along a stem. The larvae, which are creamy-white or yellowish, fleshy, legless grubs to 1.5cm long, chew into the sapwood and a bulbous, woody, irregularly shaped gall up to 30cm long and 4–5 times the diameter of the stem develops. Each gall can contain up to 100 or more larvae. The surface is roughened and marked with patches of gum, each patch indicating a chamber where a larva is active. The adults are brown weevils about 1cm long.

Climatic regions: temperate regions, mainly coastal.

Host plants: *Eucalyptus* species, including *E. viminalis*, *E. siderophloia* and *E. rudis*.

Feeding habits: the larvae feed in chambers inside the gall. The galls enlarge as feeding continues. Galls occur on stems and coppice growth, sometimes close to ground level.

Notes: an uncommon but sometimes persistent pest that causes disfigurement and stem breakage. *Strongylorhinus ochraceus* has been recorded as damaging *Eucalyptus rudis* in Western Australia. *Strongylorhinus clarki* damages eucalypts in south-eastern Australia, including Tasmania.

Control: as per introduction for gall-inducing beetles.

Stem galls on Eucalyptus *caused by gregarious weevils.* [D. Jones]

Other Gall-inducing Beetles

- A small weevil induces pale coloured galls on the growing tips of some narrow-leaved eucalypts. The gall seems to be formed by the young apical leaves fusing together to form a fleshy gall which contains the legless larvae. The adult is a small brown weevil about 0.4cm long. Attacks are mostly minor but sometimes shoot growth can be significantly disrupted by these galls.

- A few small jewel beetles (family Buprestidae) are associated with the formation of stem galls in some plants. Some of these relationships may be host specific. Members of the genus *Ethon* live and breed in small swollen galls on the stems of species of *Dillwynia* and *Pultenaea*. Similarly the jewel beetle *Dinocephalia cyaneipennis* forms rough, bump-like galls about 1cm long on the stems of *Allocasuarina distyla* and *A. littoralis*. The larvae of the jewel beetle *Diphucrania albosparsa* induce small galls on the stems of some phyllodinous wattles. The adult beetle, which is about 1.2cm long, is dark brown with various coloured metallic tints and a circle of eight small white dots on the wing-covers.

GALL-INDUCING BACTERIA – SEE CROWN GALL, CHAPTER 19.

GALL-INDUCING FUNGI OR FUNGUS GALLS

About eight species of the rust fungus genus *Uromycladium* cause woody galls to form (sometimes also witches' brooms) on numerous species of *Acacia* in Australia. These fungi, which have a very complex life cycle involving no less than three distinct types of fruiting body and up to four different types of spore, attack young stems, shoots, leaves, phyllodes, flowers and pods. Mostly the galls are hard, brown, irregularly-shaped knobs with a rust-coloured, powdery or granular appearance, however the size and type of gall formed varies with the species of fungus and the wattle involved.

Uromycladium species and host plants

- *Uromycladium tepperianum* and *U. notabile* are the most significant and widely distributed species, both causing similar damage to a range of wattles. *Uromycladium tepperianum*, for example, can infest more than 110 species of wattle. Similar fungus galls are also formed on the stems of Cherry Ballart, *Exocarpos cupressiformis*. These fungi, which are difficult to distinguish from each other outside of a laboratory, contrast with the other species of *Uromycladium* by producing large, prominent, reddish-brown galls measuring many centimetres across. The larger galls are commonly found on stems and branches; smaller galls on leaves, petioles and fruit. The galls can last for many years, increasing steadily in size. Galls up to 18cm across and weighing more than 500g have been found on stems of *Acacia mearnsii*. Extensive branching galls may also form on some species, particularly *Acacia implexa*. Mature galls produce masses of spores which cover the surface of the gall. Old galls become black and very woody. The fungus life cycle is completed during a growing season and each gall can produce millions of spores that are spread on the wind and capable of infecting other trees. In severe attacks the whole canopy of a tree can be festooned with these grotesque galls. Heavy infestations reduce flowering and seed set, greatly weaken the tree, cause branch breakage and lead to eventual death. Badly infected trees are sometimes locally tagged as 'poo trees'.

- *Uromycladium acaciae* forms small galls, including distorted swollen patches on stems and small brown powdery masses on the underside of leaves. This fungus causes little damage and is of minor significance. Hosts include *Acacia baileyana, A. dealbata, A. decurrens, A. mearnsii* and *A. paradoxa*.

- *Uromycladium alpinum* impacts on new shoots by webbing the leaves and leaflets together in a distorted sticky mass which eventually becomes brown as spores are produced. This fungus causes little damage to mature trees but can impact on seedling development. Hosts include *Acacia baileyana, A. dealbata* and *A. mearnsii*.

- *Uromycladium maritimum* forms reddish-brown to black lesions on phyllodes and stems resulting in premature shedding. This fungus causes little damage and is of minor significance. Hosts include *Acacia floribunda, A. longifolia, A. sophorae* and *A. notabilis*.

- *Uromycladium robinsonii* induces very small galls on phyllodes and sunken brown cankers on stems. Severe attacks can cause dieback in parts of the canopy. Hosts include *Acacia melanoxylon* and *A. sophorae*.

- *Uromycladium simplex* causes minute pimple-like galls on the phyllodes of *Acacia pycnantha*. Severe infestations can completely cover the phyllode surface and cause premature shedding.

Notes: A complex ecology revolves around these woody galls involving several insect groups such as weevils and moths, the larvae of which feed in the galls. These activities are in no way associated with the formation of the fungus galls.

The establishment of a successful tanning industry in New Zealand was prevented by these gall-inducing fungi, the impacts of which debilitated and killed the wattle plantations.

Uromycladium tepperianum has been successfully used in South Africa to control naturalised populations of wattles, particularly *Acacia saligna*. Some populations have been reduced by as much as 80 per cent after the fungus was introduced.

Control: weakened trees are attacked more severely than vigorous, healthy trees. Trees should be maintained in a healthy condition and galls removed, when noticed, to prevent the spread of the disease. Control of badly infected trees is impractical.

Fungus galls on stems of Acacia implexa. [D. Jones]

Old Uromycladium *gall on* Acacia pendula. [D. Jones]

Uromycladium *gall on* Acacia mearnsii. [D. Jones]

Uromycladium *gall on* Acacia podalyriifolia. [D. Jones]

Uromycladium *gall on* Exocarpos cupressiformis. [T. Wood]

Witches' Broom gall caused by Uromycladium *fungus.* [D. Jones]

Uromycladium *gall on* Acacia pycnantha. [D. Jones]

GALL-INDUCING MITES

Mites are tiny animals, some of which damage plants by piercing plant cells and sucking the sap (see chapter 6). Mites which induce galls are known as rust mites or gall mites and placed in the order Acarinae, family Eriophyidae. Commonly known as eriophyid mites (also erinose or erynium mites), they are very tiny mites that can only be readily seen under a microscope. There are hundreds of species in Australia, most undescribed. Unlike most mites which have eight legs, eriophyid mites have only two pairs of legs and an elongated sausage-like body. They are distributed by air currents and on the feet and feathers of birds. Some species of eriophyid mite cause galls on leaves, flowers and fruit. Mite galls can be formed internally within a leaf or on the surface tissue. Internal galls are often called blister galls and the mites termed blister mites. Surface galls are often hairy or with a velvety or felted texture. Some eriophyid mites are called rust mites because their feeding causes a rusty appearance on infected leaves. Attacks on leaves by gall-inducing mites usually result in unsightly leaf deformation. Within a species, some plants appear to be more susceptible to mite damage than others. Gall-inducing mites can also disrupt buds, flowering (termed blossom mites) and seed set, but leaf disfigurement is more common.

Natural control: natural predators include parasitic wasps and predatory mites.

Other control: blossom and fruit galls should be removed and burnt to reduce the build-up of the pest. Blister mites on leaves are difficult to control. Sulphur dust provides some control. Severely attacked trees are best replaced.

EXAMPLES

Bloodwood Blister Gall Mite, *Acalox ptychocarpi* and other species

Description: flat to raised, irregular, roughened blisters on the leaf surface. The galls can be much paler than the surrounding tissue (sometimes yellow), or stained red. Young shoots can be distorted by these mites and affected leaves fall prematurely.
Climatic regions: tropics, subtropics and warm temperate regions, mainly coastal.
Host plants: *Corymbia* species, particularly hybrids and cultivars of *C. ptychocarpa* and *C. ficifolia*; also *C. citriodora, C. calophylla, C. henryi* and *C. maculata*.
Feeding habits: mites shelter, feed and breed within the galls.
Notes: a very disfiguring group of pests that can have severe impacts on the establishment and shape of young trees. The mite *Acalox ptychocarpi* can have major deleterious impacts on the successful establishment of plantations of *Corymbia citriodora* subsp. *variegata* and *C. maculata*. Because of its adverse effects on the leaves, it can also severely reduce the ornamental appearance of the widely planted *Corymbia ptychocarpa* and its colourful hybrids. Usually only parts of a leaf are affected but sometimes the whole leaf is badly attacked. Attacks on young leaves can inhibit normal expansion and the leaves may develop unevenly. Some trees are attacked badly, whereas nearby trees can be virtually unscathed.

Control: as per introduction for gall-inducing mites.

Blister mite galls on Corymbia ptychocarpa. [D. Jones]

Corymbia *leaves disfigured by the Bloodwood Blister Gall Mite.* [D. Jones]

Hibiscus Erinose Mite or Hibiscus Gall Mite, *Aceria hibisci*

Description: induces flattish to raised, irregular-shaped galls that range from dark green to pale green; also leaf puckering and green velvet-textured galls on the leaves and stems; also growth distortion. The galls start off very small and increase in size as the mites continue feeding. Severe attacks can induce witches' broom-like growths.

Climatic regions: tropics, subtropics and warm temperate regions, mainly coastal.

Host plants: mainly exotic cultivars of Chinese, Indian and Hawaiian *Hibiscus* in Queensland and New South Wales; the native mallow *Abelmoschus moschatus* and some native *Hibiscus* species are susceptible to attack.

Feeding habits: mites shelter, feed and breed within the galls.

Notes: this Brazilian mite was first confirmed in south-eastern Queensland in 1992 although it is believed to have been introduced in the late 1970s.

Control: as per introduction for gall-inducing mites.

Other Gall-inducing Mites

- Many different species of native plants are damaged by eriophyid mites. A common symptom is the production of unusual growths known as witches' brooms (also see chapter 19). These structures consist of a crowded bunch of greatly reduced shoots which often take on reddish tones. Witches' brooms of this type can be common on some species of *Acacia*, *Allocasuarina*, *Eucalyptus* and *Leptospermum*.

- Eriophyid mite galls are common on the developing fruiting cones of *Banksia integrifolia* and *B. marginata*. The large misshapen galls, which start off green, become brown with a somewhat powdery or roughened texture. Although a relatively minor problem, these galls reduce seed formation.

- Several species of mites in the genus *Acadicrus* cause growth distortion on species of *Eucalyptus*. Another group of mites induce felted masses on eucalypt leaves with affected leaves often taking on bluish or purple tones.

- An unidentified eriophyid mite induces tiny, hard, green, pimple-like galls on the leaves of *Myoporum acuminatum* and *M. insulare* in temperate coastal areas. Hundreds of galls can occur on a single leaf and some plants become very heavily infested. Affected leaves fall prematurely.

- An unidentified eriophyid mite induces large, globular to cone-shaped or hard green misshapen galls on the leaves of *Banksia menziesii*. These unsightly galls, which become brown with age, reduce the value of cut flowers in the market place.

Eriophyid mite galls on fruiting cone of Banksia integrifolia. [D. Jones]

Unusual galls on cone of Banksia integrifolia *induced by eriophyid mites.* [D. Jones]

Galls on Myoporum acuminatum *induced by eriophyid mites.* [D. Jones]

Eriophyid mite galls on leaves of Banksia menziesii. [C. FRENCH]

Eriophyid mite galls on Banksia menziesii *leaves.* [C. FRENCH]

Galls on Acacia sophorae *phyllodes induced by an eriophyid mite.* [D. JONES]

- An unidentified eriophyid mite induces tiny pit galls on the phyllodes of *Acacia sophorae* in temperate coastal areas. Hundreds of pits can occur on a single phyllode and some plants become very heavily infested. Affected phyllodes turn brown over summer and fall prematurely.

- An unidentified eriophyid mite induces clusters of tiny, hard, purple, pimple-like galls on the inflorescences of *Syzygium smithii* in temperate coastal areas. Whole inflorescences can be infested on some plants.

- An unidentified eriophyid mite induces felted galls on the leaves of *Melicope elleryana* in the subtropics. Affected leaves are unsightly and fall prematurely.

- A strong russeting that is commonly seen on the leaves of *Notelaea longifolia* may be caused by rust mites.

- The Mangrove Erinose Mite, *Aceria avicenniae*, causes flattish, grey, felt-like galls on the underside of the young leaves of Grey Mangrove, *Avicennia marina*. The tiny black mites are just visible under a hand lens.

Eriophyid mite galls on inflorescences of Syzygium smithii. [D. JONES]

CHAPTER 8

LEAF-EATING CATERPILLARS

The larvae of moths and butterflies, commonly known as caterpillars, feed on the leaves, new shoots, buds, flowers and fruit of a wide range of native plants. Many of these pests are of minor importance and add an extra interesting dimension to a garden, especially considering that they metamorphose into moths and butterflies, some of our most treasured garden visitors. Some caterpillars are extremely destructive and can cause significant plant damage. Caterpillars are usually controlled by a wide range of natural predators and parasites, but occasionally the balance is upset and the pests can increase in numbers to plague proportions. In bad years some of these pests, such as the caterpillars of gum tree moth, can be present in huge numbers and may cause significant damage over large tracts of natural forest, also impacting on gardens.

Moths and butterflies are grouped in the order Lepidoptera and have two pairs of wings that are covered in tiny scales. There are approximately 21,000 species of moths and butterflies in Australia. Moths and butterflies experience complete metamorphosis with four distinct stages. The females lay eggs, usually close to the larval food source. Caterpillars hatch from the eggs and go through between four and six instars as they grow before they pupate. They spin a cocoon which can be buried in soil, hidden in leaf litter, attached to vegetation or hidden inside tunnels in wood. Adults emerge and usually feed on nectar through their sucking tube mouthparts. Some species are extremely short lived and do not feed as adults. There are a few physical differences that determine whether a caterpillar will turn into a moth or butterfly. Butterflies typically have antennae that are clubbed and hold their wings upright when at rest. Moths often have quite elaborate feathered antennae and sit with their wings flat. Many of the most voracious caterpillars, and therefore pest species, will be moths as adults.

The results of feeding by caterpillars are usually quite obvious. They either feed by grazing the leaf surface – known as skeletonising – or else by chewing the leaves. Some caterpillars eat the entire leaf, others eat the intervening tissue and avoid the main veins, leaving a ragged framework behind. Some concentrate on young tender leaves, others prefer mature leaves. Whichever way they feed damage results to the plant's foliage. Often within a large plant the damage is manageable and restricted to certain parts, such as the lower branches or the young growth. Impacts on young or newly planted shrubs and trees can, however, be much more significant. Some caterpillars, especially large ones, have a voracious appetite and can quickly defoliate a small plant. These caterpillars can cause significant problems in nurseries and their feeding can also impair the normal development of young trees and shrubs. The activities of leaf skeletonisers, which graze the leaf surface and leave behind a network of veins, can usually be readily recognised by the browning or scorched appearance of damaged tissue. Caterpillars not only eat leaves, many also feed on buds, flowers, fruit and seeds. Subterranean species mostly live on plant roots. Some species are carnivorous, cannibalising other caterpillars or eating insects such as scale or ant larvae.

Caterpillar spinning cocoon. [S.JONES]

Recognising problem caterpillars: many types of pestiferous caterpillars occur in gardens and they can often be recognised by their feeding habits and whether they build shelters or live in the open.

Solitary Caterpillars: many caterpillars occur as single individuals or feed together in small groups. Often the young caterpillars feed close together after hatching, but as they grow they spread out and feed as individuals. Their feeding habits can also change with age, the young caterpillars eating tender tissue, often grazing the leaf surface, whereas larger caterpillars feed on older material, eating chunks or whole leaves.

Hairy or woolly caterpillars that don't build shelters: Most caterpillars have hairs on their body but some species are very hairy and live in the open without the protection of any sort of shelter. Some of these hairy caterpillars have sharp hairs that can readily penetrate skin and cause severe itching, rashes and general skin irritation. A few species have barbed hairs which can be very difficult to remove. Some of these caterpillars are common garden visitors and they may feed singly or in groups (most feed in groups when young but spread out and become solitary with age). Commonly they pupate in a flimsy or sparsely spun silken cocoon. Most have a voracious appetite and feed on a wide range of plants. The moths are mainly small and in the families Arctiidae (Tiger Moths) and Lymantriidae (Tussock Moths); some larger moths of the family Anthelidae and Notodontidae also have very hairy caterpillars.

Hairy caterpillar feeding on a eucalypt. [T. Wood]

Hairy caterpillar. [T. Wood]

Hairy caterpillar that feeds on Lomandra. [T. Wood]

Hairy caterpillar that feeds on eucalypts. [T. Wood]

Hairy caterpillar. [T. Wood]

Hairy caterpillar that feeds on snow gum. [T. Wood]

Large fleshy caterpillars: most of the larger moths have conspicuous fleshy caterpillars that mainly feed on the leaves of native trees. These caterpillars, which can grow 8–12cm long, are noticeably hairy with the hairs usually arranged in tufts, sometimes on elongated structures known as turrets. Some species have nasty stinging hairs, others are quite innocuous. The caterpillars themselves are mostly an interesting addition to a garden and the large moths are very decorative, some being ornamented with startling eye-spots on the wings.

Leaf skeletonisers: these caterpillars strip off the outer skin of the leaves, exposing a skeleton-like network of cell walls and veins. Eaten areas quickly brown-off and the affected parts become very conspicuous as brown or scorched patches among normal green foliage. Commercial plantations of eucalypts can suffer badly from leaf skeletonisers with affected trees giving the appearance of devastation by fire. Damage of this type is caused by the caterpillars of several small moths. In some groups, such as cup moth caterpillars, only the smaller caterpillars feed in this way, whereas older caterpillars chew the leaves normally. Leaf skeletonisers have a ravenous appetite, feeding together in groups and are able to quickly devastate large patches of foliage.

Caterpillars that build shelters: A number of caterpillars construct shelters to protect themselves from predators. These caterpillars live within the shelters during the day, emerging mainly at night to forage and feed. Sometimes they take food back to the shelters to eat during the day. The shelters may be for a single individual or house a communal group. As well as being physical barriers to predation, the caterpillars often incorporate defence mechanisms into the shelters, including webbing and irritating hairs. Shelters vary from the flimsy protection obtained by curling the edges of a single leaf, or joining adjacent leaves together, to mobile homes, or large solid constructions that look like bags.

Leaf Rollers: a number of caterpillars, often from unrelated moth families, shelter within simple structures which they make by rolling one or more leaves and joining the edges together with silken threads. They may also form a shelter by webbing a leaf against fruit or flower buds. Those species which feed on small-leaved plants achieve the same result by joining adjacent shoots together. These caterpillars, which can be solitary or gregarious, usually live within the shelter during the day and emerge at night to feed.

Caterpillar shelter formed by rolling young leaves. [D. JONES]

Bag shelter moths: these are specialised caterpillars that construct large bag-like shelters in which they hide during the day and emerge from at night to feed. These shelters, which can be larger than a football, are constructed from leaves and twigs woven together with silk and suspended from branches by silken threads. Initially green, they quickly become brown and can be quite prominent, especially when much of the foliage around them has been eaten. Internally the shelters are very messy, containing masses of faecal pellets, dead caterpillars, cast skins and hairs. When touched, the stiff and pungently pointed hairs on many of these caterpillars can cause rashes and inflammation of sensitive areas, particularly skin and mucus membranes such as nasal passages and the eyes. The hairs are easily dislodged from the caterpillars and can be blown about on the wind. There are instances of office workers being affected by hairs originating from caterpillars feeding in nearby trees. The hairs are also present on caste skins and pupal cases. Suitable protective gloves,

Simple caterpillar shelter built by joining leaves together. [D. JONES]

Simple webbing caterpillar nest. [D. JONES]

Caterpillar nest in Angophora costata.
[D. JONES]

glasses and masks should be worn when handling these pests or their shelters. Most infestations last only a few weeks but the hairs can persist in the area after the population has declined.

Natural control: caterpillars and grubs are controlled naturally by a wide range of predators including a variety of birds, lizards, frogs, hunting spiders, assassin bugs, lacewings, hoverflies and tachinid flies; parasites include many types of small wasp, nematodes, viruses and fungi.

Other control: caterpillars can usually be readily controlled with sprays or dusts, but in recent years some of the major pests of agriculture and horticulture have become resistant to some pesticides, which is a worrying trend. Pyrethrum-based sprays kill on contact and are mostly safe for the environment because they break down upon exposure to sunlight. Dusts containing pyrethrum are also effective and safe but must be used while fresh. Preparations containing spores of the bacteria *Bacillus thuringiensis* can be very effective but are inactivated by sunlight and must be sprayed in the evenings or during cloudy weather. Spinosad is another relatively safe biological product used to control caterpillars. Severe and persistent infestations of caterpillars can also be controlled with moderately safe sprays such as malathion, but these materials should only be used after due consideration for the environment, particularly their impact on non-target beneficial species. Bag caterpillars are not only unsightly but are also very destructive and should be discouraged from becoming established in gardens. The caterpillars are best killed during the day while in the shelter. The bags should be cut from the tree while wearing suitable safety gear, sealed in a plastic bag and left in the sun to solarise or collected in a paper bag and burnt. Jetting with insecticides such as malathion also controls their activities.

SYMPTOMS AND RECOGNITION FEATURES

Caterpillar exposed, leaf surface grazed, often exposing structural veins and turning brown ......*leaf skeletonisers*

Caterpillar hidden in curled leaf, leaf surface grazed, often exposing structural veins and turning brown................ *blue gum skeletoniser*

Caterpillar camouflaged, moving with a looping motion, remaining stiff and motionless during the day................ *loopers and semi-loopers*

Fleshy hairless caterpillars that drop to the ground and curl when disturbed .... *cutworms*

Velvety caterpillars with soft hairs that hide their feet................ *...snout moths*

Small, colourful, oblong, slug-like caterpillars with retractable stinging hairs................ *cup moths*

Large, fleshy caterpillars with a prominent spine at the tail-end................ *hawk moths*

Large, colourful, fleshy caterpillars with non-stinging spines in turrets................ *emperor moths*

Hairy caterpillars with distinctive erect toothbrush-like tufts of hairs on the back................ *tussock moths*

Colourful hairless caterpillars in groups or clusters................ *cluster caterpillar, lily caterpillar*

Hairy caterpillars walking head to tail in a procession................ *procession caterpillars, white cedar moth*

Untidy mass of webbing, chewed leaves and faecal pellets .... *webbing caterpillars*

Gregarious caterpillars living communally in green or brown bags constructed from leaves .... *bag moths*

Fleshy caterpillar living singly in a mobile silken bag or case................ *case moths*

Caterpillar living within a hollow silken tube constructed along branches and stems................ *tube caterpillar*

Leaf margins curled or rolled, or adjacent leaves, and joined together by silken threads................ *leaf rollers*

MOTHS

Moths have coiled mouthparts and scales on the wings, but can usually be distinguished from butterflies by their feathery or thread-like antennae which do not end in a small club-like structure. Moths mostly fly during the night and butterflies in the day, but there are exceptions in both groups. Most moth caterpillars are plant feeders and many are serious pests. Although generally destructive, the larger and more colourful or bizarre caterpillars add interest to a garden.

HAWK MOTHS (FAMILY SPHINGIDAE)

Description: hawk moths are moderately large, fast-flying moths which have slender wings and sturdy, cylindrical bodies. They fly rapidly, their wings beating so fast that they sometimes produce an audible noise. Many fly at dusk and at night and are attracted to lights. They can often be heard banging against or fluttering on windowpanes and mesh fly screens. Although mainly night-fliers, the bee hawk moths are a distinctive group that fly during the day. Hawk moths have a long proboscis and feed on nectar while hovering in front of the flowers of a wide range of plants. The eggs are usually laid singly on the underside of a leaf of the host plant. Their caterpillars, which are large, fleshy and often very colourful, can usually be recognised by the distinctive horn (tail-horn) at the rear end (however some species lack a tail-horn). The caterpillars within a species often display variation in colour and markings. At maturity the caterpillars form a pupa amongst litter or in the soil.

Climatic regions: various species found from tropical to temperate regions.

Host plants: there are numerous species of hawk moths (about 65 species in Australia) and their caterpillars feed on a wide range of weeds and garden plants, including aroids, garden annuals, climbers, shrubs and trees, native and exotic. Often a species of plant can be the host for more than one species of hawk moth; Cunjevoi, *Alocasia brisbanensis*, for example, is attacked by six species; the Grape Vine, *Vitis vinifera*, an exotic to Australia, is also attacked by a number of species that normally feed on native grapes.

Feeding habits: these caterpillars, which mostly feed singly or in small groups, have a good appetite and can cause considerable damage to garden plants, eating the leaves down to the main veins and midrib. Attacks are usually of a sporadic nature, although some species are more destructive than others. Their presence is often advertised by masses of barrel-shaped faecal pellets which collect on leaves and on the ground under the plant.

Notes: the caterpillars of many species have prominent, often colourful false eyes near the head. When disturbed they often hunch their back, expanding the false-eyes so they look fierce to discourage possible predators. The tail-horn does not sting or have any obvious function but the whole tail region may also be lifted when the caterpillar is adopting a defensive posture. Hawk moth caterpillars are quite decorative and generally of great interest to children.

Control: we suggest you be prepared to share some leaves with them and enjoy their company.

EXAMPLES

- **Bee Hawk Moth or Gardenia Hawk Moth**, *Cephonodes kingii*, from coastal areas of Queensland and New South Wales (south to Sydney) has grey to green caterpillars with a dark lateral line on each side and a warty area behind the head. White eye-spots along the side are outlined with red. The tail-horn is curved and ornamented with black warts. The moth, with a wingspan of about 4cm, has a green body with a yellow and black-fringed abdomen and clear forewings with black tips. The moths fly during the day, hovering noisily in front of nectar-rich flowers to feed. The caterpillars feed on species of the family Rubiaceae, including the following natives – *Canthium attenuatum*, *C. coprosmoides*, *C. oleifolium*, *Gardenia ovularis*, *Larsenaikia ochreata*, *Pavetta australiensis* and *Tarenna dallachiana*. They are also very fond of the various cultivars of the popular exotic Gardenia (*Gardenia augusta*).

- **Coprosma Hawk Moth or Cissus Hawk Moth**, *Cizaria ardeniae*, from coastal areas of much of Australia has a yellowish-green to chocolate-brown caterpillar, 6–7cm long with two pale longitudinal stripes and a prominent warty tail-horn. The moth, with a wingspan of about 5cm, has lovely velvet green to brown forewings with white margins and a white stripe across each forewing that lines up with a similar stripe across the abdomen. The caterpillars feed on species of *Alloxylon*, *Cayratia*, *Coprosma*, *Cissus*, *Morinda* and *Myrmecodia*.

- **Double-headed Hawk Moth or Banksia Hawk Moth**, *Coequosa triangularis*, is the largest Hawk Moth and occurs in coastal areas over much of Australia. It has a large, stout caterpillar about 10cm long that can be wholly yellow or brilliant emerald-green with obliquely slanting yellow stripes and bands. The body is covered with short soft warts and spikes. Prominent black eye-like knobs (no tail-horn) on the anal clasper give rise to the unusual common name. The moth, with a wingspan of about 13cm, has attractively patterned brown and yellow wings. The caterpillars feed on species of *Banksia*, *Grevillea* (especially *G. robusta*), *Hakea*, *Macadamia*, *Persoonia* and *Stenocarpus*. This is an exciting caterpillar to find and one which adds interest to the garden. It can be difficult to discern the caterpillar's head because of the black knobs at the rear that resemble eyes. This species is sometimes a minor pest of commercial *Macadamia* plantations.

Double-headed Hawk Moth. [D. JONES]

Caterpillar of the Double-headed Hawk Moth. [D. JONES]

- **Emu Bush Hawk Moth**, *Coenotes eremophilae*, is widely distributed in coastal and inland areas of northern Australia. It has a black caterpillar decorated with yellow and white longitudinal stripes and red spots along the flanks. The tail-horn is black and curved. The moth, with a wingspan of about 5cm, has brown wings and yellowish markings on the abdomen. It feeds on a wide range of plants including several species of *Eremophila* and *Myoporum*; also *Albizia basaltica*, *Carissa lanceolata*, *Clerodendrum floribundum*, *Josephinia eugeniae*, *Gyrocarpus americanus*, *Hibiscus panduriformis*, *Newcastelia spodiotricha* and *Prostanthera striatiflora*. The caterpillars are voracious feeders, often defoliating one host plant before moving to another.

- **Impatiens Hawk Moth**, *Theretra oldenlandiae*, which is found in many Asian countries and on several Pacific Islands, occurs in various parts of Australia. Its readily recognisable caterpillars, which grow to about 7cm long, are velvety-black with two rows of contrasting white, yellow or orange eye-spots (a pair on each segment). They also have a slender tail-horn which the caterpillar whips or waves about when walking. The moths are pale grey-brown with a broad dark band on each forewing. Pale dorsal stripes also run the length of the body. The caterpillars feed on native plants, including *Cayraytia clematidea*, *Hibbertia scandens*, *Planchonia careya* and *Viola hederacea*; exotic garden plants include species of *Impatiens*, *Fuchsia*, *Pentas* and *Zantedeschia*; it also eats the leaves of the Grape Vine, *Vitis vinifera*.

Caterpillar of the Impatiens Hawk Moth. [T. WOOD]

- **Pale Brown Hawk Moth**, *Theretra latreillii*, which is widely distributed from Asia to New Guinea occurs in northern Australia and south to Sydney. It has yellow, green or brown caterpillars up to 7cm long with two pale dorsal lines and two prominent large eye-spots behind the head. The moths, about 6cm across, are pale brown with a small spot on each forewing. The caterpillars feed on species of *Cayratia*, *Cissus* and *Leea indica*. Exotic garden plants include species of *Fuchsia*, *Impatiens*, *Lagerstroemia*, *Parthenocissus* and *Vitis vinifera*.

- **Australian Privet Hawk Moth**, *Psilogramma casuarinae*, which is found over most of Australia lays globular greenish eggs. Young caterpillars are very thin, green and with an extended horn at the rear. Older caterpillars, which grow to 8cm long, are fleshy and green with obliquely slanting whitish bands and white (sometimes purple) dorsal patches. They have a prominent tail-horn which is covered with warts. The moth is fast-flying with a wingspan of about 10cm. It has narrow grey forewings and darker hindwings. The handsome caterpillars feed on species of *Clerodendrum*, *Jasminum* and *Olea paniculata*; also olive trees and many exotic garden plants including *Campsis*, *Spathodea*, *Tecomaria* and *Lonicera*.

Caterpillar of the Privet Hawk Moth.
[D. JONES]

Unusually coloured caterpillar of the Privet Hawk Moth. [D. JONES]

Vine Hawk Moth. [T. WOOD]

- **Vine Hawk Moth or Silver-striped Hawk Moth**, *Hippotion celerio*, which is found over most of Australia has fleshy bright green to pale green or even brown caterpillars, 6–8cm long with a prominent tail-horn, pale lateral line and a pair of large yellow and green eye-spots on the back behind the head; there are two smaller spots on the next segment. The fast-flying moth has grey-brown forewings marked with silvery streaks and dots and a prominent silvery transverse band; the hindwings have a pink patch near the body. Wingspan is about 5cm. The caterpillars feed on *Alocasia brisbanensis*, *Amorphophallus* spp., *Cayratia* species, *Cissus* spp., *Colocasia esculenta*, *Grevillea* spp., *Hibbertia scandens*, *Ipomoea* spp.; also *Fuchsia*, *Impatiens* and *Vitis vinifera*. This species, which has a worldwide distribution, shows considerable variation in the colour and patterning of the caterpillars. Its caterpillars usually cause limited damage, although they can have a severe impact on grape vines. The moth is an important pollinator of papaya. When disturbed the caterpillars hunch their back and expand the eye-spots behind the head.

- **Yam Hawk Moth or Cunjevoi Hawk Moth**, *Theretra nessus*, which is found in coastal areas from the tropics to warm temperate regions has fleshy pale green or brown caterpillars. It is 10–12cm long with a prominent tail-horn and a pale marginal stripe on each flank. The moth is fast-flying with prominent yellow bands on the abdomen and dark brownish wings. Wingspan about 9cm. The caterpillars feed on *Alocasia brisbanensis*, *Amorphophallus* spp., *Colocasia esculenta* and *Dioscorea* spp. The caterpillars of this species feed in small to large groups, often chewing large lumps out of the leaves and leaving ragged remnants. Sometimes the main veins are left as a framework. They can severely damage some plants, particularly species of *Alocasia* and *Colocasia*.

Caterpillars of the Yam Hawk Moth, Theretra nessus, *on* Alocasia brisbanensis.
[D. JONES]

CUP MOTHS, CHINESE JUNKS, SPITFIRES, STINGERS, BONDI TRAMS (FAMILY LIMACODIDAE) (SEVERAL GENERA)

Description: eggs are laid in crowded groups, sometimes covered with hairs. Young caterpillars feed communally, older caterpillars feed singly. They are flattish, fleshy, slug-like caterpillars often with groups of red, brownish or yellow retractable stinging hairs in turrets along the body. The caterpillars, which grow to 2.5–4cm long, are commonly green to bluish-green although one startling species is black with contrasting white or yellow markings. The caterpillars move like a slug because they have small legs and no prolegs. At maturity, the caterpillars build a neat egg-shaped cocoon in which they pupate. A circular lid is cut on emergence of the moth, leaving the cocoon resembling a small cup. The moths have a wingspan about 4cm across and are greyish to brownish with marbled markings. Usually Cup Moths produce two generations per year.

Climatic regions: various species found from tropical to temperate regions.

Host plants: mainly species of *Corymbia* and *Eucalyptus*, but also species of *Acacia*, *Callistemon*, *Leptospermum*, *Lophostemon confertus*, *Macadamia*, *Melaleuca*, *Planchonia careya*, *Tristaniopsis* spp. and some other rainforest plants; attacks also occur on various fruit trees including apricots.

Feeding habits: these caterpillars can cause considerable damage to garden plants. Initially they skeletonise the leaf surfaces but as the caterpillars grow they eat the whole leaf down to the midrib. Infestations can strip large trees.

Notes: the hairs on many species of cup moth caterpillars sting painfully if contacted and for this reason they are often called 'stingers'. A severe sting can be short-lasting or persist for some hours and develop into a localised ache. Outbreaks of these pests are sporadic with large numbers appearing in some years, few in others.

Control: as per introduction for caterpillars and grubs. Heavy infestations may need to be sprayed. Natural infection by virus is not uncommon, the caterpillars dying and the internal organs becoming liquid.

Young cup moth larvae skeletonising a eucalypt leaf. [T. Wood]

Cup moth caterpillar, Doratifera *sp., on* Corymbia. [D. Jones]

Cup moth caterpillar. [T. Wood]

Cup moth cocoon. [T. Wood]

Cup moth. [T. Wood]

Cup moth. [T. Wood]

EXAMPLES

Several species of cup moth produce caterpillars with a similar general appearance, but there are other species with distinctly different caterpillars.

- **Mottled Cup Moth**, *Doratifera vulnerans*, is probably the most widely distributed and commonly encountered species. The caterpillars, which grow to about 3.5cm long, are green to pink with yellowish margins, red, brown and yellow markings on the back, and four turrets of stinging hairs at each end. It ranges from the tropics to temperate regions, inland and coastal, and feeds on a wide range of eucalypts and other plants.

- **Four-spotted Cup Moth**, *Doratifera quadriguttata*, has pale green caterpillars with green and red markings, four pairs of black spots at each end of the body, brown stinging hairs in the head region and soft spikes along the side.

- **Black Slug Cup Moth**, *Doratifera casta*, from subtropics and temperate regions, has blackish caterpillars to 3.5cm long with cream to yellow margins, ornamented with soft white or yellow spikes and four turrets of stinging hairs near the head.

Cup moth caterpillar, Doratifera vulnerans. [D. BEARDSELL]

- **Painted Cup Moth**, *Doratifera oxleyi*, from temperate areas, has green caterpillars with four black or yellow turrets of yellow stinging hairs at each end, several marginal turrets and a series of outlined spots on the back and sides.

- **Pale Cup Moth**, *Doratiphora pinguis*, from temperate areas, has apple-green caterpillars boldly marked with pale blotches outlined in black, four turrets with yellow stinging hairs at each end, numerous soft turrets along the margins and two protruding at the rear like a tail.

- **Wattle Cup Moth**, *Calcarifera ordinata*, from the tropics and subtropics, has an amazingly colourful caterpillar that is decorated with aqua-blue, yellow and red markings and seven long turrets of stinging hairs as well as a series of smaller ones along the sides. This species, which has a very painful sting, feeds on species of *Acacia*.

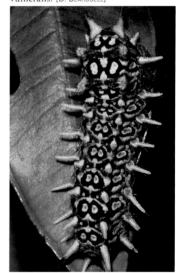

- **Macadamia Cup Moth**, *Comana fasciata*, from the tropics and subtropics has a very unusual fat caterpillar – to 3.5cm long and 1.5cm wide – that resembles a flattish green slug with numerous short marginal turrets and a yellow or white dorsal stripe that resembles the midvein of a leaf. It feeds on species of *Macadamia*.

Cup moth caterpillar, Doratifera quadriguttata. [D. BEARDSELL]

- **Striped Cup Moth**, *Anaxidea lozogramma*, also from the tropics and subtropics, has a well-armed green caterpillar measuring up to 2.5cm long with a prominent blue-margined white dorsal stripe. It has two large stinging turrets at each end and numerous short marginal turrets. It feeds on species of Proteaceae, including *Macadamia*.

- **Fern Cup Moth**, *Hedraea quadridens*, from the subtropics, has bright green caterpillars with yellow or dark lateral lines and a pale dorsal line. It has turrets of stinging hairs at each end of its body and along the sides. It feeds on bracken fern and species of *Pteris*.

- **Cocky Apple Stinger**, *Thosea penthima*, from the tropics has an unusual fattish yellow caterpillar with a darker dorsal line. It has numerous short stinging turrets in rows along the back and sides. This nasty stinging caterpillar causes problems to field workers in Darwin and other parts of the Northern Territory where it feeds on the young shoots and regrowth of *Planchonia careya* and *Terminalia ferdindandiana*.

LOOPERS, INCHWORMS OR SPANWORMS (FAMILY GEOMETRIDAE, SOME NOCTUIDAE)

Various species of moth produce caterpillars that are commonly known as loopers because of their characteristic means of locomotion – they are also known as inchworms or spanworms because of the distance they span between loops. Basically these caterpillars have a reduced number of claspers (prolegs) at the rear of their body and this reduction impacts on their means of locomotion. When moving the front part of the body is extended forward so that the caterpillar is stretched out flat, then the rear part is brought into close proximity with the front to form a characteristic loop. Another group, known as semi-loopers, do not form such a pronounced loop when moving.

Looper caterpillars are slender, often hairless or nearly so, and can be grey, green or brown with stripes, spots and other markings. They are voracious feeders and can cause considerable plant damage in a short time, especially in plant nurseries. They eat by stripping the leaves to the main veins and midrib and can defoliate a whole branch in a few days, or strip a young plant. They are mostly active at night, although the caterpillars will sometimes also feed during the day. Often they remain motionless during the day either closely clasping a stem or jutting out stiffly and resembling a small branch or twig (hence another common name, 'Twig Caterpillars'). When disturbed they often hunch their back or recurve the head in a characteristic defence pose. Some sway or thrash about wildly when threatened.

Loopers are mostly caterpillars of the moth family Geometridae, but some are found in other families such as the semi-loopers in the Noctuidae. Geometer moths mostly hold themselves flat against a surface, usually with all four wings exposed. The wings are often delicately marked and patterned to provide camouflage. They are mostly weak fliers and prefer to remain motionless, even when disturbed. They are attracted to lights at night and frequently can be observed pressed flat against windows, doors and walls.

Control: as per introduction to caterpillars and grubs. Hand picking is usually satisfactory but spraying young plants may be necessary to minimise damage.

Looper moth, Scopula perlata *(Geometridae).* [T. Wood]

Looper moth, Psilosticha absorpta *(Geometridae).* [T. Wood]

Large looper caterpillar that feeds on eucalypts. [D. Jones]

Green looper with a single pair of prolegs. [T. Wood]

Green looper with two pairs of prolegs. [T. Wood]

Looper caterpillar in camouflage pose.
[D. Jones]

Looper caterpillar on rainforest plant.
[D. Jones]

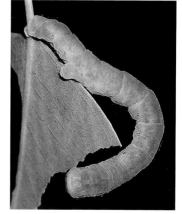

Looper caterpillar feeding on eucalypt leaf. [T. Wood]

Looper caterpillar on Corymbia torelliana. [D. Jones]

Looper caterpillar camouflaged as part of a stem. [D. Jones]

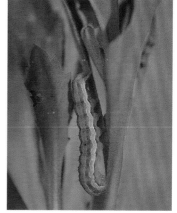

Semi-looper caterpillar on Scaevola.
[D. Jones]

EXAMPLES

Blue Gum Caterpillar or Autumn Gum Moth, *Mnesempela privata*

Description: eggs are laid in dense masses on the underside of leaves in late summer and again in late winter. Most egg-laying occurs on juvenile leaves on the lower branches of a tree. Young caterpillars are light yellow-brown with darker markings. Older caterpillars, which are fleshy, hairless and grow to 3cm long, are bluish-green with faint paler lines and red and yellow patches on each segment (two prominent yellow patches near the middle). Pupation occurs in the ground. The moths (family Geometridae) have yellowish-brown to orange-brown wings. Wingspan is about 4cm.

Climatic regions: subtropics and temperate regions, coastal and inland.

Host plants: species of *Eucalyptus* particularly those with juvenile glaucous leaves such as *E. bicostata*, *E. cinerea*, *E. crenulata*, *E. globulus*, *E. maidenii* and *E. nitens*; also *E. camaldulensis*.

Feeding habits: young caterpillars skeletonise the leaves by grazing the leaf surface and leaving a network of veins and dead tissue. Older caterpillars eat the whole leaf, often leaving the midribs intact. They are gregarious – although older caterpillars tend to be solitary or in pairs – and during the day cluster in shelters made by curling leaves and joining the margins together with silken thread. These shelters are usually found towards the end of a branch.

Notes: this is a common and serious pest of young trees in gardens and plantations, but has minor impacts on large trees. In some seasons the damage can be so severe on young trees that it impedes normal growth and development. Damage is much worse on juvenile leaves rather than mature leaves. When disturbed the caterpillars curl in a comma-shape.

Control: on small plants the caterpillars are easily removed and killed when clustered together, either when young or in the shelter. They can be killed by squashing or dropping the cluster into boiling water or kerosene covered water. Spray with malathion as a last resort (the caterpillars are protected from spray during the day by hiding within the shelter).

Mature Blue Gum Caterpillar, Mnesampela privata, *and eggs on* Eucalyptus cinerea. [D. JONES]

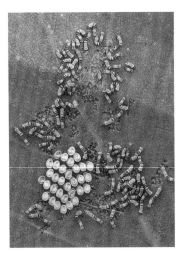

Eggs and young larvae of the Blue Gum Caterpillar, Mnesampela privata. [T. WOOD]

Caterpillars of the Blue Gum Caterpillar, Mnesempela privata. [T. WOOD]

Nearly mature caterpillars of the Blue Gum Caterpillar, Mnesempela privata. [T. WOOD]

Peppermint Looper, *Paralaea beggaria*

Description: blue-green eggs are laid in groups on the host plant. Young caterpillars are strongly marked and colourful, mainly green on the back with a black stripe and pale yellow flanks adorned with a brown stripe; white dots are also prominent. They grow to about 3.5cm long and change colour with age, usually green to blue-green with yellow stripes, but sometimes with brown or purple stripes. The moth (family Geometridae) has grey forewings with a dark central spot and paler grey hindwings. It often wraps its wings around the support when at rest. Wingspan is about 5cm.

Climatic regions: temperate regions, mainly coastal.

Host plants: species of *Eucalyptus* that are commonly known as 'peppermints', including *E. amagdylina*, *E. pulchella*, *E. radiata* and *E. robertsiana*.

Feeding habits: the caterpillars eat mature and young leaves, often defoliating twigs and patches of foliage.

Notes: although usually innocuous, this species can undergo dramatic population explosions when climatic conditions are favourable. It has been estimated that one such event in the 1970s resulted in about 1,000 hectares of peppermint forest being defoliated near Hobart. Continued defoliation of adult trees results in the growth of epicormic shoots within the canopy, altering the appearance of the trees, sometimes significantly and not for the better.

Control: as per introduction for caterpillars and grubs.

Grevillea Looper or Pink-bellied Moth, *Oenochroma vinaria*

Description: the caterpillar is a slender, hairless looper that grows to 8cm long. Colouration is an unusual grey-brown with orange mottling and a couple of orange bands. A pair of short, soft, black, thorn-like dorsal processes occur on the back towards the head. The attractive soft pink moth (family Geometridae), commonly known as the Pink-bellied Moth, is about 6cm across.

Climatic regions: tropics, subtropics and temperate regions, mainly coastal.

Host plants: mainly species and cultivars of *Grevillea*; also narrow-leaved species of *Banksia* and *Hakea*.

Feeding habits: voracious feeders that can strip patches of foliage completely. The caterpillars are usually present as individuals or in small loose groups. Feeding takes place mainly in the evenings and mornings, although the caterpillars remain active on cloudy days.

Notes: the caterpillars remain stretched out along a branch amongst the leaves during the warm part of the day. When disturbed they coil the upper part of their body inwards displaying the false spines.

Control: as per introduction for caterpillars and grubs. Spraying young plants may be necessary.

Grevillea Looper in defensive pose. [D. Jones]

Grevillea Looper feeding. [D. Jones]

Other Loopers

Fallen Bark Looper, Gastrophora henricaria, *in stick pose.* [T. Wood]

- **Brown Looper or Sinister Moth**, *Pholodes sinistraria*, from coastal parts of the tropics, subtropics and temperate regions, has shiny black young caterpillars decorated with white or yellow transverse bands, whereas older larvae are reddish brown with white spots. They grow to 5cm long. Pupation takes place in the soil without a cocoon. The moths (family Geometridae) are irregularly marked with light and dark brown patches; the wing margins are attractively scalloped. Wingspan is 5–6cm. The caterpillars eat the mature leaves of a wide variety of plants including some species of *Acacia, Macadamia integrifolia, M. tetraphylla, Senna, Syzygium floribundum* and *Syzygium smithii*. This is a relatively minor pest that can cause significant damage to young plants. When at rest they hold themselves rigidly erect; when disturbed they remain rigid but sway slowly.

- **Fallen Bark Looper**, *Gastrophora henricaria*, from temperate regions has slender, brown, stick-like caterpillars, 4–5cm long, with a pair of short soft bumps on the dorsal surface of the third abdominal segment. The attractive moths (family Geometridae), which often assume a hanging position resembling a dead leaf, are light fawn with darker dots. When the forewings are spread the contrasting orange hindwings that are marked with blackish dots and lines are revealed. The male moth has a prominent dark line that runs across each forewing. Wingspan is about 5cm. The caterpillars eat various species of *Eucalyptus* and Brushbox, *Lophostemon confertus*.

Male moth of the Fallen Bark Looper, Gastrophora henricaria. [D. Jones]

- **Green Garden Looper**, *Chrysodeixis eriosoma*, is from temperate and subtropical regions. The slender, sparsely hairy looper is bright green with faint white lines and a few dark spots. It grows to about 4 cm long and has two pairs of prolegs. Mature caterpillars pupate in a loose cocoon of curled leaves. The moths (family Noctuidae), which form a wedge-shape when sitting, are reddish-brown to coppery-brown with a pair of horn-like clusters of hairs on the head and two figure-of-eight white spots near the base of each forewing. Wingspan is about 4cm. The caterpillars, which mainly feed on the underside of leaves, eat a wide variety of garden plants including *Brachyscome, Goodenia, Pelargonium, Scaevola, Xerochrysum* and ferns. They often cut some of the main leaf veins, causing the leaf to collapse and form a curled shelter in which the caterpillar lives.

Green Garden Looper, Chrysodeixis eriosoma, *on* Goodenia *leaf.* [D. Jones]

- **Green Stick Looper**, *Chlorocoma assimilis*, from subtropical and temperate regions has a very slender, hard, green looper with a pointed head and a single pair of prolegs. Caterpillars grow to about 5.5cm long. The moths (Geometridae) sit flat with all four wings exposed. They are an attractive bright green with a marginal fringe of yellowish-brown hairs and a dorsal line of similar hairs down the body. Wingspan is about 3cm. The caterpillars, which eat at night or on cloudy days, feed on various species of phyllodinous *Acacia*. When not feeding they assume a stiff, obliquely erect posture.

- **Green Wattle Looper**, *Thalaina* sp., from temperate and subtropical regions has a very slender hard green looper with a rounded head. It grows to about 5cm long and has a single pair of prolegs. The

Grey Looper looping. [T. Wood]

Looper caterpillar damage to Macaranga tanarius *leaves.* [D. Jones]

Macaranga Looper. [D. Jones]

moths (Geometridae) are satin white and about 4cm across. The caterpillars feed on the leaflets of ferny-leaved species of *Acacia*. When not feeding the caterpillar assumes a stiff, obliquely erect posture.

- **Grey Looper**, *Cleora inflexaria*, from subtropics and temperate regions is a slender, grey looper, 3.5–4.5cm long with slight mottling on the surface. The moths (family Geometridae) are light grey with darker mottles and other markings. They sit flat on a surface with all four wings exposed. The caterpillars, which eat both young and mature leaves, feed on a range of rainforest plants in the family Lauraceae, including species of *Cryptocarya* and *Endiandra*. It is also a significant pest of avocados.

- **Macaranga Looper**, *Chlorocoma* sp., from tropical and subtropical regions is a short, fairly stout, green looper with a swollen neck-like area that partially hides the pointed head. The caterpillars, which grow 2–3cm long, often sit obliquely erect and can remain motionless for hours. The moths (Geometridae) are green with a wingspan of about 2cm. The caterpillars eat rainforest plants, including species of *Macaranga* and *Mallotus*. Damage is often very prominent on *Macaranga tanarius*. The caterpillars feed in small groups mostly on the younger leaves, sheltering by day on the leaf undersurface. They strip the leaf tissue between the veins leaving very tattered leaves.

- **Poinciana Looper**, *Pericyma cruegeri*, from tropical and subtropical regions, has an unusual slender looper with a narrow neck-like area behind the large, rounded head. Growing to about 7cm long, they range in colour from green with fine white lines when young to greyish with black spots and a narrow black dorsal band. They have two pairs of prolegs. When mature they pupate in a cocoon made among debris on the ground. The moths (family Noctuidae) have heavily mottled brown wings Wingspan is about 4cm. The caterpillars feed on native and introduced species of *Caesalpinia* and *Peltophorum pterocarpum*; also the widely grown Poinciana, *Delonix regia*. When young, the caterpillars feed in small groups, stripping the leaflets from patches of leaves. Older caterpillars tend to be solitary. This is a voracious pest that appears sporadically in large numbers. In severe attacks trees and vines can be defoliated or lose large patches of foliage, but healthy trees usually recover quickly. Caterpillars produce large quantities of droppings. When disturbed the caterpillars thrash about with the head curved back. Control is not usually practical.

- **Varied Looper**, *Capusa senilis*, from the subtropics and temperate regions has caterpillars that change colour significantly with age. Young caterpillars are yellowish with red markings; they then become green with yellow markings and finally red-brown to brown. The spiracles are white with a black outline in all stages. They grow 3–4cm long. The moths (family Geometridae), which are grey with lighter markings, are wedge-shaped when at rest, tapering from the head end to the wing tips. The caterpillars feed on

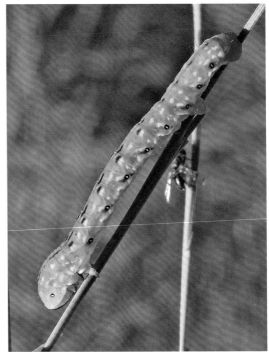

Varied Looper, Capusa senilis. [T. Wood]

Varied Looper, Capusa senilis. [T. Wood]

species of *Eucalyptus, Callitris* and various pea plants including species of *Goodia, Hardenbergia* and *Pultenaea.* They are usually solitary and feed from the leaf margins, eventually consuming the whole leaf. This species is remarkable for the colour changes which occur as the caterpillars grow.

- **Zig-zag Looper or Bizarre Looper**, *Eucyclodes pieroides*, distributed from the tropics to temperate regions, has an unusual looper caterpillar that grows 2.5–3cm long. It is brown and has flanged body segments that resemble of a twisted dead leaf. Young caterpillars attach debris in areas where the flanges will develop. Male and female moths (family Geometridae), which are about 3cm across, are different colours, the male having very pale green wings with irregular whitish marginal bands and the females green wings with a broad light brown border. The caterpillars feed on a wide range of plants including phyllodinous wattles, species and cultivars of *Callistemon, Macadamia, Melaleuca, Syzygium, Terminalia* and *Rhodamnia argentea.* They usually occur singly or in small groups. Young caterpillars graze the leaf surface, older caterpillars eat the leaf margins. In *Macadamia*, the caterpillars sometimes also eat the flowers and hulls of the nuts. This is an unusual looper which is rarely noticed because of its effective camouflage. It is a minor pest although severe attacks occur sporadically on *Acacia fimbriata*.

Zig-zag Looper. [D. Jones]

- **Flower Looper**, *Gymnoscelis subrufata*, is a minor pest of *Macadamia* in Queensland. It feeds on the buds and flowers, reducing fruit set. Its fleshy caterpillar is pale green with unusual dark zigzag-like cross-banding.

CASE MOTHS, BAG MOTHS OR BAGWORMS (FAMILY PSYCHIDAE)

These insects are so named because their caterpillars shelter inside a mobile silken case that is usually reinforced with sand grains or pieces of stick and/or leaf of the host plant on which the insect is feeding – other materials such as string, paper and weed mat have also been observed to be used. The case has an upper opening (for feeding) and a smaller lower opening (for waste disposal; also to allow the male moth to exit and the female moth to mate). Young caterpillars construct a small case and add to it as they grow. There are numerous species of case moth (many undescribed), each species constructing a unique shelter, although the case varies somewhat within a species depending on the host plant upon which the caterpillar is feeding. Usually the shelters built by male insects are about half the size of those of the females.

The caterpillars move about with the case attached. They do not leave the case but the upper parts of the body – which are well protected by an armour-like skin – emerge for feeding or when moving about. If disturbed while feeding, the head and legs are quickly withdrawn and the opening of the bag held closed by the caterpillar's front legs. The prolegs also grip firmly on the inside of the case. When resting, the caterpillar fastens the case to the trunk or branch of the tree with silken thread. Large case moths take more than a year to reach maturity. They moult and pupate inside the case. Pupation occurs

Case moth adorned with Acacia dealbata *leaf fragments.* [T. WOOD]

Case moth adorned with leaf fragments. [T. WOOD]

This tropical case moth feeds on a wide range of rainforest and garden plants. [D. JONES]

Case moth feeding on Glochidion ferdinandi. [D. JONES]

Case moth feeding on Stenocarpus sinuatus. [D. JONES]

Case moth adorned with sections of grass stem or rush. [T. WOOD]

after the opening of the case is sealed with silk – the male often pupates upside down to facilitate exit via the lower opening. Female case moths may or may not have wings, depending on the species. Wingless species (which are sometimes also legless), never leave their case. The male moths, which are small and slender-winged (often with transparent wings), leave the case from the bottom (often the pupal skin remains protruding after emergence) and fly to mate with the female. The eggs of wingless females hatch within the case and the young caterpillars lower themselves on silken threads.

Climatic regions: various species are found from the tropics to temperate regions.

Host plants: case moths feed on a wide range of plants, including many commonly grown garden species, both native and exotic. They are often seen on eucalypts and wattles. Some species seem to have a narrow host range, others feed on many different types of plant.

Notes: case moths seldom cause much damage and in fact usually add interest to a garden. They are easy to keep and can be a fascinating study for children.

Natural control: case moths are controlled naturally by a wide range of parasites and predators.

Other control: spraying is rarely needed. If severe infestations occur, individuals can be picked off by hand and squashed until the numbers are sufficiently reduced.

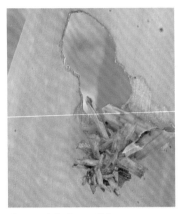

Case moth feeding on Crinum pedunculatum. [D. Jones]

EXAMPLES

- **Large Bagworm or Saunders' Case Moth**, *Metura elongatus*, is widely distributed in the tropics and temperate regions of eastern Australia. The case is an elongated silken bag, 10–15cm long, broadest in the middle and tapered to each end. It is soft, but incredibly tough, and sparsely adorned with short pieces of twig. The stout caterpillar is orange-brown with yellow bands and black markings. The spectacular male moths, which have a wingspan of about 4cm, are blackish with veined wings, yellow-orange head and thorax and an orange and black-banded abdomen. This species feeds a wide range of plants including species of *Acacia*, *Callistemon*, *Epacris*, *Eucalyptus*, *Grevillea*, *Leptospermum*, *Macadamia*, *Melaleuca* and palms. This is usually a solitary insect that chews large lumps out of leaves. Sometimes whole leaves are cut off and fall to the ground.

Saunders' Case Moth. [D. Jones]

- **Leaf Case Moth**, *Hyalarcta huebneri*, is found over most of eastern and southern Australia. The cases, which reach about 5cm long, are broadest towards the base, tapering to the top. They are adorned with pieces of leaf from the plant on which the caterpillar is feeding (sometimes also bark). The caterpillars, which grow to about 4cm long, are orange with black bands. This species feeds on a wide range of native and exotic plants, including fruit trees. Natives

Leaf Case Moth feeding on Leptospermum laevigatum. [D. Jones]

include species of *Eucalyptus*, *Corymbia*, *Grevillea*, *Leptospermum*, *Lophostemon*, *Tristaniopsis* and *Melaleuca*. A case moth of similar appearance also feeds on palms in the tropics and subtropics.

- **Conical Case Moths**, *Lepidoscia* spp. and *Narycia* spp., a group of several species widely distributed over much of Australia. They construct a small conical case up to 2cm long that tapers from a broad base to a narrow apex. The surface of the case is covered with small pieces of leaf chewed from the host plant. The cases are mainly grey to brown but are sometimes patterned with other colours as different leaf material is added. Members of this genus feed on a very wide variety of plants, ranging from species found in heathland and open forest to rainforest (some species feed on lichens growing on rainforest trees). These insects are sometimes present in substantial numbers and can damage new growth.

Ribbed Case Moth. [T. Wood]

- **Ribbed Case Moth or Silken Bag Moth**, *Hyalarcta nigrescens*, which ranges from southern Queensland to Victoria, has a case resembling a plain spindle-shaped silken bag. It is grey and unadorned with plant debris, but is reinforced by several rib-like ridges that run from top to bottom. It can reach about 4cm long. The caterpillars feed on species of *Acacia*, *Eucalyptus*, *Lophostemon* and *Tristaniopsis*. Small caterpillars graze on the leaf surface tissue which turns white. The bags are sometimes fastened by silk, completely encircling a live stem. This silk fastening can girdle the stem or twig.

- **Smooth Case Moth**, *Oiketicus herrichii*, from tropical and subtropical areas has an ellipsoid grey-brown silken case up to 5cm long that lacks any reinforcing ribs and is unadorned with plant debris. It feeds on species of *Psychotria* and some rainforest plants.

Lictor Case Moths. [D. Jones]

- **Stick Case Moth, Lictor Case Moth, Faggot Case Moth or Firewood Case Moth**, several *Clania* species, found over most of eastern and southern Australia. The cases, which reach about 5cm long, are cylindrical and completely surrounded by a palisade of closely aligned sticks. A silken bag protrudes from the lowest point. The sticks are of a similar length, but sometimes a single long one extends past the silken bag. It has been hypothesised that this longer stick assists the male moth to access the female's case. One species commonly feeds on species of *Eucalyptus* and *Corymbia*; *Clania lewinii* feeds on species of *Eucalyptus*, *Callistemon* and *Melaleuca*; another species feeds on *Allocasuarina*, *Casuarina* and *Callitris*.

- **Tower Case Moth**, *Lepidoscia arctiella*, from the tropics and subtropics, constructs a case that consists of three or four tiers, each tier broader than the previous one. Each tier is cylindrical and covered with a palisade of closely aligned sticks. This unusual species feeds on the leaves of *Brachyloma*, *Grevillea*, *Monotoca* and *Styphelia*. A similar species feeds on sheoaks.

Tower-like case moth on Casuarina cunninghamiana. [D. Jones]

OWLET MOTHS (FAMILY NOCTUIDAE)

This is the largest family of Lepidoptera in the world with some 35,000 species named and probably twice as many yet to be formally recognised. Australia has about 1,000 species. Many are significant plant pests. The caterpillars are mostly hairy but an interesting group, called semi-loopers, are easily confused with true loopers because they move in the same way.

EXAMPLES

Owlet Moth, Granny Moth, Old Lady Moth, *Dasypodia selenophora*

Description: a slender, almost hairless, mottled, grey to brown looper-like caterpillar that grows 6–7cm long. It has orange legs and head and two small soft horns near the rear. Pupation occurs in a brown parchment-like cocoon spun among leaves or in leaf litter. The decorative moths, which are brown to orange brown, are flat when at rest. Each forewing has patterns of wavy lines and a large colourful eyespot. Wingspan is 7–9cm.

Climatic regions: subtropics and temperate regions, coastal and inland.

Host plants: species of *Acacia,* both fern-leaved and phyllodinous.

Owlet Moth, Dasypodia selenophora. [D. Jones]

Feeding habits: the caterpillars, which are usually solitary, feed at night on phyllodes and leaflets. They rest during the day in a camouflaged position stretched out on the underside of a stem or branch or along the edge of a phyllode. The moths commonly enter homes in late summer-autumn, remaining at rest for hours or even days.

Notes: these caterpillars are very difficult to see during the day. It is a very minor pest that is mainly of interest for its large attractive moths.

Control: as per introduction to caterpillars and grubs. Hand picking is usually satisfactory.

Caterpillar of the Mistletoe Moth, Periscepta polysticta, *feeding on* Hibbertia. [T. Wood]

Mistletoe Moth, *Periscepta polysticta*

Description: the fleshy caterpillars, which grow 4–5cm long, are black with pale bands between each segment and numerous cream to white dots. A prominent cream stripe runs along each flank just above the legs. The moths are blackish-brown with white spots and a fringed white band on the hindwing. Wingspan is 3–4cm.

Climatic regions: subtropics and temperate regions, mainly coastal.

Host plants: species of *Hibbertia* and mistletoe.

Feeding habits: usually solitary feeding on leaves.

Notes: the caterpillars can be very difficult to see when they are at rest.

Control: as per introduction to caterpillars and grubs. Hand picking is usually satisfactory.

CUTWORMS SEVERAL GENERA

Cutworms, which feed singly, are so called because they live in the soil and chew off seedlings and small plants at ground level. When disturbed they adopt a defensive position by coiling and remaining motionless for some time. There are many different species and they can be difficult to identify accurately.

Description: cream eggs are laid singly or in small groups on plants close to soil level. The caterpillars, which are plump, fleshy and hairless, grow to 3–5cm long. Mature caterpillars pupate in the ground. The moths are relatively small (with a 4cm wingspan), grey, black or brownish and stout-bodied.

Climatic regions: various species found from tropical to temperate regions.

Host plants: cutworms favour soft seedlings and are very destructive of flower crops, vegetables and pasture plants. They can also be a significant pest in nurseries, feeding on all types of young seedlings, including wattles and eucalypts. Larger cutworms can also damage older plants, especially soft plants such as ferns.

Feeding habits: young cutworms feed on leaves and stems above ground, usually damaging tender tissue. Larger cutworms chew through the plant's stem so that it falls over. The caterpillar then often pulls all or part of the plant into a burrow to feed. They usually feed at night and hide during the day, but they are often seen on the soil surface during cloudy weather.

Notes: Cutworms congregate in weedy areas and seedbeds should be well prepared and cleared of all weeds before sowing. Several generations of cutworm can occur in a season.

Control: simple cardboard tubes can protect vulnerable plants. Handpicking at night with the aid of a torch reduces numbers. Seedlings should be sprayed or dusted for continuous protection if these destructive pests are around.

EXAMPLES

- **Brown Cutworm or Pink Cutworm**, *Agrotis munda*. This species, which is found over much of Australia (coastal and inland) has pale brown to pinkish caterpillars which grow to about 5cm long. Mature caterpillars become very dark before they pupate. Moths, which are hairy, have a wingspan of

Cutworm caterpillars: one is burrowing, the other in a defensive pose. [D. JONES]

Brown Cutworm caterpillar. [D. BEARDSELL]

Bogong Moth: its larvae are cutworms. [T. WOOD]

about 5cm, are grey to silvery with dark-brown patterns. They can secrete an unpleasant odour when disturbed. This species feeds on a wide range of garden plants, particularly fleshy annuals and perennials. It is also a pest of orchids and ferns in glasshouses.

- **Bogong Moth or Common Cutworm**, *Agrotis infusa*, is widespread in southern Australia. It is better known for its moths than its caterpillars. Bogong Moths migrate from inland areas to congregate in huge populations in the mountains of south-eastern Australia over the summer, returning in winter to lay their eggs. The Aborigines gathered the moths for food. The moths are attracted to lights and frequently enter buildings, sometimes in large numbers. The brown caterpillars feed on a wide range of plants, both exotic and native.

- **Colourful Cutworm**, *Aedia acronyctoides*, is a species from coastal parts of the tropics and subtropics. It has colourful caterpillars which grow to about 5cm long. They are speckled with mauve and white, have four or six narrow yellow to orange lines and a broad white band along the side. The moth is brownish with a wingspan of about 5cm. This species feeds on a range of plants with fleshy leaves, especially species of *Ipomoea* – it is often common on *Ipomoea pes-caprae*.

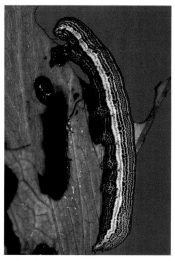

Colourful cutworm feeding on Ipomoea. [D. Jones]

Cluster Caterpillar, *Spodoptera litura*

Description: eggs are laid in a furry cluster on the underside of a leaf. Young caterpillars are fleshy, hairless and translucent green with dark marginal spots. Older caterpillars, which grow up to 5cm long, become light brown to dark brown or purplish with red or yellow lines, black marginal spots and blackish triangles along the back. Pupation occurs in the soil. The moths are stout with strongly marked grey-brown forewings. Wingspan is about 4cm.

Climatic regions: tropics, subtropics and warm temperate regions, coastal and inland.

Host plants: numerous exotic and native garden plants; natives include *Alocasia brisbanensis*, *Goodenia*, *Scaevola*, *Swainsona* and *Casuarina*; also vegetables and many field crops.

Feeding habits: young larvae feed in clusters and graze the surface leaf tissue. Larger caterpillars feed singly, eating whole leaves and boring holes in stems and petioles. They also eat flowers and fruit. Quantities of faecal pellets are obvious under the plants on which they feed.

Notes: the caterpillars are active in summer and autumn. When disturbed the caterpillars coil tightly with the head in the centre.

Control: as per control for caterpillars and grubs. Spraying is often necessary.

Cluster caterpillar, larvae of the Oriental Leaf Moth, Spodoptera litura, *feeding on* Alocasia brisbanensis. [D. Jones]

Lily Caterpillar or Crinum Caterpillar, *Spodoptera picta*

Description: eggs are laid in a furry cluster on the underside of a leaf. Young caterpillars are fleshy, hairless and greyish with darker lines and a black cross-band a short distance back from the head. Older caterpillars, which grow to 5cm long, develop a prominent yellow dorsal stripe and paler marginal stripes (still with the black cross-band). Pupation occurs in the soil. The moths have white wings with attractive reddish lines and patterns on the forewings and a brown furry body. Wingspan is about 4cm.

Climatic regions: tropics, subtropics and warm temperate regions, mainly coastal.

Host plants: these caterpillars devour fleshy native bulbs such as *Crinum angustifolium, C. pedunculatum* and *C. venosum,* in some cases eating right down through the bulbs to the roots; many different exotic bulbs are also eaten. Their voracious feeding can kill the bulbs.

Feeding habits: the caterpillars feed mainly at night or on cloudy days. They remain in groups or clusters at most stages of their development. Young caterpillars graze the surface tissue, older caterpillars eat the leaves, stems and bulbs. Often the area where they feed is heavily contaminated with their faeces, which become wet and slimy.

Notes: this destructive pest is most damaging during the summer-autumn. It is particularly prevalent in the tropics during the wet season.

Control: as per control for caterpillars and grubs. Spraying is often necessary.

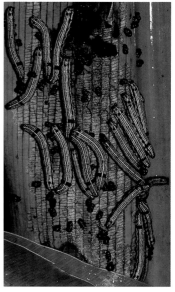

Cluster of Crinum Caterpillars, Spodoptera picta. [D.Jones]

Lily Borer or Crinum Borer, *Brithys crini*

Description: cream eggs are laid in clusters on the host plant. Caterpillars are black with prominent white spots on each segment, short sparse hairs and a reddish-brown head with black markings. Older caterpillars, which grow to 5cm long, are stout and fleshy with strongly contrasting white spots. The moths have grey-brown wings. Wingspan is about 4cm.

Climatic regions: tropics, subtropics and warm temperate regions, mainly coastal.

Host plants: these caterpillars devour fleshy native bulbs such as *Crinum angustifolium, C. pedunculatum* and *C. venosum*; many different exotic bulbs of the family Amaryllidaceae are also eaten. Their voracious feeding can severely stunt the bulbs and impede flowering.

Feeding habits: the caterpillars feed singly or in small groups, tunnelling under the outer epidermis (skin) of the leaves. Because they are black they can often be seen feeding under the intact but translucent skin. They also bore through the tissue of bulbs and pseudostems of large crinums.

Notes: this destructive pest is prevalent in the tropics during the wet season. When disturbed the caterpillars curl and form a comma-shape

Control: as per control for caterpillars and grubs. Their voracious feeding can damage a bulb in a short time.

Australian Grape Vine Moth, *Phalaenoides glycinae*

Description: green eggs are laid singly on young leaves. The fleshy caterpillars, which grow to 5cm long and have long white hairs, are black with numerous thin white or yellow bands in a cross-check pattern, red markings and a conspicuous red patch on the rear. Mature larvae pupate in the soil or in crevices. The moths are black with white and yellow markings on the wings and a tuft of orange hairs on the tip of the abdomen. Wingspan is 5–6cm.

Climatic regions: subtropics and temperate regions, inland and coastal.

Host plants: species of *Cayratia, Cissus* and *Hibbertia*; also other garden plants such as *Fuchsia, Parthenocissus* and grape vine, *Vitis vinifera.*

Feeding habits: the caterpillars feed on leaves, eating them partially or whole.

Caterpillar of the Australian Grape Vine Moth, Phalaenoides glycinae. [T. Wood]

Notes: a native moth that has adapted to feed on grape vines, where it sometimes causes significant damage. The moths fly during the day. There are two or three generations per year. Another similar species, *Phalaenoides tristifica*, also eats *Hibbertia, Pimelea* and grape vine.

Control: by hand picking and squashing, or spraying as per introduction for caterpillars and grubs.

Phalaenoides tristifica *caterpillar feeding on* Pimelea. [T. Wood]

Noctuid moths mating, possibly Phalaenoides tristifica. [T. Wood]

Native Budworm, *Helicoverpa punctigera*

Description: white to yellow eggs are laid singly on the growing tips of the plant. The fleshy caterpillars, which grow 3.5–4cm long, can be yellow, green, brown, reddish or nearly black with a dark dorsal stripe and a broad pale stripe down each side of the body. Mature larvae pupate in the soil. The nocturnal, stout-bodied moths are reddish brown with darker markings. Wingspan is 3–4cm.

Climatic regions: subtropics and temperate regions, inland and coastal.

Host plants: a commercially important native pest that attacks a large range of field crops and vegetables; this species has a very wide host range including many different exotic and native garden plants, especially soft plants such as *Brachyscome, Goodenia, Scaevola*, and legumes including *Brachysema, Crotalaria, Gastrolobium, Glycine, Hardenbergia, Hovea, Indigofera, Pultenaea* and *Swainsona*. It can also severely damage the flowers of native everlasting daisies, both in the wild and on cultivated plants.

Feeding habits: the younger caterpillars feed by grazing on young leaves, flowers and fruit; older caterpillars often burrow into fruit and pods, eating the developing seeds. The caterpillars, which are attached to the host by silken threads, often damage or destroy a very high proportion of developing seeds and can also eat the growth buds of the plant.

Notes: the moths, which are active during dusk and at night, can fly more than 500km in one night. In years of good rainfall the numbers of this pest build up during winter on native plants and weeds in inland areas before moving with the prevailing winds to coastal and near-coastal areas. The caterpillars resemble a looper but have four pairs of prolegs and do not form a loop when moving.

Control: by hand picking and squashing, or spraying as per introduction for caterpillars and grubs. Natural control is by parasitic wasps and tachinid flies.

Heliothis Caterpillar, *Heliothis punctifera*

Description: a fleshy, sparsely hairy caterpillar that grows to 4cm long with pale hairs behind its head. It is usually dark with a conspicuous broad pale band on each side, but can also be green, brown or pinkish. Pupation occurs in the soil. The moths fly at night and are stout, grey or dull-brown with a black patch on the underside of the hindwing. Wingspan is about 3cm.

Climatic region: tropics, subtropics and temperate regions, mainly coastal.

Host plants: a wide variety of plants including *Chenopodium, Solanum, Hardenbergia, Hovea, Swainsona* and members of the family Malvaceae, including *Abutilon, Gossypium, Hibiscus* and *Lavatera*; often very severe on members of the daisy family, destroying new shoots and eating flower buds before opening.

Feeding habits: feeds on tender tissue, particularly buds, flowers and fruit, but also young leaves. Larger caterpillars can also tunnel into stems, flowers and fruit, leaving a round hole.

Notes: a serious commercial pest that attacks many crops and has become resistant to pesticides. The moths of this species sometimes occur in plague proportions. The caterpillars resemble a looper but have four pairs of prolegs and do not form a loop when moving.

Control: by hand picking and squashing. Spraying as per introduction for caterpillars and grubs. Strains of pyrethrum-resistant *Heliothis* are known from some areas, especially where cotton is grown. A virus control agent has recently been developed and used successfully on commercial crops.

Joseph's Coat Moth or Painted Vine Moth, *Agarista agricola*

Description: a large, boldly-banded, fleshy caterpillar that grows to about 7cm long and has a yellow to orange patch on the rump, a yellow to orange band towards the head and the whole body adorned with unusual club-like hairs. The bands can be either black and white or black and orange. Its feet are orange. The moths are black with colourful patches on the wings (red, yellow, blue and white) and a white patch on the thorax behind the head. Wingspan is 5cm (males) to 7cm (females).

Climatic regions: tropics, subtropics and warm temperate regions, mainly coastal.

Host plants: species of the family Vitaceae, including *Ampelocissus*, *Cayratia* and *Cissus*; also the grape vine, *Vitis vinifera*.

Feeding habits: the caterpillars feed singly during the day, eating the leaves, veins and all. They strip large areas of the vine and can also eat flowers and fruit.

Notes: a large, decorative caterpillar that develops into a lovely day-flying moth. The caterpillars tuck their head inwards and arch their back when disturbed.

Control: by hand picking and moving to another host.

TUSSOCK MOTHS (FAMILY LYMANTRIIDAE) (SEVERAL GENERA)

Description: eggs are laid in large clusters. Conspicuous, very hairy caterpillars grow 3–4cm long. They are readily recognised by three or four thick conspicuous upright toothbrush-like tufts of hairs on the back; the body is covered with long hairs and a tuft of hairs also occurs on the tail (sometimes hair tufts also occur on the sides). A narrow, antenna-like tuft is found on either side of the head. Pupation occurs in a loose cocoon covered with irritant hairs which is often spun between leaves. The moths are strongly dimorphic: the males, which have large feather-like antennae, are about 2.5cm across and can fly, whereas in some species the female moths are wingless and lay clusters of eggs, usually on top of their cocoon.

Climatic regions: various species found from tropical to temperate regions.

Host plants: a wide range of native and exotic species.

Feeding habits: usually solitary in most stages of growth, the young caterpillars often skeletonising the leaves by eating the upper surface layer; larger caterpillars eating chunks from the leaves or eating the entire leaf. These caterpillars have a voracious appetite. Several generations occur and caterpillars can be found throughout the year, but are often particularly active during the cool winter months.

Notes: although generally a minor pest, these caterpillars can be very destructive. Newly hatched caterpillars disperse on the wind using silken threads to catch the breeze. The hairs on the caterpillars and cocoons can cause skin irritation and induce severe reactions in sensitive people.

Tussock moth caterpillar. [T. WOOD]

Control: as per introduction for caterpillars and grubs. Hand picking and squashing is usually satisfactory. Small Braconid wasps lay their eggs among the hairs of the caterpillar; on hatching the larvae feed on the caterpillar's body.

EXAMPLES

The moths and caterpillars of several species of tussock moth have a similar general appearance:

- **Painted Apple Moth or Wattle Tussock Moth**, *Teia anartoides*, from the subtropics and temperate regions (coastal and inland) has a yellow, brown or black hairy caterpillar with toothbrush-tufts of reddish hairs on the back. The male moths have brightly marked brown forewings and yellow hindwings with a dark brown band; the greyish female moths are wingless. This species feeds on a very wide range of plants including species of *Acacia, Callistemon, Exocarpos, Grevillea, Hardenbergia* and *Melaleuca*; also ferns, exotic pine trees and apple trees.

 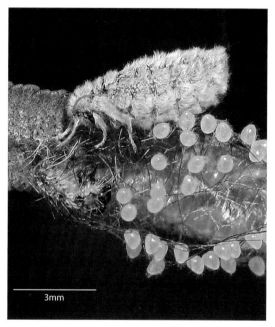

Painted Apple Moth caterpillar. [T. Wood]   *Female Painted Apple Moth and eggs.* [T. Wood]

- **Western Tussock Moth**, *Teia athlophora*, from temperate parts of Western Australia has a brown hairy caterpillar with thick toothbrush-tufts of pale hairs on the back. The male moths have grey-brown wings with lovely patterns and markings; the yellow-brown female moths are wingless. This species feeds on species of *Acacia* and native pea-flowered plants including *Brachysema, Gastrolobium* and *Hardenbergia*.

- **White-spotted Tussock Moth or Painted Pine Moth**, *Teia australis*, from subtropical and temperate regions has colourful tan and cream hairy caterpillars with toothbrush-tufts of white or yellow hairs on the back. The male moths are dark brown with an indistinct eye-spot on each forewing; females are wingless. This species feeds on species of *Acacia, Avicennia, Callitris, Grevillea* and exotic pine trees.

- **Brown Tussock Moth or Brown Tufted Caterpillar**, *Olene mendosa*, from tropical and subtropical regions has a brownish hairy caterpillar with toothbrush-tufts of white to grey-brown hairs on the back. The moths, with wingspans of 3cm for the male and 4cm for the female, have pale brown wings with darker brown markings. This species feeds on a wide variety of plants including species of *Cryptocarya, Eucalyptus, Gardenia, Grevillea, Kailarsenia, Lysiphyllum, Macadamia* and *Terminalia*.

- **Cream Tussock Moth**, *Calliteara pura*, from temperate regions has cream to yellowish caterpillars covered with very long white hairs and dense toothbrush-like tufts of white hairs on the back. The moths have white wings with patterns of fine dark lines. The wingspans are 5cm for the male and 6cm for the female. This species feeds on species of *Eucralyptus*, Gymea Lily (*Doryanthes excelsa*) and *Tristaniopsis laurina*.

White Cedar Moth, *Leptocneria reducta*

Description: a slender, very hairy, dark brown caterpillar with orange feet that grows to about 4cm long. The caterpillars are very active and move with an undulating gait, frequently travelling in procession. Pupation occurs in a hairy cocoon in litter on the ground. The moths are dull-brown with a wingspan about 4.5cm (males are smaller than females). The forewings have two dark spots, often marked with orange.

Climatic regions: tropics to temperate regions, coastal and inland (all of mainland Australia).

Host plants: restricted to the White Cedar, *Melia azedarach*, a hardy species that is an important shade tree in inland towns.

Feeding habits: a gregarious caterpillar with a voracious appetite. Infestations may completely strip even large trees of foliage. The caterpillars usually feed at night after moving in a mass up the trunk and along the branches to the leaves. In early morning they migrate to the lower parts of the tree, where they hide in crevices in the bark or gather in a community group on the bark. After stripping foliage from a tree, they move in long processions seeking other trees on which to feed. At this time they often invade homes and garages. Caterpillars ready to pupate may wander long distances.

Notes: a very destructive pest that is a common problem on White Cedar trees. The irritant hairs from the caterpillars and cocoons cause severe itching and other allergenic reactions to people and pets.

Control: banding is a simple and very effective means of controlling this pest. If a bag is tied around the trunk the larvae will seek shelter in it and can be picked out and squashed during the day. Bags or rags soaked in used motor oil will also deter them from climbing the trunk and they will then move on or starve. Spraying the foliage of small trees with a safe insecticide gives effective control but is not practical for large trees.

Cluster of caterpillars of the White Cedar Moth. [T. Wood]

Freshwater Mangrove Itchy Caterpillar, *Euproctis lutea*

Description: eggs are laid in irregular lines or masses on the host leaf, each mass covered by scales from the moth's abdomen. Young caterpillars are cream and sparsely hairy. Older caterpillars, which can grow to 1.5cm long, are brown to black with a broad white dorsal line and numerous tufts of long dark hairs – four prominent on the back and two on the rear. Pupation occurs in a cocoon under a log. The moths are yellow with wavy paler lines. Wingspan is about 3cm.

Climatic regions: tropics and subtropics, coastal and inland. A significant pest in Darwin.

Host plants: commonly *Barringtonia acutangula* and *Planchonia careya*.

Feeding habits: young caterpillars feed in groups (skeletonising leaves) beneath a shelter of silken hairs. Older caterpillars shelter during the day in a silken bag spun on the shady side of the trunk near the base.

Notes: the caterpillars and moths of this destructive pest should never be handled, as the irritant hairs are brittle and barbed. They can cause a severe reaction with sensitive tissues such as membranes of the nose, eyes or mouth. The caterpillar adopts a rigid stance when disturbed.

Control: as for bag shelter moths.

Brown-tail Gum Moth, *Urocoma baliolalis*

Description: eggs are laid in a dense scale-covered mass on the host plant. Young caterpillars are pale yellow with a dark head before becoming brown. Older caterpillars, which can grow to 4cm long, are velvety brown to black with a series of white dots and coloured hairs along the back, two short tufts of brown hairs near the head, one on the rear, and numerous tufts of spreading white hairs on the flanks and at each end. The hairs often sweep downwards like a skirt. Pupation occurs in a hairy cocoon among bark on the trunk. The hairy moths are brown with a broad white margin around each wing. Wingspan is about 5cm.

Climatic regions: subtropics and temperate regions, coastal and inland.

Host plants: species of *Eucalyptus.*

Feeding habits: young caterpillars feed in groups, skeletonising the leaves. Older caterpillars form a loose communal shelter by joining leaves together. When mature they wander, looking for a site to pupate. At this stage they can enter dwellings, causing problems because of their irritant hairs.

Notes: another species with irritant barbed hairs that can cause severe itching and rashes that last for several days. Irritant hairs not only occur on caterpillars but also caste skins, cocoons and the moths.

Control: as for bag shelter moths.

Mistletoe Itchy Caterpillar or Mistletoe Brown-tail Moth, *Euproctis edwardsii*

Description: caterpillars, which can grow to 2.5cm long, are grey to charcoal black with a broad white dorsal line, numerous red-brown dots and sparse tufts of long white hairs. Pupation occurs in a hairy cocoon among wood or under logs. The moths are pale orange-grey-brown with a tuft of yellow hairs on the tip of the black abdomen. Wingspan is about 5cm.

Climatic regions: subtropics and temperate regions, coastal and inland.

Host plants: species of Mistletoe (family Loranthaceae).

Feeding habits: caterpillars feed in groups, eating the leaves. Older caterpillars often wander, either looking for food or for a site to pupate. They may enter dwellings causing problems to people living within.

Notes: another species with irritant barbed hairs that can cause severe itching that lasts several days. The hairs readily become airborne and can enter the eyes, mouth and nose. It is best to avoid areas where this caterpillar pupates.

Control: as for bag shelter moths.

NOTODONTID MOTHS (FAMILY NOTODONTIDAE)

A small family of night-flying moths with about 2,800 species worldwide and about 90 species in Australia. One group of notodontid moths have very hairy larvae that live in silken shelters, another group has large hairless caterpillars.

EXAMPLES

Processionary Caterpillar or Itchy Caterpillar, *Ochrogaster lunifer*

Description: a grey caterpillar, densely covered with long, stiff, sharp, reddish-brown hairs. It grows to about 4cm long and has a brown head. The caterpillars hide together during the day in a large, bag-like shelter constructed from leaves and silken webbing, emerging in a procession at night to feed. Pupation occurs in a silken cocoon among soil debris. The moths are dark grey to brown with yellow bands on the abdomen and a terminal tuft of irritant white hairs. Wingspan is about 4cm.

Climatic regions: tropics, subtropics and temperate regions, very common inland but also coastal in some parts.

Host plants: attacks a wide range of phyllodinous wattles, such as *Acacia acuminata*, *A. implexa*, *A. longifolia*, *A. pendula* and *A. salicina* and some eucalypts; also *Grevillea striata*. In some years this pest can devastate large areas of wattles in inland zones, especially *A. pendula*.

Feeding habits: caterpillars shelter by day in the bags and emerge at night to feed on the foliage. On emergence they gather together and move along a pre-laid silken trail in a head-to-tail procession. They have a voracious appetite and can completely defoliate a tree. If short of food the caterpillars move in a procession across the ground to find another tree. The caterpillars move uniformly in the procession, each with its head in contact with the caterpillar preceding it. The processionary lines can be long and it is possible by directing the leading caterpillar to form a complete circle. Research has shown that cutting the rear hairs off caterpillars leads to the breakdown of the formation.

Notes: an unpleasant and destructive pest that is best controlled when first noticed. The hairs from the caterpillars are stiffly sharp and cause intense skin irritation if contacted. They can blow in the wind and still cause problems long after the caterpillars have gone. The bag-shelters constructed by the caterpillars are very conspicuous, especially in trees that have been denuded of foliage. The bags should not be handled without protective covering as they contain numerous irritating hairs and old skins cast off by the caterpillars. Aborigines will never sleep under a tree containing a bag shelter because of the danger of falling hairs. The moths also should be avoided as they also contain irritant hairs and there are recorded cases of skin irritation after moths have entered bedding or landed on washing.

In coastal areas caterpillars of this species often do not form a bag shelter but gather together on the bark at the base of the trunk.

Control: as for bag shelter moths. The caterpillars are vulnerable and easy to destroy while moving over the ground.

Processionary Caterpillar. [D. Jones]

Bag shelter built by Processionary Caterpillars in a wattle tree. [D. Jones]

Shelter built by the Processionary Caterpillar, Ochrogaster lunifer, *in* Acacia pendula. [D. Jones]

Banksia Notodontid Moth, *Psalidostetha banksiae*

Description: a spectacular, stout, fleshy, deep-chocolate-brown caterpillar, which grows to about 8cm long and has black and white spots and patches along the side, often mauve to purplish on the head and rear end. Young caterpillars have a disproportionately large head and club-like hairs. Older caterpillars lack many of the hairs. Pupation occurs in a cocoon among litter on the ground. The moths have white to grey-brown forewings with black and white bands and spots, cream to pale orange hindwings and an orange body. Wingspan is 6–7cm.

Climatic regions: tropics, subtropics and temperate regions, coastal and inland.

Host plants: species of *Banksia, Dryandra, Grevillea* and *Hakea.*

Feeding habits: this species, which is active during the day, usually feeds singly or in small groups, chewing large lumps out of the leaves.

Notes: a striking addition to the garden. It is of sporadic occurrence and rarely causes significant damage. The caterpillar reacts to disturbance by jerking the head backwards and extruding a red forked appendage (osmeterium) from the underside of the body near the head.

Control: as per introduction for caterpillars and grubs. If needed, hand-picking and moving to another host is usually satisfactory.

Wattle Notodontid Moth, *Neola semiaurata*

Description: an unusual, stout, fleshy, pinkish-brown caterpillar, which grows to about 6cm long, with spotted prolegs and white spots and patches along the side and a few black, club-like hairs scattered along the body. The head has two blackish lines, the tail a projecting horn and the whole rear end is shaped like an anvil. Older caterpillars have fewer hairs. Pupation occurs in a cocoon among litter on the ground. The moths have brown forewings and pale orange hindwings and body. Wingspan is about 6cm.

Climatic regions: tropics, subtropics and temperate regions, mainly coastal.

Host plants: species of *Acacia* (phyllodinous and ferny-leaved)*, Brachychiton, Dodonaea, Leptospermum* and *Grevillea*, especially *G. robusta.*

Feeding habits: this species, which is active during the day, usually feeds singly or in small groups, chewing large lumps out of the leaves.

Caterpillar of the Wattle Notodontid Moth, Neola semiaurata. [T. Wood]

Caterpillar of Wattle Notodontid Moth, Neola semiaurata, *in defensive pose.* [T. Wood]

Notes: a striking addition to the garden. It is of sporadic occurrence and rarely causes significant damage. The caterpillar reacts to disturbance by jerking the head backwards and extruding a double-forked red appendage (osmeterium) from the underside of the body near the head; it also raises the rear end exposing four electric blue eye-spots.

Control: as per introduction for caterpillars and grubs.

Variegated Notodontid Moth, *Aglaosoma variegata*

Description: a colourful, stout, fleshy caterpillar, which grows to about 8cm long. Colours on the body include brown prolegs, black flanks and silvery grey along the back. Prominent blue dots occur along the sides, on some prolegs and in three bands across the back behind the head. Long hairs are scattered along the body and distinct tufts of short brown hairs along the back. Pupation occurs in a cocoon among litter on the ground. The moths have white forewings with black markings and the hindwings are light brown. The plump body has black and orange hairs. Wingspan is about 5cm.

Climatic regions: subtropics and temperate regions, mainly coastal.

Host plants: species of *Acacia, Banksia, Casuarina* and *Grevillea.*

Feeding habits: this species feeds singly during the day.

Notes: a colourful caterpillar that is a striking addition to the garden. It is of sporadic occurrence and rarely causes significant damage.

Control: as per introduction for caterpillars and grubs.

Colourful caterpillar of the Variegated Notodontid Moth, Aglaosoma variegata, *on* Grevillea asplenifolia. [D. Jones]

Sparshall's Moth or Porcupine Caterpillar, *Trichiocercus sparshallii*

Description: eggs are laid in clusters on the host plant. Newly hatched caterpillars are yellowish with darker markings and long white hairs. Mature caterpillars, which grow to 4cm long, are brown to black with conspicuous brush-like tufts of red hairs (sometimes yellow) along the back, the rest densely covered with long pale hairs. A single 'antennae' of black or brown hairs arises behind the head. Pupation occurs in a cocoon spun among litter on the ground. The moth is grey or white with black hairs on the head. Wingspan is about 4cm. Male moths have a tuft of long hairs protruding from the tip of the abdomen.

Climatic regions: subtropics and temperate regions.

Host plants: species of *Eucalyptus,* including *E. cephalocarpa, E. cinerea, E. leucoxylon, E. melliodora* and *E. polyanthemos*; also *Lophostemon confertus.*

Feeding habits: young caterpillars cluster in groups, sometimes moving in a head-to-tail procession. Older caterpillars are mainly solitary.

Notes: usually a relatively minor pest.

Control: as per introduction for caterpillars and grubs.

Hairy caterpillar, possibly of Sparshall's Moth, Trichiocercus sparshalli. [T. Wood]

EMPEROR MOTHS (FAMILY SATURNIIDAE)

A small family of large, furry moths with decorative wings that are often adorned with prominent eye-spots. There are about 1,500 species worldwide with only 15 species in Australia. The caterpillars are large, often colourful, and with tufts of hairs sometimes carried on long turrets.

EXAMPLES

Emperor Moth, *Syntherata janetta*

Description: white eggs are laid in rows on leaves. Newly hatched caterpillars are yellow with stiff hairs. They become greener as they grow and have numerous tufts of non-stinging pinkish hairs on the body. Mature caterpillars are about 8cm long, plump, fleshy and highly decorative. Pupation occurs in a tough dark brown cocoon. The moths, which have a wingspan of 12–14cm, are yellow, orange, pinkish brown or dark brown with small eye-spots on all four wings (but inconspicuous on the hindwings).

Climatic regions: tropics, subtropics and warm temperate regions (south to Sydney).

Host plants: various mangroves and rainforest trees including *Alphitonia, Avicennia, Barringtonia, Callistemon, Ceriops, Euodia, Evodiella, Glochidion, Melicope, Planchonia, Podocarpus, Terminalia* and *Timonius*; also the introduced Pepper Tree (*Schinus molle*).

Feeding habits: the caterpillars feed singly or in small groups during the day. They chew large lumps out of the leaves sometimes leaving only the main veins.

Notes: the unusual colourful caterpillars always create interest and the large moths, which are attracted to light at night, are especially decorative. They cause minimal damage and are a delightful addition to gardens.

Control: by hand-picking and moving to another host.

Caterpillar of Emperor Moth, Syntherata janetta, *on* Melicope elleryana. [D. JONES]

Emperor Moth, Syntherata janetta. [D. JONES]

Emperor Gum Moth, *Opodiphthera eucalypti*

Description: white eggs are laid on leaves in small clusters. Newly hatched caterpillars are black for about a week with tiny turrets on their back. They become bluish-green as they grow and have numerous tufts of non-stinging reddish hairs on the body. Mature caterpillars are about 10cm long, fleshy and highly decorative. Pupation occurs in a tough brown cocoon (sometimes grouped together). The moths, which have a wingspan of 12–15cm,

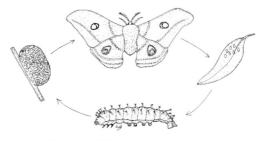

Emperor Gum Moth life cycle. [D. JONES]

are grey to pinkish-brown or reddish with eye-spots on all wings (very prominent on the hind wings).

Climatic regions: tropics, subtropics and temperate regions.

Host plants: many species of *Eucalyptus* – two favoured hosts are *E. leucoxylon* and *E. doratoxylon*; also *Lophostemon confertus* and the introduced Pepper Tree, *Schinus molle*.

Feeding habits: the caterpillars feed singly or in small groups during the day. They chew large lumps out of the leaves, sometimes leaving only the bare midribs. They strip small trees completely and can damage bigger trees when present in large numbers.

Notes: this insect is familiar to many people, both as the handsome moth and the large colourful caterpillar. Despite their destructiveness the caterpillars are very decorative and add interest to a garden. Their numbers can be easily checked by handpicking and transferring to other host plants. Large birds such as cuckooshrikes and currawongs, eat the caterpillars.

Control: by hand-picking and moving to another host.

Emperor Gum Moth. [R. Farrow]

Cocoons of the Emperor Gum Moth on a eucalypt stripped of foliage. [D. Jones]

Caterpillar of the Emperor Gum Moth. [T. Wood]

Helena Gum Moth, *Opodiphthera helena*

Description: newly hatched caterpillars are light brown with white lines, yellow spots and seven black turrets on their back (one on the tail). They become bright green as they grow, with tiny rough hairs all over and six purple and red or blue turrets on the thorax bearing non-stinging hairs. Mature caterpillars are about 8cm long, fleshy and highly decorative. Pupation occurs in a tough brown cocoon, sometimes spun between leaves. The moths, which have a wingspan of 12–17cm, are pinkish-brown with large eye-spots on all wings (very prominent on the hind wings).

Climatic regions: subtropics and temperate regions (including Western Australia).

Helena Gum Moth, Opodiphthera helena. [T. Wood]

Host plants: many species of *Eucalyptus*.

Feeding habits: the caterpillars feed singly or in small groups during the day. They chew large lumps out of the leaves, sometimes eating the veins and all. They strip small trees completely and can damage bigger trees when present in large numbers.

Notes: a large, attractive caterpillar that grows into a lovely moth (very similar to the Emperor Gum Moth). When disturbed the caterpillars tuck their head inwards, arch their back and project the turrets.

Control: by hand-picking and moving to another host.

SNOUT MOTHS (FAMILY LASIOCAMPIDAE)

A small family of moths consisting of about 2,000 species worldwide, with about 70 species in Australia. They are furry moths that hold the wings in a V-shape like a gable roof when at rest. Most fly at night, some during the day. The caterpillars are very hairy with the hairs along the side spreading widely and providing camouflage.

Caterpillar of the moth family Lasiocampidae that feeds on eucalypts.
[D. Jones]

Caterpillar of the family Lasiocampidae that feeds on eucalypts. [T. Wood]

Caterpillar of the Gum Snout Moth, Entometa fervens, *on* Callistemon.
[D. Beardsell]

EXAMPLES

Gum Snout Moth, *Entometa fervens*

Description: ovoid mottled brown eggs are laid in small groups on leaves. The fleshy grey to brown caterpillars, which grow 6–7cm long, are wrinkled and the sides covered with soft velvety hairs that mask the feet. There is a knob-like appendage near the rear and two soft horns on the back behind the head. Pupation occurs in a white parchment-like cocoon spun among leaves. The moths, brown to reddish brown, are wedge-shaped when at rest and with a prominently pointed head. Wingspan is 4–5cm.

Climatic regions: subtropics and temperate regions, mainly coastal.

Host plants: species of *Callistemon* and *Eucalyptus*; also apple trees.

Feeding habits: the caterpillars, which are usually solitary, feed at night on leaves. They rest during the day in a camouflaged position stretched out along a branch. When disturbed the caterpillar lifts its rear and coils its head inwards exposing a broad black band. At the same time it projects the horns behind its head.

Notes: these caterpillars can be very difficult to see when they are at rest. They can defoliate young apple trees and apple rootstocks, hence are a pest in orchards and fruit tree nurseries.

Control: as per introduction to caterpillars and grubs. Hand-picking is usually satisfactory.

Wattle Snout Moth, *Paraguda nasuta*

Description: globose white eggs are laid in rows on twigs. The slender caterpillars, which grow 5–6cm long, change colour with age. Young caterpillars are green, older caterpillars green or brown, and when about to pupate they all become brown. The caterpillars, which look soft and velvety from a covering of hairs (especially on the sides), have a dark dorsal line. Pupation occurs in a white cocoon spun among leaves. The moths, which are brown with darker lines and spots, are wedge-shaped when at rest, with a prominently pointed head. Wingspan is 3–4cm.

Wattle Snout Moths, Paraguda nasuta, mating. [T. Wood]

Climatic regions: tropics, subtropics and temperate regions, mainly coastal.

Host plants: species of *Acacia* (phyllodinous and ferny-leaved) and *Exocarpos*; also *Pinus radiata*.

Feeding habits: usually solitary, feeding at night on leaves and leaflets. The caterpillars rest during the day stretched out along a twig or branchlet. When disturbed the caterpillar coils its head inwards exposing two broad black bands.

Notes: the caterpillars can be very difficult to see when they are at rest.

Control: as per introduction to caterpillars and grubs. Hand-picking is usually satisfactory.

Casuarina Caterpillar or Sheoak Moth, *Pernattia pusilla*

Description: a slender, hairy brown caterpillar growing to 3cm long with a disproportionately large and prominent head. It is predominantly brown but often has reddish markings on the back and each segment has two small brownish dorsal knobs. When mature it spins a small white parchment-like cocoon on a stem. The moths are stout-bodied. Females are about 3cm across and sluggish with brown wings. Males are smaller, about 2cm across, and faster with attractive patterns on the forewings.

Climatic regions: tropics, subtropics and temperate regions, inland and coastal.

Host plants: many species of *Allocasuarina* and *Casuarina*, including the prostrate or cascading selection of *Casuarina glauca*.

Feeding habits: a very destructive gregarious pest which feeds on the stems and leaf scales of species of sheoaks. The caterpillars leave white patches along the stems where they have eaten and may also ringbark branches. Severe infestation during dry periods or on weakened trees can result in plant death. This pest can cause severe damage to nursery plants and young trees.

Notes: the caterpillars drop to the ground when disturbed and this can be used as an aid in their control.

Control: If plastic sheets are spread on the ground and the branches beaten with sticks the caterpillars can be easily collected and destroyed.

Casuarina Caterpillar. [D. BEARDSELL]   *Casuarina Caterpillar cocoon.* [D. JONES]

LAPPET MOTHS OR HAIRY MARYS (FAMILY ANTHELIDAE)

A small family of large, furry moths with about 74 of the 90 or so species found only in Australia. Most species fly at night. The moths have broad wings and often exhibit colour variation. The caterpillars are very hairy and usually can be recognised by a vertical band of pale colour on their face.

White-stemmed Gum Moth, *Chelepteryx collesi*

Description: brown eggs are laid in groups or clusters. Small caterpillars are blackish with long hairs. They grow and become grey with red, white or blackish bands and yellow warty areas, each wart carrying a tuft of long stiff sharp reddish hairs. Mature

Typical hairy anthelid caterpillar. [T. WOOD]

caterpillars are stout, fleshy and up to 12cm long. Pupation occurs in a brown silken cocoon 8–12cm long, drawn out at each end and armed with numerous protruding, irritant hairs. The moth is grey to dull-brown moth with conspicuous blackish zig-zag bands on the wings. Wingspan is about 14cm (males) to 16cm (females).

Climatic regions: tropics, subtropics and temperate regions, chiefly coastal.

Male White-stemmed Acacia Moth,
Chelepteryx chalepteryx, *feeding on*
Hakea *flowers.* [D. JONES]

Caterpillar of the White-stemmed Gum Moth. [S.JONES]

Host plants: species of *Angophora*, *Corymbia*, *Eucalyptus* and *Exocarpos*; also broad-leaved species of *Melaleuca* and *Lophostemon confertus*.

Feeding habits: a solitary species that chews large lumps out of leaves, sometimes leaving only the bare midribs.

Notes: a large, handsome caterpillar that should not be handled because of its sharp hairs that readily penetrate skin and cause a painful reaction. The cocoons, which are also formidably armed with sharp irritant hairs, are a familiar sight in some districts, usually in sheltered positions such as under eaves or on shed and garage walls. Mature caterpillars ready to pupate often wander long distances seeking a suitable site.

Another species, *Chelepteryx chalepteryx,* known as the White-stemmed Acacia Moth, has caterpillars of the

same shape but smaller and reddish-brown with a pale dorsal line and white or yellowish warts carrying sharply pointed hairs. It spins a similar well-armed cocoon to the White-stemmed Gum Moth. The moths are slightly smaller (wingspan about 10cm) and have attractive red patches and black castellated markings on the hindwings. The caterpillars feed on *Acacia*, *Exocarpos* and *Doryanthes excelsa*.

Control: by hand-picking and squashing. Spraying is rarely successful because of the caterpillar's nomadic habits.

Male and female White-stemmed Gum Moths. [D. Jones]

Hairy Mary Caterpillar or Variegated Caterpillar, *Anthela varia*

Description: white eggs are laid in rows on the host plant. Newly hatched caterpillars are blackish with long white hairs. Mature caterpillars, which grow 6–7cm long, are brown to black and densely covered with long hairs which can be a mixture of black, white and brown. Pupation occurs in a cocoon spun between dead leaves or among leaf litter on the ground. The moth is yellow to brown with darker wavy lines on the wings. Wingspan is about 7cm (male) or 9cm (female).

Climatic regions: tropics, subtropics and temperate regions, coastal and inland.

Host plants: species of *Eucalyptus* and some Proteaceae, including *Macadamia* and *Stenocarpus*.

Feeding habits: a solitary caterpillar that eats large chunks from leaves.

Notes: usually a relatively minor pest.

Control: as per introduction for caterpillars and grubs. Hand-picking and squashing is usually satisfactory.

Hairy Mary Caterpillar, Anthela varia. [T. Wood]

Large Woolly Bear, *Anthela canescens*

Description: mature caterpillars, which grow 4–5cm long, are grey to black and densely covered with long hairs – a mixture of grey or brown hairs and tufts of white hairs; there are also two erect pencil-like tufts of hairs on the thorax and two rows of yellow dots along the back. Pupation occurs in a hairy cocoon spun between dead leaves. The moth is pale yellow to pinkish brown with one or more dark lines on the wings and a thick furry body. Wingspan is about 5cm (male) or 7cm (female).

Climatic regions: tropics, subtropics and temperate regions, coastal and inland.

Host plants: species of *Corymbia* (*C. citriodora*, *C. henryi*, *C. torelliana*) and *Eucalyptus* (*E. deanei, E. dunnii, E. grandis, E. saligna, E. scoparia*).

Feeding habits: a solitary caterpillar that eats large chunks from leaves.

Notes: usually a relatively minor pest, but can cause significant damage to young plants. Sometimes a persistent pest of nursery plants.

Control: as per introduction for caterpillars and grubs. Hand-picking and squashing is usually satisfactory.

Wattle Woolly Bear, a compex of species including *Anthela repleta*

Description: mature caterpillars, 4–5cm long, are densely hairy with each body segment carrying tufts of spreading long grey, brown or black hairs; they also have rows of prominent white spots among the hairs. The head is reddish-brown with a prominent vertical white central band. Pupation occurs in a brown cocoon spun between phyllodes. The moth is rusty brown with a thick furry body and a wingspan of about 4cm.

Climatic regions: temperate regions, coastal and inland.

Host plants: species of *Acacia* (*A. binervata, A. dealbata, A. floribunda, A. mearnsii, A. melanoxylon*).

Feeding habits: a solitary caterpillar that eats large chunks from leaves and leaflets.

Notes: usually a relatively minor pest, but can cause significant damage to young plants. A whole complex of species are included in this group.

Control: as per introduction for caterpillars and grubs. Hand-picking and squashing is usually satisfactory.

Hairy caterpillar of Anthela repleta. [T. Wood]

Moth of the Anthela repleta *complex.* [D. Jones]

Rose Anthelid, *Chenuala heliaspis*

Description: caterpillars, which grow 4–5cm long, are grey to yellowish and densely covered with long hairs which can be a mixture of grey or black. Two tufts of dense hairs occur on the back behind the head. Pupation occurs in a cocoon spun between leaves. The moths are dimorphic, the males rusty brown with red hindwings, the females much paler. Wingspan is about 6cm (male) or 7cm (female).

Climatic regions: subtropics and temperate regions, mainly coastal.

Host plants: species of *Acacia* and *Eucalyptus*.

Feeding habits: feeds singly or in small groups.

Notes: usually a relatively minor pest.

Control: as per introduction for caterpillars and grubs. Hand-picking and squashing is usually satisfactory.

Female Rose Anthelid moth, Chenuala heliaspis. [T. Wood]

CONCEALER MOTHS (FAMILY OECOPHORIDAE)

A large family that is diverse in Australia with about 2,300 species named and a similar number yet to be formally recognised. They are small moths, many with fringed or scaly hindwings. The caterpillars live in simple shelters made by joining leaves together with silk or constructing silk-lined tunnels.

Opened shelter of Smooth Bag Shelter Moth, Heteroteucha dichroella. [D. JONES]

Smooth Bag Shelter Moth, *Heteroteucha dichroella*

Description: a smooth, almost hairless brown caterpillar, 1.5–2cm long with two pale dorsal lines and pale bands between each segment. A stiff transparent bristle protrudes from the side of each segment. Pupation occurs within the bag. The attractive moths have yellow forewings with brown tips. Wingspan is about 2cm.
Climatic regions: subtropics to temperate regions, mainly coastal.
Host plants: species of *Angophora*, *Corymbia* and *Eucalyptus*.
Feeding habits: a gregarious caterpillar that forms a loose bag shelter by joining patches of leaves together. The caterpillars move onto the foliage at night to feed, retreating in early morning to the shelter of the bag.
Notes: a relatively minor pest that occurs sporadically.
Control: as for bag shelter moths.

Gumtree Leaf Roller, *Wingia aurata*

Description: plump, fleshy, hairless cream caterpillars with a dark brown head. They grow to about 2cm long. The moth is bronze-yellow to orange with a central dark line on the forewing. It is unusual in that the pointed wing-tips curve upwards when at rest. Wingspan is about 2cm.
Climatic regions: tropics, subtropics and temperate regions, coastal and inland.
Host plants: species of *Angophora*, *Corymbia* and *Eucalyptus*.
Feeding habits: the caterpillar lives singly, joining adjacent leaves together to form a silken tube. The silk is very tough and the caterpillars probably emerge at night to feed.
Notes: a widespread and common species that causes limited damage.
Control: as per introduction for caterpillars and grubs.

BELL MOTHS OR LEAF ROLLERS (FAMILY TORTRICIDAE)

Damage caused by leaf-rolling caterpillars to new growth of Pomaderris. [D. JONES]

Leaf-rolling caterpillar feeding on young eucalypt leaves. [D. JONES]

A family of small to medium-sized moths consisting of about 5,000 species worldwide with more than 1,200 species in Australia. A number of exotic species are economically important pests of fruit trees and grape vines. The caterpillars often live in simple shelters made by joining adjacent leaves together. A few are borers.

Light Brown Apple Moth or Leaf Roller, *Epiphyas postvittana*

Description: cream eggs are laid in dense clusters on the leaf of a host plant. Young caterpillars are green, older caterpillars are green with narrow yellow bands between the segments; when mature they often become pinkish. They grow to about 2cm long. Pupation occurs within the leaf shelter. The moth is velvety and brown with a pale brown area on each forewing above a dark brown basal patch. The moth sits with the forewings folded over the hindwings in a bell-shaped arrangement. Wingspan is about 2cm.

Climatic regions: tropics, subtropics and temperate, coastal and inland.

Host plants: feeds on a tremendous range of plants, including weeds, forbs, shrubs and trees; also ferns, orchids and palms.

Feeding habits: the caterpillars curl leaves and join adjacent leaves together with silk to form a shelter. They commonly attack soft new foliage and young growing tips. They feed within the shelter and also emerge at night to feed on adjacent plant material. The caterpillar can often be found hiding within the shelter by day. When disturbed they wriggle actively and frequently drop to the ground to avoid capture.

Notes: a very familiar, persistent and destructive pest that is common wherever plants are grown (including in greenhouses). It is very destructive of soft new growth. Grey Mould, *Botrytis cinerea*, often grows on tissue damaged by this pest.

Control: as per introduction for caterpillars and grubs. Regular spraying of commercial crops is often needed to control the activities of this persistent pest.

Caterpillar of the Light Brown Apple Moth. [D. BEARDSELL]

Damage caused by Light Brown Apple Moth on Thryptomene. [D. JONES]

Green Wattle Leaf Roller, *Epiphyas ashworthana*

Description: caterpillars are very slender and green with a few short hairs on the back. They grow to about 2cm long. Pupation occurs within the leaf shelter. The moth is light grey-brown with darker patches. Wingspan is about 2cm.

Climatic regions: temperate regions, coastal and inland.

Host plants: fern-leaved species of *Acacia*.

Feeding habits: the caterpillars join adjacent young leaves together with silk to form a shelter. They feed within the shelter, probably also emerging at night to feed. When disturbed they wriggle actively and may drop to the ground.

Notes: a minor pest that causes limited damage.

Control: as per introduction for caterpillars and grubs.

Other Bell Moths

• **Fern Spore Caterpillar:** these tiny caterpillars, 0.1–0.2cm long, feed in the sporing patches of the elkhorn fern, *Platycerium bifurcatum*, on the tips of the fertile fronds. They eat the spores and the underlying frond tissue causing the tips of the fronds to turn brown and die. Fine webbing is often noticeable where they are feeding. The satiny white moths are about 0.3cm long.

TUFT MOTHS (FAMILY NOLIDAE)

A small family of drab to colourful moths consisting of about 1,400 species worldwide with about 170 species in Australia. Some caterpillars of this family are quite colourful, others much less so and well camouflaged.

EXAMPLES

Gum Tree Moth or Gum Leaf Skeletoniser, *Uraba lugens*

Description: eggs are laid in large groups (up to 200 at a time) in lines or parallel rows, usually on the underside of a leaf , but sometimes on the upper surface. Most egg-laying occurs on the lower branches of a tree. Young caterpillars graze together, first on the leaf on which the eggs were laid, moving to adjacent leaves as they grow. They feed as a group for the first four moults then move about singly, feeding as solitary caterpillars that eat whole leaves rather than skeletonising. Mature caterpillars, which are about 2.5cm long, are greyish with yellow and brown markings and abundant long white hairs. They have the unusual habit of piling cast-off skins on their head, creating a hat-like structure. Pupation occurs in hairy cocoons hidden in the bark or among leaf litter. The moth has patchy light and dark grey wings. Wingspan is about 3cm.

Climatic regions: tropics, subtropics and temperate regions, coastal and inland.

Host plants: various species of *Eucalyptus*, particularly those with glaucous leaves such as *E. bicostata*, *E. cinerea*, *E. crenulata*, *E. globulus*, *E. macrocarpa*, *E. maidenii*, *E. morrisbyi* and *E. nitens*; also green-leaved species such as *E. argophloia*, *E. calophylla*, *E. camaldulensis*, *E.crebra*, *E. delegatensis*, *E. ficifolia*, *E. marginata*, *E. obliqua*, *E. pilularis*, *E. saligna*, *E. tereticornis*, *E. viminalis* and *E. wandoo*. *Corymbia citriodora* and *Lophostemon confertus* are also host plants.

Feeding habits: young caterpillars congregate in groups and graze the surface of the leaf, leaving the veins which turn brown. Older caterpillars feed singly, eating the leaves whole.

Notes: often the outbreaks of this pest are relatively minor, impacting on individual trees or small patches within a single tree. Young trees, however, can be severely impacted by this pest. In particular, repeated attacks can severely weaken young trees and affect later development and shape. Major outbreaks of Gum Tree Moth occur in some years and these can devastate large areas of forest. This was exemplified in March 2011 when a devastating outbreak defoliated large tracts of jarrah forests in south-west Western Australia. The caterpillars drop on a silken thread if threatened. Contact with the hairs on the caterpillars can cause stinging followed by severe itching, rashes and other allergenic reactions.

 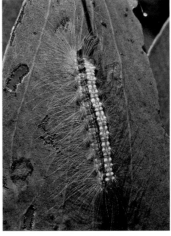 

Young caterpillars of the Gum Tree Moth, Uraba lugens. [T. Wood]

Mature caterpillar of the Gum Tree Moth, Uraba lugens. [T. Wood]

Caterpillar of the Gum Tree Moth, Uraba lugens, *showing 'hat'.* [T. Wood]

Research has shown that three geographical races of Gum Tree Moth have been recognised; coastal and inland races which have summer-autumn and winter generations and a highland race which has a single generation each year.

Control: look for lines of eggs which are quite visible and easily removed. Watch for early stages of leaf damage during early summer and again in late autumn. For small trees cut off affected foliage and burn or squash caterpillars.

Rough Bollworm, *Earias huegeliana*

Description: a stout, sparsely hairy caterpillar that grows to 2cm long. It is patchy grey and brown with rows of yellow-orange spots on the back. Pupation occurs in a papery cocoon spun on the host plant. The moths are cream to dirty white with a darker green or brown stripe on the forewings. Wingspan is about 2cm.

Climatic region: tropics and subtropics, mainly coastal.

Host plants: mainly plants of the family Malvaceae including native species of *Abelmoschus*, *Abutilon*, *Alyogyne*, *Gossypium* and *Hibiscus*; also *Adansonia*.

Feeding habits: the caterpillars tunnel in developing buds and fruit, affecting flowering and fruit set. They also bore into stems and damage new growth.

Notes: a sporadic pest that fluctuates considerably in numbers. It can be a nuisance in gardens, severely impacting on flowering in some seasons.

Control: by hand-picking and squashing.

Blue Quandong Moth, *Pisara hyalospila*

Description: a slender, bright green caterpillar that grows to 2cm long. It has a distinct pale line along its back edged with small red patches. Transparent hairs arising in tufts spread widely from along its flanks. Pupation occurs in a papery cocoon spun on twigs of the host plant (often in groups). The moths are light grey to dark grey with darker patches and lines. Wingspan is about 3cm.

Climatic region: tropics and subtropics, mainly coastal.

Host plants: Blue Quandong, *Elaeocarpus angustifolius*; possibly also White Beech, *Gmelina leichhardtii*.

Feeding habits: young caterpillars skeletonise the leaves, older caterpillars eat patches out of the leaves or consume whole leaves.

Notes: a sporadic pest that fluctuates considerably in numbers. Whole trees can be defoliated in some years. Successive attacks can result in dieback and possible death.

Control: by hand-picking and squashing. Spraying young trees may be necessary.

PYRALID MOTHS (FAMILY PYRALIDAE)

A very large and diverse family of moths consisting of about 16,000 species worldwide with about 1,100 named species in Australia and many more awaiting formal recognition. It includes many large and colourful moths. The caterpillars mostly live hidden within shelters made by tying leaves together with silk, or in silken tubes or cases. Some are borers.

EXAMPLES

Tube Caterpillar, *Poliopaschia lithochlora*

Description: a slender, fleshy brown, sparsely hairy caterpillar which grows about 2cm long. It lives in a hollow tube constructed from waste products. The moths are dull pale grey with a wingspan of about 3cm.

Climatic regions: tropics and subtropics, mainly coastal.

Host plants: species and hybrids of *Grevillea*; also broad-leaved species of *Melaleuca* (*M. leucadendron*, *M. quinquenervia* and *M. viridiflora*).

Feeding habits: the caterpillars graze and chew leaves within a hollow tube which they build from webbing, waste products and chewed leaves. These tubes are found over the leaves and stems where each caterpillar has eaten and they become larger as the caterpillar grows. Eventually the area of dead tissue and brown tunnels gives notice of their activities.

Notes: the general external appearance of this pest's shelters resemble those of the web-covering borers, however no boring or bark-feeding activity takes place within the tube. This species is being considered as a biological control agent for naturalised populations of *Melaleuca* in Florida.

Control: this pest's activities are usually of a minor nature and hand-picking is adequate for its control.

Webbing Caterpillar or Teatree Web Moth, *Orthaga thyrisalis*

Description: slender, fleshy, almost hairless caterpillars about 2.5cm long. They are grey-brown with a broad pale green dorsal stripe bisected by a narrow dark line. These caterpillars spin protective webbing over the stems and leaves of the plants on which they feed. The moths have unusual bulging eyes and strongly patterned wings in shades of light and dark brown.

Climatic regions: tropics, subtropics and temperate regions, coastal and inland.

Host plants: a wide range of garden plants but especially small-leaved Myrtaceae (species of *Babbingtonia*, *Baeckea*, *Beaufortia*, *Kunzea*, *Leptospermum*, *Melaleuca* and *Regelia*).

Feeding habits: the caterpillars congregate in the webbing during the day and emerge at night to feed, sometimes taking leaves back to the webbing. The webbing, which becomes filled with their droppings, caste skins, chewed leaves and other plant debris, eventually becomes an unsightly mess. The size of the

Webbing Caterpillar nest in Melaleuca armillaris. [D. Jones]

Webbing Caterpillar damage to Acacia cardiophylla. [D. Jones]

Webbing Caterpillar nest on wattle. [D. Jones]

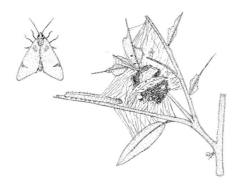

Webbing Caterpillar. [D. Jones]

Webbing Caterpillars on Labichea nitida. [S. Jones]

webbed mass increases as the caterpillars grow and large areas of the plant become denuded of leaves.
Notes: this pest is so persistent and damaging that gardeners often avoid growing susceptible species and cultivars in areas where it is active. Another unknown webbing caterpillar feeds on species of *Acacia*. It can cause severe damage to prostrate wattle cultivars such as *A. cardiophylla* 'Gold Lace' and *A. pravissima* 'Golden Carpet'
Control: by spraying malathion onto the stems and foliage around the web. Small webs can be dragged out or cut off and the contents crushed.

Kurrajong Bag Moth, *Dichocrosis clytusalis*

Description: a slender, pale green, sparsely hairy caterpillar that grows about 2.5–3cm long. The caterpillars live together in groups and construct bags by rolling leaves and joining them together with silken threads until a large mass is formed. Pupation occurs inside the bag. The moth is yellow with black wavy lines on all four wings. Wingspan is 2.5–3cm.
Climatic regions: tropics, subtropics and temperate regions, inland and coastal.
Host plants: species of *Brachychiton*, particularly *B.acerifolius, B. populneus* and *B. rupestris;* also some *Brachychiton* hybrids.
Feeding habits: caterpillars shelter by day in the bags and emerge at night to feed. They concentrate on relatively young leaves, eating the tissue and leaving a framework of veins. Most attacks occur during late summer and autumn.
Notes: a common species that can completely defoliate young trees. Affected trees usually recover but their growth rate is impeded and their final shape can be affected as some side branches can die after the defoliation. The impact of the caterpillars is much worse in dry years and on trees that have been lopped for forage.
Control: as for bag shelter moths.

Shelter of Kurrajong Bag Moth. [D. JONES]

Shelter built by the Kurrajong Bag Moth on Brachychiton acerifolius. [D. JONES]

Shelter built by the Kurrajong Bag Moth on Brachychiton populneus. [D. JONES]

Gardenia Leaf Roller, *Parotis marginata*

Description: the hairless, fleshy caterpillars are cream to pale green with darker hues on the back and a series of raised shiny black knobs. They grow to about 2.5cm long. Pupation occurs within the leaf shelter. The attractive moth is bright green with a notched brown basal margin on each wing. Female moths exert a flower-like tuft from the abdomen when seeking a mate. Wingspan is about 3cm.
Climatic regions: tropics and subtropics, mainly coastal.
Host plants: native species of *Alstonia* and *Gardenia*; possibly other species of the family Rubiaceae.
Feeding habits: the caterpillars create a simple shelter, either by curling a single leaf and holding the

edges together with silken threads, or joining adjacent leaves together. They feed within the shelter, probably also emerging at night to feed. Young caterpillars can skeletonise leaves; older caterpillars eat whole leaves or leave a framework of veins.

Notes: a minor pest that usually causes limited damage, although it can significantly damage young nursery plants.

Control: as per introduction for caterpillars and grubs.

FORESTER MOTHS (FAMILY ZYGAENIDAE)

A small family of small, shiny moths consisting of about 1,000 species worldwide with around 43 species found in Australia. The stout, hairy caterpillars feed on species of *Cayratia*, *Cissus* and *Hibbertia*.

Satin Green Forester, *Pollanisus viridipulverulenta*

Satin Green Forester moth, Pollanisus viridipulverulenta. [T. WOOD]

Description: stout brown caterpillar growing to about 1cm long with rows of warts, each wart containing a clump of short hairs. Pupation occurs in a small cocoon spun between leaves or among leaf litter on the ground. The moth has brilliant metallic green forewings and brown hindwings. Wingspan is about 3cm.

Climatic regions: subtropics and temperate regions, mainly coastal.

Host plants: species of *Hibbertia*, including *H. diffusa*, *H. linearis*, *H. obtusifolia*, *H. rufa*, *H. sericea* and *H. stricta*.

Feeding habits: a solitary caterpillar which eats grooves in the upper leaf surface.

Notes: usually a relatively minor pest. Often brown leaf remnants indicate its presence.

Control: as per introduction for caterpillars and grubs. Hand-picking is usually satisfactory.

WOOLLY MOTHS (FAMILY EUPTEROTIDAE)

A small family of medium-sized to large woolly moths which is poorly represented in Australia (about 8 species). One species lives in groups within a silken shelter.

Lewin's Bag Shelter Moth, *Panacela lewinae*

Shelter built by caterpillars of Lewin's Bag Shelter Moth on Exocarpos cupressiformis. [D. JONES]

Description: a very hairy, dark brown caterpillar, 4–5cm long with a wavy or zig-zag yellowish line along each side and a black head with a round yellow or white patch. The caterpillars hide together during the day in a loose bag-like shelter constructed from leaves and silken webbing, emerging at night to feed. Pupation occurs in a silk cocoon among ground litter. The attractive moths are reddish-brown with a wingspan of about 3cm. The male moths have a dark band on each forewing.

Climatic regions: subtropics to temperate regions, mainly coastal.

Host plants: species of *Angophora*, *Corymbia*, *Eucalyptus*, *Syncarpia* and *Exocarpos cupressiformis*.

Feeding habits: a gregarious caterpillar with a voracious appetite that can cause significant damage to young trees. The shape of the bag shelter varies considerably from structures as wide as long, to long draw-out narrow bags. The caterpillars move onto the foliage at night to feed, retreating in early morning to the shelter of the bag. The bags become messy with accumulations of faecal pellets, cast skins, dead caterpillars and leaf fragments.

Notes: a destructive pest that can be a problem on young trees. The irritant hairs from the caterpillars and bag shelters (even old shelters) can cause severe itching, rashes and other allergenic reactions. The hairs can become airborne if the shelter is moved and can be breathed into the lungs. Caterpillars living within the shelter often make scratching or scraping noises when disturbed.

Control: as for bag shelter moths.

Caterpillars of Lewin's Bag Shelter Moth, Panacela lewinae.
[D. Jones]

PALM MOTHS (FAMILY AGONOXENIDAE)

A very small family of moths which has only about four species recognised worldwide.

Palm Moth or Coconut Flat Moth, *Agonoxena phoenicea*

Description: slender, small, hairless grey caterpillar growing to 2cm long. The moth, which is about 1cm across, has yellow wings with a red-brown line on each forewing. It has been recorded that when threatened the moths, rather than flying, prefer to run with the wings wrapped tightly around the body.

Climatic regions: tropics.

Host plants: *Archontophoenix alexandrae* and *Cocos nucifera*.

Feeding habits: the caterpillars feed on the underside of leaflets while hidden beneath a flimsy, silken web.

Notes: a minor pest of limited significance.

Control: as per introduction for caterpillars and grubs.

BUTTERFLIES

Butterflies share many features with moths, such as scales on the wings and coiled mouthparts, but can usually be distinguished by their antennae ending in a small club-like structure. Butterflies mostly fly during the day and moths at night, but there are some day-flying moths. The caterpillars of butterflies are plant feeders but few are serious pests. Their presence in a garden provides interest with the butterflies adding colour, beauty and movement.

WHITES AND YELLOW BUTTERFLIES (FAMILY PIERIDAE)

A family of medium-sized butterflies consisting of about 1,100 species worldwide with 36 species occurring in Australia. The butterflies are mainly in shades of white, cream or yellow. The slender, fleshy caterpillars are well camouflaged on their food source.

Cassia Caterpillar, four species of *Catopsilia*

Description: white spindle-shaped eggs are laid singly on a leaf. The slender, smooth, yellow and green caterpillars, with a dark dorsal stripe and pale lateral band, grow to about 4cm long. The butterflies are fairly large, white or yellow (wingspan 5–6cm), and are commonly known as emigrants.
Climatic regions: tropics and subtropics, coastal and inland.
Host plants: species of *Cassia* (*C. brewsteri*, *C. marksiana* and *C. queenslandica*) and *Senna* (*S. aciphylla*, *S. barclayana*, *S. leptoclada* and *S. surattensis*).
Feeding habits: caterpillars are solitary or feed in loose groups chewing large lumps out of the leaflets. They often centre their activities near the tips of shoots and can defoliate some stems completely. The caterpillars rest along the midrib during the day, merging in well with the leaflet tissue.
Notes: these insects are most active during the warm summer months. The caterpillars are usually regarded as a minor pest. The attractive butterflies fly actively during the day.
Control: as per introduction for caterpillars and grubs.

Common Grass Yellow Butterfly, *Eurema hecabe*

Description: caterpillars are slender and covered with short hairs. Young caterpillars are yellowish green, older caterpillars are green with a conspicuous white lateral band. Mature caterpillars grow to 3cm long. Pupae are green, about 2cm long with a pointed head. The butterflies are bright yellow with a black marginal band.
Climatic regions: tropics and subtropics, coastal and inland.
Host plants: many feather-leaved wattles including *Acacia baileyana*, *A. muelleriana*, *A. rubida* and *A. spectabilis;* also *Abrus precatorius*, *Albizia lebbeck*, *Breynia oblongifolia* and species of *Aeschynomene*, *Indigofera*, *Phyllanthus*, *Senna* and *Sesbania* .
Feeding habits: the caterpillars feed in small loose groups. They eat at night, resting on the underside of leaves during the day. They are voracious feeders despite their small size and can defoliate even quite large plants.
Notes: The butterflies are most active during March-April and congregate in large numbers around the food plants to lay their eggs.
Control: as per introduction for caterpillars and grubs.

SWALLOWTAIL BUTTERFLIES (FAMILY PAPILIONIDAE)

A small family of large, colourful butterflies consisting of about 550 species worldwide with 18 species in Australia. When disturbed, the large fleshy caterpillars rear up and extend an unusual forked appendage (termed osmeterium) that emits a strong, unpleasant smell from the second segment behind the head.

Dainty Swallowtail Butterfly or Dingy Swallowtail Butterfly, *Papilio anactus*

Description: yellow globular eggs are laid on the margins of young leaves. Young caterpillars, which feed mainly on younger leaves, are dark with broad yellowish bands and two rows of bristly hairs. Mature caterpillars, which are about 3.5cm long, are green to blackish with cream to bold yellow or orange markings, blue and white dots and short bristles. The greenish or greyish chrysalis, about 3cm long, is held to the support by centrally attached silken threads. The butterflies have a wingspan of about 7.5cm. The body is black; forewings black with numerous variably-shaped white blotches; hindwings with large red spots towards the margins.

Climatic regions: tropics, subtropics and temperate regions of the eastern states, coastal and inland.

Host plants: native species of *Acronychia, Citrus, Geijera parviflora* and *Melicope*; also cultivated citrus.

Feeding habits: the caterpillars feed singly or in small groups, mainly on younger more tender leaves. They are voracious feeders and quickly strip the foliage from the stems where they are active.

Notes: this is a relatively minor pest whose activities can be tolerated because of the beautiful butterflies which have a gliding flight. When disturbed the caterpillars evert an orange horned osmeterium from just behind the head. This emits a smell like rotten citrus. Often the caterpillar also thrashes its head about in an attempt to smear the predator with the acidic material.

Control: the caterpillars are attacked by a range of predators including small birds, especially Silvereyes. Hand-picking and transferring to another host is usually an effective way to reduce numbers.

Orchard Swallowtail or Large Citrus Butterfly, *Papilio aegeus*

Description: cream globular eggs are laid on the upper surface of young leaves and shoots. Young caterpillars are black to brown with three white patches. Mature caterpillars, which are about 6cm long, are greasy-green with white, yellow or brown markings (resembling a bird's dropping). The greenish or greyish chrysalis, about 4cm long, is held to the support by centrally attached silken threads. The butterflies have a wingspan of 12cm (males) to 14cm (females). The body is black; the forewings black to greyish with a white stripe; the hindwings black with red, blue and white crescent-shaped spots (female), or mainly black and white (male).

Climatic regions: tropics, subtropics and temperate regions over much of Australia, coastal and inland.

Orchard Swallowtail Butterflies mating (female top). [T. Wood]

Young caterpillar of the Orchard Swallowtail Butterfly. [T. Wood]

Nearly mature caterpillar of the Orchard Swallowtail Butterfly. [T. Wood]

Host plants: native species of *Acronychia, Citrus, Cryptocarya, Flindersia, Geijera parviflora, Halfordia, Melicope, Murraya, Zanthoxylum* and *Zieria*; exotics include cultivated citrus and *Choisya ternata*.

Feeding habits: the caterpillars feed singly or in small groups, mainly on younger more tender leaves. They are voracious feeders and quickly strip the foliage from the stems where they are active.

Notes: this is often a relatively minor pest whose activities can be tolerated because of the beautiful butterflies. Significant damage is sometimes caused. When disturbed the caterpillars evert a red horned osmeterium from just behind the head. This emits a smell like rotten citrus. Often the caterpillar also thrashes its head about in an attempt to smear the predator with the smelly acidic material. Eggs of this species are commercially available and are often raised in school projects to educate students on insect metamorphosis.

Control: the caterpillars are attacked by a range of predators including small birds and wasps.

GOSSAMER BUTTERFLIES (FAMILY LYCAENIDAE)

A very large family of small butterflies consisting of more than 5,000 species worldwide with 141 species known from Australia. The caterpillars tend to be broad and flattish with the head tucked under. They attach themselves by a pad of silk. Many species produce secretions that both attract and subdue ants.

Imperial Hairstreak Butterfly, *Jalmenus evagoras*

Description: white to pale green spindle-shaped eggs are laid in clusters or rows. The flattish red-brown caterpillars, which grow to to 4cm long, have numerous short blackish turret-like growths on the back and sides. They also have short hairs and short white lines on the flanks. The butterflies, which have a wingspan of about 4cm, are shiny blue with black margins and orange spots on the base of the hindwing. The undersides of the wings are a lovely satiny grey with black and orange markings. The hindwings each have a short curled tail and two pairs of yellow or orange spots.

Climatic regions: tropics, subtropics and temperate regions, coastal and inland.

Host plants: species of *Acacia*, both ferny-leaved and phyllodinous, including *A. falcata, A. dealbata, A. decurrens, A. harpophylla, A. irrorata, A. leiocalyx, A. mearnsii, A. melanoxylon, A. parramattensis* and *A. spectabilis*.

Feeding habits: the caterpillars feed in groups during the day, always attended by ants of the genus *Iridomyrmex*. Female butterflies will only lay eggs on host plants that support suitable species of ant. The ants protect them from predators and parasites and in return obtain secretions from the

Caterpillars of Imperial Hairstreak Butterfly attended by ants. [T. Wood]

Imperial Hairstreak Butterfly feeding. [T. Wood]

caterpillars. The caterpillars chew large lumps out of the phyllodes and leaflets. They are voracious feeders and can defoliate a small tree, often concentrating on specific trees in an area, leaving nearby trees of the same species untouched. Even more remarkably the same trees will be eaten year after year. When mature the caterpillars pupate together in a silken web spun between petioles and branchlets. The ants also attend the pupae.

Notes: these insects are most active during late spring and summer. The caterpillars often attack moderately young wattles and although they can cause significant damage, healthy plants usually regrow strongly. The medium-sized attractive butterflies fly during the day and add beauty and movement to a garden. The genus *Jalmenus*, consisting of about 10 species, is widely distributed in Australia. The caterpillars feed on a wide range of wattles, both fern-leaved and phyllodinous; also some species of *Cassia* and *Alectryon*.

Control: as per introduction for caterpillars and grubs. Although the larvae are destructive, the plant damage is temporary, the life cycle a fascinating study and the butterflies very decorative.

Pupae of Imperial Hairstreak Butterfly. [T. WOOD]

Cycad Blue Butterfly, *Theclinesthes onycha*

Description: white ovoid eggs are laid singly or in small groups on new fronds. Young caterpillars are green, later becoming brown to blackish. They grow to 1cm long, are flattish, strongly segmented and can have a darker dorsal line. The butterflies, which have a wingspan of about 3cm, have a shiny metallic pale blue central patch bordered with grey-brown. The hindwings each have a short, thin tail and yellow or orange spots.

Climatic regions: tropics, subtropics and temperate regions, coastal and inland.

Larva of Cycad Blue Butterfly on Macrozamia secunda. [M. CLEMENTS]

Damage to leaflets of Cycas media *caused by caterpillars of the Cycad Blue Butterfly.* [D. JONES]

Host plants: cycads, particularly native species of *Cycas* and *Macrozamia*; also garden-grown exotic cycads including *Ceratozamia robusta* and *Cycas revoluta*.

Feeding habits: the caterpillars are usually attended by ants. They hide during the day and feed mainly at night, grazing the surface tissue of the leaflets.

Notes: these insects are most active in spring and summer. They concentrate on newly emerging fronds and can cause considerable damage in a short period. Damaged areas quickly become white and papery and this can last for years, rendering affected plants unsightly. This is the only native butterfly that feeds on cycads. The medium-sized, attractive butterflies fly during the day and add beauty and movement to a garden. Two subspecies have been described, subsp. *capricornia* from northern Queensland and subsp. *onycha* from southern areas. It is suspected that the northern subspecies mainly feeds on species of *Cycas* and the southern subspecies mainly on *Macrozamia,* however, the commonly grown Sago Palm, *Cycas revoluta*, is frequently attacked in Sydney gardens, the realm of subsp. *onycha*.

Control: as per introduction for caterpillars and grubs. Spraying newly emerging fronds is often necessary to prevent damage.

BRUSH-FOOTED BUTTERFLIES (FAMILY NYMPHALIDAE)

A very large family of medium-sized to large butterflies consisting of about 6,000 species worldwide with 82 species known from Australia. Many species have the upper surface of the wings brightly coloured and the lower surface dull for camouflage when resting. The fleshy caterpillars are often boldly marked, some adorned with soft horn-like processes.

Tailed Emperor Butterfly, *Polyura pyrrhus*

Description: this lovely butterfly produces large fleshy caterpillars up to 8cm long with a rough skin as if covered with tiny warts. The colour is dark green with yellow longitudinal bands and two broad transverse yellow stripes across the back. The caterpillar's head has a startling ornament of four backward-projecting horns – two long and two short. The chrysalis is green. The handsome, fast-flying butterfly has white wings edged with black and short tails on the hindwings. Wingspan is about 8cm.

Climatic regions: tropics, subtropics and temperate regions, chiefly coastal.

Host plants: feather-leaved species of *Acacia*, also *Paraserianthes lophantha, Brachychiton spp., Cassia* spp. and *Pararchidendron pruinosum*.

Feeding habits: solitary or in small groups on mature leaves. The caterpillars rest on a pad of silken hairs, moving away to feed and returning when full.

Notes: an exciting caterpillar that adds interest to a garden. The butterfly usually flies high in the tree canopy.

Control: we suggest you be prepared to share some leaves with them and enjoy their company.

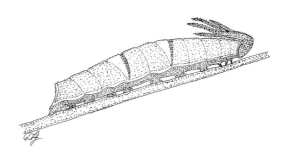

Tailed Emperor Butterfly caterpillar. [D. JONES]

SKIPPERS (FAMILY HESPERIIDAE)

A large family of butterflies consisting of more than 3,500 species worldwide with 124 species known from Australia. These small butterflies can be recognised by their distinctive fast darting flight.

Palm Dart Butterflies, two species of *Cephrenes*

Description: eggs are laid singly on the underside of palm fronds. The caterpillars, which grow to about 4cm long, are slender and more or less translucent with a prominent head and a tuft of white hairs on the rear segment. Those of the Yellow Palm Dart, *Cephrenes trichopepla*, are yellow with a black and white striped head; those of the Orange Palm Dart, *C. augiades*, are green with a white head marked with a black V-shape. They reside on the underside of a leaflet, in a shelter made by pulling the leaflet margins together with silk. Pupation occurs within the shelter, the pupa covered with chalky secretions. The butterflies are colourful skippers or darts, with a wingspan of 3–4cm and brown and orange colours in the wings and abdomen.

Climatic regions: tropics, subtropics and temperate zones, mainly coastal.

Host plants: native and exotic palms, particularly species of *Archontophoenix, Livistona* and *Ptychosperma.*

Feeding habits: the caterpillars live by day in a flimsy silken shelter made by sewing edges of the leaflets together, emerging at night to feed. They eat the palm leaflets to the midrib. Severely damaged palms appear sparse with ragged fronds.

Notes: a significant pest of palms, especially young plants and nursery stock. The Yellow Palm Dart is distributed from Queensland to Victoria and South Australia, the Orange Palm Dart from the coastal areas of all states.

Control: as per introduction for caterpillars and grubs. Spraying is often necessary to control these pests in nurseries. Noisy Miners are a useful predator of this species.

Palm dart butterfly. [D. JONES]

CHAPTER 9

LEAF-EATING BEETLES

Beetles are a group of commonly-encountered insects that are mostly easy to recognise. They share the common feature of having two hardened wing-covers (known as elytra) that cover and protect the membranous wings that are used for flying. The delicate flying wings are folded beneath these covers when not in use. Millions of beetles exist in numerous habitats scattered around the world. Australia has about 60,000 species of beetle, many yet to be formally recognised. Scientists place beetles in the order Coleoptera.

Beetles feed on a wide range of materials, including plants. A number of beetles eat the leaves of native plants. Some of these beetles are quite innocuous and cause limited damage, others are much more conspicuous and quite destructive. In some species it is the beetle only that feeds on leaves, whereas in others the beetles and larvae may share the food source and feed in close proximity to each other. Often the beetle larvae may feed on different parts of the plant to the adult beetle, including roots, branches and stems. Those beetles that have larvae which bore into trunks, branches and stems are dealt with in chapter 13.

Life cycle: beetles, which are the adult stage of the life cycle, mate and lay eggs. The larvae –usually known as grubs – bear little resemblance to the beetles, but look more like a caterpillar. They feed continually, increasing in size and eventually becoming large enough to develop into an adult insect. At this stage the grub changes to a pupa and after some time a beetle emerges from the pupa.

Symptoms of plant damage: beetles feed on plants in different ways and the remnants left after feeding can provide clues to the identity of the culprits involved. Many species, especially weevils, graze the surface of the leaf, leaving the veins and structural elements intact. Often those beetles which feed in this way will graze a narrow band or strip of tissue from the leaf surface. These insects may also graze strips from the green bark found on branchlets and young stems. Areas damaged by this grazing dry out, becoming white and papery, later brown and necrotic. By contrast other leaf-eating beetles eat the leaves partially or whole, often starting from the margins and feeding inwards towards the midvein. Typically leaf tissue eaten by beetles has ragged or sawtooth margins. Sometimes the midvein is left intact. Some beetles eat holes in a leaf, leaving most of the leaf tissue uneaten. Photos of beetle-chewed leaves are included in the prelude.

Beetles can feed in different ways to their larvae. For example *Paropsis* beetles often chew leaves inwards from the margins, whereas the larvae feed either by eating small areas from leaf margins or eating holes in the leaf. Swarming beetles, which feed in loose communal groups, eat large patches of foliage, moving from branch to branch or plant to plant. Sometimes the canopies of substantial trees can be defoliated by this type of communal feeding. Successive severe defoliation of the upper parts of trees can impact on their shape and lead to a condition known as 'broom topping', whereby the growths resemble the strands of a straw broom.

Notes: leaf-eating beetles are mainly a problem during spring and the summer months. The fact that many beetles can fly means that they are highly mobile and can move from plant to plant. It also means that outbreaks of beetle damage, which are often very sporadic, can appear suddenly and without warning. Some beetles are voracious and cause significant damage to plants within a short time of their appearance. They can also move on after only a short period of feeding, leaving few clues as to the cause of the damage. Healthy trees usually recover satisfactorily after beetle attacks, but persistent attacks can be quite debilitating and may weaken plants and cause abnormal branching. Attacks on seedlings and young plants can cause significant setbacks, including death.

Natural control: the beetles and larvae are eaten by a wide range of animals, including marsupials, rodents, birds, frogs, lizards, spiders, bugs and lacewings; predatory beetles also prey on other beetles and their larvae; parasites include many types of wasps, fungi, viruses and bacteria, and these affect both adults and larvae.

Other control: beetles can be difficult to kill because of their nomadic habits. Also, some species are known to be resistant to chemicals. Chemicals effective against them include sprays containing malathion and pyrethrum, but these materials should only be used after due consideration for the environment.

SYMPTOMS AND RECOGNITION FEATURES

Leaf margins chewed in a ragged manner..*beetles in general*
Leaf margins nibbled ....*eucalypt leaf beetles*
Leaves eaten completely or leaving the midribs only..*large swarming beetles*
Leaf surfaces heavily grazed, quickly taking on a brown scorched appearance..
....*small swarming beetles*, *fig leaf beetle*
Short irregular patches of surface tissue of eucalypt leaves grazed, turning white...........*eucalypt weevils*
Very soft new shoots and leaves of eucalypts and wattles eaten..*small weevils*
Small holes chewed in leaves to give a 'shot-hole' appearance...*flea beetles*
Small patches of bark grazed or rasped..*weevils*
Young palm fronds emerging already chewed...*palm leaf beetle*
Small circular brown areas eaten on elkhorn fern fronds...*elkhorn fern beetle*
Coils of white toothpaste-like wax on orchid plants ....*dendrobium beetle*
Patches grazed out of orchid leaves..*dendrobium beetle*
Surface of wattle leaflets grazed and turning brown ....*wattle blight*
Narrow strips grazed out of wattle phyllodes..*weevils*
Small oval to round domed beetles feeding on eucalypt or wattle leaves...................*eucalypt leaf beetles*
Sturdy beetles with antennae at the end of a snout ....*weevils*
Large colourful beetles that fly noisily..*christmas beetles, flower scarabs*
Small leaf-feeding beetles that hop when disturbed ....*flea beetles*

LEAF BEETLES (FAMILY CHRYSOMELIDAE)

A very large group of beetles consisting of about 33,000 species currently recognised worldwide, plus many more that await formal recognition. Australia has about 2,250 species. This family is not only large but also very complex, with 11 subfamilies recognised (10 of these occur in Australia), covering an extensive range of habitats and insect morphology. Most of the Australian species of Chrysomelidae feed on eucalypts and wattles.

CHRYSOMELIN LEAF BEETLES OR EUCALYPT LEAF BEETLES (FAMILY CHRYSOMELIDAE, SUBFAMILY CHRYSOMELINAE)

These beetles are very common in Australia, feeding mainly on eucalypts and wattles.

Description: eggs are laid in groups on young leaves or towards the tip of a shoot. The larvae are fleshy grubs that initially feed in groups, but may disperse as they get older. The grubs, which grow to about 1cm long, often have dark heads and pale bodies. They have three pairs of legs on the thorax but none on the abdomen. The beetles are small, commonly 0.5–1.2cm long, often oval and tortoise-shaped (paropsine beetles or tortoise leaf beetles) but sometimes more elongate and longer than wide. There are many species in this common group of beetles and they occur in a wide range of colours. The grubs are fleshy and grow to about 1cm long. Some are cylindrical in shape, others plump and nearly globular.

Climatic regions: various species found from tropical to temperate regions, coastal and inland.

Host plants: species of *Acacia*, *Angophora*, *Casuarina*, *Corymbia* and *Eucalyptus*; also numerous rainforest plants.

Feeding habits: adults and larvae often feed together in close proximity. When young, the larvae usually congregate in clusters and may be mistaken for sawflies (they often lift their rear end when disturbed). They also often jerk their tails in the air as they feed. As they grow they usually spread out and feed singly. The larvae feed on young leaves, sometimes eating them right to the midrib. They can defoliate whole sections of foliage leaving just the stems. Severe infestations can result in complete defoliation of the upper part of the canopy. Irregularly eaten leaves are often a signature of the beetles which usually nibble from the margins or eat large scallop-like chunks from the edge.

Notes: there are numerous species of chrysomelin beetles (several genera and some 700 named species in Australia). A distinctive group of nearly round, dome-shaped beetles, they are sometimes called eucalypt tortoise beetles because of their shape. When disturbed they attach themselves tightly to a leaf by gripping with specialised pads on their feet. They also withdraw their legs and antennae under their

A leaf beetle, Paropsis *sp.* [T. Wood]

Nearly mature leaf beetle larvae. [T. Wood]

body to make predation difficult. Some are quite colourful and attractive beetles. When disturbed they commonly fly or drop to the ground and play dead. The larvae of some species have special glands on their rear end that are extruded when disturbed, releasing fluids that smell strongly of eucalyptus oil. Others can apparently release toxins such as hydrocyanic acid.

Some species of native leaf beetles have become established overseas and cause damage far in excess of that usually found in Australia. For example, the beetle *Paropsis charybdis*, which is relatively uncommon and harmless in its native region of south-eastern Australia, has become a serious defoliator of several species of *Eucalyptus* in New Zealand since its introduction in the early 1900s. Its attacks have been so severe that they have prevented the successful establishment of wood fibre plantations.

Control: mostly these beetles, which add variety and interest to a garden, cause limited damage. If a nuisance they can be discouraged by hosing, use of sticks or hand picking and squashing. Spraying with malathion or pyrethrum should only be used as a last resort and with due consideration for the environment. The predatory shield bug *Oechalia schellenbergii* seems to specialise in feeding on the larvae of some of these beetles.

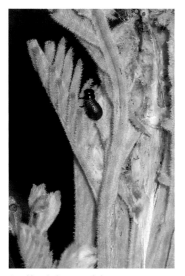

Leaf beetle larva and damage to new shoot of Acacia dealbata. [D. Jones]

Leaf beetle larvae feeding on Acacia phyllodes. [D. Jones]

Eucalypt leaf beetle larvae, Paropsis *sp.* [D. Jones]

A leaf beetle, Paropsisterna *sp.* [T. Wood]

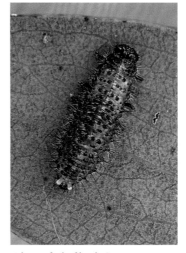

A larva of a leaf beetle, Paropsisterna *sp.* [T. Wood]

A leaf beetle, Paropsis *sp.* [T. Wood]

EXAMPLES OF CHRYSOMELIN LEAF BEETLES

- **Eucalyptus Leaf Beetle**, *Paropsis atomaria*, has pinkish eggs which are laid in rings on petioles and twigs. The grubs, which grow to about 0.5cm long, are yellowish green with thick black stripes. The tortoise-shaped beetles, which are about 1cm long, are light tan to pale brown with brown or yellow specks. This species, which is known to feed on a wide range of eucalypts, can be especially damaging to *Corymbia citriodora* subsp. *citriodora*, *C. maculata*, *Eucalyptus cloeziana*, *E. dunnii*, *E, grandis*, *E. saligna* and *E. tereticornis*.

Eucalyptus Leaf Beetle, Paropsis atomaria. [T. Wood]

Eucalyptus Leaf Beetle, Paropsis atomaria. [T. Wood]

Larvae of Eucalyptus Leaf Beetle, Paropsis atomaria. [T. Wood]

Larva of the Variable Leaf Beetle, Paropsis variolosa. [D. Jones]

- **Variable Leaf Beetle**, *Paropsis variolosa*, has cylindrical greenish eggs which are laid on young leaves. The grubs, which grow to about 0.8cm long, range from light brown to dark brown, becoming blackish as they mature. They have a black head and tail and are also adorned with white spots. The tortoise-shaped beetles, which are about 1cm long, range from orange-brown to dark brown with paler spotting. This species, which extends from the subtropics to temperate regions, feeds on a number of species of *Angophora*, *Corymbia* and *Eucalyptus*; also *Lophostemon confertus*.

- **Spotted Leaf Beetle**, *Paropsis maculata*, has cylindrical greenish eggs which are laid on twigs and young leaves. The grubs, which grow to about 0.6cm long, are pale green with a black head and blackish rear which is draw out into a tail-like extension. The tortoise-shaped beetles, which are about 0.8cm long, are reddish-brown with prominent golden-yellow spots and markings. This species feeds on eucalypts.

- **Red-spotted Leaf Beetle**, *Paropsis beata*, is a distinctive brown or blackish beetle about 1cm long with about six large spots on the wing-covers. The larvae grow to about 0.8cm long and are pale yellow and brown with long black hairs. This species, which is common in eastern Australia, was found in New Zealand in 2012.

- **Blue Gum Leaf Beetle**, *Cadmus excrementarius*, is a serious pest of blue gum plantations in south-west Western Australia. The beetles eat the young tips and juvenile leaves of seedlings and young trees, defoliating all or parts of the trees and altering normal growth and development. The beetles, which are 0.6–0.8cm long,

Red-spotted Leaf Beetle, Paropsisterna beata. [D. Jones]

are brownish-yellow with about six prominent black spots on the wing-covers. They swarm at times, causing significant damage in a short period of feeding. The larvae cause no damage to the trees but feed on fallen leaves. This pest, which is native to eastern Australia, was first recorded in Western Australia in 1904. It also feeds on *Eucalyptus marginata*, *E. rudis* and *Corymbia calophylla*. Several other species of *Cadmus* from the eastern states, including *C. litigiosus*, have also become established as significant pests of eucalypt plantations in Western Australia.

- **Silver Wattle Leaf Beetle**, *Calomela vittata*, is common on the east coast and its range extends to Tasmania. It feeds on pinnate-leaved wattles, including *Acacia dealbata*, *A. mearnsii* and *A. decurrens*, and also on species of *Eucalyptus*. The plump larvae often feed in the new shoots of the wattles, eating them from the inside before they expand. The adults are attractive green to yellowish beetles with a dark band on each wing-cover. The wing-covers are covered with jewel-like spots that sparkle in the sunlight.

Leaf beetle, Cadmus litigiosus. [T. WOOD]

Yellowish variant of Silver Wattle Leaf Beetle, Calomela vittata. [T. WOOD]

Silver Wattle Leaf Beetles, Calomela vittata, *feeding on* Acacia dealbata *(alongside a leafhopper nymph).* [D. JONES]

Silver Wattle Leaf Beetle, Calomela vittata. [T. WOOD]

- **Brown Paropsine Leaf Beetle**, *Paropsisterna variicollis*, is a species which is widely distributed from Queensland to Tasmania, feeding on the leaves of several species of *Eucalyptus*. It lays groups of pale yellow eggs on leaves and the larvae are creamy white with black heads. The beetles are pale brown to dark brown with light coloured dots on the wing-covers.

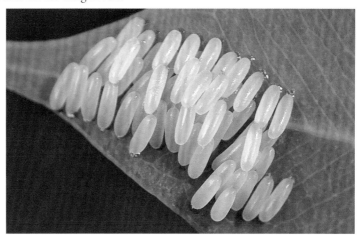

Female Brown Paropsine Leaf Beetle, Paropsisterna variicollis, with eggs. [T. Wood]

Eggs of Brown Paropsine Leaf Beetle, Paropsisterna variicollis. [T. Wood]

Newly hatched larvae of Brown Paropsine Leaf Beetle, Paropsisterna variicollis. [T. Wood]

First-instar larvae of Brown Paropsine Leaf Beetle, Paropsisterna variicollis. [T. Wood]

Second-instar of Brown Paropsine Leaf Beetle, Paropsisterna variicollis. [T. Wood]

Third-instar larvae of Brown Paropsine Leaf Beetle, Paropsisterna variicollis. [T. Wood]

- **Northern Leaf Beetle**, *Paropsisterna cloelia*, lays cylindrical cream to yellowish eggs in rows on young leaves. The grubs, which grow to about 0.7cm long, are cream to yellowish-green with a black head and narrow black stripes. The tortoise-shaped beetles, which are about 0.8cm long, are commonly yellowish-orange to reddish-brown, but can be black with an orange head and margins. This species feeds on eucalypts and can cause extensive damage to the crowns of eucalypts grown in plantations.

- **Shiny Leaf Beetle**,

    Paropsisterna liturata, is common in the ACT and surrounding areas. It feeds on eucalypts. The shiny beetles, which can be red or brown, are about 1cm long and have a black patch on each wing-cover.

Shiny Leaf Beetle, Paropsisterna liturata, *with paropsis larvae in background.*
[D. JONES]

- **White Leaf Beetle**, *Dicranosterna circe*, has cylindrical bluish eggs which are laid in rows on leaves. The grubs, which grow to about 0.7cm long, have a fattened abdomen. They are greyish to pale green with black cross-bands and black wart-like spots. The tortoise-shaped beetles, which are about 1cm long, are cream with a dark brown line between the wing-covers and three dark brown spots (one on each wing-cover and one in the middle). This species feeds on phyllodinous species of *Acacia*.

Shiny Leaf Beetle, Paropsisterna liturata.
[T. WOOD]

Fig Leaf Beetle, *Poneridia semipullata*

Description: eggs are laid in clusters on the underside of leaves. Very young grubs are yellow and look like clumps of sawdust on the leaf. Older larvae become black and fleshy. Widest at the head end and growing to about 1cm long, the grubs are hairy with a roughened surface. Pupation occurs in the soil. The beetle (family Chrysomelidae), which is about 1cm long, is dull-brown with a yellow head and thorax and a dark tip on the wing-covers. It also has a black spot on the thorax and prominent antennae.

Climatic regions: tropics, subtropics and temperate regions, chiefly coastal.

Host plants: species of *Ficus* including the cultivated fig, *Ficus carica*; Fig Leaf Beetle damage is often prominent on sandpaper figs.

Feeding habits: both beetles and larvae feed on the leaves. The larvae, which resemble sawfly larvae, usually feed in groups. They graze the leaf surface and leave white papery tissue behind. The edges of badly eaten leaves often curl upwards.

Larvae of Fig Leaf Beetle, Poneridia semipullata, *on* Ficus coronata. [R. ELLIOT]

Notes: a sporadic and persistent pest that is of periodic significance in summer rainfall regions. Badly affected plants look unsightly. Two other native species of *Poneridia*, *P. australis* and *P. ficus*, also feed on figs.

Control: hand-picking and squashing or by spraying with pyrethrum or malathion. The grubs are often attacked by a predatory grey bug about 1cm long.

Native Palm Beetles, species of *Anadastus, Hemipeplus, Plesispa*

Three small, slender, native beetles in the family Chrysomelidae feed on the leaves of palms, attacking natural populations, nursery plants and palms in gardens.

Description: *Hemipeplus australasicus* is an elongated beetle about 0.4cm long which is pale yellow with a reddish-brown spot on the thorax. *Anadastus* species is about 1cm long, orange and black, with conspicuous knobs on the end of the antennae. *Plesispa* species, which grows about 1cm long, has a brown head, yellow thorax and black wing-covers.

Climatic regions: tropics.

Host plants: palms.

Feeding habits: beetles chew the surface of young fronds, leaving sunken areas that turn brown.

Notes: these are relatively minor pests that can cause problems in some seasons, especially on young palms.

Control: as per introduction for leaf-eating beetles.

Palm Leaf Beetle or Coconut Leaf Beetle, *Brontispa longissima*

Description: flat brown eggs are laid in rows on unopened leaflets. The slender, plump, white larvae, which grow to about 0.8cm long, have a series of short, soft marginal spines and curved earwig-like pincers on the rear end. The beetles, about 1 cm long, are narrow and flat. They are black in the lower half with an orange head and orange in the upper half of the wing-covers and thorax.

Climatic regions: tropical and subtropical regions. This introduced pest from Indonesia and some Pacific Islands was originally known only from Darwin, Cooktown and the Torres Strait Islands. It has since been found in Broome and is now well established in north-eastern Queensland with an isolated outbreak on the Sunshine Coast in south-eastern Queensland.

Host plants: about 20 species of palms including several native species (*Archontophoenix alexandrae, Carpentaria acuminata, Caryota albertii, Livistona muelleri, Ptychosperma elegans, P. macarthuri* and *Wodyetia bifurcata*); the Coconut Palm, *Cocos nucifera*, is also badly affected.

Feeding habits: the beetles and larvae concentrate on emerging young leaves, particularly the spear leaf. They shelter within the folds of the leaf and chew the soft surface tissue. The beetles chew narrow strips, the larvae eat out larger areas. The damaged area curls, turns brown and, as the leaves expand, takes on a scorched appearance. The insects move onto the new unexpanded spear leaf (mature or expanded leaves are not touched).

Notes: this pest, first noticed in Darwin in 1979, has spread fairly quickly in northern Australia. It has had a significant impact on palms grown in tropical areas and probably also affects natural populations. Damaged leaves lose the ability to photosynthesise and prolonged attacks can result in all the leaves in a palm's crown being affected. These attacks weaken mature palms and severe persistent attacks can readily kill young palms.

Control: the beetles are weak flyers and most distribution of this pest is by movement of infested palms. Inspection of the youngest leaf should indicate the presence of the pest. Drenching young

Damage to young fronds of coconut palm by Palm Leaf Beetle, Brontispa longissima. [D. Jones]

Damage caused by Palm Leaf Beetle, Brontispa longissima. [D. Jones]

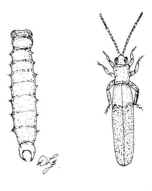

Palm Leaf Beetle, Brontispa longissima. [D. Jones]

leaves with a mixture of malathion and wetting agent will kill the insects but not the eggs. Repeated applications will be necessary as new leaves emerge, until the pests are eliminated. Control is not practical on large palms. Severely affected palms are best destroyed. An entomopathogenic fungus, *Metarhizium anisopliae*, known as the Green Muscadine Fungus, has been found to kill the beetles during the wet season in Darwin. A small parasitic wasp, *Tetrastichus brontispae*, has also been released to control the pest.

Clerodendrum Leaf Beetle, *Phyllocharis cyanicornis*

Description: grub-like larvae grow to 1cm long and feed communally in groups. The beetles, about 1.2cm long, are bright orange with nine large black spots on the thorax and wing-covers.

Climatic regions: tropics, subtropics and warm temperate regions, chiefly coastal.

Host plants: species of *Clerodendrum*, especially *C. floribundum* and *C. tomentosum.*

Feeding habits: the larvae graze the leaf surface. The beetles chew large irregular lumps out of the leaves. Beetles and larvae may feed together.

Notes: although usually of minor occurrence, this species can cause significant damage in some seasons, leaving eaten plants looking very tattered. The beetles, which shelter on the undersides of leaves during the day, quickly drop to the ground when threatened or disturbed.

Control: as per control for leaf-eating beetles. Spraying is rarely necessary.

Larva of Clerodendrum Leaf Beetle, Phyllocharis cyanicornis. [D. JONES]

Pittosporum Leaf Beetle, *Lamprolina aeneipennis*

Description: ugly, grub-like, spotted larvae grow to 1cm long, feed communally in groups and are covered in soft spines. The beetle has dark, shiny, blue-black metallic wing-covers and a yellow-orange thorax and head. Male beetles are about 1cm long, females about 1.5cm.

Climatic regions: tropics, subtropics and temperate regions, chiefly coastal.

Host plants: *Pittosporum* species, particularly *P. undulatum* and *P. venulosum*; possibly also species of *Auranticarpa* and *Bursaria.*

Feeding habits: the larvae graze the leaf surface or mine tunnels beneath the epidermis. The beetles

Damage caused by Pittosporum Leaf Beetle, Lamprolina aeneipennis. [D. JONES]

Pittosporum Leaf Beetle, Lamprolina aeneipennis. [D. JONES]

Pittosporum Leaf Beetle, Lamprolina aeneipennis. [D. JONES]

chew large irregular lumps out of the leaves. Beetles and larvae may feed together. The beetles quickly drop to the ground if disturbed.

Notes: although usually of minor occurrence, this species can cause significant damage in some seasons. A related beetle, *L. impressicolis*, in which the adults have similar colouration to those of Pittosporum Leaf Beetle, feeds on the leaves of *Bursaria* and *Citriobatus*.

Control: as per control for leaf-eating beetles. Spraying is rarely necessary.

Elkhorn Fern Beetle, *Halticorcus platyceri*

Description: eggs are laid on a true frond where it forks. Young larvae are yellow; mature larvae, which grow to 0.7cm long, are bright orange to red fleshy grubs that feed singly or in groups. The small, round, tortoise-shaped beetle, about 0.35cm long, is shiny bluish-black with four prominent dull-red spots on the wing-covers.

Climatic regions: tropics and subtropics, chiefly coastal.

Host plants: mainly the Elkhorn Fern, *Platycerium bifurcatum*; also *Asplenium australasicum*.

Feeding habits: the larvae tunnel into the tissue of the fronds, feeding on the internal tissues and forming mines. They often tunnel close to the veins. Affected areas become brown and crowded with frass. The beetles chew small round to oval areas in the surface of the true fronds towards the tips, feeding on the upper cell layers but leaving the lower surface intact. These areas become brown and leave a distinctive and easily recogniseable pattern of damage. Both types of fronds, the nest leaves and the true fronds, can be damaged. Sometimes the eaten areas overlap and the damaged areas coalesce. Severely eaten parts of the fronds may die.

Notes: a sporadic pest that can be very damaging in some seasons. The attacks, which occur in autumn and early summer, are worse on plants that have been neglected, especially drought-stricken or underwatered plants. The beetles jump or drop readily when disturbed.

Control: as per introduction for leaf-eating beetles. On some occasions a regular spray with pyrethrum or malathion may be necessary to achieve control. The larvae are difficult to kill because they are protected within the leaf tissue.

Elkhorn Fern Beetle, Halticorcus platyceri, *and damage to Elkhorn Fern frond.* [D. Jones]

Vitex Leaf Beetle or Spotted Leaf Beetle, *Chalcolampra 18-guttata*

Description: the larvae are small grubs that graze on the leaf surface. The beetle, 0.7–0.8cm long, is mostly black with 18 prominent white spots on the wing-covers; the thorax and head are orange-red. The beetle has prominent, beaded antennae.

Climatic regions: tropics and subtropics, mainly coastal.

Host plants: species and cultivars of *Vitex ovata* and *V. trifolia*; occasionally observed on *Clerodendrum floribundum*.

Feeding habits: the beetles chew large lumps out of the younger leaves and also the bark of young stems.

Notes: this handsome beetle can cause significant damage to *Vitex* plants in some years. Young plants are susceptible to persistent attacks which can impede their establishment. Badly-chewed plants look very unsightly. The beetles drop to the ground when threatened. The unusual scientific name refers to the 18 spots on the beetle's wing-covers.

Control: as per introduction for leaf-eating beetles. Spraying young plants may be necessary until they become established.

Vitex Leaf Beetle. [D. Jones]

Lily Leaf Beetle or Bowenia Beetle, *Lilioceris nigripes*

Description: the larvae are small grubs that chew on young leaves. The beetle, 0.7–0.8cm long, is reddish-yellow to reddish-brown with black legs. The beetle has prominent black beaded antennae.
Climatic regions: tropics and subtropics, mainly coastal.

Bowenia Beetles, Lilioceris nigripes, *feeding on young* Bowenia *leaflets.* [G. WILSON]

Larvae of Bowenia Beetle on young leaflets of Bowenia. [G. WILSON]

Host plants: bulbs and other monocotyledons; also cycads, especially species of *Bowenia*.
Feeding habits: the beetles chew large lumps out of the younger leaves.
Notes: this beetle can cause significant damage to some bulbs and *Bowenia* plants in gardens and in the wild.
Control: as per introduction for leaf-eating beetles. Spraying young growth may be necessary to minimise damage.

Iridescent Leaf Beetle, *Edusa glabra*

Description: the larvae are small grubs that feed in groups, grazing the leaf surface. The beetles, 0.4–0.5cm long, are iridescent in blackish-green or blackish-purple tones.
Climatic regions: tropics and subtropics, mainly coastal.
Host plants: many species of rainforest plants, including species of *Alectryon*, *Cryptocarya*, *Cupaniopsis*, *Endiandra*, *Mallotus*, *Psychotria* and *Randia*.
Feeding habits: the beetles, which also feed in groups, chew large lumps out of the margins of younger leaves; they also feed on new shoots.
Notes: although small, this beetle can cause significant damage in a short period. Because the attacks concentrate on new growth, the damage is visible for many months and badly-chewed plants look unsightly. The beetles fly when threatened.
Control: as per introduction for leaf-eating beetles.

Wattle Blight, Pyrgoides orphana, *on* Acacia baileyana. [D. BEARDSELL]

Wattle Blight or Fire Blight Beetle, *Pyrgoides orphana*

Description: The elongated beetles, 0.6–0.8cm long, are pale green with cream and red-brown to brown stripes on the wing-covers. The grubs grow to a similar length and are stout, greenish and taper to a point at the tail.
Climatic regions: temperate regions.
Host plants: fern-leaved wattles such as *Acacia dealbata*, *A. decurrens* and *A. mearnsii*.
Feeding habits: the beetles and larvae feed together in groups. They graze the surface of the leaflets and the damaged tissue turns brown. The larvae, which cling closely to the stems, are well camouflaged.
Notes: a serious pest that was of particular importance in the early 1900s when plantations of wattles were established to provide bark for the tanning industry. Affected trees look as if they have been scorched. Whole trees and colonies of wattles may be attacked and under suitable conditions the insects increase in numbers very rapidly. Severe infestations can defoliate large trees and repeated attacks can cause tree death.
Control: as per introduction for leaf-eating beetles. Spraying is often necessary.

ORCHID BEETLE OR DENDROBIUM BEETLE
(FAMILY CHRYSOMELIDAE, SUBFAMILY CRIOCERINAE)

This beetle, *Stethopachys formosa*, is a serious pest for orchid growers.

Description: cylindrical green eggs about 0.15cm long are laid singly on new growth and on the base of young leaves. After hatching the greenish larvae bore into stems and new growths. Mature larvae are slug-like orange grubs about 1cm long. Pupation occurs in a waxy cocoon surrounded by strands and coils of white toothpaste-like wax. The beetle, which is about 1.2cm long, is shiny and orange with four prominent large black spots on the wing-covers and prominent antennae.

Climatic regions: tropics, subtropics and warm temperate regions, chiefly coastal. This pest can also become established in glasshouses in temperate areas.

Host plants: epiphytic orchids in general but particularly species in the *Dendrobium* alliance; often very damaging on *D. speciosum* and its relatives and hybrids; also *Cymbidium canaliculatum* and *C. suave*. Orchid beetles cause damage to naturally occurring native orchids as well as those grown in gardens, verandahs, shadehouses and glasshouses.

Feeding habits: larvae tunnel in new shoots and graze on young leaves. They can also feed on flowers and young fruit. Their major impact

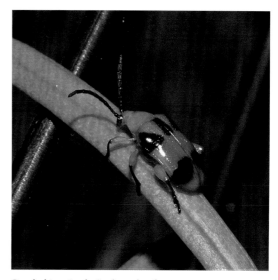

Dendrobium Beetle, Stethopachys formosa. [A. Stephenson]

Dendrobium Beetle cocoons on stem of Cymbidium suave.
[D. Jones]

Dendrobium Beetle damage to Dendrobium speciosum *leaf.*
[D. Jones]

follows their tunnelling in young pseudobulbs where they eat the internal tissue, often feeding in groups within the pseudobulb. Affected pseudobulbs often die back from the top, losing all the apical buds, Very young shots commonly become watery and rot. The beetles graze on most parts of the orchid plant, including new shoots, leaves, buds, flowers and seed pods. They excavate irregular patches of tissue, leaving the veins which initially turn white, then become brown and unsightly.

Notes: a persistent, serious and annoying pest for orchid-growers. The beetles are present all year in the tropics and several generations occur in a season. Further south the beetles are most active during the summer months, although some beetles survive winter, especially mild winters. The beetle drops to the ground at the slightest sign of disturbance or threat. Newly emerged beetles are pale yellow-orange and fly readily compared with older, darker orange beetles that are reluctant to fly, but drop quickly when disturbed.

Control: regular observation, hand picking (hold a container underneath the beetle) and squashing are useful for eliminating the beetles. Sprays containing pyrethrum and malathion will kill the beetles, but the larvae are almost impossible to control in their tunnels.

FLEA LEAF BEETLES OR SHOT-HOLE BEETLES (FAMILY CHRYSOMELIDAE, SUBFAMILY GALERUCINAE, TRIBE ALTICINI)

There are about 450 species in this group of beetles, feeding on a wide range of plants. A distinctive group jump when disturbed.

Description: eggs are laid on the host plant or in the soil. The larvae feed in a variety of ways, including stem borers, leaf miners and root chewers. The beetles, which are mostly small, shiny and metallically coloured, commonly feed in groups and can jump actively when threatened. Most species have stoutly developed hindlegs to aid in their jumping, although they can walk normally when not threatened.

Climatic regions: various species found from tropical to temperate regions, coastal and inland.

Host plants: a large range of plants, particularly shrubs and herbaceous species, but also eucalypts and *Acacia*. These beetles can be damaging to many plants from inland areas, including species of *Eremophila (E. altemifolia, E. denticulata* and *E. glabra)*; also numerous rainforest plants.

Feeding habits: the beetles feed on leaves, characteristically chewing a series of small holes in the upper surface of the leaf or completely through the leaf. Damaged leaves look as if they have been blasted by shot from a shotgun. With continued feeding the eaten areas often overlap or coalesce.

Notes: these beetles, of which there are numerous Australian species, were previously placed in their own subfamily (Alticinae).

Control: as per introduction for leaf-eating beetles. Hosing, disturbing with sticks or squashing helps to move them on. Spraying with malathion or pyrethrum should only be used as a last resort and with due consideration for the environment.

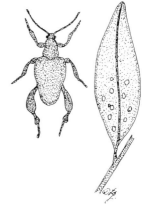

Flea leaf beetle. [D. Jones]

Flea leaf beetle damage to leaves of Correa alba. [D. JONES]

Flea leaf beetle damage to leaf of Alyogyne. [D. JONES]

Flea leaf beetle damage to leaves of Lagunaria patersonii. [D. JONES]

EXAMPLE

- **Small Blue Leaf Beetle**, *Nisotra breweri*, has larvae which feed on roots or bore in stems. The beetles, which are about 0.5cm long, are dark metallic blue with an orange thorax and head. They feed singly or in small groups and jump actively when disturbed. Occurring in the subtropics and temperate regions, this beetle feeds on plants in the family Solanaceae (particularly *Solanum*) and Malvaceae (*Hibiscus* and *Lagunaria*).

SMALL SWARMING CHRYSOMELID BEETLES

In tropical and subtropical regions there are a few species of small beetles which congregate together in swarms and feed sporadically, moving rapidly onto another tree when disturbed or feeding is completed. These beetles are very destructive, usually congregating on young shoots and eating the surface of the leaves, leaving a lacy, ragged appearance. They also attack green bark. Within a few hours the damaged tissue browns or blackens, giving the plants a blasted or scorched appearance. Attacks are most prevalent during the wet season from December to March. Sometimes the beetles shelter by day on the underside of leaves or in curled and damaged leaves.

Natural control: natural enemies include lizards, frogs, spiders and assassin bugs.

Other control: as the beetles are easily disturbed the swarms can be dispersed by beating with sticks or hosing. Most attacks are over by the time the damage is noticed and chemical sprays should not be applied unless the beetles are still active. Chemicals used in control are pyrethrum-based contact sprays and sprays containing malathion. Because they are highly mobile these pests can be very difficult to control.

EXAMPLES

Monolepta Beetle or Red-shouldered Leaf Beetle, family Chrysomelidae, subfamily Galerucinae

This beetle, *Monolepta australis*, which appears sporadically in large swarms, can be very damaging.
Description: eggs are laid singly or in small groups just beneath the soil surface. The small white larvae feed on the roots of grasses, clover and other plants. Pupation occurs in the soil, the beetles emerging following heavy rain ending a dry spell. The beetles are plump, about 0.6cm long and yellow or dull-orange with a bright red band at the top of the wing-covers and two red spots towards the base. One to three generations occur per season.
Climatic regions: tropics, subtropics and warm temperate regions, mainly coastal.

Host plants: a wide range of native and exotic garden plants including *Ceratopetalum gummiferum*, *Corymbia spp.*, particularly *C. torelliana* (a favourite host), *Eucalyptus* spp., *Ficus* spp., *Leptospermum obovatum*, *L. petersonii*, *Macadamia* spp. and *Melaleuca decora*; also avocados, lychees, peaches, nectarines, mangos.

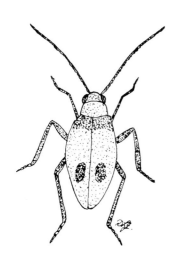

Monolepta australis, *known as Monolepta Beetle or Red-shouldered Leaf Beetle.* [D. Jones]

Feeding habits: beetles gather in swarms, congregating on the leaves and shoots and feeding voraciously; they also eat buds and flowers. They graze the leaf surface leaving most of the veins intact. When disturbed they fly or drop to the ground.

Notes: a major pest that often gathers in huge swarms that move regularly. A swarm can appear from nowhere and devastate garden plants within a short time. Attacks are usually short-lived with the swarms moving on within a few hours or days. Swarms commonly occur in spring and late summer/autumn, but can appear at other times. Severe attacks on small plants may result in complete defoliation and death.

Control: as per introduction for leaf-eating beetles. Spraying is often necessary, although this pest is resistant to many chemicals. Yellow sticky traps attract the beetles and can be useful for early detection. The beetles are also attracted to white and can be drowned in white containers filled with water. *Corymbia torelliana* is also sometimes planted as a windbreak and to provide an early warning system for the presence of the beetles.

Swarming Rhyparida Beetles, family Chrysomelidae, subfamily Eumolpinae

There are about 111 species of *Rhyparida* in Australia, some of which congregate in swarms and damage plants.

Description: eggs are laid in the soil and the small white grubs feed on roots – often those of grasses and other pasture plants. The beetles, 0.3–0.5cm long, are mostly orange-brown to brown or black, plain or with darker markings, sometimes shiny.

Climatic regions: mainly coastal areas of the tropics and subtropics; a few species occur in temperate regions.

Host plants: species of *Acacia, Corymbia and Eucalyptus*; also some rainforest trees and many garden plants and fruit trees including avocado, hibiscus, lychee, mango and mangosteen; also sugarcane.

Feeding habits: swarms of beetles descend on a tree eating young and recently mature leaves. Most feeding occurs after dark, the insects hiding on the undersides of the leaves during the day. Terminal parts of stems and branches become leafless and appear as if skeletonised or scorched. Stem dieback and growth retardation can follow such attacks.

Notes: a group of beetles that appear sporadically, swarming in large numbers soon after the first heavy rains of the summer. About 16 problem species occur in Australia, mostly in the tropics. The swarms are often localised with the worst damage occurring in tropical coastal areas, especially on sites adjacent to sugarcane or pasture. Attacks can be short-lasting but the damage incurred can be severe and debilitating.

Control: as per introduction for leaf-eating beetles. Despite often occurring in large numbers, the beetles hide during the day and can be difficult to see. Shaking branches may reveal their presence. These beetles are very fond of *Corymbia torelliana,* which is often planted as a windbreak. Regular monitoring of the windbreak can provide an early warning system for the presence of the beetles.

JEWEL BEETLES (FAMILY BUPRESTIDAE)

These beetles, which are generally elongated and often brightly coloured, are an interesting addition to gardens but their larvae damage native plants. The beetles are generally found feeding in flowers, whereas the larvae are usually borers. Jewel beetles are treated in Chapter 13.

SCARAB BEETLES (FAMILY SCARABAEIDAE)

A very large group of beetles consisting of more than 30,000 species currently recognised worldwide plus many more that await formal recognition. Australia has more than 2,200 species. This family of beetles is extremely diverse and difficult to characterise. Their distinctive antennae, which consist of three or four flattened leaf-like segments, is the most reliable feature for recognition. Many of these beetles feed on plants in some part of their life cycle. Some beetles do not feed at all.

CHRISTMAS BEETLES
(FAMILY SCARABAEIDAE, SUBFAMILY RUTELINAE)

Christmas beetles, which are large beetles of the Scarab family, get their name from their abundance at or close to Christmas time. In all there are 36 species of Christmas beetle native to Australia, most in the genus *Anoplognathus*. Growing 2–3.5cm long, the beetles are clumsy and noisy fliers. The common species have a shiny appearance and are buff or a pale-brown colour, whereas, others may have a metallic appearance. The beetles feed and fly during the day and are also attracted to lights at night, often gathering in large swarms. The larvae live in the soil for about 12 months.

Christmas beetles prefer eucalypt trees growing among pasture and may swarm in huge numbers on relict trees left after clearing, often resulting in complete defoliation. Attacks usually occur with devastating rapidity. Because of their voracious appetite and attraction for trees left in paddocks, they have been implicated in eucalypt decline. Even healthy trees can be weakened by sustained attacks. Christmas beetles also attack eucalypt plantations, sometimes feeding in populations of mixed species. Repeated attacks on young trees can inhibit normal growth and development and may also result in tree death. They are often accompanied in their feeding activities by larvae and beetles of *Paropsis* beetles.

Climatic regions: tropics and temperate regions, coastal and inland.

Host plants: their main diet consists of both young and mature eucalypt leaves, although sometimes they also eat the foliage of related plants such as bloodwoods, melaleucas and leptospermums. Some species of *Corymbia* and *Eucalyptus* seem to be more attractive to them than others. Favoured species include *Corymbia citriodora*, *C. maculata*, *Eucalyptus argophloia*, *E. cloeziana*, *E. dunnii*, *E. grandis* and *E. tereticornis*. Their larvae feed on grass roots and possibly the fine roots of other plants.

Feeding habits: beetles cluster on shoots, eating irregular strips and ragged patches from the leaves. They usually begin feeding at the margins, often towards the leaf tip. Large populations result in the leaves being devoured right down to the midribs; sometimes even the midribs and petioles are eaten. Mature leaves as well as young and recently expanded leaves are all consumed. Usually the area beneath their feeding site is littered with their faecal pellets, and these can often be heard falling through the foliage and onto the ground.

Notes: it has been suggested that these handsome and familiar beetles may be declining in number. Certainly they are more common in some years than others, but anecdotal evidence suggests that they may be declining in some of our larger cities due to habitat loss. For example, it has been reported huge numbers drowned in Sydney Harbour in the 1920s and the numbers of beetles were so prodigious that tree limbs were weighed down by their mass.

Christmas beetle, possibly Anoplognathus rugosus. [T. Wood]

Christmas beetle, Anoplognathus porosus. [D. Jones]

Christmas beetle, Anoplagnathus viriditarsis. [D. Jones]

Christmas beetle. [D. Jones]

Christmas beetle. [T. Wood]

Christmas beetle. [T. Wood]

Natural control: large birds such as boobook owls, nightjars, frogmouths, currawongs, drongos and magpies feed on Christmas beetles.

Control: as per introduction for leaf-eating beetles. At low population levels these beetles are rarely troublesome in gardens and in fact make an interesting and decorative addition. However, control measures may be necessary if they reach plague proportions. Swarms can be disrupted by beating with sticks or hosing with jets of water. When disturbed during the day the beetles usually drop straight to the ground and hide in the grass. Individuals can be collected and destroyed if plastic sheets are spread under infested trees before disturbance. Individuals can be squashed or dropped in tins containing a mixture of water and kerosene. The beetles can also be collected when swarming around lights at night. Spraying is not recommended as the beetles move when disturbed.

EXAMPLES OF CHRISTMAS BEETLES

- **Brown Christmas Beetle or Golden Christmas Beetle**, *Anoplognathus chloropyrus*, is a common species from subtropical and temperate regions. It grows to about 2cm long and is orange-brown to yellow in colour with smooth or slightly roughened wing-covers.

- **Bronze Christmas Beetle**, *Anoplognathus punctulatus*, is a sturdy brownish to bronze beetle with smooth wing-covers that is common in south-eastern Australia. The beetles are about 2cm long.

- **Washerwoman or Common Christmas Beetle**, *Anoplognathus porosus*, from subtropical and temperate regions, is mostly creamy orange with a darker brownish thorax. The wing-covers have faint lines irregularly adorned with short dark grooves and small sunken pits.

- **King Beetle**, *Anoplognathus viridiaeneus*, is an uncommon species from temperate regions which grows to about 3cm long. The smooth, highly glossy beetles are greenish yellow or brownish yellow.

- **Iridescent Christmas Beetle**, *Anoplognathus montanus*, from the subtropics grows about 2.5cm long. The shiny smooth beetles have a bronze-coloured thorax and head and cream-brown wing-covers.

- **Glossy Christmas Beetle**, *Anoplognathus viriditarsus*, from the subtropics is bronze-coloured with an iridescent green sheen.

- **Green Christmas Beetle**, *Anoplognathus punctulatus*, is an attractive bright green beetle that grows about 2.5cm long and is found in the coastal forests of eastern Queensland and northern New South Wales.

- **Pitted Christmas Beetle**, *Anoplognathus boisduvali*, from subtropical and temperate regions is a brownish beetle about 2.5cm long with rows of brown to black pits on the wing-covers.

SWARMING SCARAB BEETLES
(FAMILY SCARABAEIDAE, SUBFAMILY MELOLONTHINAE)

These small beetles gather in swarms, sometimes causing serious damage.

Description: several species of small beetle, 0.5–1.5cm long, in three or four genera.

Climatic regions: tropics, subtropics and temperate regions, coastal and inland.

Host plants: mainly species of *Acacia, Angophora, Corymbia* and *Eucalyptus*, but a wide range of plants can be hosts.

Feeding habits: active beetles which swarm over trees, stripping bare large areas of foliage. They fly during the day or night and, when feeding, usually concentrate on the younger leaves and growing tips.

Notes: a group of pests that can cause considerable damage to young trees. Large swarms can defoliate small to moderately sized trees in a short period of feeding. They can be very debilitating in newly established eucalypt plantations where repeated swarming can result in tree death. Damage is often also severe on young trees regenerating in paddocks, especially those adjacent to larger, established trees. Often swarms consist of more than one species of beetle.

Control: as per introduction for leaf-eating beetles.

EXAMPLES

- **Spring Beetles or Liparetrus Beetles**, about 12 *Liparetrus* species. These are significant pests of young eucalypts. Eggs are laid in the soil and the whitish larvae feed on fine roots, pupating in the soil. After emergence from the pupa, the beetles await rain to soften the soil, emerging en masse, usually in spring. They fly in swarms during the day, attacking eucalypt seedlings and trees up to three years old, eating buds, leaves and young stems. They also feed on coppice growth. The stout, hairy beetles, which range in length from 0.5cm to 1.2cm, can be recognised by their abbreviated wing-covers which fall well short of covering their bulging abdomen. They fly readily when disturbed and also burrow into the soil in areas where they are feeding.

Spring beetles, Liparetrus *sp., feeding on* Eucalyptus nichollii. [D. JONES]

- **Heteronyx Beetles**, several *Heteronyx* species. Eggs are laid in the soil and the whitish larvae – cockchafers with an orange-brown head – feed on roots, pupating in the soil. The larvae of at least one species, *H. elongatus*, can kill eucalypt seedlings by eating their roots. After emergence from the pupa, the beetles await rain to soften the soil, emerging at night to feed (the beetles of some species do not feed at all), usually in spring and summer. They feed on young growth, sometimes defoliating the upper canopy of moderately large trees. Some species also eat juvenile foliage. The beetles return to the soil to hide during the day, and holes around the trees are an indication of the presence of these pests. The stout, clumsy brown or black beetles, which range in length from 0.5cm to 3cm, can be recognised by their long, uniformly coloured, pitted wing-covers (with no grooves or lines) which cover most of their abdomen (with usually only the tip protruding). Swarming often occurs on warm summer evenings.

- **Metallic Beetles or Diphucephala Beetles**, about 69 *Diphucephala* species. Native species distributed from the tropics to Tasmania and over much of the inland. Eggs are laid in the soil and the larvae feed on roots and organic matter. Pupation occurs in the soil and the beetles emerge from rain-softened soil. The small beetles, 0.5–1cm long, have attractive iridescent or metallic tones. They often gather in groups to feed, sometimes in swarms. They feed during the day, eating the leaves of a wide range of plants, not only in open habitats but also in wet forests and rainforest. They also feed on pollen and can often be seen on wattle flowers. One metallic green species, *D. colaspidoides*, from south-eastern Australia including Tasmania, is known to cause significant damage to species of *Acacia*, *Eucalyptus* and *Nothofagus cunninghamii*. Another metallic green species appears in large numbers in summer in south-eastern Queensland, feeding on rainforest plants, especially *Alphitonia excelsa*.

- **Automolus Beetles**, *Automolus* species. These small, hairy, brown beetles are 0.4–0.5cm long and feed on eucalypts during the day. The wing-covers extend over most of the body. The beetles eat young and recently mature leaves and can cause extensive damage to plantation trees, including *Corymbia citriodora* subsp. *variegata*, *Eucalyptus cloeziana*, *E. dunnii* and *E. grandis*.

- **Tropical Swarmer**, *Epholcis bilobiceps*, is a swarming scarab beetle from north-eastern Queensland. The larvae probably feed on grass roots but the beetles, which feed at night, eat the foliage of eucalypts. They are most active during and soon after the wet season and cause significant damage to plantation trees, including *Eucalyptus grandis* and *E. pellita*. Even mature trees can be defoliated by large swarms of this beetle.

BROWN EUCALYPT BEETLES OR LOUSY BEETLES (*LEPIDIOTA* SPECIES)

Description: larvae are plump white grubs that feed on the roots of grasses. The beetles are stout, dark brown to brown and about 2.5cm long, with conspicuously hairy legs.

Climatic regions: tropics.

Host plants: species of *Eucalyptus.*

Feeding habits: the beetles congregate on young shoots and recently matured leaves, eating them to the midrib. Periodically they swarm in large numbers and can strip trees of their foliage.

Notes: The activities of this beetle closely parallel those of Christmas beetles (*Anoplagnathus* species). The adult beetles feed on a range of eucalypts while the larvae are curl grubs which feed on roots in the soil. The beetles are active just before the wet season and fly in late afternoon making a loud buzzing noise with their wings.

The larvae of this nectar scarab beetle, Phyllotocus *sp., are cockchafer grubs.*
[T. Wood]

Control: as the beetles will not fly after dark they can be caught by shaking or knocking them out of the tree onto sheets of plastic and squashing them or tipping them into a tin of water topped with kerosene.

COCKCHAFER GRUBS OR CURL GRUBS (FAMILY SCARABAEIDAE, SUBFAMILY MELOLONTHINAE)

These soil-living grubs are the larvae of chafer beetles. They feed on the roots of a wide range of plants and are often found by gardeners when digging in soil. They are treated in chapter 14.

FLOWER CHAFERS (FAMILY SCARABAEIDAE, SUBFAMILY CETONIINAE)

These very common beetles feed among flowers of the Myrtaceae family. They are active during the day and fly quickly and noisily when disturbed. They feed actively on nectar and pollen and can often be seen feeding in the flowers of eucalypts and angophoras. They are sometimes thought of as pests. Any damage to the flowers during feeding is greatly offset by their contribution as pollinators. Their larvae eat decaying leaf litter and other organic matter. Some large colourful common species are attractive and interesting garden visitors.

Description: stout, active, fairly broad beetles that grow 1.5–2cm long. Colours range from brown to yellow, often with colourful bands or spots.

Climatic regions: tropics, subtropics and temperate regions, coastal and inland.

Host plants: mainly flowering Myrtaceae.

Feeding habits: they eat nectar and pollen, often causing minor damage to the flowers.

Notes: the beetles fly noisily during the day. If disturbed while feeding they fly freely or sometimes drop to the ground and pretend to be dead.

Control: enjoy their activities, noise and colours.

Banded Flower Chafer, Chondropyga dorsalis. [T. Wood]

Brown Flower Chafer, Aphanestes gymnopleura. [T. Wood]

Spotted Flower Chafer, Neorrhina punctatum, *feeding in* Melaleuca huegelii. [D. Jones]

Fiddler Beetle, Eupoecila australasiae. [T. Wood]

WEEVILS, SNOUT BEETLES (FAMILY CURCULIONIDAE)

There are about 40,000 species of weevils in the world with about 6,500 species known in Australia. All weevils have an elongated snout (often resembling a trunk), known scientifically as a rostrum, with the jaws at the apex and the distinctive 'elbowed' and clubbed antennae extended on either side. The eyes, situated at the base of the snout, are often large and prominent. Weevils, which range from tiny insects to large stout insects, are often slow moving. Often the wing-covers are hardened and covered with bristles and/ or knobbly projections. Some species can fly, others are wingless. The larvae, which range from maggot-like grubs to fleshy grubs, feed on a wide range of materials; many are legless. The beetles and larvae of numerous weevils feed on plant organs, often on different parts, with the larvae of many species tunnelling in roots (see also Root Feeders, chapter 14) and stems while the adult weevils feed on foliage, flowers or fruit.

Damage to young growth of Eucalyptus parvula *caused by Snout Weevils.*
[D. Jones]

EXAMPLES

Botany Bay Weevil or Diamond Weevil, *Chrysolopus spectabilis*

Description: eggs are laid in bark near the base of a wattle tree. The larvae are fleshy white grubs that tunnel in the wood. The weevils are black with patches of pale green to blue-green scales on the wing-covers and larger colourful areas on the head, legs and underside of the body. This is a sturdy insect, about 2.5cm long with roughened surface and a long snout (rostrum).
Climatic regions: subtropics and temperate regions of eastern Australia, mainly coastal.
Host plants: fern-leaved and phyllodinous wattles. At least 28 species are known host plants of this

Botany Bay Weevil, Chrysolopus spectabilis. [T. Wood]

Botany Bay Weevils, Chrysolopus spectabilis*, preparing to mate.* [T. Wood]

weevil, including *Acacia aulacocarpa, A. concurrens, A. leiocalyx, A. longifolia, A. mearnsii, A. pycnantha* and *A. sophorae.*

Feeding habits: the weevils graze narrow strips from the surface of wattle phyllodes. The larvae feed as borers in wood and roots.

Notes: this common, attractive-looking weevil was discovered during the first voyage by James Cook and was the first Australian insect to be formally named and described. The weevils drop to the ground when disturbed and play dead.

Control: usually only occurs in very low numbers. Although the damage done by the adult weevils is usually minor, the larvae can cause considerable damage (even death) to wattles and other plants. Sometimes severe outbreaks occur, causing damage to plants over a large area. The beetle is an iconic insect to be enjoyed rather than destroyed.

Elephant Weevil, *Orthorhinus cylindrirostris*

Description: eggs are laid on the host plant. The larvae are stout fleshy white grubs that tunnel in the wood. The weevils, which range from 1.2–2cm long (females are smaller than males), are sturdy with a very rough, uneven (knobbly) surface on the wing-covers, long forelegs, prominent black eyes and a very long trunk-like snout (rostrum). Colours are brown to dark brown with small black warts and patches of grey and white scales.

Climatic regions: tropics, subtropics and temperate regions, coastal and inland.

Host plants: species of *Eucalyptus* are the primary hosts of this insect (less commonly *Lophostemon confertus* and *Castanospermum australe*), but it has spread its boundaries and is now a significant pest of citrus trees and grape vines.

Feeding habits: the weevils eat young growth, soft new leaves and buds. They can also eat young green bark, sometimes ringbarking a shoot. The larvae feed on wood and roots.

Notes: the weevils either drop to the ground when disturbed and play dead or fly away.

Control: as per control for leaf-eating beetles. Spraying is often necessary to control their activities in orchards and vineyards.

Fuller's Rose Weevil, *Asynonychus cervinus*

Description: yellow eggs are laid in clusters in a fold or crevice. The larvae, which grow to about 0.9cm long, are hairy legless white grubs. The weevils, which are about 1cm long, are grey to light brown with slightly roughened wing-covers.

Climatic regions: subtropics and temperate regions, coastal and inland.

Host plants: an extensive range of garden plants, including natives: also crops, vegetables, succulents, bulbs, ferns and orchids.

Feeding habits: the weevils feed on leaf margins, leaving a series of scoop-like indentations. Badly affected leaves become very ragged. The larvae feed in the soil on the fine roots of a wide range of plants.

Notes: an introduced species from South America that, once established, is a persistent pest.

Control: as per control for leaf-eating beetles. Spraying may be necessary.

Garden Weevil, *Phlyctinus callosus*

Description: eggs are laid in soil litter. The larvae, which grow to about 0.6cm long, are hairy legless white grubs with an orange head and black jaws. The weevils, which are about 0.7cm long, are grey brown with a broadly V-shaped, faintish white stripe towards the end of the wing-covers. The surface of the body is unevenly bumpy, each bump containing bristles.

Climatic regions: subtropics and temperate regions, coastal and inland.

Host plants: a huge range of garden plants, crops, vegetables, succulents, bulbs, ferns and orchids; also a pest of grape vines and apple trees.

Feeding habits: the weevils either graze the surface of soft tissue such as buds, fruit and flowers, leaving narrow damaged areas, or eat characteristic saucer-shaped depressions in leaves and soft green stems. Initially green, these eaten areas turn brown and unsightly. The larvae feed in the soil on the roots and tubers of a wide range of plants. They can also ringbark seedlings.

Notes: an introduced species that is native to South Africa, where it is known as the Banded Fruit Weevil. It is a persistent pest and its activities can significantly impact on the appearance of prized plants. The weevils drop to the ground when disturbed and play dead.

Control: as per control for leaf-eating beetles. Spraying may be necessary.

Garden Weevils, Phlyctinus callosus, *feeding on* Crinum *buds.* [D. Jones]

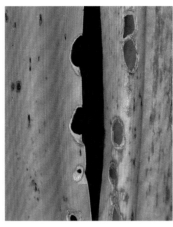

Damage to Crinum *leaves by Garden Weevil,* Phlyctinus callosus. [D. Jones]

Melaleuca Leaf Weevil or Melaleuca Snout Beetle, *Oxyops vitiosa*

Description: cylindrical yellow eggs are laid singly or in small groups on shoot tips and covered with a secretion that dries hard. The small (about 0.15cm long) greyish to blackish slug-like larvae feed on leaves. Pupation occurs in the ground. The grey and brown marbled weevils, which range from 0.7–0.9cm long, have a few knobbly bumps on the wing-covers, shiny black eyes, a trunk-like nose and prominent antennae.

Climatic regions: tropics.

Host plants: broad-leaved *Melaleuca* species, particularly *M. quinquenervia.*

Feeding habits: both the weevils and larvae feed on leaves and shoots. The weevils eat holes in the leaves and buds. They also feed on young shots, grazing green bark, ringbarking the shoot or biting completely through the stem. The larvae graze irregular strips from the surface tissue, often chewing right through to the epidermis on the other side of the leaf. The strips initially turn white but later become brown. The larvae trail a black faecal string behind them as they feed, and as they grow they become covered in an oily secretion that becomes black and shiny. Initial feeding is in narrow patches but in bad infestations most of the leaf surface is destroyed

Notes: this is a relatively minor pest that causes limited damage in Australia. It has been introduced successfully into Florida to help control naturalised populations of *Melaleuca quinquenervia.* In some sites its feeding has reduced flowering by 80 per cent or more, thus reducing recruitment.

Control: as per control for leaf-eating beetles.

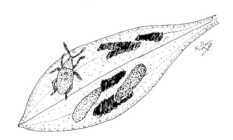

Melaleuca Leaf Weevil. [D. Jones]

Eucalyptus Weevils or Eucalyptus Snout Beetles, several *Gonipterus* species

Description: eggs are laid in batches, with each batch in a greyish capsule. The fleshy larvae, which grow to about 1cm long, are yellowish or greenish with three darker stripes, sometimes becoming wholly blackish with age. Pupation occurs in the ground. The weevils, 0.8–1.3cm long, are grey-brown to brown with paler patches; the underside is wholly pale coloured. They have a short, trunk-like nose, black eyes and short but prominent antennae.

Snout weevil, Gonipterus scutellatus. [D. Jones]

Larva of the snout weevil, Gonipterus scutellatus. [D. Jones]

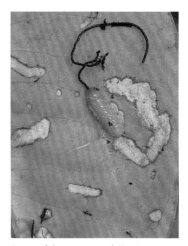

Larva of the snout weevil, Gonipterus gibberus. [D. Jones]

Snout weevil larvae feeding on Eucalyptus parvula. [D. Jones]

Snout weevil damage to eucalypt leaves. [D. Jones]

Damage to young Corymbia *leaves caused by larvae of the snout weevil.* [D. Jones]

Climatic regions: subtropics and temperate regions, coastal and in the ranges and tablelands.

Host plants: many species of *Eucalyptus*; also *Corymbia maculata* and probably other *Corymbia* species.

Feeding habits: both the weevils and larvae feed on leaves and shoots. The larvae graze irregular narrow patches from the surface of leaves, the eaten areas becoming white. Sometimes they concentrate on very young soft leaves at the shoot tips, often destroying the leaves before they expand. Badly eaten leaves often curl inwards, turn brown and fall prematurely. The larvae trail a black faecal string behind them as they feed. The weevils usually rest on stems during the day and feed at night. Feeding weevils eat patches out of the leaves, sometimes with areas of grazed tissue on the margins. They also chew lumps from leaf margins.

Notes: *Gonipterus* is a genus of about 20 species of native weevils and includes a complex of about 10 similar-looking species that are currently under study. These weevils can cause significant damage to seedlings, small trees and coppice growth. Persistent attacks by these weevils can be a major factor preventing successful seedling establishment of eucalypts in paddocks and among pasture. They are also a very significant cause of stunting and decline of seedlings regenerating naturally around mature trees. One species, *G. platensis*, which was an accidental introduction from Tasmania, has become a serious pest of *Eucalyptus globulus* plantations in Western Australia. Attacks on mature trees are usually of minor significance in Australia, but some of these weevils have become established in other countries (in Africa, America, Europe and New Zealand), where they cause major economic losses in eucalypt plantations. Introduction of an egg-parasitising wasp from Australia, *Anaphes nitens*, has partially reduced their impact.

The weevils often have a characteristic resting pose, with their legs reaching right around the supporting stem or petiole (see photo). They often drop and fly when disturbed, but sometimes also cling very tightly to the stem on which they are sitting.

Control: as per control for leaf-eating beetles. Spraying seedlings and young plants may be necessary for successful establishment in areas where these pests are persistent.

Sugarcane Weevil Borer, *Rhabdoscelus obscurus*

Description: eggs are laid in a small cavity excavated in the stems or leaf bases by the female. The legless larvae, which are plump fleshy white grubs with a brown head, tunnel in the stem. Pupation occurs in a fibrous cocoon. The weevils, which are 1.2–1.5cm long, are brown to black with reddish areas on the thorax and wing-covers. They also have a longish slender snout (rostrum).

Climatic regions: tropics, subtropics and warm temperate regions, mainly coastal.

Host plants: although sugarcane is the primary host of this insect, which has become well established in northern and eastern Australia after its introduction from Papua New Guinea, it also commonly attacks palms (natural populations, as well as those grown in nurseries and gardens). Several native species are susceptible to attack, including *Carpentaria acuminata*, *Caryota albertii*, *Normanbya normanbyi*, *Ptychosperma elegans* and *Wodyetia bifurcata*.

Feeding habits: the larvae, which can feed communally, eat the soft internal tissue of the stems of palm seedlings, concentrating on areas towards the growing point. They can also bore into the leaf bases and trunks of older plants (plants up to about one metre tall). Seedlings often collapse and die; older plants suffer trunk damage, including splitting and bore-holes. Usually a slimy jelly-like substance oozes from areas damaged by this pest. The weevils apparently cause little damage.

Notes: this is a very serious pest of palms that has become distributed into subtropical areas by the movement of infested plants. It can cause serious losses in nurseries, especially to seedlings and young potted plants. The weevils are attracted to sugarcane waste, a material which has been commonly included in potting mixes. The impact of this pest on natural populations of native palms is largely unknown but cannot be considered as beneficial.

Control: as per control for leaf-eating beetles. Spraying is necessary to control this pest but the chemicals used are hazardous to humans and the environment. The chemicals authorised for control are broad-spectrum pesticides that can impact non-target organisms, are known to leave residual levels within food crops and are considered too hazardous for domestic use.

White-fringed Weevil, *Naupactus leucoloma*

Description: white eggs are laid in clusters on the base of plants or in soil litter. The curved, legless larvae, which grow to about 1.3cm long, are sparsely hairy, creamy-white grubs with a brown head and black jaws. The female weevils (males are absent in Australia), which are about 0.8–1.2cm long, are dark grey with two whitish lines down each side of the body. The abdomen is densely covered with short pale hairs.

Climatic regions: subtropics and temperate regions, coastal and inland.

Host plants: a huge range of garden plants, crops, vegetables, succulents, bulbs, ferns and orchids; also a pest of grape vines and apple trees.

Feeding habits: the weevils chew lumps out of leaf margins, often concentrating on the basal parts of the leaf. The larvae, which cause the most damage, actively feed on plant roots and tubers, causing significant damage (furrows, tunnels, chewed roots). Leaf yellowing and wilting are signs of their presence.

Notes: an introduced species that is native to South America. Males are unknown in Australia, all reproduction being by parthenogenic females. The wing-covers of this persistent pest are fused, meaning the weevils cannot fly, but they can walk actively. Because the weevils cannot fly, numbers build up in localised sites, causing significant damage. The weevils drop to the ground when disturbed and play dead.

Control: as per control for leaf-eating beetles. Spraying may be necessary.

Other Weevils

- **Palm Weevil Borer or Lesser Coconut Weevil**, *Diocalandra frumenti*, is an exotic species that has become established in Queensland and the Northern Territory. The larvae bore tunnels in the roots, trunks, petioles, inflorescences and fruit of palms. Pupation occurs within a tunnel. The shiny black weevils, 0.6–0.8cm long, are very slender and have four large reddish to yellowish spots on the wing-covers. Gum leaks from the damaged sites and affected leaves become yellow and die. Death of mature palms has been recorded. Native species attacked by this pest include *Archontophoenix alexandrae* and *Howea belmoreana*. Cut or damaged surfaces should be covered with acrylic paint to prevent egg-laying by the weevils.

- **Vine Weevil**, *Orthorhinus klugii*, is a native weevil that feeds on wattles and has also adapted to grape vines. The brown weevils, which range from 0.6–0.8cm long (females are smaller than males), are densely covered with soft brown bristles. The surface of the body is knobbly with numerous small warts (two large bumps on the wing-covers), long forelegs, prominent black eyes and a long trunk-like nose. The larvae are stout fleshy white grubs that tunnel in the wood.

Vine Weevil, Orthorhinus klugii. [T. Wood]

- **Wattle Pig Weevils**, several *Leptopius* species, are a group of native weevils found over much of Australia that feed on various species of wattle and some other native trees. The weevils are sturdy, 1.5–2.5cm long, grey to brown, and their surface often wrinkled and ornamented with bumps, warts and tubercles. The snout resembles a trunk and the antennae are prominent. They feed by grazing strips and patches from the foliage and may also ringbark stems. The larvae of some species feed on roots (see also chapter 14).

- **New Growth Weevils**, *Merimnetes* species, are small grey weevils, 0.3–0.4cm long, that eat the soft new leaves of some species of *Angophora* and *Eucalyptus*. They appear as the new shoots emerge and, although small, they can be very destructive and reduce the leaves to ribbons. The larvae feed on roots. A small black species, possibly a species of *Euops*, is also very destructive of new the growth of *Angophora floribunda* and *Eucalyptus camaldulensis*. Eaten tissue turns black and leaks sap.

Weevil, Merimnetes *sp., on young leaf of* Angophora floribunda. [D. Jones]

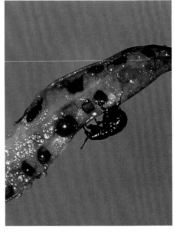

This tiny black weevil, possibly a species of Euops, *causes severe damage to new shoots and leaves of young eucalypts.* [D. Jones]

- **Acacia Weevils**, *Melanterius* species, at least one species attacks the phyllodes and developing pods of wattles in semi-arid areas. The weevils eat out sections of tissue, which turns black. Known to occur on *Acacia anceps* from Eyre Peninsula, South Australia.

Acacia weevil, Melanterius *sp., feeding on* Acacia anceps. [S. Jones]

Acacia weevil damage on phyllodes of Acacia anceps. [S. Jones]

Acacia weevil damage on pods of Acacia anceps. [S. Jones]

Miscellaneous Weevils

Unknown weevil that feeds on Grevillea alpina. [T. Wood]

Weevil of the genus Aeonychus *feeding on* Grevillea alpina. [T. Wood]

Weevil, probably of the genus Perperus, *feeding on* Acacia dealbata, *the larvae feed on wattle roots.* [T. Wood]

These weevils, Aterpodes tuberculatus, *feed on* Allocasuarina verticillata. [D. Jones]

CHAPTER 10

THE GRASSHOPPER AND PHASMID FAMILIES

CRICKETS, SANDGROPERS, LOCUSTS, GRASSHOPPERS, KATYDIDS AND STICK INSECTS

Most of the insects grouped together in this chapter belong to the insect order Orthoptera; the stick insects are placed in the order Phasmida. Orthoptera is a very diverse group of insects that embraces more than 2,800 named native species with numerous others awaiting formal recognition. Most species have their hindlegs modified for jumping, a significant feature that helps identify members of the order. The majority of species feed on plants, although only a few are significant pests in gardens.

SYMPTOMS AND RECOGNITION FEATURES

Brown-winged insects with shovel-like front legs with four claws, other legs of similar length............ *mole crickets*

Brown-winged insects with shovel-like front legs with two claws, other legs of similar length.........*changa cricket*

Brown wingless insects with shovel-like front legs, other legs shorter..*sandgropers*

Black insects with back legs modified for hopping and long antennae...*field crickets*

Green, grey or brown insects with back legs modified for hopping and short antennae.......*grasshoppers and locusts*

Green, grey or brown insects with back legs modified for hopping and long antennae ....*katydids*

Long, slender insects with short antennae and no legs modified for hopping..*phasmids*

CRICKETS

There are several groups of crickets, all belonging to the superfamily Gryllacridoidea. The members of two groups of crickets, which are commonly found in gardens, cause problems on occasions.

FIELD CRICKETS

Field crickets belong to the family Gryllidae. Field crickets, which live in holes in the ground or under rocks, are often black, have a flattish body and can hop. These crickets commonly chirp or trill on warm evenings and are easily distinguished from grasshoppers by their long antennae. At least two species of field crickets are common garden residents. They remain hidden during the day and feed at night, eating plant roots and leaves. They also feed on dead insects and the eggs of grasshoppers, moths and flies.

   Field crickets pass through several nymph stages, moulting between each stage before they become an adult. These nymphs are miniature versions of the adult except that they are wingless. Nymphs and adults are commonly present in gardens during the warm months of the year and often become active in the late summer and autumn. Some species of field crickets undergo periodic population explosions and swarm in large numbers.

Natural control: birds, lizards, frogs and spiders feed on field crickets.

Other control: pieces of cloth or hessian on the ground and night inspections by torchlight can be used to monitor field cricket numbers. Catching and squashing is a simple control method, except that they can be difficult to catch. Baiting with commercial snail preparations is successful. Spraying sensitive plants with malathion can also be effective as a protective agent.

EXAMPLES

Black Field Crickets, two species of *Teleogryllus*

Description: stout, flattish, blackish crickets, 2.5–3.5cm long, with a broad shiny black head and powerful jumping hindlegs. The females have a long ovipositor which projects back from the abdomen.
Climatic regions: tropics, subtropics and temperate regions, coastal and inland.
Host plants: feeds on the tender shoots of many plants. These crickets can be significant pests of plants in glasshouses and bushhouses where they feed on orchid root tips, young shoots and developing fern fronds. They are also important pests of pastures and many field crops.
Feeding habits: field crickets usually hide by day and feed at night. They can be solitary or gather in groups.
Notes: these insects frequently congregate in glasshouses because of the warmth and constant humidity. In some years they are present in plague proportions and may invade houses at night. They live for only a few months. *Teleogryllus commodus* is mostly jet black, whereas *T. oceanicus* has brown on the wing-covers.
Control: as per introduction for field crickets. Removing obstacles from the ground eliminates hiding places.

Black Field Cricket. [D. Jones]

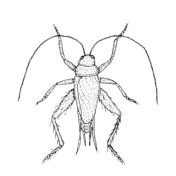

Black Field Cricket. [T. Blake]

MOLE CRICKETS

Mole crickets belong to the family Gryllotalpidae. They live underground and have cylindrical bodies with short front legs that are highly modified for digging. Their hindlegs are not enlarged like field crickets and they cannot hop. Mole crickets, which are common garden residents, are easily distinguished from grasshoppers by their body shape, very soft abdomen and shovel-like front feet. They are subterranean, living in well-built horizontal and vertical tunnels and galleries. Both sexes call loudly on moist or wet warm evenings, with males advertising their presence to females from the entrance of their burrows with a continuous trilling song. They eat plant roots, leaves, earthworms and ground-dwelling grubs and caterpillars.

Mole crickets pass through several nymph stages, moulting between each stage before they become an adult. These nymphs are miniature versions of the adult. Female mole crickets remain in their burrows with the eggs and nymphs. Nymphs and adults are commonly present in gardens during the warm months of the year and often become active in the autumn. They are often seen on the ground surface at night, especially after significant rainfall.

Natural control: females of the wingless metallic blue solitary wasp *Diamma bicolor*, commonly known by the misleading name of 'blue ant', sting and paralyse mole crickets, laying an egg on each victim. After hatching the wasp larva feeds on the paralysed cricket.

Other control: examination of potted plants containing mole crickets shows definite root damage, however, the level of damage they cause in gardens is largely unknown since they live underground. Control in a garden environment is impractical because of the species' subterranean habits. Mole crickets can be flushed out of pots by drenching with soapy water. Pots should be elevated above the ground level by using pot feet or something similar to deter crickets from entering through drainage holes.

EXAMPLES

Mole Crickets, several species of *Gryllotalpa*

Description: strong, cylindrical, brown crickets 3–4cm long. The shovel-like front legs are modified for digging, with four specialised claws to help move the soil. The males are wingless or with very short wings; females have gauzy wings folded beneath short covers.

Climatic regions: tropics, subtropics and temperate regions, coastal and inland.

Host plants: a range of native and exotic plants.

Feeding habits: feeds on roots and leaves.

Notes: mole crickets live in burrows and are rarely seen above ground. Estimates suggest there are at least 20 native species described in Australia, with others awaiting recognition. They occur in many different habitats. Some common species have adapted well to gardens. The significance of their feeding activities is difficult to estimate, however they have a hearty appetite and damage is likely to be appreciable. Two species that probably originated in eastern Australia have become established as pests in south-west Western Australia where they cause significant damage to vegetable gardens, potted plants and lawns. When held in the hand mole crickets try to prise their way between the fingers.

Control: as per introduction for mole crickets. Control is often not practical.

Mole cricket. [D. Jones]

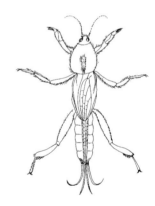

Mole cricket. [T. Blake]

Changa Mole Cricket, *Scapteriscus didactylus*

Description: strong, cylindrical, brown crickets, 3–4cm long. The shovel-like front legs are modified for digging but have only two specialised claws to help with the digging. Strongly veined wing-covers hide gauzy wings folded beneath.

Climatic regions: naturalised in the Newcastle area of New South Wales.

Host plants: seedlings and lawn and pasture grasses, including native species.

Feeding habits: feeds on roots during the day and emerges at night to eat above-ground plant organs. Its extensive burrowing activities cause subsidence and have impacted significantly on lawns and golf courses.

Notes: an introduced species that is native to north-eastern parts of South America and Caribbean islands. It has become well-established around Newcastle, Maitland and Williamstown in northern New South Wales since its introduction sometime around 1982. It has been reported that captured animals often 'play dead'.

Control: as per introduction for mole crickets. Chemical control has been recommended by authorities.

SANDGROPERS

Sandgropers, which are sometimes confused with mole crickets, are unusual insects that belong to the family Cylindrachetidae. Fourteen species are native to Australia and they seem to favour sandy soils. They feed on fungi, termites and plants. At least one species is sometimes a pest of crops and gardens.

EXAMPLE

Sandgroper, *Cylindraustralia kochii*

Description: sturdy, cylindrical, wingless brown crickets about 5cm long. The head and thorax are darker brown than the paler abdomen. The shovel-like front legs are flattened for digging and the other two pairs of legs are short.

Climatic regions: temperate regions of Western Australia, coastal and inland.

Host plants: a pest of crop plants including wheat, oats, barley and lupins; in gardens it may also feed on a range of native and exotic plants.

Feeding habits: feeds on roots, stems, leaves and flowers. Forms characteristic raised trails when burrowing close to the soil surface.

Notes: a slow-growing species that may take several years to complete its life cycle. It lives in burrows and is rarely seen.

Control: not practical. Farmers control this insect by ploughing late in the season when the insects are active and planting alternate crops of legumes.

GRASSHOPPERS AND LOCUSTS

Grasshoppers and locusts, of which about 700 species are native to Australia, are familiar garden insects. All grasshoppers and locusts eat plants, and although they are generally regarded by gardeners as pests, researchers have found that the majority of these insects only feed on a limited range of plants – some species, for example, will eat plants of a single genus or within a particular family. Although the group includes some very important economic pests, most grasshoppers cause limited plant damage and indeed can add interest to a garden. Grasshoppers and locusts belong to the family Acrididae; a specialised group of grasshoppers with cone-shaped heads, known as Pyrgomorphs, are placed in the family Pyrgomorphidae – see entry for Pointed Grasshopper below.

Grasshoppers are generally regarded as solitary insects, whereas locusts are solitary when in low numbers but as a population increases they become gregarious and swarm. In certain years locusts gather together in large swarms and migrate, eating all plant material they encounter as they travel. Some locusts, such as the Australian Plague Locust, *Chortoicetes terminifera*, reach plague proportions in years when a particular pattern of climatic conditions occurs.

Most grasshoppers and locusts are short-lived, lasting less than a year. Some species hatch in autumn, the majority in spring. The majority of species are active during the day in the warmer months of the year. They pass through nymph stages before they become an adult. The nymphs are miniature versions of the adult except that they are wingless (the adults of some species are also wingless) and may be a different colour. The nymphs also feed on plants.

Solitary grasshoppers are not usually as great a problem as the swarming types, although some

The Southern Pyrgomorph, Monistria concinna, *is an attractive solitary grasshopper.* [T. Wood]

Any gardener would be happy to have Leichhardt's Grasshopper, Petasida ephippigera, *in their garden.* [D. Jones]

The Striped Grasshopper, Macrotona australis, *is an active garden resident.* [T. Wood]

Grasshopper damage to young palm leaf. [D. Jones]

species, if present in sufficient numbers, can cause appreciable damage. Grasshoppers feed by chewing large lumps out of leaves, commonly starting from the margins and working towards the midrib. Often the feeding is erratic, the insects moving on and leaving partially eaten leaves behind. Plagues of locusts feed voraciously, stripping all foliage from any plants in their path. They even chew green bark and stems when hungry enough.

Natural control: grasshoppers and locusts are eaten by birds, lizards, frogs, marsupials, ants, predator bugs and robber flies. Specialised maggots of some tachinid flies and sarcophid flies eat grasshoppers. They are also killed by parasitic wasps, internal worms, nematodes, bacteria and fungi. Wasps of the genus *Scelio* parasitise grasshopper and locust eggs.

Other control: solitary grasshoppers are best controlled by hand-picking and squashing. They are generally easier to catch in the morning, especially following a cold night. They are attracted to the colour yellow – yellow buckets containing water can be used as a trap to drown them. Domestic poultry eat them. Some sprays, such as garlic spray, may repel them from foliage. Valuable plants should be covered to deny access to the foliage. Control of locust swarms is extremely difficult and is usually carried out by farmers and authorities on an area-wide basis. Egg-laying areas are monitored annually to check on populations. Best control is obtained by spraying in the hopper stage. Continual spraying of registered chemicals is necessary as the adults are nomadic and can fly considerable distances. Aerial spraying is the only effective method of controlling adult locusts. Experience in the USA shows that a protozoan parasite, *Nosema locustae*, can suppress grasshopper populations for several seasons, but this organism is not available in Australia.

EXAMPLES

Australian Plague Locust, *Chortoicetes terminifera*

Description: a brown to green locust that grows 2.5–4.5cm long. When in flight it can be recognised by a prominent black patch on the tip of each clear hindwing (the mark is also visible when at rest). An X-shaped mark on the top of the thorax is also useful for identification and its hopping legs usually have some red to scarlet colour on the shanks. The males are smaller than the females.
Climatic regions: tropics, subtropics and temperate regions, mainly inland but also coastal.
Host plants: almost anything.

Feeding habits: a gregarious species that eats large chunks out of leaves. Whole plants, even trees, can be defoliated during swarming.

Notes: a widespread major pest which assumes plague proportions in years of good rainfall, hatching in large numbers and migrating in huge swarms. The locusts travel on the prevailing wind, denuding pastures, crops, orchards and natural vegetation. Very large and dense swarms can contain millions of insects, eating tonnes of vegetation each day and causing huge economic losses. When disturbed these insects fly and on landing have the characteristic habit of turning back towards the point from where they began the flight.

Control: as per introduction for locusts and grasshoppers.

Australian Plague Locust, Chortoicetes terminifera. [P. Cochard]

Giant Grasshopper or Hedge Locust, *Valanga irregularis*

Description: a large, stout, creamy-brown to pinkish-brown locust with dark spots and mottles. Females grow to 7.5cm long, males to 5.5cm. The forewings are mottled and the hindwings transparent grey. Spines on the jumping legs are orange or red with black tips. Nymphs can be bright-green and look vastly different from the adults.

Climatic regions: tropics and subtropics, coastal and inland.

Host plants: a wide range of exotic and native species; very damaging to palms, epiphytic orchids, gingers and *Hibiscus tiliaceus.*

Feeding habits: usually a solitary insect but sometimes present in groups. Chews large lumps out of leaves and also eats most plant organs including flowers and fruit.

Notes: this is Australia's largest grasshopper and it is a very destructive species. The bright green nymphs are mostly seen from spring to late summer and the adults from autumn through winter to spring. Although this species is a strong flyer its movements are greatly influenced by the prevailing winds. The adults kick strongly with their spiny hind legs when caught.

Control: as per introduction for locusts and grasshoppers. Hand picking early in the morning when the insects are sun-baking is often successful. The eggs are parasited by a tiny wasp, *Scelio flavicornis.* Sarcophid flies of the genus *Blaesoxipha* lay their eggs on grasshoppers and the maggots eat them as food.

Giant Grasshopper, Valanga irregularis. [D. Jones]

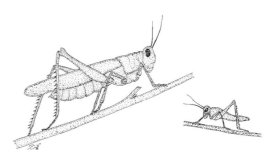

Giant Grasshopper and nymph. [D. Jones]

Giant Green Slantface or Long-headed Grasshopper, *Acrida conica*

Description: a long, slender, green or brown, striped grasshopper with a characteristic elongated head and flat antennae. The upper part of the abdomen is pink or red. Adults grow 7–8cm long and when in flight emit an unusual crackling or buzzing sound.

Climatic regions: tropics, subtropics and temperate regions, chiefly coastal.

Brown variant of the Giant Green Slantface grasshopper, Acrida conica.
[T. Wood]

Green variant of the Giant Green Slantface grasshopper, Acrida conica.
[D. Jones]

Host plants: grasses and a range of garden plants.

Feeding habits: a solitary, well-camouflaged grasshopper that chews large lumps from leaves and may even eat young shoots entirely.

Notes: a slow-moving, distinctive species which excites interest among children. It hops poorly and flies for only short distances.

Control: as per introduction for locusts and grasshoppers. Control is rarely necessary.

Migratory Locust, *Locusta migratoria*

Description: a stout locust with a large, powerful head and spiny hopping legs. Females grow to about 6.5cm long, males are smaller. Their colours vary with the growth phase; solitary adults are brown and green whereas gregarious adults are bluish-green and yellow; solitary nymphs are green or brown but gregarious nymphs are yellow with black spots.

Climatic regions: tropics and subtropics, coastal and inland.

Host plants: a range of native and exotic plants; also pastures and many crops.

Migratory Locust, Locusta migratoria. [R. Farrow]

Feeding habits: chews large lumps out of leaves.

Notes: a destructive pest which also occurs in Africa, Asia and New Zealand. Swarming populations mainly occur in the Central Highlands of Queensland. This locust migrates south into subtropical regions during the summer. Although a strong flyer, the insects are mainly distributed by prevailing winds.

Control: as per introduction for locusts and grasshoppers. Hand-picking early in the morning can be successful.

Pointed Grasshopper or Pyrgomorph, *Atractomorpha crenaticeps*

Description: a slender, green grasshopper with an elongated head and short, stiff, purplish antennae. The abdomen has a pink flush which is mostly hidden by the wings. Females grow to about 4cm long, males to about 3cm long. The tip of the wing-covers is pointed. Males are often seen riding on the female's back.

Climatic regions: tropics, subtropics and temperate regions, mainly coastal.

Host plants: a wide range of native and exotic plants.

Feeding habits: a slow-moving species which feeds singly or in small, loose groups. Individuals are well camouflaged and can be difficult to discern.

Notes: although mostly a minor pest in gardens, this species can be a persistent nuisance in nurseries and vegetable patches.

Control: as per introduction for locusts and grasshoppers. Hand-picking is often successful with this species.

Spur-throated Locust, *Austacris guttulosa*

Description: a large, stout locust which is pale brown to dark brown with white stripes. Females can grow to 8cm long, males to 6.5cm. There is a characteristic peg-like spur projecting on the underside between the first pair of legs. The forewings are transparent with dark blotches while the hindwings are transparent. Young nymphs are bright green, older nymphs yellowish brown.

Climatic regions: mainly tropics and subtropics, commonly inland, less common on the coast.

Host plants: a wide range of native and exotic shrubs, trees, crops and grasses; can severely damage palms; also very fond of *Hibiscus tiliaceus*.

Feeding habits: The adults congregate together in swarms during winter and often roost in trees close to water. In some years they form huge swarms, defoliating large trees and even breaking branches by their sheer weight. They migrate at night during the wet season, moving into tall grassland with the aid of prevailing winds.

Notes: eggs are laid in the wet season and nymphs develop over summer and autumn. The adults live for about a year.

Control: as per introduction for locusts and grasshoppers.

Spur-throated Locust, Austacris guttulosa. [P. COCHARD]

Wingless Grasshopper, *Phaulacridium vittatum*

Description: nymphs, which are grey-black with tiny wing-covers, range from about 0.2cm to 1cm long. Adults grow to 1.5–2cm long, the males are smaller than the females. They are brown with orange-brown hindlegs and sometimes have two prominent white stripes on the thorax. Mostly the adults have short, stubby wings and cannot fly, but a small proportion develop wings and can fly short distances.

Climatic regions: subtropics and temperate regions, coastal and inland.

Host plants: a wide range of native and exotic plants, especially forbs and herbaceous plants,

Wingless Grasshopper, Phaulacridium vittatum. [P. COCHARD]

including vegetables (but not grasses). They also eat the leaves of young shrubs, trees and grape vines.

Feeding habits: this species congregates in large swarms during late summer and autumn. It is an efficient climber and can be a significant pest of newly planted trees and vines. It sometimes also causes economic damage to nursery plants.

Notes: eggs laid in late summer-autumn hatch the following spring. The nymphs feed near where they hatched and develop into adults in late spring-early summer. They move in search of food, roaming widely as the season dries off and food becomes scarce.

Control: as per introduction for locusts and grasshoppers.

Yellow-winged Locust, *Gastrimargus musicus*

Description: a large, stout locust which is usually green, less commonly pale brown. Females can grow to 5cm long, males to 3.5cm. The thorax is ridged and the forewings have blackish mottles. The hindwings have a large bright yellow patch subtended by a black band. Adult males produce a characteristic loud clicking sound when they fly.

Climatic regions: tropics, subtropics and temperate regions, mainly inland but also coastal.

Host plants: a wide range of native and exotic plants.
Feeding habits: this mostly solitary locust gathers together in swarms in some years. When swarming, most of the insects are brown rather than green.
Notes: a very distinctive species which is easily recognised by the bright yellow patch on the hindwings and the noise the males make when in flight. It is most noticeable during summer. This locust is sometimes a pest of gardens in inland areas.
Control: as per introduction for locusts and grasshoppers.

Yellow-winged Locust, Gastrimargus musicus. [P. Cochard]

KATYDIDS

Katydids, sometimes called long-horned grasshoppers, belong to the family Tettigoniidae. Mostly nocturnal, katydids are often well camouflaged in their surroundings and are generally difficult to see. The males sing, sometimes quite loudly, to attract females for mating. Katydids feed on a range of materials; some are predators, others feed on plants.

Plant-feeding katydids lay their eggs in living or dead plant tissue. After hatching the small nymphs move onto their host. Often the nymphs are of very different colours to their parents. Some take on bright colours to warn off predators, some resemble ants, others adopt the colours of their background as camouflage. They pass through nymph stages as they grow, moulting between each stage.

Natural control: katydids are eaten by birds, marsupials, lizards, frogs, spiders, ants, predatory bugs, praying mantids, predatory katydids and tree crickets.

Other control: katydids are rarely a pest in gardens and most people would happily adopt them for the musical interest they add. If they do become a pest, hand collection and squashing is usually sufficient to reduce numbers. Spraying is not necessary in home gardens.

Katydid on Syzygium. [D. Jones]

Katydid nymphs often resemble an ant. [D. Jones]

EXAMPLES

Garden Katydid, *Caedicia simplex*

Description: adults are bright green and 4–5cm long, with antennae up to 6cm long. The forewings form a narrowly pointed tent-like cover over the abdomen which has pink and yellow markings.
Climatic regions: mainly temperate regions, coastal and inland.
Host plants: several species of *Eucalyptus*; also many other garden plants.
Feeding habits: a solitary insect that chews lumps out of leaves.
Notes: a common garden insect that is not easily seen because of its efficient camouflage. Occasionally it assumes pest proportions. The nymphs are often brightly coloured (pink to red or purple).
Control: as per introduction for katydids.

Garden Katydid, Caedicia simplex.
[D. Jones]

Giant Torbia, *Torbia perficita*

Description: adults are bright green, about 5cm long with antennae up to 6cm long. The forewings, which are marked with numerous fine veins, form a narrowly pointed tent-like cover over the abdomen.
Climatic regions: tropics, subtropics and temperate regions, mainly coastal.
Host plants: several species of *Eucalyptus*; also other plants.
Feeding habits: a solitary insect that chews lumps out of leaves.
Notes: an interesting species which camouflages itself well among eucalypt leaves. It is attracted to lights at night.
Control: as per introduction for katydids.

Male Giant Torbia, Torbia perficita. [T. Wood]

Nymph of the Giant Torbia, Torbia perficita. [T. Wood]

Female Giant Torbia, Torbia perficita. [T. Wood]

Gumleaf Torbia, *Torbia viridissima*

Description: black eggs are laid in rows along leaf margins. The nymphs are pale green with darker markings. Adults are bright green, about 5cm long with antennae up to 6cm long. The forewings form a narrowly pointed tent-like cover over the abdomen. They are green with a prominent thick white vein.

Climatic regions: tropics, subtropics and temperate regions, coastal and inland.

Host plants: several species of *Corymbia* and *Eucalyptus*.

Feeding habits: a solitary insect that chews lumps out of leaves.

Notes: a widely distributed species that occurs in most habitats where eucalypts grow. The nymphs have a remarkable resemblance to ants.

Control: as per introduction for katydids.

The Olive Green Coastal Katydid eats orchids and ferns. [D. Jones]

Olive-green Coastal Katydid, *Austrosalomona falcata*

Description: Adults are brownish-green to olive-green, about 6–7cm long with antennae up to 8cm long. The translucent forewings form a narrowly shiny tent-like cover over the abdomen.

Climatic regions: tropics, subtropics and temperate regions, mainly coastal.

Host plants: succulent tissue of many garden plants including buds, flowers and new shoots; can also be a significant pest of orchids and ferns in greenhouses and shadehouses.

Feeding habits: a solitary insect that eats lumps out of plant organs.

Notes: a widely distributed species that is common in coastal gardens. The males sing loudly on warm summer nights. Adults can bite strongly when handled.

Control: as per introduction for katydids.

Palm Katydid, *Segestidea queenslandica*

Description: adults are light grey to brown, about 9cm long with antennae up to 9cm long. The forewings, which are marked with numerous darker veins, form a narrowly pointed tent-like cover over the abdomen.

Climatic regions: tropics, mainly coastal.

Host plants: several species of palms, native and exotic.

Feeding habits: a solitary insect that chews large lumps out of palm leaflets leaving a ragged appearance.

Notes: a common species that feeds at night. It can be damaging to young palms and is a nuisance in nurseries. The females are commonly parthenogenetic, laying eggs in the ground. The nymphs, which remain still during the day, resemble bird droppings

Control: as per introduction for katydids. Spraying may be necessary in nurseries.

Other Katydids

- A very interesting and colourful addition to the garden is the Crested Katydid or Superb Katydid, *Alectoria superba*. It grows to about 7cm long and has a rounded bright-yellow dorsal crest with a black border. It is mainly found in subtropical regions.

STICK AND LEAF INSECTS (PHASMIDS)

These unusual insects, which are mainly nocturnal, belong to the insect order Phasmida (or Phasmatodea). There are about 100 named species in Australia, all of which feed on plant organs. Stick insects are so called because of their slender, stick-like appearance and their ability to remain motionless for long periods, passing themselves off as a dead stick or small branch. Leaf insects on the other hand, camouflage themselves as leaves, either living (green) or dead (brown). Stick insects can become very large, and some species stretch to 50–60cm long with their legs extended. They usually move slowly with an ungainly gait, although the nymphs of some species run like ants.

Stick insects are voracious feeders and eat the leaves completely, including the veins and midrib. Some species also feed on flowers and green bark. Small populations cause limited damage but a few species are known which have the ability to impact dramatically on natural forests in highland areas of the eastern states. These destructive phasmids occasionally undergo huge population explosions, the insects quickly reaching plague proportions and devastating forests, resulting in branch dieback and tree death. Phasmids seldom cause much damage in home gardens and in fact are usually welcomed for their unusual and fascinating appearance. Children in particular find them a fascinating addition to the garden fauna. Several species, such as the Goliath Stick Insect, *Eurycnema goliath*, can be purchased as pets from commercial outlets.

Female stick insects lay numerous eggs (the record is more than 2,000) over several months. When laying the females of many species drop or flick the eggs which then fall to the ground. The eggs, which can look like small seeds, often have an apical knob (termed a capitulum) on which ants feed after they have carried them to their nest. After hatching the small nymphs climb quickly to the top of shrubs or trees. Stick and leaf insects pass through nymph stages as they grow, moulting between each stage. The wingless nymphs are usually very slender and also feed on leaves. In some species adults of one sex only are capable of flight, while in others both sexes can fly. Although stick insects mate normally, some species can also reproduce parthenogenetically.

Natural control: stick and leaf insects are eaten by birds, marsupials, lizards, frogs, spiders, nematodes, ants, predatory bugs, praying mantids, predatory katydids, tree crickets and mites. Small flies and wasps also parasitise the eggs.

Other control: stick and leaf insects are rarely a pest in gardens and most people would happily adopt them for the interest they add.

A well-camouflaged stick insect. [T. Wood]

EXAMPLES

Tessellated Stick Insect, *Anchiale austrotessulata*

Description: a slender, pale brown to dark brown or green stick insect. The males, which have larger wings and can fly, grow to 9cm long and are more slender than the females. Females, which are poor flyers (mainly gliding), grow to 15cm long. The wings are strongly tessellated.

Climatic regions: subtropics and warm temperate regions of the east coast.

Host plants: many species of *Corymbia* and *Eucalyptus*; also *Allocasuarina littoralis, Lophostemon confertus, L. suaveolens* and *Syncarpia glomulifera.*

Feeding habits: feeds singly or in small to large groups, stripping leaves from branches.

Notes: a common species that becomes abundant in some years when it can devastate large areas. In some locations it has been linked with eucalypt dieback. The females can reproduce parthenogenetically. The males mate frenetically, even mating with other species.

Control: as per introduction for stick and leaf insects.

Spur-legged Stick Insect, *Didymuria violescens*

Description: a sturdy brown or green stick insect. The males, which grow to 10cm long, have dark purple wings which are opened and displayed when disturbed. The females, which are of similar size but often stouter, have short pale wings.

Climatic regions: subtropics and temperate regions of the east coast, including South Australia and Tasmania.

Host plants: species of *Eucalyptus*, especially in highland areas.

Feeding habits: feeds singly or in small groups, stripping leaves from branches and shoots.

Notes: a widespread and often common species that undergoes population explosions in some years, resulting in significant damage to forests. Damaging cycles, which last two seasons, often follow a cool summer. The females can reproduce parthenogenetically.

Control: as per introduction for stick and leaf insects.

Ringbarker Stick Insect, *Podacanthus wilkinsoni*

Description: a fairly stout, yellowish-brown or greenish-brown stick insect. The males, which grow to about 9cm long, have a reddish band on the forewing. The females, which are of similar size to the males, have gauzy pink hindwings. The wings sometimes have an orange blotch.

Climatic regions: subtropics and temperate regions of the east coast.

Host plants: many species of *Eucalyptus*, especially *E. bicostata, E. dalrympleana, E. radiata, E. robertsonii, E. mannifera, E. pauciflora, E. stellulata* and *E. viminalis.*

Feeding habits: a gregarious species which gathers in colonies, stripping large trees completely of leaves. Green bark is sometimes also eaten. Trees can be completely defoliated. Branch dieback and even tree death can follow these infestations.

Ringbarker Stick Insects, Podacanthus wilkinsoni. [T. Wood]

Ringbarker Stick Insect in camouflage pose. [T. Wood]

Spiny Leaf Insect, Extatosoma tiaratum. [T. Jones]

Notes: in some years this stick insect appears in plague numbers and large areas of forest can be ravaged.

Control: as per introduction for stick and leaf insects.

Spiny Leaf Insect, *Extatosoma tiaratum*

Description: a large, leaf-mimicking, pale green or brown leaf insect with a prominent conical head. The males, which grow to 11cm long, are slender with large mottled dark brown and pale wings. The females, which grow 15cm long, are much broader and with very small wings. In the female, the edges of the abdomen and the legs have flattened, leaf-like plates and short, sharp spines.

Climatic regions: disjunctly distributed in the tropics, subtropics and temperate regions of the east coast.

Host plants: mainly species of *Eucalyptus*; also *Alphitonia petriei*, *Brachychiton acerifolius*, *Buckinghamia celsissima* and *Callicoma serratifolia*; also exotic garden plants including roses, *Calliandra* and guava.

Feeding habits: feeds singly or in small groups. A slow-moving insect that chews lumps out of leaves. Often curls its abdomen when at rest and can secrete smelly liquid from the mouth.

Notes: an exciting insect that always attracts attention when discovered. It rarely causes significant damage in gardens.

Control: rarely necessary.

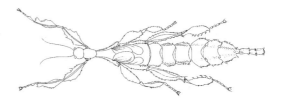

Spiny Leaf Insect, Extatosoma tiaratum. [T. Blake]

Other Phasmids

- **Titan Stick Insect**, *Acrophylla titan*, is widely distributed in the subtropics and temperate parts of eastern Australia. It feeds on several species of *Eucalyptus* as well as some *Acacia*, *Callistemon* and *Callitris columellaris*. It is sometimes found in gardens. Mostly greyish brown, the males grow to about 15cm long and the females to 25cm long.

- **Wülfing's Stick Insect**, *Acrophylla wuelfingi*, from north-eastern Queensland, feeds on a variety of rainforest plants. It has been recorded as a pest of the kauri pine *Agathis robusta*, defoliating even large trees in some years. The males, which are brown with a green thorax, grow to about 14cm long; the pinkish-brown females grow to more than 20cm long.

- **Strong Stick Insect**, *Anchiale briareus*, is common in north-eastern Queensland and feeds on species of *Acacia*, *Casuarina*, *Corymbia*, *Eucalyptus* and some rainforest plants. Adults are either brown or green, the males grow to 10cm long and the females to 17cm long.

- **Margin-winged Stick Insect**, *Ctenomorpha marginipennis*, is a dark brown species that is widespread in south-eastern Australia. It is often found in gardens and feeds on a wide range of native and exotic plants. The males grow to about 12cm long and the females to 20cm long.

- **Goliath Stick Insect**, *Eurycnema goliath*, is often found in gardens in New South Wales and Queensland feeding on eucalypts and phyllodinous species of *Acacia*. The adults are green and white with a yellow head and transparent wings with a bold red band on the upper margin.

- **Red-winged Stick Insect**, *Podacanthus viridiroseus*, is very widely distributed in a narrow band from north-eastern Queensland to South Australia. It feeds on several species of *Eucalyptus* and is often found in gardens. Commonly green, the males are slender insects which grow up to 8cm long and the plump females, which have bright pink wings, grow to about 11cm. The colourful wings are often displayed when the insect is disturbed.

Margin-winged Stick Insect, Ctenomorpha marginipennis.
[T. Jones]

CHAPTER 11

OTHER LEAF-CHEWING PESTS

COCKROACHES, EARWIGS, LEAF MINERS, SAWFLIES

SYMPTOMS AND RECOGNITION FEATURES

Slender insects with a pair of forceps on the rear end...*earwigs*

Fast-running insects with long antennae, repulsive smell when squashed.............................*cockroaches*

Slender, convoluted tunnels eaten between upper and lower leaf surfaces ....*leaf miners*

Broad blisters eaten between upper and lower leaf surfaces..............................*leaf blister sawflies*

Leaf surface grazed, often exposing structural veins..*sawflies*

Leaves completely eaten leaving only the structural veins..*sawflies*

Leaves completely eaten leaving only the petiole as a stub..*sawflies*

Small patches of bark grazed...*sawflies*

Small patches eaten out of soft tissue especially root tips and young shoots ....*cockroaches and earwigs*

Ugly tapered grubs congregating in large colonies during the day ....*sawflies*

Grubs regurgitating oily yellow, green or brown fluid when disturbed...........................*sawflies*

COCKROACHES (ORDER BLATTODEA)

There are about 4,500 species of cockroach in the world, of which about 30 are pests that invade homes. Cockroaches can be recognised by their flattened bodies, long antennae and mouthparts situated on the underside of the head. Some species are slow movers but many of the problem species can run very fast when disturbed.

Native cockroaches: it is not generally appreciated that Australia has more than 400 species of native cockroach. Most natural habitats have populations of native cockroaches that feed on organic material such as rotting wood, decaying leaves and bark. Some are pollen feeders, others have adapted to highly specialised habitats including blind and wingless species that live in caves and sink holes. Most are nocturnal but some venture out during the day. A few are even useful predators of other insects and in recent times some large species have been kept as novelty pets. Very few damage living plants. Few if any native species enter houses regularly and none contaminate food.

Introduced Cockroaches: household cockroaches are familiar and much detested pests that are loathed for their dirty habits, in particular the fouling of cupboards and contamination of food. Several introduced species have been linked with allergic reactions in humans and they are also vectors of microbes involved in infections such as salmonella and dysentery. There are about a dozen species of cockroaches that invade Australian houses and cause other urban problems. All are invasive pests from other countries that have become established in urban environments. A couple of these introduced species are significant pests of plants, especially those grown in containers. Many species produce an obnoxious smell as a defence mechanism when threatened. This odour, especially from squashed cockroaches, can linger and contaminate clothing and food.

Life cycle: after mating the females of many species deposit eggs in a prominent protruding case attached to the abdomen. This structure, known as an ootheca, has a hard shell that protects the eggs after laying. The nymphs, which appear as miniature versions of the adults, moult as they increase in size. Not all cockroaches produce ootheca; some species, such as the Surinam Cockroach, produce eggs that hatch internally within the female's body.

Natural control: cockroaches are eaten by birds, small marsupials, small rodents, lizards and frogs. They may also be eaten by centipedes and predatory insects such as beetles and parasitised by wasps. Cockroach oothecae can also be subject to parasitism by wasps.

Other control: inspection at night by torchlight is useful for determining numbers. Rubbish which provides shelter should be cleaned up and the area dusted with lime. Eucalyptus spray or spray made with garlic and cayenne pepper can be used as a deterrent. Boric acid spread in cracks and crevices will deter cockroaches as long as it remains dry. Sticky traps are a useful means of control. Baiting or spraying with commercial preparations containing pyrethrum may be necessary to protect valuable plants.

EXAMPLES

Australian Brown Cockroach, *Periplaneta australasiae*

Description: a large cockroach, 3–3.5cm long, that is brown with bright yellow marginal bands on the thorax. It has spiky legs and long antennae. It enters houses and runs quickly, hiding when disturbed. The adults are winged and can fly actively.

Climatic regions: although mainly a pest of the tropics, this species can survive in warmer parts of temperate regions where it is mostly active in the summer months.

Host plants: eats decaying organic matter and the soft tissue of living plants. It can be a significant pest of a range of plants grown indoors and in glasshouses. Colonies can establish readily in artificially heated greenhouses and glasshouses.

Feeding habits: these pests shelter by day and feed at night. They gnaw soft tissue such as new roots, shoots and buds.

Notes: a cosmopolitan species that is well-established in the tropical areas of many countries. The common name is misleading as the species is not an Australian native but originates in Asia and has been introduced to Australia. Adults fly on warm nights and are attracted to lights.

Australian Brown Cockroach, Periplaneta australasiae. [S. JONES]

Control: as per introduction for cockroaches.

Smoky Brown Cockroach, *Periplaneta fuliginosa*

Description: a large cockroach, 3–3.5cm long, that is uniformly brown to almost black with no obvious paler markings. It lives and breeds in gardens, feeding on organic matter and plant material, often entering houses. Adults are winged and can fly actively.

Climatic regions: tropical to temperate regions (mainly coastal). It is most active during the warm months.

Host plants: destructive of indoor plants, ferns, bulbs and epiphytic orchids in glasshouses and bush-houses.

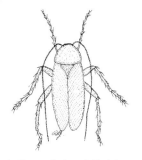

Smoky Brown Cockroach, Periplaneta fuliginosa. [D. JONES]

Feeding habits: these pests shelter by day and feed at night. They gnaw soft tissue such as roots and young shoots. They often eat the green growing tips of the roots of epiphytic orchids, effectively curtailing normal root growth.

Notes: this introduced pest may be troublesome during the summer months. Adults fly on warm nights and are attracted to lights.

Control: as per introduction for cockroaches.

Surinam Cockroach, *Pycnoscelus surinamensis*

Description: a plump cockroach, 2–2.5cm long, that has relatively short legs and very thin antennae. The winged adults (males and females) are uniformly brown with a shiny black head. Wingless females can also be common. These are black and shiny all over apart from the dull posterior segments. In many countries the females can produce young without mating. Squashed insects do not release a strong smell.

Climatic regions: mainly tropical regions with a warm to hot humid climate.

Host plants: eats decaying organic matter and the soft tissue of living plants. Attacks a wide range of indoor and glasshouse plants including ferns, bulbs and orchids; also attacks susceptible plants in the garden.

Feeding habits: these insects congregate in piles of garden litter, organic garden mulch and compost heaps. They burrow readily and breed quickly in ideal conditions. They can also shelter and breed in the coarse materials used in potting mixes for orchids and ferns. They mainly feed at night although they can sometimes be found feeding during the day.

Notes: a very destructive pest that can swarm in large numbers, causing considerable damage to soft plant tissue. Roots and new growth are often completely destroyed and sustained attacks usually result in plant death.

Control: as per introduction for cockroaches. The removal of rubbish and litter is of major importance for the control of this pest. Organic garden mulches provide an ideal environment where they can live and breed. Spraying is often necessary to protect valuable plants.

Orchid roots damaged by Surinam Cockroach, Pycnoscelus surinamensis. [D. JONES]

EARWIGS (ORDER DERMAPTERA)

There are about 2,000 species of earwigs in the world. They are distinctive insects that have a pair of forceps at the rear end of the abdomen. They hide in crevices and under logs, stones and rocks. They are common in weedy areas, among litter, under containers (especially plant pots) and anywhere there is an abundance of decaying organic matter.

Native earwigs: there are about 100 species of earwig native to Australia. They feed on plant and animal material including decaying plant matter, pollen and dead invertebrates. Some species of native earwigs are beneficial because they feed on caterpillar eggs and small soft-bodied insects such as aphids. Some of the larger native species may use their forceps to catch small soft caterpillars on which they feed. The introduced European Earwig (see below) is a significant pest in some areas.

Life cycle: after mating, the adult female lays batches of eggs in a chamber beneath the soil surface during autumn. She cares for the eggs by keeping them free of fungus and parasites and guarding them from predators. The eggs hatch in spring and the young nymphs, which are paler than the adults, remain with their mother up to the second or third moult. Earwigs reach maturity after the fourth moult.

Natural control: many native birds feed on earwigs. Poultry also eat them and are especially efficient at locating them when scratching for food. Earwigs are also eaten by lizards, frogs, spiders, ants and mantids.

Other control: earwig numbers can be reduced by luring them into sites where they are attracted for shelter. Balls of crumpled newspaper work well, also pots containing shredded newspaper. The paper is collected at intervals and burnt. Earwigs can also be trapped in dishes of vegetable oil or beer traps. Dusting with lime acts as a deterrent. Plants in containers should be raised off the ground to deter earwigs from entering through drainage holes. Baiting with commercial snail preparations (iron based) is a useful technique for protecting valuable plants in glasshouses. Spraying with a product containing malathion is also effective when large numbers occur.

EXAMPLE

European Earwig, *Forficula auricularia*

Description: a slender, dark brown insect, 1–1.3cm long, with a pair of forcep-like pincers at the rear end. Gauzy wings are folded under small shield-shaped covers on the thorax. Female earwigs have relatively small straight pincers, whereas the males have larger curved pincers with small teeth on the inner margins.

Climatic regions: temperate and cooler subtropical regions.

Host plants: earwigs attack seedlings of many species, especially vegetables, and also eat the flowers of a wide range of garden plants. European Earwigs seem to find the warm humid conditions of glasshouses and greenhouses congenial and can cause damage to a range of plants, especially those with soft new growth. They are a significant pest of epiphytic orchids, causing damage to new shoots, the tips of actively growing roots, buds and flowers. They also damage very young developing fern fronds.

European Earwig, Forficula auricularia.
[T. Wood]

The European Earwig also has its good points, feeding as it does on insect eggs and aphids.

Feeding habits: earwigs hide in dark places during the day and emerge at night to feed. They often shelter in the type of coarse potting mix used for epiphytic orchids. They can release an obnoxious smell when annoyed.

Notes: an introduced pest that is native to Europe and western Asia. It is common in southern Australia and occurs in plague proportions in some years. The beneficial Australian Brown Earwig, *Labidura truncata*, is a similar-looking native species that is easily confused with the European Earwig. It is of comparable size but can be distinguished by its overall light tan to fawn colouration with dark brown areas on the wing-covers and abdomen.

Control: as per introduction for earwigs.

OTHER EARWIGS

- **The Black Field Earwig**, *Nala lividipes*, is widely distributed in Europe and Asia and common as an introduced pest in Australia, where it is frequently found in inland and tropical gardens. It feeds on roots and soft stem tissue and can cause considerable damage to seedlings. The black shiny adults, about 15mm long, also feed on caterpillars and wireworms.

LEAF MINERS

The larvae of a wide range of different insects feed on the sap obtained from the internal tissues of leaves. These insects, known as leaf miners, live and feed inside the leaf, eating layers of cells between the upper and lower surface and forming tunnels or blisters. The leaf epidermal cells and cuticle are left intact and provide protection against predation by other insects such as ants. The larvae often feed only on certain types of cells within the leaf, avoiding those rich in cellulose or defensive cells containing chemicals such as tannins. Some leaf miners pupate inside a leaf, others emerge from the leaf and pupate in the soil. Many leaf miners are host-specific, attacking only a single species of plant, others are less selective.

Most leaf miners are the larvae of moths (Lepidoptera – especially in the family Gracillariidae), leaf blister sawflies (Hymenoptera family Pergidae – treated at the end of this chapter) and true flies (Diptera), but some beetles (Coleoptera) and at least one native wasp genus (Hymenoptera) also feeds in this way. The larvae are generally flattened so that they can feed in the confined areas of the inner leaf tissue. Caterpillars of the moth family Gracillariidae are flattened in the early stages of development but later change to a more normal caterpillar shape. Some larvae are so specialised that they have lost their prolegs.

The areas eaten by leaf miners can be seen from outside the leaf, especially after the damaged tissue changes colour and dies. They show up either as narrow lines (tunnels or mines), or broadly eaten areas which appear as pale or papery blisters or blotches. These blisters can be mistaken for the damage caused by some leaf spotting fungi. Often a tunnel either includes or finishes in a broadly eaten area of this type. The tunnels often meander, twist and convolute (termed serpentine) in an apparently random manner. The pattern of tunnel meandering is, however, not random, but is often unique to a species and can sometimes, in conjunction with the species of plant being attacked, be used to identify the insect involved. Not all tunnels are serpentine, some leaf miners construct star-like mines. Close examination of the tunnels over a period of time will show an increase in tunnel width as the larvae grows.

Leaf miner damage on Scaevola taccada*: note convoluted tunnels of different widths ending in pupation chambers.* [D. Jones]

Leaf miner damage, with tunnels ending in large blisters. [D. Jones]

Leaf miners can be a serious pest in forestry situations but are not usually a major problem in gardens, although they will make affected plants look unsightly. Leaf miners are commonly encountered on grevilleas, bottlebrushes, eucalypts and wattles but are usually a minor pest in gardens, although occasionally severe damaging outbreaks occur. For example, studies during an outbreak of Wattle Leaf Miner on relatively large trees of *Acacia prominens* showed 30–56 per cent of the phyllodes to have been damaged.

Natural control: leaf miners are controlled by parasitic wasps which lay eggs through the tunnel wall into the body of the larvae. They are sometimes also targeted by small birds such as pardalotes.

Other control: leaf miners are very difficult to effectively control without resorting to highly toxic chemicals because the larvae are shielded by the leaf surfaces. Fortunately, serious outbreaks are sporadic and most attacks are of a minor nature. Affected shoots or individual leaves should be removed and destroyed to reduce numbers. Spraying new growth with a pest oil will deter egg-laying. These sprays should not be used in hot weather or foliage damage may result.

EXAMPLES OF LEAF MINING MOTHS

Banksia Leaf Miner, *Stegommata sulfuratella*

Description: a small, pale green caterpillar, up to 5mm long, that forms conspicuous tunnels and blisters within the leaf tissue. These become brown and unsightly. Attacks occur mainly on young leaves and often concentrate around the leaf apex. Pupation occurs in a silken cocoon among the leaves. The adult is a small, shiny white moth with fringed wings in the family Lyonetidae. Wingspan 1.5cm.

Climatic regions: subtropics and temperate regions, mainly coastal.

Host plants: *Banksia integrifolia* and *B. serrata*.

Feeding habits: this insect mainly damages leaves in the lower parts of the tree. Affected leaves can become distorted and fall prematurely, imparting a bare look to the lower branches.

Notes: attacks are usually sporadic and mostly cause minor damage, but sometimes young plants can be severely affected. This pest has recently been found in New Zealand.

Control: as per introduction for leaf miners.

Blackwood Leaf Miner, *Acrocercops alysidota*

Description: a tiny pale caterpillar, 0.5–0.6cm long, that excavates narrow tunnels lengthwise in the phyllodes. The adult is a small dark moth in the family Gracillariidae with a wingspan of about 0.8cm. It has narrow wings with pale chevron–like markings and delicately fringed hindwings

Climatic regions: subtropics and temperate regions.

Host plants: a range of phyllodinous wattles including *Acacia longifolia*, *A. melanoxylon*, *A. pycnantha*, *A. sophorae*, and *A. saligna*.

Feeding habits: feeds mainly in young phyllodes which can become badly mined. The tunnels are often close to the midrib but usually twist and convolute. They end in a blister. Badly affected phyllodes fall prematurely. Later stage larvae can also tunnel into stems causing dieback and impacting on shape and leader development

Notes: this pest causes significant damage to Blackwood, *Acacia melanoxylon*, in New Zealand where this wattle is grown commercially for its attractive wood.

Control: as per introduction for leaf miners.

Blackbutt Leaf Miner, *Acrocercops laciniella*

Description: a small, pale caterpillar, up to 0.4cm long, that forms conspicuous blisters within the leaf tissue. The adult is a small, grey-brown moth in the family Gracillaridae with a wingspan of about 1cm.

Climatic regions: subtropical and temperate areas of eastern Australia.

Host plants: several eastern eucalypts, especially Blackbutt, *Eucalyptus pilularis*; also *E. agglomerata*, *E. eugenoides*, *E. globoidea*, *E. microcorys*, *E. muelleriana*, *E. piperita*, *E. saligna* and *E. triantha*.

Feeding habits: the larvae form convoluted tunnels and large irregular blisters in young leaves.

Notes: a serious pest of forestry plantations which often significantly impairs the growth of young trees. This pest has recently been found in New Zealand.

Control: as per introduction for leaf miners.

Citrus Leaf Miner, *Phyllocnistis citrella*

Description: a tiny pale green caterpillar that excavates tunnels within the leaf tissue of *Citrus* species and cultivars. The adult, which is a small moth about 0.2cm long in the family Gracillaridae, flies at night. It has wings with fringed margins and a black dot on the tip of each forewing.

Climatic regions: tropics, subtropics and temperate regions, coastal and inland.

Host plants: species and cultivars of *Citrus*, including the native species; also other members of the family Rutaceae.

Feeding habits: eggs are laid on new shoots and the larvae tunnel in immature leaves creating silvery convoluted (serpentine) tunnels. Affected leaves twist and curl or become cupped. They are often pale green, contrasting with the normal dark green foliage.

Notes: most attacks by this Asian pest occur in late summer and autumn. It was first recorded in northern Australia in 1912 and since then has spread to most areas where citrus is grown.

Control: as per introduction for leaf miners. Removing badly affected shoots is recommended, followed by spraying with a pest oil.

Bloodwood Leaf Miner or Corymbia Leaf Miner, unknown Lepidoptera

Description: a small, thin, pale caterpillar, 0.2–0.4cm long, that forms slender straight or convoluted tunnels within the leaf tissue of flowering bloodwoods. Pupation occurs in a small expanded loop at the end of a tunnel. The adult moth emerges through a hole cut in this region.

Climatic regions: tropics, subtropics and temperate areas of eastern Australia.

Host plants: *Corymbia ficifolia*, *C. ptychocarpa* and hybrids involving these species.

Feeding habits: eggs are laid in young leaves and the tunnels formed by the larvae cause blistering and distortion.

Notes: a serious pest of nurseries which impacts on the appearance of plants grown for sale. This pest seriously impacts on the growth of stock trees used to produce grafting material and on any grafted plants produced.

Control: as per introduction for leaf miners; a difficult pest to control.

Corymbia leaves damaged by Corymbia Leaf Miner. [D. Jones]

Grevillea Leaf Miner, *Peraglyphis atimina*

Description: a tiny caterpillar, 0.5–1cm long, that excavates tunnels within the leaf tissue of *Grevillea* species and cultivars. The adult is a brown moth, 1.2–1.5cm long, in the family Tortricidae.

Climatic regions: tropics, subtropics and temperate regions, coastal and inland.

Host plants: *Grevillea* species and cultivars, particularly those with broad leaves, including *G. banksii* and *G. robusta*.

Feeding habits: larvae form tunnels and blisters in the leaves; these eventually turn grey-brown and become unsightly.

Notes: a sporadic pest which can cause significant problems in some years. Sustained attacks can seriously debilitate plants. Leaves can be tunnelled so badly that they fall prematurely from the plant.

Control: as per introduction for leaf miners.

Silky oak leaves damaged by Grevillea Leaf Miner. [D. Jones]

Jarrah Leaf Miner, *Perthida glyphopa*

Description: a small cream caterpillar, up to 0.4cm long, that forms conspicuous blisters within the leaf tissue of jarrah. Holes are cut in the leaves by mature larvae and used to form a cocoon. Pupation occurs in the ground. The adult is a small grey-brown moth about 0.6cm long in the family Incurvariidae.

Climatic regions: temperate areas of south-western Western Australia.

Host plants: Jarrah, *Eucalyptus marginata*, and Flooded Gum, *E. rudis*, are often severely attacked; about 10 other western eucalypts are affected to varying degrees.

Feeding habits: the larvae form convoluted tunnels and large grey blisters. Most feeding occurs during winter.

Notes: a serious pest of forestry plantations which often significantly impairs the growth of young trees and can permanently influence tree shape. Leaves can be fully mined by this pest and badly affected shoots appear as if fire-scorched. The moths emerge from the soil in autumn and can often be seen running on twigs during sunny days.

Control: as per introduction for leaf miners; a difficult pest to control.

Macadamia Leaf Miner, *Acrocercops chionosema*

Description: a tiny white to yellow caterpillar, 0.4–0.6cm long, that excavates tunnels and blisters within the leaf tissue. The mature stage of the larvae, which develops red bands, cuts a hole through the epidermis and pupates in a cocoon. The adult is a small moth in the family Gracillariidae with a wingspan of about 0.6cm. It has brown forewings with three silver-white bands and narrow hindwings edged with silky hairs.

Macadamia *leaves damaged by Macadamia Leaf Miner.* [D. Jones]

Climatic regions: tropics, subtropics and temperate regions, chiefly coastal.

Host plants: species and cultivars of *Macadamia* and *Stenocarpus*, particularly *S. salignus* and *S. sinuatus*.

Feeding habits: the larvae form silvery convoluted tunnels and blisters sometimes covering a whole leaf. New leaves are mainly attacked and badly infested shoots appear as if scorched by a fire. Affected leaves are often shed prematurely.

Notes: a pest of commercial significance causing growth retardation in young trees. Damage is worst in areas adjacent to natural forest. Fully grown larvae can often be seen through the skin of a blister.

Control: as per introduction for leaf miners. Young trees will require spraying, but mature trees tolerate the damage without major setback.

Wattle Leaf Miner, *Acrocercops plebeia*

Description: a tiny, pale green to whitish caterpillar, 0.6-0.8cm long, that excavates tunnels in the phyllodes, eventually forming a blister. The mature stage of the larvae, which becomes dark red, cuts a hole through the epidermis and pupates in a cocoon or drops to the ground. The adult is a small moth in the family Gracillariidae.

Climatic regions: subtropics and temperate regions of the east coast.

Host plants: a wide range of phyllodinous wattles including *Acacia buxifolia*, *A. implexa*, *A. podalyriifolia*, *A. pravissima*, *A. prominens* and *A. rubida*.

Acacia podalyriifolia *phyllodes damaged by Wattle Leaf Miner.* [D. Jones]

Feeding habits: damages mature phyllodes which become blotched, discoloured and eventually covered by a large blister. Badly affected phyllodes fall prematurely.

Notes: severe and persistent attacks by this leaf miner on Queensland Silver Wattle, *Acacia podalyriifolia*, have led to it being rejected as a suitable ornamental in some areas.

Control: as per introduction for leaf miners.

Other Leaf Mining Moths

- The larvae of the moth *Pectinivalva anazona*, in the family Nepticulidae, attack the leaves of *Lophostemon confertus*.

- The larvae of *Porphyrosela aglaozona*, in the family Gracillaridae, feed on species of *Desmodium*, *Glycine*, *Kennedia* and *Phaseolus*.

Phyllode damage in Acacia implexa *caused by Wattle Leaf Miner.* [D. Jones]

EXAMPLES OF LEAF MINING FLIES

Pittosporum Leaf Miner, *Phytoliriomyza pittosporophylli*

Description: a tiny pale grub about 0.3cm long that excavates short curved tunnels close to the leaf midrib, these appearing as pale coloured circular areas 0.1–0.3cm across within the leaf tissue. The adult is a tiny fly about 0.3cm long in the family Agromyzidae.

Climatic regions: subtropics and temperate regions, mainly coastal.

Host plants: species of *Pittosporum*, particularly *P. undulatum*.

Feeding habits: larvae feed within small areas of leaf tissue.

Notes: this leaf miner is very common and widely distributed. Although the leaves may become unsightly when badly attacked, overall this is a fairly minor pest which causes limited damage. *Pittosporum undulatum* has become established as a significant naturalised invasive weed in South Africa and Jamaica and the Pittosporum Leaf Miner may have potential for biological control.

Control: rarely necessary as the trees cope well with the damage.

Pittosporum undulatum *leaves damaged by Pittosporum Leaf Miner.* [D. Jones]

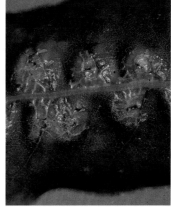

Pittosporum Leaf Miner. [D. Beardsell]

Rainforest Leaf Miners, family Agromyzidae

Description: the larvae are tiny cylindrical or tapered maggots that bore convoluted tunnels in leaves and stems. Adults are tiny flies, 0.2–0.3cm long, often with a distinctive patch of colour on the head. There are about 150 species of Agromyzid flies in Australia, most being leaf miners.

Climatic regions: tropics, subtropics and temperate regions, mainly coastal.

Host plants: numerous plants, especially those growing in rainforest and heavy forest.

Feeding habits: larvae form pale coloured, convoluted tunnels, sometimes ending in a blister.

Notes: these flies are mostly minor pests which cause limited damage. Usually they attack young growth, feeding as the leaves mature and often impeding normal leaf development on that shoot.

Control: rarely necessary as the plants mostly cope well with the damage.

Leaf miner damage on a rainforest plant.
[D. Jones]

Leaf miner damage to Cissus hastata.
[D. Jones]

Leaf miner damage to Elaeocarpus holopetalus. [D. Jones]

Leaf miner damage to Eupomatia laurina. [D. Jones]

Leaf miner damage to a rainforest plant.
[D. Jones]

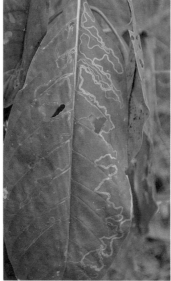

Leaf miner damage: convoluted trails on a rainforest tree leaf. [D. Jones]

Leaf miner damage: radiating patterns on a rainforest tree leaf. [D. Jones]

Other Leaf Miners

* The larvae of the Elkhorn Fern Beetle, *Halticorcus platycerii*, mine the fronds of *Platycerium bifurcatum* and *Asplenium australasicum* (more details in chapter 9).

Leaf Mining Sawflies – see Leaf Blister Sawflies at the end of this chapter.

CLUSTERING SAWFLIES (ORDER HYMENOPTERA)

Sawflies are not actually true flies but belong to a primitive group of wasps classified scientifically in the order Hymenoptera, family Pergidae. They get their name from a saw-like ovipositor on the underside of the female's abdomen which is used to slit open leaves and bark prior to egg-laying. Sawflies have caterpillar-like larvae (grubs) which taper from the chest to the tail. They have three pairs of true legs on the thorax and either no prolegs (subfamily Perginae) or six to eight pairs of prolegs (subfamily Pterygophorinae) on the abdomen (true caterpillars have up to five pairs of prolegs).

The larvae are commonly known as spitfires by children because when disturbed they jerk their heads and tails upwards, at the same time regurgitating (but not spitting) a sticky yellow-green to brown fluid which usually smells strongly of an aromatic oil. Commonly it is eucalyptus oil but the particular smell of the oil depends on the type of plant being eaten. Some species pupate in a cocoon in the ground, others in holes bored in soft wood or bark. Adult sawflies, which have four wings and resemble wasps, can be recognised by the thick area between the thorax and abdomen (thin and waist-like in a true wasp) and they cannot sting.

There are about 200 species of native sawflies. Several species feed on eucalypts, others attack *Callistemon, Callitris, Ficus, Leptospermum, Melaleuca* and some rainforest plants. Some specialised sawflies are parasitic. Sawflies can be very destructive to young plants and if unchecked can defoliate whole trees or impact significantly on a large section of a tree or shrub. Defoliation is common in trees up to about six years old and the larvae will migrate to adjacent plants if they run out of food. They are regarded as significant pests of eucalypt plantations and also commonly attack trees along road verges. Although damage by these pests often appears to be limited, research shows that attacks by large groups of sawflies significantly slow the growth of young trees and sustained attacks can cause plant death.

Life cycle: adult sawflies do not feed. Male sawflies are quite rare in some species; most adults are parthenogenic females that can produce eggs without mating. The female lays her eggs in groups (known as 'pods') or lines ('rafts') in specialised slits cut between the upper and lower surfaces of a leaf, or slits in bark. In some species of sawfly the adults guard their eggs until hatching. The eggs can take up to two months to hatch. Young larvae graze the leaf surface in a similar manner to leaf skeletonisers, leaving the veins. As they increase in size, however, they eat large areas of a leaf or whole leaves with the petiole left as a protruding stub. The larvae pass through six stages as they increase in size, each stage lasting three or four months. When fully grown the larvae pupate. Some species pupate in the soil, forming a cocoon of cemented soil particles, other species bore a hole in soft bark. The adults emerge sporadically, some taking a few months to develop, others a year or two.

Notes: Sawflies are classified into several subfamilies. The larvae of two of these subfamilies are commonly encountered and can be readily identified by morphological features (see below).

Natural control: small birds such as pardalotes feed on the eggs. Large birds, including currawongs, ravens, Gang-gang Cockatoos and cuckooshrikes feed on the larvae while they are on the tree, in the process knocking many to the ground where they can be attacked by other predators – the larvae of some species are helpless on the ground or when isolated from the cluster. They are also eaten by lizards and frogs, and are parasitised by wasps and tachinid flies. Bacterial and fungus diseases also kill many, especially in wet weather. Small groups of larvae are less able to survive adversity than large groups.

Other control: sawflies frequently cause consternation when found on garden plants. The adults do not sting and the larvae are harmless, although the fluids they regurgitate can irritate the eyes. Although destructive, their occurrence is usually sporadic and they are readily controlled by picking off the clusters and squashing them, or by dropping them in boiling water or a mix of water and kerosene. They can also be knocked to the ground with sticks or jets of water. Once on the ground they are not mobile and are subject to attack by predators. Clusters of larvae can also be killed by spraying with a product containing malathion or pyrethrum.

SAWFLIES OF SUBFAMILY PERGINAE

Larvae of the subfamily Perginae are large and fleshy with no prolegs on the abdomen and no protruding spine-like tail. They commonly feed on species of *Angophora*, *Corymbia* and *Eucalyptus*. The larvae congregate in groups on stems or leaves during daylight hours. Often there are several groups on a tree but eventually, as the larvae grow, the groups merge together to form one huge cluster that can contain hundreds of individuals. At night the larvae move in procession to the young shoots which they systematically strip of all leaves. When mature the larvae move en masse down the tree trunk into the soil, pupating in a cluster of fused cocoons made from cemented soil particles.

Sawfly larvae exude irritating fluid when disturbed. [T. Wood]

Sawfly larvae feeding on Rhodamnia *in rainforest.* [D. Jones]

Sawfly larvae. [D. Jones]

Sawfly larvae. [T. Wood]

Sawfly larvae. [T. Wood]

Sawfly larvae. [T. Wood]

Sawfly larvae. [T. Wood]

Sawfly larvae. [T. Wood]

EXAMPLES

Cypress Pine Sawfly, *Zenarge turneri*

Description: pale yellow eggs are laid singly near the base of a leaf scale. Young larvae feed singly on young growth, often resulting in the tips of shoots withering and falling off. Young larvae are yellow, older larvae become green to glaucous-green depending on the host plant. The larvae grow 2–2.5cm long. The end of the abdomen is often curled like a short trunk. Mature larvae travel down the trunk in groups to pupate in a white cocoon in the soil. Adults are pale orange sawflies with prominent black eyes.

Climatic regions: subtropics and temperate regions, coastal and inland.

Host plants: species of *Callitris* and exotic species of *Cupressus*.

Feeding habits: grubs feed singly, sometimes congregating in clusters. Often individual branches within a tree will be defoliated, resulting in dieback.

Notes: this species is of sporadic occurrence. In some years it occurs in large numbers, causing significant damage to stands of native pine, especially in inland districts. Although attacks of this pest are most severe on young trees, complete defoliation of large trees has been recorded. Stressed trees, such as are found on sites of poor drainage or growing in heavy soil, are more severely attacked than healthy trees. Attacks by Cypress Pine Beetle, *Diadoxus erythrurus* (see chapter 13), may follow sawfly attacks, often resulting in tree death. Sawfly larvae feeding on plants of golden cypress exhibit stronger yellow colouration than those feeding on green cypress plants and *Callitris*. Inland populations of this sawfly have been described as a distinct subspecies (subsp. *rabus*).

Control: as per introduction for sawflies. Spraying may be necessary.

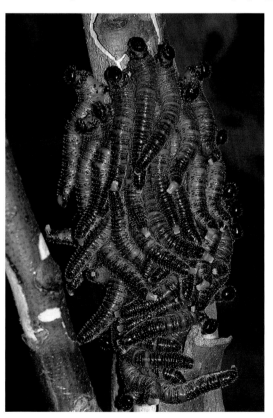

Larvae of the Eucalyptus Sawfly. [T. Wood]

Eucalyptus Sawfly, *Perga kirbii*

Description: fleshy pale-brown tapering grubs, 3–4cm long, with a conspicuous black head and yellow tip on the tail. The adult is a brown sawfly about 2cm long.

Climatic regions: subtropics and temperate regions of eastern Australia from Queensland to Victoria.

Host plants: *Corymbia maculata, Eucalyptus camaldulensis, E. grandis, E. leucoxylon.*

Feeding habits: larvae congregate in huge clusters on the trunk by day and disperse at night to feed.

Notes: a common pest which can defoliate trees.

Control: as per introduction for sawflies.

Large Green Sawfly, *Perga affinis*

Description: distinctive black grubs to about 4cm long with short white bristly hairs. The grubs taper from the front to the back and the lower abdomen curls like an elephant's trunk. Mature larvae pupate in massed groups in the soil. The adult is a metallic green sawfly about 2.5cm long.

Climatic regions: subtropics and temperate regions of eastern Australia (Queensland to Tasmania).

Host plants: *Corymbia ficifolia* and many species of *Eucalyptus*, including *E. amygdalina, E. bicostata, E. globulus, E. lansdowneana, E. leucoxylon, E. ovata E. polyanthemos, E. sideroxylon* and *E. viminalis*

Feeding habits: the grubs systematically strip young trees from the top down. The stems of defoliated trees often take on reddish tones. After stripping a tree of leaves, the grubs move en masse across the ground to an adjacent tree.

Notes: attacks occur from autumn to early summer. This species is an important pest of plantation-grown eucalypts and also commonly damages trees in farm windbreaks and along road verges.

Control: as per introduction for sawflies.

Larvae of the Large Green Sawfly. [T. Wood]

Red-brown Sawfly or Brown Sawfly, *Peragrapta polita*

Description: fleshy, light brown to reddish-brown, tapering grubs, up to 3cm long with a black head, short bristly hairs on the body and a short black tip on the tail. Mature larvae often become bright red before pupating. The adult is a brown sawfly about 2.5cm long with yellow antennae.

Climatic regions: tropics, subtropics and temperate regions of eastern Australia.

Host plants: *Angophora floribunda, Corymbia citriodora, C. gummifera,* many species of *Eucalyptus,* including *E. bicostata, E. botryoides, E. crebra, E. deanei, E. elata, E. globulus, E. grandis, E. saligna, E. scoparia* and *E. sideroxylon;* also *Melaleuca quinquenervia.*

Feeding habits: the grubs congregate in small to moderately large clusters during the day, dispersing at night to feed.

Notes: when disturbed the larvae lift their tails together as a group.

Control: as per introduction for sawflies.

Small Brown Sawfly, *Pseudoperga lewisii*

Description: fleshy, grey-brown to light brown grubs, up to 2.5cm long with a black head, short bristly hairs on the body and a blackish tail. The adult is a small pale brown to yellow-brown sawfly about 1.8cm long.

Climatic regions: tropics, subtropics and temperate regions of eastern Australia from Queensland to Tasmania.

Host plants: *Corymbia gummifera, Eucalyptus dives, E. nitens, E. obliqua, E. pauciflora, E. tereticornis* and *E. viminalis.*

Feeding habits: the grubs congregate in small to moderately large clusters during the day, dispersing at night to feed.

Notes: when disturbed the larvae lift their tails together as a group.

Control: as per introduction for sawflies.

Adult female Small Brown Sawfly guarding young larvae. [T. WOOD]

Mature larvae of the Small Brown Sawfly. [T. WOOD]

Steel-blue Sawfly, *Perga dorsalis*

Description: distinctive blue-black to black grubs which grow up to about 7cm long and have prominent bristly white hairs. The grubs taper noticeably from the front to the back and the lower abdomen curls like an elephant's trunk. Mature larvae migrate in groups, burrowing in soil at the base of the tree, pupating in a dark brown cocoon made from cemented soil particles. The cocoons are often joined together in a mass. The adult is a colourful sawfly (brown with yellow and red markings) about 2.5cm long.

Young larvae of the Steel-blue Sawfly. [T. WOOD]

Larvae of the Steel-blue Sawfly. [T. WOOD]

Climatic regions: subtropics and temperate regions of eastern Australia.

Host plants: *Corymbia citriodora*, *C. gummifera* and many species of *Eucalyptus*, including *E. camaldulensis*, *E. globulus*, *E. grandis*, *E. leucoxylon* and *E. obliqua*. Older larvae often can be seen in large prominent clusters on the trunk and larger branches of white-trunked eucalypts, such as *E. haemastoma* and *E. racemosa* subsp. *rossii*. Swamp gum, *E. ovata*, is also a favourite. This sawfly has also been reported to feed on *Callistemon*.

Feeding habits: this species clusters in crowded groups, the older larvae aggregating into very large clusters (sometimes with hundreds in a cluster). They communicate by tapping their tails. When disturbed they exude a sticky yellow liquid from their mouth that smells of eucalyptus oil. At night they disperse to the outer branchlets, eating the leaves. Older larvae may also eat bark.

Notes: a repulsive looking grub that is a major pest of plantation-grown eucalypts. Whereas damage by this pest often appears to be limited, research shows that attacks by this sawfly significantly slow the growth of young trees and sustained attacks can cause mortality. *Eucalyptus globulus* is especially susceptible.

Control: as per introduction for sawflies. Usually controlled by hosing and beating with sticks to knock larvae to the ground, but young trees may need to be sprayed.

SAWFLIES OF SUBFAMILY PTERYGOPHORINAE

Larvae of the subfamily Pterygophorinae are slender with six or more prolegs on the abdomen and a distinctive spine-like tail on the tip of the abdomen. Often their bodies are adorned with jewel-like warts. They feed on a wide range of plants, including eucalypts. Feeding takes place day and night, with the larvae grouped side by side in small clusters. Mature larvae either pupate in the soil or in holes bored in bark.

EXAMPLES

Bottlebrush Sawfly or Ringed Sawfly, *Pterygophorus cinctus*

Description: eggs are laid in a raft inside the edge of a leaf. The larvae are fleshy pale brown to grey-brown tapered grubs up to about 3cm long with a black head, black legs and a warty black spine-like tail. Small jewel-like warts cover the surface of the thorax. The adult sawfly, with a wingspan of about 2cm, has a yellow body with black bands.

Climatic regions: subtropical and temperate regions.

Host plants: species and cultivars of *Callistemon* and *Leptospermum*.

Feeding habits: the grubs congregate in small groups, feeding side by side on a leaf, commonly leaving the main veins as a skeleton. Large larvae eat whole leaves.

Notes: when disturbed the larvae lift their heads and tails and exude a dollop of sticky liquid.

Control: as per introduction for sawflies.

Ironbark Sawfly or Green Long-tailed Sawfly, *Lophyrotoma interrupta*

Description: fleshy, tapering grubs, up to 3cm long, greenish in colour with a black head, jewel-like warts on the thorax and a long black warty spine-like tail. The adult is a yellow and black sawfly about 2cm long with black wings.

Climatic regions: tropics, subtropics and temperate regions of eastern Australia.

Host plants: *Angophora floribunda*, *Corymbia gummifera*, *C. maculata*, and many species of *Eucalyptus* including *E. camaldulensis*, *E. crebra*, *E. grandis*, *E. melanophloia*, *E. moluccana*, *E. obliqua*, *E. ovata*, *E. siderophloia* and *E. viminalis*.

Feeding habits: grubs congregate in dense clusters during the day and disperse at night to feed on the leaves.

Notes: in some seasons this pest devastates large areas of forest in Queensland, attacking a wide range of trees. Individual trees of Silver-leaved Ironbark, *Eucalyptus melanophloia*, which are isolated in paddocks are much favoured and may be completely defoliated. Before pupating the caterpillars congregate en masse at the base of the trunk. In some inland districts of Queensland, browsing cattle eat the clusters of grubs causing significant stock losses (the cost of deaths in 1981 was estimated at more than one million dollars). Dead caterpillars are also toxic to cattle if eaten. Impacts that have resulted from the activities of this pest include clearing ironbarks from paddocks and restricting access by cattle to infested paddocks between July and October.

Control: as per introduction for sawflies.

Paperbark Sawfly, *Lophyrotoma zonalis*

Description: fleshy, greenish, tapering grubs to 3cm long with a black head, jewel-like warts on the thorax and a long, black, warty, spine-like tail. The adult is a yellow and black sawfly about 2cm long.

Climatic regions: tropics, subtropics and temperate regions of eastern Australia.

Host plants: broad-leaved species of *Melaleuca*, including *M. dealbata*, *M. leucadendra*, *M. quinquenervia* and *M. viridiflora*.

Feeding habits: the grubs congregate in small groups feeding side by side on a leaf, commonly leaving the main veins as a skeleton. Large larvae eat whole leaves.

Notes: this species has been investigated as a possible biological control agent to control naturalised populations of *M. quinquenervia* in Florida. When disturbed the larvae lift their heads and tails and exude a dollop of sticky liquid. Mature larvae bore into soft wood or the papery bark of the tree to pupate, sometimes causing significant bark damage in the process.

Control: as per introduction for sawflies.

Long-tailed Sawfly, *Pterygophorus insignis*

Description: greenish, brownish or reddish grubs, up to 3cm long with jewel-like warts over much of the surface, including the spine-like tail. The adult is a small orange and black sawfly.

Climatic regions: tropical to temperate regions.

Host plants: species and cultivars of *Callistemon* and *Leptospermum*; also *Melaleuca armillaris* and *M. quinquenervia*.

Feeding habits: young grubs skeletonise tissue leaving the veins while larger grubs eat the leaves entirely, sometimes leaving only the midrib. They may feed day and night, although sometimes they cluster on leaves in small groups during the day.

Notes: a very destructive pest of bottlebrushes, often defoliating whole plants. In the early stages the insects are hardly noticeable, but the results of their ravenous appetite very soon become obvious. Their feeding can impact the growth habit of the plant and interfere with flowering.

Control: as per introduction for sawflies. Spraying is often necessary.

Larvae of the Long-tailed Sawfly.
[D. Jones]

Long-tailed Sawfly larvae on Callistemon. [D. Jones]

Other Sawflies

- The Red Ash Sawfly, *Philomastix xanthophylax*, attacks the trees of Red Ash, *Alphitonia excelsa*, in winter, often causing considerable damage, sometimes defoliation. The slender green larvae, which grow to about 2cm long, have two spine-like tails at the base. Mature larvae pupate in the ground.

- The Blueberry Ash, *Elaeocarpus reticulatus*, is attacked by larvae of the sawfly *Pteryperga galla*. The larvae congregate in groups on the trunk and larger branches eating patches of bark. They also pupate in groups.

- Some species of *Syzygium*, notably *S. francisii*, are attacked by the larvae of *Pterygophorus turneri*. They are greenish with a black warty spine-like tail and two erect black spikes on the thorax behind the head.

- Fronds of the semi-aquatic fern *Marsilea drummondii* are eaten by the larvae of the sawfly *Warra froggatii*.

- In New South Wales and Victoria the Narrow-leaved Peppermint, *Eucalyptus radiata*, is attacked by the larvae of *Lophyrotoma cyanea*.

- The leaf blister sawflies are a specialised group of sawflies that feed on the tissues between the leaf surfaces – see next entry.

Sawfly larvae feeding on Syzygium *in rainforest.* [D. JONES]

LEAF BLISTER SAWFLIES

The larvae of the leaf blister sawflies (order Hymenoptera, family Pergidae) feed on the internal tissues of leaves in a similar way to the leaf miners. Some species concentrate on the upper internal cells of the leaf, others are less specific. The larvae create large blisters that eventually turn white to brown and become unsightly. Most attacks concentrate on growth that is just hardening and commonly take place on branches in the lower parts of the tree (usually less than 4–6m from the ground). Attacks are worst on young trees and the insects can be present all year round in some regions. The eggs are laid singly or in rows on the underside of a leaf near the midrib. The larvae, which resemble small, flattish grubs, spend all their feeding life within the blister. Each blister in its early stage contains a single larva, but as the feeding continues the blisters enlarge, overlap and merge, eventually containing several larvae. Pupation occurs within silken cocoons built inside the blister. The larvae are able to continue growth and pupate inside fallen leaves. The adult sawflies, which are actually a highly specialised wasp, are small, being 0.4–0.7cm long, and often orange in colour. They move

Eucalyptus nichollii *badly damaged by leaf blister sawflies.* [D. BEARDSELL]

about actively and are winged. The females have an ovipositor on the underside of the abdomen that resembles a sawblade. The female uses this blade to cut slits in leaves when depositing the eggs.

Natural control: leaf blister sawflies are attacked by parasitic wasps which lay eggs through the blister wall into the body of the larvae. They are also targeted by small birds such as pardalotes.

Other control: leaf blister sawflies are very difficult to effectively control without resorting to highly toxic chemicals, because the larvae are shielded by the leaf surfaces. Most attacks are sporadic and of a minor nature and mechanical control such as regularly removing affected growth can be effective at reducing insect numbers. Spraying new growth with pest oil will deter egg-laying. These sprays should not be used in hot weather or foliage damage may result.

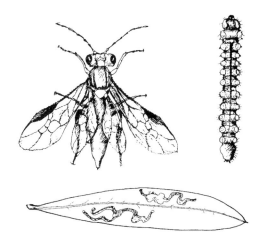

Leaf blister sawfly adult, damaged leaf and larva. [T. BLAKE]

EXAMPLES

Eucalyptus Leaf Blister Sawfly, *Phylacteophaga froggatii*

Description: small, nearly legless pale yellow grubs, 0.6–1cm long, that graze the upper layers of cells within the leaves, forming a large papery brown blister and leaving the cuticle intact. The tunnels of larvae that will become adult female insects are larger than those which will become adult males. Adjacent tunnels coalesce to form a large blister containing several larvae. The adult insect is a small orange sawfly, 0.6–0.8cm long with a thick waist and prominent black eyes.

Climatic regions: subtropics and temperate regions (including Western Australia), coastal and inland.

Host plants: *Corymbia citriodora, C. ficifolia, C. maculata, Eucalyptus botryoides, E. cinerea, E. globulus, E. nitens, E. perriniana, E. rudis, E. sideroxylon, E. saligna* and *E. viminalis;* also *Lophostemon confertus* and occasionally *Agonis flexuosa*. This pest is often severe on *Eucalyptus tereticornis* and *E. nicholii* planted in inland towns.

Feeding habits: a gregarious species that causes large grey papery blisters on leaves, which eventually become brown or white and unsightly.

Notes: a serious pest which can defoliate young trees, impede growth and influence final tree shape. Sustained attacks can cause dieback and even death. A biological control program in New Zealand using the parasitic wasp *Bracon phylacteophagus* was highly successful in controlling this pest.

Control: as per introduction for leaf miners. Spraying young trees may be necessary to aid their establishment.

Damage to eucalypt leaf by leaf blister sawfly. [D. JONES]

Leaf blister sawfly damage on Eucalyptus botryoides. [D. JONES]

Other Leaf Blister Sawflies

- The larvae of *Phylacteophaga eucalypti*, another widespread and common species with a similar range to that of *P. froggatii*, feed on the leaves of numerous *Eucalyptus* species and some species of *Corymbia*.

- The larvae of the Western Leaf Blister Sawfly, *Phylacteophaga occidens*, feed on the leaves of *Eucalyptus marginata* and *E. rudis*.

- The larvae of the Queensland Leaf Blister Sawfly, *Phylacteophaga rubida*, feed on the leaves of *Tristaniopsis conferta*.

- The larvae of the Tasmanian Leaf Blister Sawfly, *Phylacteophaga amygdalina*, feed on the leaves of *Eucalyptus amygdalina*.

Leaf blister sawfly damage to Corymbia maculata. [D. Jones]

Leaf blister sawfly damage on Leptospermum laevigatum. [D. Jones]

Leaf blister sawfly damage on Acacia melanoxylon. [D. Jones]

CHAPTER 12

SNAILS AND SLUGS

There are about 1,000 species of snails and slugs native to Australia, but most of these are not pests and very few are ever found in gardens. Some of these harmless slugs and snails feed only on dead material, algae, fungi or animal wastes, while a few are actually carnivorous, eating other slugs, snails, worms and other soft-bodied creatures. Some native species live high in the canopy of trees, eating fungi or lichens, and are rarely seen.

Red Triangle Slug: a harmless native species. [D. Jones]

A native slug. [D. Jones]

More than 65 species of slugs and snails have been introduced accidentally or deliberately from overseas and have become established in various parts of Australia. Many of these are serious pests of horticulture and agriculture and can quickly build up in numbers and run riot in gardens unless checked. Some of these are familiar and common pests that attack a wide range of plants. They are very destructive of seedlings and plants with soft or tender growth such as lilies, bulbs, kangaroo paws, ferns, gingers and orchids. Sometimes susceptible plants can be eaten right back to ground level. Even the surfaces of bulbs and rhizomes are sometimes eaten. Slugs and snails also eat the green bark on the stems of shrubs, less commonly trees. When in sufficient numbers these pests can cause severe bark damage, sometimes ringbarking the trunk and killing the plant. Susceptible native plants include species of *Atriplex, Chorizema, Chrysocephalum, Correa, Eremophila, Goodenia, Grevillea, Hovea, Hymenosporum, Melia, Myoporum, Prostanthera, Rhagodia, Scaevola, Westringia* and *Xerochrysum*.

Snail and slug damage is fairly distinctive. The animals graze on the upper layers of the plant tissue leaving strips of eaten tissue and they can also eat chunks out of leaves, especially after repeated feeding. In severe infestations these pests congregate on succulent tissue and the destruction of tissue can be rapid. Repeated browsing on soft tender growth results in the surface layers being stripped off and the sappy cells exposed. These damaged areas often become slimy from the exudation of sap and are a point of entry for disease. The grazed areas are at first pale-green and then become papery and die. Because of some basic structural differences, these two groups of pests are dealt with separately.

SNAILS

Snails have a hard external shell that protects the body of the animal. Lines of growth are visible on the shell, these added to as the animal grows. Snails have four retractile tentacles at the head end; the lower pair act as feelers, the upper pair carry the eyes. They also have very efficient mouthparts consisting of a rasp-like tongue and thousands of teeth. Snails glide on the slime they produce from the underside of their foot. They thrive in coastal areas, especially those with limestone and limey soils where there is abundant calcium for shell development. Snails shelter in cool protected sites during the day, become active after rain and mainly feed at night. They become dormant over summer or during long dry periods, sealing off their shells with a thin mucus layer to prevent water loss. Some species climb on structures before entering dormancy. All snails are hermaphrodites and can produce large numbers of eggs – up to 1,000 eggs each season from a single mated snail.

Natural control: snails are eaten by birds, rats, large lizards and predatory slugs; also ducks and other poultry.

Other control: regular baiting with commercial snail preparations is the best means of controlling these pests. Baits containing metaldehyde, methiocarb or those based on iron chelate are most commonly used. Baits based on iron chelate are slower to kill the snails but still very effective, and less toxic to children, wildlife, domestic pets and beneficial animals. Poisonous baits are best scattered rather than placed in heaps. Baits can be applied to prepared areas a few days before planting to assist with early control. Baits should be applied just before rain to coincide with the maximum activity of these pests. Baiting after the first autumn rains is recommended for some types of snail as they are foraging after emerging from dormancy. Soft tender plants should be protected regularly with baits, especially when they are making new growth. Young plants and seedlings should be baited until established. Searching by torchlight at night, handpicking and squashing is a useful technique to reduce numbers, especially in glasshouses and ferneries. Shallow containers of stale beer are often used to lure and drown snails. Another useful attractant is bran soaked in vinegar. Finely-ground tobacco dust is an effective killer of slugs and snails and works well if spread around tender plants. Copper bands on pots and trees provide a barrier as they will not be crossed by snails.

EXAMPLES

Common Garden Snail or European Snail, *Helix aspersa*

Description: a large snail with a hard, globose shell 2–4cm across. The shell is yellowish-brown with five dark brown bands interrupted with yellowish streaks, mottlings and blotches. The soft, slimy body, which is brownish grey to glistening purple-black, retracts fully into the shell when inactive. The head has four slender retractile tentacles, the upper pair carrying the eyes.

Climatic regions: tropics (mainly highland areas), subtropics and temperate regions, chiefly coastal.

Host plants: feeds on the leaves and green bark of a very wide range of plants including some crop species. These snails are very destructive of plants with soft tissue such as ferns, orchids, bulbs and kangaroo paws.

Feeding habits: grazes long patches of tissue from the surface layers. Damaged areas leak sap and become slimy. Repeated grazing may result in holes eaten right through the tissue. The snails are most active at night but also feed during drizzly or rainy weather.

Notes: native to the Mediterranean region and western Europe, but now widespread around the world. This is the most serious and destructive snail introduced into Australia, where it has been established for more than 120 years. Its most obvious impacts are in gardens, nurseries, greenhouses, orchards and vegetable crops. Less well known is its ability to establish in bushland near houses and in tussock grassland, especially in near-coastal sites. In these native habitats the snails' feeding quickly eliminates

Snail damage to Hymenosporum flavum. [R. Elliot]

Common Garden Snail. [T. Blake]

tender plants such as ground orchids, ferns and lilies. Its eggs, which are laid in a concealed nest among loose litter, are clusters of small white spheres. Common Garden Snails survive dry periods by sealing the opening of their shell and remaining dormant.

Control: as per introduction for slugs and snails.

Green Snail, *Cantareus apertus*

Description: juveniles have a greenish to yellowish-green shell and a cream body. Adults have a uniformly coloured olive-green to brown shell (with no mottles or lighter markings), measure 2–2.5cm across and a have a cream to white body, often with a grey dorsal band and grey eye-stalks.

Climatic regions: well established around Perth, with an isolated outbreak near Capel in the southwest of Western Australia. In 2011 about six infested properties were found at Cobram East in northern Victoria.

Host plants: feeds on the leaves and green bark of a range of native plants as well as vegetables, crop plants and pasture.

Feeding habits: grazes surface tissue leaving translucent patches. Only leaf veins may be left after severe attacks.

Green Snail and Common Garden Snail compared. [A. Everett]

The Green Snail is an exotic environmental pest recently found in Western Australia and northern Victoria. [A. Everett]

Notes: this recently introduced pest is native to southern Europe and northern Africa. It feeds mainly from autumn to spring and becomes dormant over late spring-summer, burrowing into soil and surviving underground. Eggs are laid beneath the soil surface. This snail is a prolific breeder, with up to 1,000 young snails recorded per square metre. It can survive in native bush and is a serious threat to native ground orchids, lilies and other plants with fleshy tissue. It appears that this pest is not considered an important environmental threat in Victoria because insufficient funds have been made available to support an attempt at its eradication.

Control: as per introduction for slugs and snails. Baits must be applied when the snails are actively feeding (late autumn-spring).

Asian Tramp Snail or Bradybaena Snail, *Bradybaena similaris*

Description: adults have a circular brownish or yellowish shell, 1.2–1.6cm across, usually with a narrow brown stripe on the edge of the whorl. The shell flares at the mouth where it is usually white. The body is tan to brown and smooth with brown eyestalks. Juveniles are miniatures of the adults.

Climatic regions: well established on the east coast from Brisbane south to near Bega, NSW.

Host plants: feeds on a large range of soft plants including orchids, lilies and ferns; also a serious pest of citrus.

Feeding habits: grazes surface tissue leaving exposed strips; also eats leaves from the margins inwards and feeds on green bark.

Notes: originating from eastern Asia, it is now widespread in the tropics and is a serious pest of crop plants, greenhouses and gardens. It establishes quickly in an area and can become the dominant snail pest, out-competing others. The snails begin egg-laying when about six months old and can produce up to 100 eggs in a single batch.

Control: as per introduction for slugs and snails. Baits must be applied when the snails are actively feeding.

White Garden Snail, Sandhill Snail or White Italian Snail, *Theba pisana*

Description: a small to medium sized snail with a somewhat flattened shell measuring 1.5–2.5cm across, which is white to yellow-brown and usually ornamented with narrow brown spiral bands.

Climatic regions: temperate regions, mainly coastal.

Host plants: a wide range of garden plants and crops including citrus and grape vines; it especially attacks native legumes in gardens and natural habitats.

Feeding habits: grazes surface tissue of green stems and leaves.

Notes: an introduced snail from the Mediterranean region which was brought to coastal areas of Perth in the 1890s and is now widely distributed in coastal districts of southern Australia, including Bass Strait islands and northern Tasmania. It climbs readily and in dry times congregates in huge numbers on trees, weeds, crop plants, fence posts, walls and under leaves.

Control: as per introduction for slugs and snails. This pest is killed by heavy frosts and burning.

White Garden Snails on grass tree leaves.
[S. Jones]

Shells of the White Garden Snail (the brightly coloured ones are alive). [D. Jones]

Garlic Snail, *Oxychilus alliarius*

Description: a small snail with a somewhat flattened, coiled (4–5 whorls), dark brown, glossy shell, 0.5–0.6cm across. The body is shiny blue-grey and emits a strong garlic smell when squashed.

Climatic regions: temperate regions (southern New South Wales, southern Victoria, Tasmania).

Host plants: feeds mainly on the tender growth of fleshy plants such as bulbs, ferns and orchids; it is particularly fond of the soft new growth and growing root tips of orchids and ferns.

Feeding habits: grazes small patches or chews lumps out of very soft tissue.

Notes: a European snail that was first recorded in Australia in 1852. It is found mainly in moist, humid situations and establishes readily in glasshouses and shadehouses. It feeds on other snails including native species and is associated with a significant decline of native snails in Hawaii. At least two other species of *Oxychilus* are naturalised in Victoria.

Control: as per introduction for slugs and snails.

Garlic Snail. [D. JONES]

Bark Snail or Orchid Snail, *Zonitoides arboreus*

Description: a tiny snail with a flattish, coiled (4.5 whorls), light brown to creamy brown, glossy shell measuring 0.4–0.6cm across. The body is shiny blue-grey. When sliding the body extends from the shell for 0.6–0.7cm. No smell is noticeable when squashed.

Climatic regions: subtropical and temperate regions (Queensland, New South Wales, possibly Victoria).

Host plants: feeds mainly on the tender growth of fleshy plants such as orchids and ferns.

Feeding habits: grazes small patches from the surface of leaves, buds and petals; also new shoots and new roots. Feeding snails concentrate on the soft growing tips of roots which either cease growth or continue growing and develop a narrowed or constricted area in the region where feeding occurred.

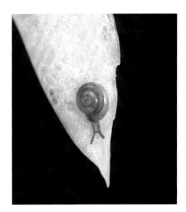

Orchid Snail. [D. JONES]

Notes: a North American snail that was apparently first collected in Sydney in 1868 and described as *Helix lyndhurstensis*, a purported new species of native snail. It favours moist, humid conditions and establishes readily in greenhouses, glasshouses and shadehouses. It burrows into coarse, open potting mixes, such as those based on pine bark, and emerges to feed after dark and in moist weather. It has been widely distributed through orchid collections by passing on live plants containing the pest. It also establishes readily in garden litter and will invade natural habitats. This snail is readily confused with the previous species but is smaller overall and does not smell of garlic when squashed.

Control: as per introduction for slugs and snails.

Orchid Snail damage to roots. [D. JONES]

Tiny Spiral Snail, *Cochlicella ventricosa*

Description: a small snail with a thin-textured almost transparent conical shell about 1cm long. The colour is whitish-grey with a brown spiral band.
Climatic regions: temperate regions of south-eastern Australia.
Host plants: feeds on a wide range of garden plants; also ferns and orchids in glasshouses.
Feeding habits: grazes small patches of tissue.
Notes: an introduced species from the Mediterranean region which feeds mainly at night, hiding during the day. It congregates on sheltered walls and other structures in dry conditions.
Control: as per introduction for slugs and snails.

Other Snails

- **Giant African Snail**, *Achatina fulica*, does not occur in Australia although there have been a couple of localised populations found in Queensland which have been subsequently eradicated. This African snail, one of the largest and most damaging land snails in the world, is widely naturalised in the Indo-Pacific region and feeds on a huge range of plants. Large individuals can weigh up to 1kg with a brown conical shell up to 30cm long. They reproduce fast and can quickly colonise an area. They pose a serious threat to gardens and commercial plant growers and also natural populations of many types of native plants including ferns and ground orchids. Any suspicious sightings should be reported immediately to a biosecurity department.

SLUGS

Slugs lack the external hard shell of snails – although some have a much-reduced internal shell – but otherwise are basically similar in structure. They have tentacles on the head identical to a snail and the part of the back behind the head has a swelling known as the saddle or mantle. They breathe through a hole known as the pneumostome. Slugs are hermaphrodites and both members of a mating couple can lay eggs, which are often hidden in potting mix. Slugs like mild to warm wet weather. They can only tolerate limited dryness and rehydrate through their skin. Some slugs can burrow in soil and feed on plant roots. Slugs favour heavy soils and those rich in organic matter.

Natural control: slugs are eaten by birds, frogs, lizards, predatory slugs, assassin bugs, native earwings, carabid beetles, ducks and other poultry.

Other control: as for snails.

Slug damage on a leaf of kangaroo paw. [R. ELLIOT]

EXAMPLES

Common Garden Slug, *Deroceras invadens*

Description: a slender grey to brown smooth slug with a lighter-coloured saddle behind the head. Grows to 2–3.5cm in length. Releases clear mucus when sliding.

Climatic regions: tropics, subtropics and temperate regions, coastal and inland.

Host plants: feeds on the leaves, soft stems, green bark, flowers and soft fruit of a very wide range of plants. It is very destructive of plants with soft tissue including ferns, orchids, forbs, lilies, bulbs and kangaroo paws.

Feeding habits: grazes strips and patches of tissue from the surface layers. Affected tissue becomes slimy. This slug is most active at night and during rainy weather.

Notes: a common, serious horticultural pest that probably originates from the Mediterranean region and was first found in Australia in 1967. It establishes readily in any moist area where there is food, especially on disturbed sites. It can survive in native bush and grassland, and is a serious threat to ground orchids, lilies and other plants with fleshy tissue. Populations increase quickly under suitable conditions.

Control: as for slugs and snails.

Grey Garden Slug or Reticulated Slug, *Deroceras reticulatum*

Description: a slender pale yellow to cream or whitish smooth slug with a dark brown to grey mottled pattern. Grows 4–5cm long. Releases milky white mucus when sliding.

Climatic regions: subtropics and temperate regions of eastern Australia, mainly coastal and adjacent inland areas.

Host plants: feeds on the leaves, soft stems, green bark, flowers and soft fruit of a very wide range of plants, including field crops and vegetables.

Feeding habits: grazes strips and patches of tissue from the surface layers; eats lumps out of leaves; completely destroys seedlings. Affected tissue becomes slimy. This slug is most active at night and during rainy weather.

Notes: this slug, which is from Europe and North Africa, is one of the most serious garden pests. Its numbers can build up quickly under suitable conditions.

Control: as per introduction for slugs and snails.

Great Yellow Slug, *Limax flavus*

Description: a stout, yellow, smooth slug with orange-brown or greenish markings. Grows 8–10cm long and has bluish tentacles. Releases thick yellow mucus when sliding.

Climatic regions: temperate regions, mainly coastal.

Host plants: feeds on tender growth of a range of plants, especially seedlings.

Feeding habits: a large slug which grazes patches of tissue from the surface layers. This slug is most active at night and during rainy weather.

Notes: a European slug that was first recorded in Australia in 1852.

Control: as per introduction for slugs and snails.

Hedgehog Slug, *Arion intermedius*

Description: a plump, small, cream to pale yellow, smooth slug with a dark grey to black head and tentacles. Grows to 2cm long and has the breathing pore in front of the mantle. Releases clear mucus when sliding. Raised bumps appear on the upper surface when the slug is disturbed.

Climatic regions: temperate regions (New South Wales, Victoria, Tasmania), mainly coastal and adjacent inland areas.

Host plants: feeds on tender growth of a range of plants,

Feeding habits: grazes strips and patches of tissue from the surface layers.

Notes: a European slug that lives in gardens and also invades native bushland and grassland.

Control: as per introduction for slugs and snails.

Black Slug, *Arion ater*

Description: a large black slug (sometimes brownish) that grows 10–15cm long. The body is distinctively roughened with rows of bumps or tubercles – these bumps are often in lines on the upper side of the foot. The foot is striped with orange or yellow. Releases thick, viscous mucus when sliding. This mucus is difficult to wash off and apparently foul-tasting. When disturbed this slug curls and rocks from side to side.
Climatic regions: temperate areas (South Australia, Victoria, New South Wales); populations are established in the Dandenong Ranges and Otway Ranges, Victoria.
Host plants: feeds nocturnally on fungi and plants (living and dead); apparently damaging to seedlings.
Feeding habits: grazes strips and patches of tissue from the surface layers; eats lumps out of leaves. The slugs are most active during rainy warm weather in spring and summer.
Notes: a large European slug with a distinctive bumpy appearance. It is potentially a serious invader.
Control: as per introduction for slugs and snails.

Striped Slug, *Lehmannia nyctelia*

Description: a slender, yellowish to light-brown, smooth slug that grows 4–6cm long. It has two darker stripes on the tail and three stripes on the mantle. Releases clear mucus when sliding.
Climatic regions: temperate regions (New South Wales, Victoria, Tasmania, South Australia, Western Australia), coastal and inland.
Host plants: feeds on a wide range of plants.
Feeding habits: grazes strips and patches of tissue from the surface layers. The slugs, which are most active during rainy weather, often congregate in groups when sheltering.
Notes: a European slug first recorded in Australia in 1881. It is easily recognised by the stripes on its back.
Control: as per introduction for slugs and snails.

Striped Slug. [D. JONES]

Black Keeled Slug or Greenhouse Slug, *Milax gagates*

Description: a slender, dark grey to blackish, smooth slug with a ridge or keel on its back. Grows 4–5cm long. Releases clear mucus when sliding.
Climatic regions: subtropics and temperate regions, coastal and inland.
Host plants: feeds on soft shoots, flower buds and root tips of a wide range of plants.
Feeding habits: grazes strips and patches of tissue from the surface layers. It burrows readily into soil, eating roots, tubers and bulbs.
Notes: this species, native to North Africa and Europe, readily populates cool moist sites, including ferneries, greenhouses and shadehouses.
Control: as per introduction for slugs and snails.

Other slugs

- **Leopard Slug**, *Limax maximus*, a native of Europe, is well established in temperate parts of Australia. It is a large slug, 10–20cm long, which is greyish or brownish with prominent dark spots. It eats fungi, plants (living and dead), carrion, pet food, animal faeces and is usually considered beneficial since it kills and eats other slugs. It does however, eat plants. This slug mates in a distinctive manner, the mating pair entwining while hanging suspended on a string of thick mucus.

Leopard Slug. [D. JONES]

CHAPTER 13

BORERS, TWIG GIRDLERS AND BARK-FEEDING PESTS

A range of specialised insects attack the bark and wood of plants causing injuries ranging from minor to severe and even death. Most species included in this chapter attack living plants and need moist wood to complete their life cycle. A number of these insects take many years to become adult, sometimes emerging long after a tree has died or been cut for timber. Insects which feed on dry wood and thus are major pests of stored timber, buildings and furniture, are not dealt with in this book.

Whereas wood-feeding insects can attack healthy shrubs and trees, their major activities occur in weakened or stressed plants. Eucalypts with vigorous coppice growths following storm damage, bushfires or pollarding are frequently a favourite target. Other common examples of stressed and susceptible plants include trees retained in paddocks after clearing thick forest and scrub; those damaged and scarred by bushfires; elderly plants nearing the end of their life; and trees or shrubs with root damage, as commonly occurs following activities such as house construction, renovations, footpath and road construction, and any activity that involves ditch digging. Plants grown well away from their natural habitats (such as natives from the sandplains of Western Australia grown in eastern state gardens) and plants that have been weakened by nutrient starvation and inadequate watering can also become the targets of borer attack. Situations like these are far more common in gardens than is often realised and normal good gardening practices such as watering, light fertilising and mulching all serve to keep the plants healthy and help resist attack.

BORERS

As the common name suggests, these insects bore holes and tunnels through the bark and wood of plants. Commonly they attack the trunk and larger branches, but some smaller borers tunnel down the centre of young shoots causing wilting. Mostly the borers are grubs of specialised beetles, or the caterpillars of some large moths. Numerous native plants are subject to borer attack, however the most susceptible are species of *Acacia*, *Allocasuarina*, *Angophora*, *Corymbia* and *Eucalyptus*. Other commonly attacked plants include species of *Banksia*, *Callistemon*, *Callitris*, *Leptospermum*, *Melaleuca*, *Olearia*, *Pomaderris*, *Prostanthera* and *Grevillea*.

This Exocarpos *trunk broke at the site of borer activity.*
[D. Jones]

Accumulation of frass at the base of a wattle trunk is an indication of borers. [D. JONES]

Gallery of borer tunnels in old wattle trunk. [D. JONES]

Bark removed to show a gallery of borer tunnels in the sapwood. [D. JONES]

Borer damage is often found in the vicinity of a branch junction. [D. JONES]

Frass covering borer tunnels on Hakea *stems.* [S. JONES]

Borer hole surrounded by a large area of eaten bark. [D. JONES]

Borer attacks are most severe on weakened and stressed plants. Damaged tissue is a common point of entry for borers. Borers are also attracted to areas already damaged by borer attack, resulting in compounded attack. The exudation of gum from the trunk and branches and accumulations of sawdust and frass produced by the insects are good indications of borer activity, although an entry hole chewed by the grub is the most obvious feature of their presence. Often new cambial growth is produced by the plant around the entry hole. Some larvae tunnel just below the bark, others concentrate on the sapwood, and some of the larger species tunnel deep into the heartwood. Some borers construct simple tunnels, others form complex galleries.

Healthy plants are normally able to tolerate borer attack. Damaged sites are generally sealed off by the exudation of gum or kino which greatly restricts the borers' activities and may eventually cause their death. Borers can reduce the life expectancy of a plant and sustained attacks commonly lead to plant death. Borer activity also weakens the mechanical strength of young trunks and branches and it is not uncommon for breaks to occur at sites of borer activity. Severe damage by borers can also lead to the entry of fungal diseases.

While it is apparent that borers mostly attack plants which are weakened or have lost vigour, this is not always the case. The authors have witnessed numerous borer attacks on healthy and vigorous garden plants. Damage like this can be especially common in gardens recently established in estates that support

Borers in Hakea *seed pods.* [D. Jones]

Gum sealing site of borer damage in Acacia. [D. Jones]

Yellow-tailed Black Cockatoo with beetle larva taken from Melaleuca armillaris. [D. Jones]

large eucalypts and wattles weakened as a result of development and clearing. Stressed trees like these commonly support large populations of borers which spread into the nearby gardens.

Natural control: predators include birds, assassin bugs and predacious beetles; borers are also parasitised by wasps, viruses and fungi. Large cockatoos often feed on borer larvae and in the process cause considerable damage to the host tree.

Other control: control measures should begin as soon as borer activity is noticed. Probing the holes with soft pliable wire will kill any nearby larva. Squirting solutions of a mild contact insecticide into the hole may also be effective. Soap solutions or methylated spirits may cause the larvae to emerge and it can then be destroyed.

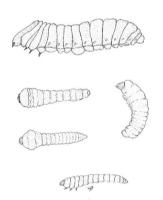

Examples of borer larvae. [D. Jones]

Severely affected branches should be removed and burnt. Dying or dead trees are best removed to reduce infestation of healthy plants. Bluestone paste will deter female beetles from laying eggs if applied in spring (see deterrents, chapter 2). Persistent borer attacks indicate that the plant is weakened or stressed and steps to improve its health should be taken quickly.

TRUNK AND LARGE BRANCH BORERS

Some borers, mostly larvae of jewel beetles and longicorn beetles, concentrate their attacks on tree trunks and the larger stems and branches. Some species of large wood moths and swift moths are also involved. The damage shows up as extrusions of gum, piles of frass and chewed sawdust, and split or peeling bark which lifts easily. The tunnel entrances can be seen when the covering material is removed. Sometimes an actively chewed area of bark or bark regrowth is present around the tunnel entrance. In the early stages of damage by these borers, the young leaves become yellow and die. The yellowing extends into older leaves and the tips die back and eventually the whole branch can die. Severely attacked trees look very unhealthy, usually with a number of dead branches and

Staining of eucalypt trunk due to gum exudation from borer damaged site. [D. Jones]

much of the trunk damaged. Because the trees are weakened they can shed branches during strong winds and be a potential hazard. Young trees suffering trunk damage can also break off at the weakened area during strong winds.

BEETLE BORERS OF TRUNKS AND LARGE BRANCHES

A couple of beetle families contain important borers. Sometimes they can be recognised by their activities, including the type of wood that is tunnelled, the pattern of the galleries formed and the shape of the holes. Most damage is caused by the larvae which are commonly flattish, fleshy grubs with a distinctive tapered shape. The adults often use damaged tissue as a point of entry. Such areas, including damaged or split bark, fire scars, storm-damage and pruning cuts, are attractive to

Severe damage to Eucalptus *trunk caused by persistent borer attack over several years.* [D. Jones]

female beetles which lay eggs on the exposed and damaged tissue. The young larvae may feed on the bark around the damaged area before tunnelling into the wood. Some larvae tunnel just below the bark, others concentrate on the sapwood, whereas some of the larger larvae are able to tunnel deep into the heartwood. Some borers construct simple tunnels, others form complex galleries. Often a specialised pupal chamber is associated with the tunnel. Borer attacks can be cyclic, with new generations of beetles attracted to the damaged areas to lay eggs and the growth of new bark is unable to seal the damaged areas.

JEWEL BEETLES (FAMILY BUPRESTIDAE)

This is a large family of beetles numbering more than 15,000 species worldwide; about 1,200 species are native to Australia. Known commonly as jewel beetles because of the combinations and patterns of attractive colours, these handsome beetles are widespread throughout Australia. They are also often known as 'metallic wood-boring beetles' because of the shiny metallic sheen or tinting on the wing-covers of some species. The larvae of most species are wood borers, attacking the stems, trunks and/ or roots of a very wide range of plants. Typical jewel beetle larvae are plump, creamy grubs with small rounded heads, much enlarged thoracic segments, often then tapering suddenly to an extended narrow abdomen. They tunnel into the sapwood of the host and form a series of interconnecting oval or flattish tunnels and galleries. The old tunnels are usually packed with

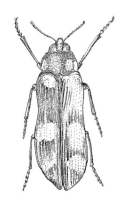

Adult jewel beetle. [T. Blake]

Larvae of a jewel beetle in Emmenosperma alphitonioides. [D. Jones]

Jewel beetle, Castiarina crenata. [T. Wood]

Jewel beetle, Castiarina decemmaculata. [T. Wood]

Jewel beetle, Castiarina sexplagiata. [T. Wood]

Jewel beetle, Castiarina subvicina. [T. Wood]

sawdust and frass (waste products). The beetles, which can fly actively during warm days, congregate and feed on flowers, chiefly those of the family Myrtaceae, sometimes also pea flowers (Fabaceae). The beetles of some species also feed on leaves. A few species of jewel beetles induce galls (see chapter 7). Typically the beetles fly rapidly when disturbed or drop to the ground. Jewel beetle numbers can fluctuate significantly from year to year. They are especially adapted to the drier regions, often using eucalypts, sheoaks, native pines and wattles as their host plants, but also occur in many other habitats. Species from higher rainfall areas use many other plant genera as hosts. Some species are apparently attracted to trees burnt during bushfires.

EXAMPLES

Banksia Jewel Beetles, two species of *Cyrioides*

Description: eggs are laid on the bark of *Banksia* species. The larvae are thickset, legless, white grubs, broadest at the head end. They bore an elliptical hole about 1cm wide in the trunk. The handsome slender beetles are about 3cm long.

Climatic regions: eastern Australia, subtropics and temperate regions, mainly coastal.

Host plants: *Banksia integrifolia, B. marginata* and *B.serrata.*

Feeding habits: the larvae bore deep into the wood of the trunk forming characteristic flattish tunnels.

They can also apparently tunnel below ground into the main roots. The beetles are easily disturbed, quickly flying or dropping to the ground.

Notes: the beetles of *Cyrioides imperialis* are black with large yellow patches on the wing-covers, whereas *C. australis* is dark blue with faint lines on the wing-covers. These beetles are often destructive pests that periodically appear in significant numbers. The larvae are quite capable of killing weak or unhealthy *Banksia* plants.

Control: as per introduction for borers.

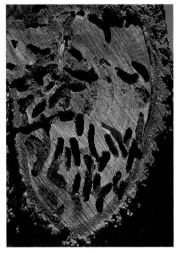

Sawn dead branch of Banksia integrifolia *showing borer tunnels, probably caused by jewel beetles.* [D. Jones]

Impact of sustained borer attack on Banksia integrifolia. [D. Jones]

Hoop Pine Jewel Beetle, *Prospheres moesta*

Description: eggs are laid on the bark or on damaged areas of Hoop Pine. The larvae are creamy-white legless grubs broadest at the head end. They tunnel into the trunk. The slender shiny black beetle is about 2cm long.

Climatic regions: subtropics, mainly coastal.

Host plants: Hoop Pine, *Araucaria cunninghamii.*

Feeding habits: the larvae bore deep into the wood of the trunk, forming a series of interconnecting oval tunnels.

Notes: these pests, which can be very destructive of Hoop Pines, are important in commercial softwood forestry operations. They often enter through injury sites and can cause severe damage to weak or unhealthy trees.

Control: as per introduction for borers.

Cypress Pine Jewel Beetles or Murray Pine Borers, three species of *Diadoxus*

Description: eggs are laid on the bark or on damaged areas of native pine. The larvae are creamy-white, legless grubs, up to 2.5cm long, broadest at the head end. The handsome slender, shiny, green and brown beetles are about 2cm long.

Climatic regions: subtropics and temperate regions, mainly inland.

Host plants: species of *Callitris*, particularly *C. glaucophylla, C. huegelii* and *C. endlicheri*; also cultivated plants of the introduced exotic cypress, *Cupressus macrocarpa* var. *lambertiana.*

Feeding habits: the larvae bore interconnecting galleries into the sapwood and deep tunnels into the wood of the trunk and large branches. The tunnels are oval in cross-section.

Notes: two species of insect, *Diadoxus erythrurus* and *D. scalaris*, commonly feed on species of *Callitris* in inland areas, with *D. erythrurus* being the most damaging. Another species, *D. jungi*, bores into *Callitris* growing in coastal and near-coastal areas of southern Australia. These borers appear sporadically and cause serious damage in some years. Attacks can be very severe following prolonged droughts and after bushfire.

Control; as per introduction for borers.

Sheoak Jewel Beetle, *Temognatha suturalis*

Description: eggs are laid on the bark of sheoaks. The larvae, which are creamy-white legless grubs, broadest at the head end, tunnel into the trunk. The sturdy greenish-black beetle, which is 4–5cm long, has shiny yellow to rusty-brown wing-covers.

Climatic regions: temperate regions, mainly coastal.

Host plants: species of *Allocasuarina* including *A. littoralis*, *A. torulosa* and *A. verticillata*.

Feeding habits: the larvae bore into the sapwood of the trunk, forming a series of interconnecting oval tunnels.

Notes: a relatively widespread species that is sometimes locally common.

Control: as per introduction for borers.

Sheoak Jewel Beetle, Temognatha suturalis. [D. Jones]

Wattle Jewel Beetle, *Agrilus australasiae*

Description: eggs are laid on the bark or on damaged areas of wattles. The larvae, which are creamy-white legless grubs, broadest at the head end, tunnel into the trunk. The small, very slender, bronze to brownish metallic-coloured beetle is about 1.2cm long with pale flanks and a couple of pale stripes.

Climatic regions: eastern Australia, subtropics and temperate regions, mainly coastal.

Host plants: species of *Acacia*, including *A. dealbata*, *A. decurrens*, *A. paramattensis*, *A. pycnantha* and *A. sophorae*.

Feeding habits: the larvae bore into the wood of the trunk forming a series of interconnecting oval tunnels. Exit holes are more or less circular but strongly convex on one side of the circle, the other side flatter.

Notes: the beetles feed on the leaves of the host plants. Wattle Jewel Beetle will sometimes infest a single tree feeding in a group while leaving nearby trees untouched. Severe attacks can result in dieback and death.

Control: as per introduction for borers.

Other Jewel Beetles

- The larvae of *Prospheres aurantiopictus* can bore into damaged areas on the trunks of weakened trees of Hoop Pine, *Araucaria cunninghamii*, but mainly infest fallen logs. Cases are known where the beetles have emerged from timber sawn 10–20 years previously, sometimes appearing after chewing through wall furniture, panelling or even books.

- The larvae of *Cisseis fascigera* bore into the stems of *Eucalyptus wandoo* causing significant dieback in some regions of Western Australia. Large populations of the beetles appear following particular climatic patterns, especially dry years that follow wet years.

- The larvae of *Chrysobothris subsimilis* bore into wattles growing in semi-arid and arid areas of inland Australia, including the Waddy Tree, *Acacia peuce*. The larvae are equipped with massive jaws to help deal with the very hard dense wood of these wattles.

- The larvae of *Temognatha heros* bore into the stems and roots (including the 'mallee roots') of several species of mallee eucalypts. The beetles feed on the flowers of *Melaleuca uncinata* and several eucalypts.

- The larvae of *Melobasis azureipennis* bore into the pithy wood of *Grewia latifolia*, which is widely distributed across northern Australia and has a range that extends south along the east coast. The small beetle, about 1cm long, has an orange-red thorax and metallic blue wing-covers with characteristic saw-tooth edging.

- The larvae of at least three species of jewel beetles in the genus *Anilara* attack species of *Flindersia*. The larvae and beetles of *A. obscura* and *A. olivia* feed on *Flindersia xanthoxyla*. An undetermined species of *Anilara* bores into the inland Leopard Tree, *Flindersia maculosa*, sometimes causing extensive damage, weakening trees and resulting in dieback.

- A remarkable charcoal-coloured species, *Merimna atrata*, commonly known as the 'fire beetle' appears in large numbers following bushfires. It is widely distributed in Australia and probably uses numerous species of Myrtaceae as host plants.

LONGICORN BEETLES (FAMILY CERAMBYCIDAE)

Members of this large group of beetles, numbering more than 20,000 species (1,200 native species), can be recognised by their long and conspicuous antennae which are often adorned with prominent spines. Some species are small and slender, but many are large beetles which range from dull to quite colourful. These beetles, which have powerful jaws, are frequently common during summer and are often attracted to lights at night. Some species squeak when handled. The larvae of most species are borers which attack trees, shrubs and herbaceous plants. Typical longicorn beetle larvae are white fleshy grubs which taper downwards from the thorax. Their legs are weak and poorly developed or absent. The larvae can bore deep into the heartwood, but commonly excavate tunnels in the sapwood and just below the bark. The tunnels are usually more or less circular in cross-section, although some are oval. Attacks on trees by longicorn beetles are common and they frequently cause considerable damage. They mainly enter through wounds and their impacts may be cumulative, with female beetles laying eggs in tissue damaged by active borer larvae. Longicorn beetles that ringbark twigs are dealt with later in this chapter.

Longicorn borer damage to base of Eucalyptus *trunk.* [D. JONES]

Larva of a longicorn beetle. [D. BEARDSELL]

Small longicorn beetle. [D. JONES]

Longicorn beetle, Phacodes obscurus.
[D. Jones]

Flower longicorn beetles in the genus Pempsamacra *feeding in* Xerochrysum.
[T. Wood]

Longicorn beetle, Pempsamacra pygmaea. [T. Wood]

Brown longicorn beetle. [D. Jones]

Powdery Longicorn Beetle, Phacodes obscurus. [D. Jones]

The larvae of this longicorn bore into wattles. [D.Beardsell]

EXAMPLES

Banksia Longicorn Beetle, *Paroplites australis*

Description: a slender, reddish-brown to dark brown, somewhat shiny beetle, about 5cm long with antennae of a similar length. The larvae are large, fleshy, yellowish, legless grubs, broadest at the head end.

Climatic regions: temperate regions of south-eastern Australia, mainly coastal.

Host plants: a wide range of native plants including *Allocasuarina verticillata, Angophora floribunda, Banksia integrifolia, B. marginata, B. serrata, Eucalyptus longifolia, E. ovata* and *E. saligna.* It has also been reported to attack exotic trees including elms and oaks.

Feeding habits: larvae tunnel into the wood, forming large circular tunnels packed with woody frass. The beetles hide in fissures and crevices in the bark.

Notes: a widespread and common insect that can cause significant damage. It is often found on large old *Banksia* plants.

Control: as per introduction for borers.

Old borer damage in Banksia: *note shape of tunnels.* [D. Jones]

Bullseye Borer, Karri Borer or Marri Borer, *Phoracantha acanthocera*

Description: a slender, pale yellow beetle, 3–4.5cm long with red-brown markings on each wing-cover. The thorax has a curved spine on each side. Antennae are about 4 cm long. The larvae are fleshy, cream, legless grubs which grow to 6cm long.

Climatic regions: tropics, subtropics and temperate regions, mainly coastal.

Host plants: *Corymbia calophylla* and several species of *Eucalyptus*, including *E. diversicolor, E. grandis* and *E. marginata*.

Feeding habits: larvae bore circular tunnels extending for several metres up the sapwood of the trunk. They also cut air holes or vents in the bark to eject excess frass. A structure that resembles a 'bullseye' is formed close to the bark prior to pupation. This is sometimes associated with bark splitting. Pupation occurs in a chamber near the 'bullseye'.

Notes: this species can cause severe damage to young plantation trees – trees as young as two years old have been attacked by this borer. Damage is also frequent in regrowth following forestry operations. The presence of vent holes, frass, kino and dead or dying branches are useful indicators of the larvae.

Control: as per introduction for borers.

Fig Tree Longicorn Beetle, *Acalolepta vastator*

Description: a grey to grey-brown beetle, about 3.5cm long with antennae about 5cm long. A prominent spine is present on each side of the thorax. The larvae are fleshy, shiny, cream-coloured, legless grubs, broadest at the head end. The head is prominent, with large jaws.

Climatic regions: tropics, subtropics and warm temperate regions, mainly coastal.

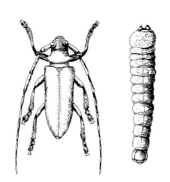

Host plants: mainly species of *Ficus* but also other natives including *Hymenosporum flavum* and *Toona ciliata*; exotic species include elm, grape vine, poplar, mulberry and wisteria.

Feeding habits: the larvae, which grow to about 4cm long, bore large round tunnels in the trunk, larger branches and roots.

Notes: a very destructive native borer of commercial significance. Serious attacks on large figs planted in parks have required the removal of affected trees. Large limbs have been known to die or break because they are structurally weakened by the borer's activities.

Control: as per introduction for borers. Spraying may be necessary with this species.

Fig Tree Borer adult and larva. [T. Blake]

Tropical Fruit Tree Longicorn Beetle, *Acalolepta mixtus*

Description: a dark grey beetle, 2–3cm long with red-brown markings on the wing-covers. Antennae are 1.5–2.5cm long. The larvae are cream-coloured, legless grubs which grow to 4cm long and are broadest at the head end.

Climatic regions: tropical regions, mainly coastal.

Host plants: a wide range of native species including *Adansonia gregorii, Adenanthera pavonina, Ficus virens, Planchonia careya* and species of *Terminalia*; also a significant pest of many tropical fruits including cashew, citrus, coconut, mango and papaya.

Feeding habits: the larvae bore round tunnels in the trunk, often close to the ground, causing the bark to crack and lift. Exit holes are often surrounded by areas where the bark has fallen off.

Notes: a very destructive native borer that has become a significant pest to tropical fruit growers.

Control: as per introduction for borers. Spraying may be necessary with this species.

Giant Longicorn Beetle, *Eurynassa odewahni*

Description: a large, stout, brown beetle which grows to 8cm long with antennae about 6cm long. The larvae are large fleshy cream legless grubs, broadest at the head end.
Climatic regions: tropics, subtropics and temperate regions, mainly coastal.
Host plants: species of *Eucalyptus*.
Feeding habits: larvae bore circular tunnels in the wood of the trunk and large branches.
Notes: the beetles of this species are large and distinctive insects.
Control: as per introduction for borers.

Leopard Longicorn Beetle, *Penthea pardalis*

Description: a dumpy pale yellow beetle about 2cm long and spotted with darker dots and blotches. Banded antennae are about 2cm long. The larvae are fleshy, cream-coloured, legless grubs which grow to about 2.5cm long.
Climatic regions: tropics and subtropics, mainly coastal.
Host plants: several *Acacia* species, including *A. aulacocarpa*, *A. holosericea*, *A. leptostachya* and *A. mangium*.
Feeding habits: the larvae chew broad tunnels in the sapwood of the trunk and larger branches. Branch dieback and breakage is common in severe infestations. Frass collects where branches join, in bark crevices and on the ground.
Notes: this species, which is also found in New Guinea, has become established as a serious pest of *Acacia mangium* plantations in several countries. Dead or dying branches are a common indicator of the activities of the larvae.
Control: as per introduction for borers.

Musk Longicorn Beetle, *Eurynassa australis*

Description: a large, shiny brown beetle measuring 5–6cm long with antennae about 3cm long. The larvae are fleshy, cream-coloured, legless grubs, broadest at the head end.
Climatic regions: subtropics and temperate regions, mainly coastal.
Host plants: *Acacia decurrens*, *Casuarina glauca*, *Eucalyptus acmenioides*, *E. punctata* and *E. squamosa*.
Feeding habits: larvae bore circular tunnels in the wood of the trunk and large branches.
Notes: the beetles discharge a musky odour when handled. The larvae were eaten by the Aboriginal people, both fresh and after cooking.
Control: as per introduction for borers.

Poinciana Longicorn Beetle, *Agrianome spinicollis*

Description: a brown beetle, 4–5cm long with pale yellow-brown wing-covers. Antennae are 4–5cm long. The larvae are cream-coloured legless grubs which grow to 5cm long and are broadest at the head end.
Climatic regions: tropics, subtropics and temperate regions, coastal and inland.
Host plants: a wide range of native species including *Angophora floribunda*, *Acacia maidenii* and other wattles, *Brachychiton populneus*, *Eucalyptus acmenioides*, *E. moluccana*, *E. saligna*, *E. umbra*, *Ficus macrophylla*, *F. watkinsiana*, *Flindersia schottiana*, *Grevillea robusta*, *Hibiscus tiliaceus*, *Howea forsteriana* and *Melaleuca quinquenervia*; also a significant pest of many exotic plants including apples, bauhinia, citrus, poinciana, poplars and willows.
Feeding habits: the larvae bore round tunnels in the trunk and larger branches.
Notes: a very destructive native borer with a wide host range.
Control: as per introduction for borers.

Tiger Longicorn Beetle, *Phoracantha semipunctata*

Description: a slender, dark reddish-brown beetle, 2–3cm long with a prominent yellow and black pattern on the wing-covers. Antennae are 3–4.5cm long. Eggs are laid beneath loose bark or in crevices. The larvae are fleshy white to cream, legless grubs to about 2.5cm long. Pupation occurs in a chamber sealed of with a plug of closely packed frass.

Climatic regions: tropics, subtropics and temperate regions, coastal and inland.

Host plants: species of *Angophora*, *Corymbia* and *Eucalyptus*.

Feeding habits: larvae tunnel under the bark and in the sapwood of trunks and large branches, usually forming galleries. Very young larvae sometimes tunnel the bark surface before chewing into the sapwood.

Notes: the beetles squeak noisily when handled and chewing noises made by feeding larvae can sometime be heard. This species has been accidentally introduced to several overseas countries resulting in significant economic loss in eucalypt plantations. Biological control has been achieved in California by introducing a chalcid wasp, *Avetianella longoi*, that parasitises the eggs.

Control: as per introduction for borers.

Tiger Longicorn Beetle. [D. Beardsell]

Longicorn Beetle in the genus Phoracantha. [T. Wood]

Wattle Longicorn Beetle, *Uracanthus triangularis*

Description: a slender, grey beetle about 3cm long with a dark brown, shiny triangular patch on each wing-cover. Antennae are about 4cm long. The larvae are fleshy, cream-coloured, legless grubs about 3cm long.

Climatic regions: tropics, subtropics and temperate regions, mainly coastal.

Host plants: *Hakea gibbosa*, *Eriostemon australasius* and *Acacia* species.

Feeding habits: larvae bore circular tunnels in the wood of the trunk and large branches.

Notes: dead or dying branches are a common indicator of the activities of the larvae.

Control: as per introduction for borers.

Other Longicorn Beetles

- The larvae of the Yellow Longicorn Beetle, *Phoracantha recurva,* tunnel into the bark, sapwood and heartwood of species of *Angophora*, *Corymbia* and *Eucalyptus*, often causing considerable damage. The beetles are smaller versions of the Tiger Longicorn Beetle (see above) but paler and with dense yellow hairs on the underside of the antennae.

- The larvae of the Two-hole Longicorn Beetle, *Phoracantha solida*, damage the stems many species of *Eucalyptus*, including plantations of *E. globulus* and *E. grandis*. They cut oval tunnels in the heartwood with twin openings surrounded by clean areas where the damaged bark has fallen off.

- The feather-horned Yellow Box Longicorn Beetle, *Distichocera macleayi*, is a handsome insect which has very prominent feather-like antennae. Its larvae feed on eucalypts and can be very destructive of Yellow Box trees, *Eucalyptus melliodora*.

- Native species of *Ficus* are tunnelled by the larvae of at least two longicorns other than the Fig Tree Longicorn Beetle (see entry above). The beetles may also feed on fig sap oozing from wounds. Wallace's Longicorn Beetle, *Batocera wallacei,* is a large, decorative species found in north-eastern Queensland. The attractively marked beetles (shiny green to brown with white markings), which are about 8cm long, have antennae 20–23cm long. *Batocera boisduvali,* which ranges from north-eastern Queensland to north-eastern New South Wales, is about 5cm long and grey to light brown with white to yellow blotches on the wing-covers. Its antennae are 8–10cm long.

WEEVILS (FAMILY CURCULIONIDAE)

Weevils are dealt with in some detail in Chapter 9. These insects can be recognised by the front part of the head being drawn into a snout with the antennae at the tip. The larvae are fleshy, legless grubs. The larvae of many weevils attack a range of living and unhealthy trees. Many species also riddle fallen logs with holes.

EXAMPLES

Kurrajong Weevil, *Axionicus insignis*

Description: a plump, grey-brown and white mottled weevil about 1.5cm long. The larvae are fleshy, white, legless grubs.

An unidentified weevil. [T. Wood]

Climatic regions: tropics, subtropics and temperate regions, mainly inland.

Host plants: species of *Brachychiton,* particularly the Kurrajong, *B. populneus.*

Feeding habits: the larvae enter through wounds left by storm damage or when plants are pollarded for stock feed. They bore into the wood, excavating large round tunnels.

Notes: this pest can be very destructive in weak, unhealthy or badly damaged trees. Fungal infections of the damaged sites can lead to rotting and tree death.

Control: as per introduction for borers.

Other Weevils

- The large, fleshy, white, legless larvae of Australia's largest weevil, *Eurhampus fasciculatus,* bore into the trunks of dying trees and fallen logs of species of *Agathis* and *Araucaria,* creating large tunnels deep into the wood. The weevils, which are about 6cm long, have tufts of hair on the wing-covers.

Belid beetle, Rhinotia haemoptera. [T. Wood]

- The larvae of numerous species of small slender weevils in the family Belidae bore into the wood of stressed or unthrifty wattles, but they do not usually damage healthy trees. A common species in eastern Australia, *Rhinotia suturalis,* has been found damaging *Acacia dealbata, A. decurrens, A. leiocalyx* and *A. sophorae.*

- Root-boring weevils are included in chapter 14.

Belid beetle, Rhinotia suturalis. [T. Wood]

AUGER BEETLES (FAMILY BOSTRICHIDAE)

Auger beetles, also known as shot-hole borers, are mainly pests of seriously weak, dying or recently fallen trees. Australia has about 45 species of these beetles, which are especially common in the tropics. The beetles are mostly elongated with the head hidden from above by a large and strongly humped thorax. The beetles bore small circular tunnels in branches and stems to lay eggs. Tiny beetles that bore very small tunnels are referred to as pinhole borers. Some species live in colonies in chambers in the wood. The larvae tunnel in the sapwood.

Pinhole borers in rainforest tree bark.
[D. JONES]

EXAMPLES

Large Auger Beetle, *Bostrychopsis jesuita*

Description: stout, dark-coloured beetle, 1.5–2cm long. The thorax continues as an outgrowth over the head, which is permanently downturned. The males have roughened wing-covers, whereas the tips of the female wing-covers are smooth. The larva is a thickset, C-shaped, white grub which grows to 1.2cm long and has true legs.

Climatic regions: tropics, subtropics and temperate regions, coastal and adjacent inland.

Host plants: a wide variety of native plants including many species of wattle, eucalypts, kurrajong, *Brachychiton populneus,* white cedar, *Melia azederach* and silky oak, *Grevillea robusta*; also a pest of citrus trees and grape vines.

Feeding habits: the larvae bore circular tunnels in the sapwood of the trunk and large branches, producing a soft powdery frass. The tunnels are filled with droppings and undigested wood particles.

Large Auger Beetle. [D. BEARDSELL]

Notes: a common and sometimes serious pest, although most attacks occur on weakened, stressed or dying trees. This pest is also frequent on recently fallen timber.

Control: as per introduction for borers.

Other Auger Beetles

• **The Common Auger Beetle**, *Xylopsochus gibbicollis*, feeds on native plants and has become a pest of grape vines. The beetles, which fly actively during spring and summer, are small, brown and about 0.2cm long with a rounded head and truncate tip to the abdomen. The cream-coloured, legless larvae have black jaws and grow to about 0.3cm long. The larvae feed internally on the sapwood, often encircling a stem and causing breakage.

• **The Particoloured Auger Beetle**, *Mesoxylion collaris*, also feeds on a range of native plants, including species of *Acacia*, *Banksia* and *Eucalyptus*. The beetles are black with a rusty brown thorax. The larvae bore tunnels in the stems leaving frass-filled galleries and tiny exit holes. Massed feeding can cause general debilitation of the affected tree.

CATERPILLAR BORERS OF TRUNKS AND LARGE BRANCHES

The larvae (caterpillars) of some moths are important borers that can cause significant damage to trees. Two families of large moths, the caterpillars of which bore into trunks and large branches, are of particular significance.

SWIFT MOTHS OR GHOST MOTHS (FAMILY HEPIALIDAE)

Australia has about 120 species of moths in the family Hepialidae. These are mostly large moths that emerge and fly on rainy nights in autumn, winter and spring. The moths do not feed and only live a single night. The larvae of some species live in the soil, feeding on roots; a few others bore tunnels in the stems of shrubs and trees; others feed on litter. The tunnels can persist in the tree for many years, but those where the larvae are active have fresh sawdust near the entrance. The larvae of some of these borers feed on the bark around the tunnel entrance and on callus tissue that grows as a result of the damage. After pupation the pupal case is often seen projecting from the tunnel.

EXAMPLES

Bent-wing Ghost Moth or Bent-wing Swift Moth, *Zelotypia stacyi*

Description: the large, grey to orange or light brown moths are fast-flying 20–25cm across with relatively narrow wings and a relatively long thin body. The smaller male is orange-brown with whitish or silvery markings (some in intricate patterns), the female less colourful. The forewings of the male have a prominent eye-spot. There are shaggy hairs on the hindwings. The larvae are brownish caterpillars that grow to about 15cm long.

Climatic regions: subtropics and warm temperate regions of the east coast, mainly coastal.

Host plants: several species of *Eucalyptus* that grow in wetter forests, including *E. grandis*, *E. saligna*, *E. tereticornis* and *E. tessellaris*.

Feeding habits: the larvae tunnel into stems and branches of saplings, regrowth shoots and older trees. A vertical shaft, in which the larvae live, is formed at the end of the entrance tunnel, the opening of which is camouflaged by a mixture of webbing and frass. At maturity the covering over the tunnel entrance is replaced by a tight-fitting plug constructed to block the entrance while pupation occurs. The moths emerge in late summer.

Notes: at times this species is very destructive, especially of saplings. The moths can sometimes be found hanging suspended from twigs, giving the appearance of a dead leaf.

Control: as per introduction for borers.

Banded Swift Moth, *Aenetus ligniveren*

Description: the moths have bright green forewings adorned with white diagonal stripes and shiny, creamy-grey hindwings. Wingspan is 5–7cm. The slender cream to yellowish larvae grow to about 5cm long and are sparsely hairy with a dark head.

Climatic regions: tropics, subtropics and temperate regions, mainly coastal.

Host plants: a common garden visitor that feeds on a wide range of native plants including species of *Acacia*, *Babbingtonia*, *Baeckea*, *Callistemon*, *Cassinia*, *Dodonaea*, *Eucalyptus*, *Leptospermum*, *Melaleuca*, *Pomaderris, Prostanthera, Syzygium*, and *Westringia*; also some deciduous trees and exotic shrubs.

Feeding habits: initially the larvae bore horizontally, usually at a site on the underside of a leaning branch or at a branch junction, then create a vertical tunnel downwards through the centre of a stem. Vertical tunnels can be one metre or more long. The entrance of the horizontal tunnel is covered with a mixture of webbing and frass. The caterpillar emerges at night and feeds on bark close to the tunnel entrance. The plant responds to the damage by growing callus tissue on which the larva also feeds.

Notes: this species causes shoot dieback and breakage. Its impacts are usually of a minor nature but the site where the larva feeds can become unsightly.

Control: as per introduction for borers.

Spotted Swift Moth, *Aenetus eximius*

Description: the moths have bright green spotted forewings and orange-brown hindwings. Wingspan is about 5cm. The slender, pale larvae grow to about 5cm long and are sparsely hairy with a dark head.

Climatic regions: tropics, subtropics and temperate regions, mainly coastal.

Host plants: feeds on a wide range of plants from wet forest and rainforest including species of *Acacia, Atherosperma, Cassinia, Daphnandra, Diploglottis, Dodonaea, Doryphora, Eucalyptus, Glochidion, Lophostemon, Pomaderris, Prostanthera* and *Syzygium*.

Feeding habits: as for the previous species.

Notes: as for the previous species.

Control: as per introduction for borers.

WOOD MOTHS OR CARPENTER MOTHS (FAMILY COSSIDAE)

These are large grey to brownish moths, the larvae of which tunnel in shrubs and trees. The large larvae, which can be up to 15cm long, bore circular tunnels up to 3cm across into the heartwood of the tree. Attacks are usually sporadic, which is just as well because the tunnels are large enough to significantly reduce the mechanical strength of a tree trunk or branch, resulting in breakage. These tunnels are usually very neat, as if they had been made with a drill or auger. The tunnels persist in the tree for many years, but those where the larvae are active have fresh sawdust near the entrance. Frequently the larva may feed on the bark around the tunnel entrance. In some species there is little sign of the presence of the borer (apart from wastes ejected through a small opening in the bark) until

Cossid larva. [CSIRO Ecosystem Sciences]

Damage caused by Yellow-tailed Black Cockatoos seeking borer larvae. [CSIRO Ecosystem Sciences]

Plantation trees destroyed in strong winds following trunk damage by cossid larvae. [CSIRO Ecosystem Sciences]

Yellow-tailed Black Cockatoo feeding in wattle. [M. Sutcliffe]

Caterpillar of a wood-boring moth. [D. Jones]

a hole is cut through the bark to allow emergence of the moth.

The Cossidae, which are known as wood moths, include Australia's largest and heaviest moths – large females of *Endoxyla cinereus* can weigh up to 30g and have a wingspan of up to 25cm. They are generally dull-coloured, heavy-bodied moths with narrow scaly wings. The thick, fleshy larvae have horny plates on the surface of the thoracic segments.

EXAMPLES

Giant Wood Moths, *Endoxyla cinereus* and *E. magnifica*

Description: the large grey moths have a dark abdomen. Female moths, which have a heavy, fat body full of thousands of eggs, have a wingspan of 20–25cm, about twice that of the males. Eggs are laid in crevices in the bark. Young larvae feed on roots and when about 12 months old they climb the tree and bore a hole in the trunk. Mature larvae are 15–17cm long and cream with darker bands.

Climatic regions: tropics and subtropics.

Host plants: species of *Eucalyptus*.

Feeding habits: the larvae tunnel vertically upwards in the heartwood of the trunk. When mature they bore a horizontal tunnel to the outside, sealing the opening with a plug of webbing mixed with frass. Pupation occurs upside-down in the vertical chamber which is sealed at the base by a sticky repellent secretion. The moth emerges through the plugged hole which is separate from the tunnel entrance.

Giant Wood Moth, Endoxyla magnifica, *from Eden, New South Wales.* [D. Jones]

Notes: large and mature trees are able to cope well with the impacts of the larvae, but saplings and younger trees can suffer trunk weakening that can lead to breakage. Significant damage is often inflicted on affected trees by Yellow-tailed Black Cockatoos, which tear open trunks and branches while seeking the larvae as food. This species, one of Australia's largest moths, is a national treasure. The moths, which can be mistaken for small birds, fly at night in late spring and early summer.

Control: as per introduction for borers.

Australian Goat Moth, *Zyganisus caliginosus*

Description: the slaty-grey moths have a wingspan of 5cm, the forewings are streaked and marked with black lines. The larvae are plump fleshy red caterpillars that grow to 5cm long.

Climatic regions: subtropics and temperate regions of eastern Australia, mainly coastal.

Host plants: species of *Angophora* and *Eucalyptus*.

Feeding habits: the larvae, which feed in groups, bore in the larger stems and trunks, feeding between the inner surface of the bark and the sapwood. When mature they tunnel into the sapwood to pupate.

Notes: the larvae cause structural damage and weaken trees. The moths fly during rainy weather from late spring to mid-summer.

Control: as per introduction for borers.

Wattle Goat Moth, Endoxyla eucalypti.
[D. Jones]

Wattle Goat Moth, *Endoxyla eucalypti*

Description: the large, grey to light brown moths have a wingspan of 10–12cm and the forewings are marked with white streaks and patches. The top of the thorax has a prominent bald patch outlined with a narrow blue band. The larvae are large, fleshy, greyish or pinkish caterpillars that grow 10–15cm long. They have horny plates on the upper parts of the body. Pupation occurs in a tunnel in the soil. The moth emerges in rainy weather, leaving the upper part of the pupal case protruding from the ground.

Climatic regions: tropics, subtropics and temperate regions of eastern Australia, mainly coastal.

Host plants: species of *Acacia*.

Feeding habits: the larvae attack wattles, usually boring the lower trunk near ground level and into the roots. They bore large, round tunnels about 3cm across deep into the heartwood. Each larva may live within a tree for four to five years

Notes: these insects are of significance because they can weaken the structural strength of saplings and large trees. Several similar species are known as wattle goat moths, the name arising from the strong smell of the larvae. The moths fly during rainy weather from late spring to midsummer.

Control: as per introduction for borers.

Speckled Goat Moth, *Endoxyla lituratus*

Description: the large grey moths have a wingspan of 6–8cm, the forewings are heavily speckled and marked with a few dark blotches and lines. The plump abdomen is banded with brown and grey. The top of the thorax has a prominent shiny bald patch outlined with a black border. The larvae are large, fleshy, cream-coloured caterpillars that grow 8–10cm long.

Climatic regions: tropics, subtropics and temperate regions of eastern Australia, mainly coastal.

Host plants: species of *Acacia*.

Feeding habits: the larvae bore in the larger stems and trunks of wattles, forming a network of tunnels. The larva may live within a tree for a few years

Notes: the larvae can cause structural damage and weaken trees. The moths fly during rainy weather from late spring to mid-summer.

Control: as per introduction for borers.

Other Wood Moths

- The larvae of the Culama Wood Moth, *Culama australis*, tunnel between the sapwood and bark of eucalypts. The damage is mostly minor but can reduce the value of timber. The moth, which is grey with narrow black reticulated patterns, rests with the wings wrapped around its body. This common species is widely distributed in south-eastern Australia.

- The larvae of the Grass Tree Wood Moth, *Cossodes lyonettii*, feed on the roots of species of *Xanthorrhoea* with subterranean trunks in south-west Western Australia. The moths are beautifully patterned with black and white forewings and iridescent purple hindwings.

SMALL BRANCH AND TWIG BORERS

These borers concentrate their activities on the twigs and smaller branches of the outer canopy of a tree or shrub. Some species attack soft shoots before the growth hardens, others feed on mature growth. The larvae of some species bore down the centre of shoots causing wilting, others feed around the stem under the bark, cutting the sapwood and causing the branch to snap off where it has been weakened. These pests include specialised longicorn beetles that feed on bark, small weevils and the larvae of some moths. Damage caused by these pests usually shows up when the affected shoot wilts, dies or breaks off in the wind.

Stem breakage in Acacia sophorae *due to borer damage.* [D. JONES]

Dead branch in Acacia sophorae *caused by borer.* [D. JONES]

Stunted young plant of Corymbia torelliana *weakened by trunk borer.* [D. JONES]

Severe borer damage in trunk of Corymbia torelliana. [D. JONES]

Borer damage on the stem of a young wattle: note exudation of gum. [D. JONES]

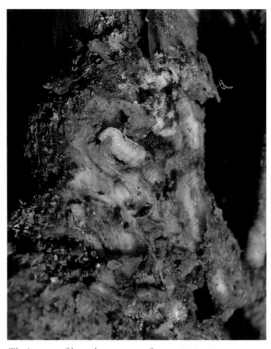

Sap exudation due to borer damage in Acacia *stem.* [D. JONES]

The impacts of borer larvae on a soft stem. [D. JONES]

WEB-COVERING BORERS

These are moths of the family Xylorictidae, the larvae of which are common borers which have the characteristic of covering their tunnels by an accumulation of sawdust-like material enclosed in a fine web. Sometimes the web includes partly eaten leaves, frass, faecal pellets and detritus. Removing the covering material reveals the tunnels and galleries. The tunnel is a slanting hole about 0.5 cm across, frequently with the green bark under the webbing completely eaten away. The tunnel extends through the centre of branches, considerably weakening them. These are very destructive insects which should be controlled when first noticed. Affected limbs become yellowish and sickly and usually snap off in a strong wind. Some moths in the family Hepialidae also feed in a similar manner.

Larva of a Xylorictid moth borer in Callistemon. [D.BEARDSELL]

Larva of a Xylorictid moth in Callistemon. [D.BEARDSELL]

Stem borer damage to Prostanthera nivea *covered with webbed frass.* [D. BEARDSELL]

Stem borer damage to Prostanthera nivea, *with frass removed.*
[D. BEARDSELL]

Web-covering borer damage in Stenocarpus sinuatus. [D. JONES]

EXAMPLES

Banksia Web-covering Borer, *Xylorycta strigata*

Description: the pale brown satiny moths have a wingspan of about 4cm. The hindwings have a fringe of silvery hairs. The slender and cream to greenish larvae grow to about 2.5cm long and have darker markings.

Climatic regions: subtropics and temperate regions, mainly coastal.

Host plants: species of *Banksia*, particularly *B. aemula*, *B. integrifolia* and *B. serrata*; also *Lambertia formosa*.

Feeding habits: larvae tunnel down the centre of shoots near the tip. The tunnel entrance is covered with large blobs of webbing and brown faecal material. The caterpillar emerges at night and drags leaves back to the tunnel for food.

Notes: a common but fairly minor pest.

Control: as per introduction for borers.

Banksia Web-covering Borer in Banksia integrifolia – *note harvested food leaf in web.* [D. JONES]

Banksia Web-covering Borer in Banksia serrata. [D. JONES]

Fruit Tree Borer or Pecan Stem Girdler, *Maroga melanostigma*

Description: the satiny white moths, with a wingspan of about 4cm, have a characteristic black spot on each forewing. The sparsely hairy, fleshy caterpillars, which grow to about 3.5cm long, are pale brown with a dark brown head.

Climatic regions: tropics, subtropics and temperate regions, coastal and inland.

Host plants: a wide range of native plants including species of *Acacia, Albizia, Allocasuarina, Banksia, Cassinia, Eucalyptus, Jacksonia, Leptospermum, Melaleuca, Ozothamnus, Pithecellobium, Prostanthera, Santalum* and *Senna*; also a serious pest of fruit trees, especially *Prunus* (apricot, cherry, plum, peach), fig, citrus, grape, pecan and apple; also many deciduous ornamental trees including species of birch, elm, oak, gingko and maple.

Feeding habits: young larvae feed on the bark and as they grow they tunnel into the sapwood forming extensive shallow galleries covered with a mass of brown chewed wood fragments and webbing. The larvae emerge at night to feed on bark surrounding the tunnel entrance. Large patches of bark are often eaten on the trunk and larger branches. Smaller branches, even the trunk, can be ringbarked. Older larvae can tunnel deep into the heartwood, greatly weakening even large branches.

Notes: a widespread and common pest that can be very damaging. When disturbed the caterpillars wriggle rapidly and when threatened the moths fall to the ground and lie on their back with the abdomen curved upright.

Control: as per introduction for borers. Squash any larvae exposed when the webbing is removed.

Small Fruit Tree Borer, *Cryptophasa albacosta*

Description: the moths, with a wingspan of 4–5cm, have whitish satiny forewings and brown hindwings. The greyish-blue, sparsely hairy, fleshy larvae, attractively marked with transverse black bands, grow to about 3cm long.

Climatic regions: subtropics and temperate regions, coastal and inland.

Host plants: several species of *Banksia*, especially *B. serrata*; also *Callicoma serratifolia, Ceratopetalum gummiferum* and commercial *Macadamia* plantations; this species also attacks exotic plants including apricot, plum, poplars and *Tamarix*.

Feeding habits: larvae tunnel in the bark and sapwood, often entering at a branch junction. The entrance hole is covered with a webbed brown mass consisting of chewed wood, faecal pellets and webbing. The caterpillar emerges at night to chew off whole leaves, which it then lodges petiole-end first in the tunnel entrance as food. The caterpillar is protected within the tunnel while it feeds, drawing the leaf into the tunnel as it is eaten.

Notes: the larvae tunnel into the sapwood and heartwood. It is common for branches to break at the point where these borers have been feeding.

Control: as per introduction for borers.

Wattle Web-covering Borer, *Cryptophasa rubescens*

Description: the moths are light brown and satiny, about 5cm long, with a crest of hairs behind the head. The hindwings are cream to pale yellow. The greenish, sparsely hairy, fleshy larvae grow to about 5cm long.

Climatic regions: subtropics and temperate regions, coastal and inland.

Host plants: several species of *Acacia*.

Feeding habits: larvae tunnel in the bark and sapwood, forming shallow tunnels. The entrance hole is covered with a webbed brown mass consisting of chewed wood, faecal pellets and webbing. The caterpillar emerges at night to chew off phyllodes, which it then lodges in the tunnel entrance as food, drawing them into the tunnel as they are eaten.

Notes: the larvae mainly feed on small branches but they also tunnel into the heartwood. It is common for branches to break at the point where these borers have been feeding.

Control: as per introduction for borers.

Other Web-covering Borers

- The larvae of at least two other species in the genus *Cryptophasa* bore in shoots and twigs and cover their tunnels with a webbing of silk, wood particles and waste products, drawing in leaves as food. *Cryptophasa immaculata* feeds largely on *Banksia serrata* and *Cryptophasa spilonota* on *Banksia integrifolia*.

- The pinkish larvae of some species of small moths in the genus *Scieropepla* tunnel into the buds, flower spikes and developing cones of banksias, producing masses of brown sawdust-like frass.

- The larvae of the unusual moth *Eschatura lemurias* bore into the sapwood and emerge from the tunnels to feed on the bark of some subtropical rainforest trees, including *Eleaocarpus angustifolius* and *Syzygium floribundum*. Other species of *Eschatura* occur in tropical rainforest further north. The moths are unusual in having pointed tips on the forewings.

- The larvae of the moth *Xyloricta luteotactella* bore into the twigs and fruit of *Macadamia* plants. The tunnel entrances are covered with webbing containing brown sawdust. The bark is commonly encircled by their feeding and the twigs die and break off. For more information see Macadamia Twig Girdler (this chapter).

Web-covering borer in Banksia integrifolia. [D. Jones]

SOFT SHOOT AND TIP BORERS

A few specialised borers attack the new shoots of plants causing them to wilt and die. Such outbreaks are usually of a minor nature and help promote branching. Persistent attacks, however, are damaging and can impact on plant development and shape. The symptoms of tip borers are usually obvious with wilting and yellowing being the major features. Eucalypts may produce small globules of clear gum which appear similar to drops of rain or dew. The tips of eucalypt shoots may also blacken. Plants which produce flushes of growth such as *Melaleuca* and *Callistemon* may have only one or two shoots in each cluster affected. These show up fairly quickly by curling their leaves and eventually turning brown and dying. Damaged tips frequently break off in heavy rain or wind.

Control: if necessary by cutting off and burning affected tips or by spraying new shoots with pyrethrum or malathion when the pests become active.

EXAMPLES

Cedar Tip Moth, *Hypsipyla robusta*

Description: fleshy, red or greenish, sparsely hairy caterpillars with shiny black spots which grow to 2.5cm long. Pupation occurs in cocoons spun in tunnels or among litter at the base of the tree. The moths (family Pyralidae) have prominent dark veins on their fringed wings, the forewings are brown, the hindwings paler. The wingspan is about 3cm.

Climatic regions: tropics, subtropics and temperate regions (south to Eden, NSW), chiefly coastal.

Host plants: in Australia, mainly Australian Red Cedar, *Toona ciliata*; also mangroves in the genus *Xylocarpus*. Cedar Tip Moth is widespread from Africa to Asia and the Pacific region. It causes significant economic losses to a number of valuable plantation-grown timber trees in the family Meliaceae, including species of *Cedrela*, *Khaya* and *Swietenia*.

Feeding habits: larvae tunnel into the growing tip of the main shoot or enter at a nearby leaf axil, causing shoot death followed by a proliferation of side-shoots. Young trees affected in this way become stunted and crooked from an early age, greatly reducing their value as timber trees and for ornamental purposes. Large quantities of faecal pellets held in webbing indicate the presence of the borer. The larvae can also eat the flowers (amidst clumps of webbing) and fruit of Australian Red Cedar. They sometimes bind the fruit in webbing to prevent it falling. They may also feed on the bark around the tunnel entrance.

Notes: a very persistent and debilitating pest that commonly causes stunting or even death of cultivated Australian Red Cedar plants. Seedlings planted in full or partial sun are attacked more readily and frequently than are seedlings planted in the shade or under nurse trees (fast-growing and often short-lived species planted to provide shelter to young seedlings). The moths are strong fliers and can seek out host trees over long distances.

Control: by regularly spraying young trees with pyrethrum or malathion. Control is impractical in larger trees.

Cedar Tip Moth damage. [D. Jones]

Callistemon Tip Borer, unidentified moth

Description: the borers, which are fleshy, cream caterpillars, about 0.3cm long, develop into an unknown species of small, blackish, metallic moth which measures about 0.5cm long.

Climatic regions: tropics, subtropics and temperate regions, mainly coastal.

Host plants: species and cultivars of *Callistemon* and some *Melaleuca* species.

Feeding habits: eggs are laid near the top of young shoots and the larvae bore down the centre of a shoot, which usually dies or breaks off.

Notes: although a relatively minor pest, this borer can be a persistent problem. Several cycles of the moth are produced in a season. Some individual plants seem to be more prone to attacks by this pest than others. Similarly some species and cultivars are more prone to attack.

Control: as per introduction for borers. Removing and burning affected shoots will reduce population levels.

Thin Strawberry Weevil or Eucalypt Seedling Borer, *Rhadinosomus lacordairei*

Description: eggs are laid near the tips of seedlings and young shoots. The larvae, which are pale-yellow, legless grubs about 0.6cm long, tunnel down the centre of a stem. The thin brown weevils (family Curculionidae) are about 1cm long. The head is drawn out with antennae at the very tip and the deeply furrowed wing-covers are notched and pointed at the apex.

Climatic regions: tropics, subtropics and temperate regions.

Host plants: seedlings of a wide range of eucalypts. Often severe on tropical eucalypts including *Eucalyptus miniata*, *E. phoenicea* and *E. shirleyi*. Also attacks the young shoots of trees in October to December. Also an occasional pest of strawberry plants.

Feeding habits: larvae tunnel down the centre of young shoots near the tips, which die causing premature branching and stunting. The tunnel entrance is a small hole which is often visible on the stem in the region where the shoot collapses. The weevils feed on young leaves.

Notes: this species is sometimes a nuisance in nurseries.

Control: as per introduction for borers. Spraying may be necessary to prevent damage in nursery plants.

Macaranga Tip Borer, *Cryptophasa* species

Description: fleshy, greenish, sparsely hairy caterpillars which grow to 2.5cm long, with a brown or black head. The moths (family Xylorictidae) are silky and brownish with a wingspan of about 4cm.

Climatic regions: tropics and subtropics, mainly coastal.

Host plants: *Macaranga tanarius*.

Feeding habits: larvae tunnel in shoots just below the tip and feed in the centre of the stem. The entrance hole is covered with a mat of webbing and brown faeces. The larvae also feed on young leaves.
Notes: it is common for a few larvae to feed together in a shoot. Affected shoot tips die and the damaged stems turn black.
Control: a minor pest rarely worth controlling.

Other Soft Shoot and Tip Borers

- **Elkhorn Borer** – the caterpillars of an unidentified species of moth bore circular holes in the pads and sterile fronds of the Elkhorn Fern, *Platycerium bifurcatum*. The fleshy grey caterpillars grow to about 1.2cm long and the moth is grey with a wingspan of about 2cm. The tunnels are hidden by webbing and the caterpillars bring back pieces of frond as food. The tunnel entrances become exposed on older pads. This pest can severely damage weak or debilitated clumps of the fern.

- **Fern Borers** – the larvae of two native species of small weevils (family Curculionidae) tunnel into the young fronds of ferns causing them to shrivel and die. Sometimes they only attack the upper part of a developing frond, causing the tip to die back. The brown weevils, 3–6mm long, have rounded bumps on the wing-covers. *Syagrius fulvitarsus*, known as the fern weevil, attacks native ferns including species of *Blechnum* and *Doodia*. It was accidentally introduced into Hawaii in 1903 where it spread quickly, damaging and killing several species of fern over significant areas until introduced wasp parasites provided a degree of control. Another weevil, *Neosyagrius cordipennis,* is commonly known as the Maidenhair Fern Weevil because of its favoured food item.

- **Sandpaper Fig Borers** – the young shoots of some sandpaper figs (in particular *Ficus coronata* and *F. fraseri*) are often tunnelled by the larvae of the weevil, *Hylescinus fici*. Affected new shoots turn white and papery.

Damage caused by the Elkhorn Borer. [D. JONES]

Tip borer damage in Macaranga tanarius. [D. JONES]

TWIG GIRDLERS AND RINGBARKERS

These pests are prominent in tropical and subtropical regions, less so in temperate areas. Their activities are noticeable mainly during the hot humid months of December to March. The most recognisable symptom of a twig girdler is the rapid browning and death of twigs and branches within a shrub or tree without apparent cause. Close examination of the affected branch often reveals its death can be traced to patches of bark removed from a site towards the base of the necrotic tissue. The damaged area consists either of a series of neat concentric rings eaten right through the bark to the wood or larger areas of bark completely eaten out. Damaged sites quickly turn from green to brown. Some tunnelling caterpillars can also girdle branches and twigs but their activities are revealed by masses of webbing.

Girdled twigs or branches commonly break off in the wind, usually snapping at the damaged site. Twig girdlers range from innocuous to damaging pests and attack a wide range of native and exotic plants. Usually by the time the dead twigs and branches become noticeable it is too late to take remedial action because the insects have departed. Twig girdling is mainly carried out by various species of beetle, particularly longicorns, and it is a deliberate and important part of their life cycle. The girdling restricts the movement of photosynthetic materials back to the roots and starch accumulates in the tissue above the damaged site. The insect lays its eggs here and the larvae feed on the enriched tissue.

Control: these pests are difficult to control as they have usually departed by the time the results of their feeding become obvious and spraying is not justified. Removing and burning damaged shoots will kill any eggs or larvae that are still present.

Twig girdler damage to Acacia pycnantha. [D. JONES]

Site of twig girdler damage in wattle stem. [D. JONES]

Twig girdler damage on Alphitonia. [D. JONES]

Longicorn twig girdler feeding on Acacia dealbata. [T. WOOD]

Longicorn twig girdler feeding on the bark of Acacia dealbata. [D. JONES]

Twig girdling longicorn beetle. [D. BEARDSELL]

EXAMPLES

Twig Girdling Longicorn Beetle, *Platyomopsis humeralis*

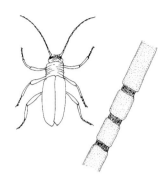

Twig girdler. [T. BLAKE]

Description: a handsome beetle (family Cerambycidae), about 2cm long with antennae of a similar length. Beetles are dark grey with various spots and stripes and may have pubescent areas on the abdomen. The head, with its prominent mandibles, is set at right angles to the body. The larvae are whitish legless grubs, 2.5–3cm long.

Climatic regions: tropical and subtropical regions.

Host plants: a wide range of native plants from open forest and rainforest, including several species of *Acacia, Allocasuarina littoralis, Alphitonia excelsa, A. petriei. Cassia marksiana, Casuarina cunninghamiana, Commersonia bartramia*; this longicorn also attacks the serious introduced tropical weed *Mimosa pigra.*

Feeding habits: the beetles graze concentric rings of bark from twigs and branches which subsequently die above the damaged site. The trunks of young plants can also be girdled, causing premature branching. Attacks on plantations of the tropical wattles *Acacia auriculiformis* and *A. mangium* can be very damaging, especially in the first year after planting.

Notes: a destructive and annoying pest. Symptoms of twig girdling are usually not obvious until the insects have left the scene. Ringbarked branches die or break off readily in storms.

Control: as per introduction for twig girdlers.

Dieback Borer, *Platyomopsis armatula*

Description: a fairly stout, grey-brown beetle (family Cerambycidae) about 2cm long. The wing-covers have a few lumps and roughened projections. The antennae are about 2cm long. Larvae are fleshy, legless, cream grubs, about 1.5cm long.

Climatic regions: tropics, subtropics and warm temperate regions.

Host plants: species of *Callistemon, Eucalyptus, Leptospermum* and *Melaleuca.*

Feeding habits: larvae tunnel under the bark, feeding on the sapwood. Shoots and small branches frequently die when they have been ringbarked by the larvae. It is not uncommon for damaged but still living branches to break off in a storm.

Notes: a very destructive species. Removing damaged bark can frequently expose the tunnels of the larvae.

Control: as per introduction for borers.

Pittosporum Longicorn Beetle, *Strongylurus thoracicus*

Description: a stout, brown beetle (family Cerambycidae) about 3cm long with a row of large white spots down each side of the thorax. Antennae are about 4cm long with prominent spines. The larvae are creamy legless grubs about 3cm long.

Climatic regions: tropics, subtropics and temperate regions, chiefly coastal.

Host plants: *Melia azedarach* and species of *Pittosporum.*

Feeding habits: young larvae feed on the bark, whereas larger larvae bore round tunnels in the sapwood just below the bark. Often complete patches of bark are eaten, especially on small to medium-sized branches which die back or break off at the damaged site.

Notes: dead twigs and branches are commonly seen on plants where this borer has been active. Broken stems may litter the ground.

Control: as per introduction for twig girdlers.

Wattle Ringbarking Beetle, *Ancita marginicollis*

Bark-feeding longicorn beetle of the genus Ancita *on* Acacia dealbata. [T. Wood]

Description: a stout, dark grey-black, mottled beetle (family Cerambycidae), about 1.5cm long with prominently banded antennae about 3 cm long.

Climatic regions: tropical and subtropical regions.

Host plants: species of *Acacia, Alphitonia* and other rainforest plants.

Feeding habits: the beetles graze patches of bark from twigs and small branches, frequently ringbarking them.

Notes: a very destructive pest, most active in spring and early summer. Ringbarked branches frequently die or readily break off in storms.

Control: as per introduction for borers.

Fig Bark Beetle, *Aricerus eichoffi*

Description: the stout, reddish-brown weevil (family Curculionidae) is tiny, about 0.2cm long. Larvae are tiny fleshy grubs with true legs.

Climatic regions: subtropics and warm temperate regions.

Host plants: species of *Ficus*.

Feeding habits; the larvae bore tunnels down the centre of branchlets and twigs. The weevils eat patches of bark.

Notes: although this is a fairly minor pest, the twigs and branchlets affected by the borers usually die.

Control: as per introduction for borers. Cut off and burn affected branchlets.

Macadamia Twig Girdler or Waratah Bud Borer, *Xyloricta luteotactella*

Description: the satiny white moths, family Xyloryctidae, have yellow legs and a wingspan of 2.5–3cm. Eggs are laid in leaf axils on terminal shoots. The larvae are pale green caterpillars with darker rows of dots and a black head. They are somewhat fleshy, sparsely hairy and grow to about 2.3cm long. They are very active when disturbed.

Climatic regions: tropics and subtropics, mainly coastal.

Host plants: several members of the family Proteaceae, including species of *Alloxylon, Banksia, Buckinghamia, Grevillea, Hakea, Lambertia. Macadamia, Persoonia, Stenocarpus, Telopea* and *Xylomelum.*

Feeding habits: the larvae attack young shoots, often concentrating at branch junctions, in leaf axils or at other girdled sites. Young larvae feed on the bark and as they grow they tunnel, girdling the twigs.

Damage caused by the Waratah Bud Borer. [D. Jones]

They also skeletonise leaves. Affected sites, which are covered with a mass of brown chewed leaf and wood fragments and webbing, often contain several larvae in different stages of growth. The larvae feed amid the webbing and also emerge at night to gather food. Leaves are frequently nipped off in the region where the larvae are feeding and may be included in the webbing.

Notes: a serious pest of *Macadamia* plantations which is active over most of the year. Persistent attacks can impact on the shape of affected plants. Damaged shoots frequently become swollen and misshapen or die, and new growths appear from lower down the stem. This insect is also a serious pest of waratahs, both commercially and in home gardens, the caterpillars commonly destroying flower buds and stem tips, resulting in shorter branches and loss of production.

Control: as per introduction for borers. Natural control is by several species of small wasps. Cutting and burning affected shoots is effective on small trees.

Other Twig Girdlers and Ringbarkers

Many longicorns of the genera *Ancita* and *Platyomopsis* are active twig girdlers in tropical and subtropical regions.

- The pale grey-brown Teatree Longicorn Beetle, *Platyomopsis obliqua*, eats large patches of the flaky bark of some species of *Leptospermum* and *Melaleuca*. It is about 3.5cm long and has spiky bumps on the wing-covers and thorax.

- The unusual Green Longicorn Beetle, which feeds on the soft growth of many plants, including *Acacia* and *Passiflora*, has bright emerald green wing-covers and long hairy antennae. It is about 2cm long with antennae about 3cm long.

- The weevils of *Ancita crocogaster* graze the bark of developing wattle shoots, causing dieback. This sturdy beetle, which is about 1cm long, is blackish with white markings on the wing-covers. Its hairy antennae are about 2cm long.

- Small weevils of the genus *Auletobius* ringbark the terminal shoots of many native plants after laying eggs among the leaves. The larvae feed on the dying and withered tissue.

BARK FEEDERS

Damage caused by bark-chewing caterpillar. [D. JONES]

Bark-chewing caterpillar on Acacia doratoxylon*; note covered tunnel and chewed bark.* [D. JONES]

The larvae of many different types of insect make tunnels through the bark of trees and shrubs, eating along the surface of the sapwood but not entering it. Attacks are usually of a sporadic and minor nature but occasionally the bark of an individual tree may be riddled. Severe attacks like this may be an indication that the tree has been weakened by some other factor. Control of bark feeders is not usually practical.

EXAMPLES

Pine Bark Weevil, *Aesiotes notabilis*

Description: a dull-grey weevil (family Curculionidae), about 1.5cm long with numerous conspicuous bumps and protruberances on its wing-covers. The larvae are fleshy white legless grubs.
Climatic regions: tropics, subtropics and warm temperate regions, mainly coastal.
Host plants: species of *Agathis* and *Araucaria*.
Feeding habits: larvae enter through deep wounds in the bark and feed actively around the point of entry. They also enter through wounds induced by pruning in all seasons except winter. Attacks can persist and secondary pests often enter the damaged site, sometimes resulting in ringbarking and death. The larvae also feed on the root system of Hoop Pine seedlings, causing stunting and sometimes death.
Notes: a serious pest of cultivated plants of *Agathis* and *Araucaria*, especially trees grown in plantations where pruning is important for commercial tree development.
Control: as per introduction for borers.

Scribbly Gum Moths, about 15 species of *Ogmograptis*

Description: the tiny moths (family Bucculatricidae) have a wingspan of 0.3–0.8cm and long hairs on the lower parts of the wings. The female moths lay eggs in depressions in the bark and the tiny cream to brown larvae (about 0.2–0.4cm long) feed specifically on the cork cambium layer (also known as phellogen) within the bark. Pupation occurs in a distinctive ribbed cocoon at the base of the tree or in litter.

Climatic regions: subtropics and temperate regions, mainly coastal.

Host plants: scribbles have been recorded on the smooth bark of 26 species of *Eucalyptus*. They are particularly common and prominent on *E. haemastoma, E. pauciflora, E. racemosa* subsp. *racemosa, E. racemosa* subsp. *rossii* and *E. stellulata*.

Feeding habits: the larvae produce distinctive zig-zagging or sinuous tunnels. Each tunnel, which contains a single larva, becomes wider as the caterpillar grows. When about half-grown the larva does a reverse turn and tunnels back the way it came, either enlarging parts of the existing tunnel or creating a new one on a similar course to the earlier one. The early instars of a larva, which are legless, feed on the bark eaten while the tunnel is being formed. By contrast, the last larval-instar (which has legs), feeds on the callus growth produced as a growth response by the tree in the second half of the very tunnel it created by its earlier feeding. The shape and form of the track depends on the species of moth involved. The characteristic brown tunnels of theses insects, which are so obvious on some smooth-barked eucalypts, can only be seen when the outer layer of bark is shed.

Notes: these moths are not pests as their activities do not harm the tree and in fact the scribbles add considerable interest to the eucalypts. Recent studies have shown that host-specific interactions occur between species of moths and eucalypts. Thus *O. scribula* occurs only on *Eucalyptus pauciflora*; *O. fraxinoides* occurs on *E. fraxinoides* and also on *E. pauciflora*, sometimes both species together on the same tree; *O. racemosa* occurs on *E. racemosa* subsp. *racemosa* and *E. racemosa* subsp. *rossii*; *O. pilularis* occurs on *E. pilularis*.

Control: unnecessary and impractical.

Bark scribble moth trails on Eucalyptus haemastoma. [R. Elliot]

Other Bark Feeders

- Scarab Beetles – some large scarab beetles feed on the bark of many different plants in tropical and subtropical regions. Native plants include cassias and some rainforest species such as Australian Red Cedar, *Toona ciliata*. The insects congregate and lift and tear the bark with their strong mouthparts. Branches may die from this treatment. Attacks are usually sporadic and can be easily disrupted.

- Moth larvae – the larvae of many species of moth tunnel in the bark of smooth eucalypts, usually causing minor damage. Their activities may be increased on stressed eucalypts left standing after vegetation has been cleared, or on trees retained in housing estates and other developments. See also the various species of web-covering borers many of which feed on bark.

- Proteaceae – unidentified caterpillars chew on the bark of some species of the Proteaceae family, including *Alloxylon flammeum* and waratahs. The eaten areas are covered with coarse chewed

Tube Caterpillar. [D. Jones]

frass particles which become brown with age. Damage ranges from minor attacks to large areas which can result in dieback or breakage.

Bark-chewing caterpillar damage to Alloxylon *stem.* [D. JONES]

Significant bark damage to Alloxylon flammeum *caused by an unknown caterpillar.* [D. JONES]

Weevil, Aterpodes tuberculatus, *rasping the bark of* Allocasuarina littoralis. [D. JONES]

- The adults and larvae of several species of weevils and longicorn beetles graze or rasp young bark (see Leaf-eating Beetles, chapter 9, and Twig Girdlers, this chapter).

BANKSIA FLOWER SPIKE BORERS

Damage caused by a Banksia *flower borer.* [S. JONES]

- Caterpillars of specialised moths in the genus *Arthrophora* (family Tortricidae) bore into the developing flower spikes of various species of *Banksia*. The caterpillars live in the stem and eat sections of buds. This feeding impairs the appearance of the spike as it often develops unevenly and its symmetry is ruined. The main impact is on cut flowers as the damaged spikes are unsaleable. Spraying in the early stages of spike emergence may offer some control.

CHAPTER 14

ROOT-FEEDING PESTS AND OTHER ROOT PROBLEMS

Plant roots provide the important roles of anchorage in the soil and the uptake of water and nutrients for normal growth. Factors which interfere with root growth and development affect the general health and growth of plants. These include root-feeding pests, root diseases and soil factors such as pH, nutrient availability, aeration and drainage.

SYMPTOMS AND RECOGNITION FEATURES

Seedlings toppling over .... *fungus gnats, damping-off (chapter 19)*
Potting mix loose, appearing as if ploughed or extra wet .... *fungus gnats, curl grubs*
Tiny flies running over soil surface, tiny white maggots present .... *fungus gnats*
Sturdy beetles with antennae on the end of a 'nose', drop to ground when disturbed .... *weevils*
Tunnels in roots .... *moth larvae, weevil and other beetle larvae*
White grubs in soil, usually curled in a C-shape .... *curl grubs*
Tiny plump white insects among roots, usually associated with white waxy threads .... *root coccoids*
Stunted roots, abnormally branched roots, galls or swellings on roots .... *nematodes*

ROOT-FEEDING PESTS

The activities of root-feeding pests are less obvious than other pests because they primarily attack plants at ground level or below ground, including the crown, base of the stem and root system (some of these pests can also damage above ground organs). Feeding activities by these pests include sucking sap, eating root hairs and soft root tips, chewing grooves on the surface of roots and boring tunnels into roots. The impacts of root-feeding pests include sudden wilting, complete plant collapse (in the case of seedlings, herbaceous plants and forbs), leaf yellowing, stem dieback and stem death. Often the death of a branch or a particular region of a shrub or tree can be traced to the destruction of the roots immediately below. Affected plants can linger and die slowly following root damage. Root-feeding pests include fungus gnats, root coccoids, some beetle larvae (particularly weevils and scarabs) and moths in the family Hepialidae.

FUNGUS GNATS

Fungus gnats (genera *Bradysia* and *Sciara* in the family Sciaridae) are tiny flies which are often noticeable in moist, sheltered areas where they breed in the soil. The winged flies often gather together in colonies and can be frequently seen running actively over the soil surface or flying close to the ground. Fungus gnats are often common in the warm, humid conditions of shadehouses, greenhouses and glasshouses. Fungus gnats do not feed at all, but the larvae, which are tiny maggots, are a serious pest for gardeners, hobbyists, nurserymen and hydroponic growers. The larvae feed on fungi, soil microbes, decaying plant matter and tender plant tissue, including root hairs, fine roots, callus growth, young stems and other soft growth. The larvae and adults can both transport fungal spores and increase the impact of fungal diseases such as *Botrytis*, *Fusarium* and *Pythium*.

Seedlings are a common target of fungus gnats. Seedlings damaged at or close to ground-level sometimes topple over, this symptom being easily confused with 'damping off' – a disease which is caused by soil fungi. The larvae also damage cuttings in propagation nurseries, feeding on callus growth and young roots. They are also devastating on soft plants recently removed from sterile conditions, such as orchid seedlings, mericlones, fern prothalli, fern sporelings and tissue-cultured explants. Fungus gnats can be a significant problem for fern growers by damaging tender roots and young fronds. The tubers and roots of terrestrial orchids can also be damaged.

Fungus gnats find modern soil-less potting mixes and composts very attractive and are a common problem in plants potted into these mixes. Most damage occurs to the crown and roots growing near the surface. The maggots burrow freely in the top layers and the potting mix commonly appears wet and the soil surface disturbed as if it had been ploughed by a miniature tractor. The presence of small white maggots in the mix, or small flies emerging from the soil and running over the surface is a sure indication of the presence of fungus gnats.

Life cycle: minute white eggs laid on the surface of soil or potting mix hatch into tiny legless white maggots with a prominent black head. The maggots can grow to about 0.5cm long in 2–3 weeks before pupating. The flies are small dark gnats, 0.2–0.4cm long, with long legs and one pair of wings. The life cycle takes three to four weeks and an adult female can lay up to 300 eggs. In warm weather it is usual for eggs, maggots of various sizes and adults to be present at any one time.

Control: good management is the best way to avoid attacks by fungus gnats. Fungus gnat outbreaks can be sporadic but quick control is necessary because their numbers increase so quickly. Thorough sterilisation of all propagation mix is essential, followed by hygienic growth practices and constant monitoring. Ensure good drainage by using well-aerated potting mix; avoid overwatering; ensure good ventilation in greenhouses; don't overcrowd plants; don't let algae, mosses and liverworts colonise the potting mix. Repot regularly as fungus gnats are attracted to degraded mix. Adult fungus gnats are attracted to yellow sticky traps (identify with a hand lens). Repotting can save an affected plant but it may be necessary to first wash off the old potting mix and dip the root system in a solution of pyrethrum.

Fungus gnats can be controlled by natural predators. According to Industry & Investment NSW, three predators are available from biological control firms. The mite *Stratiolaelaps scimitus* can be applied to the potting media at planting time and again two weeks later (it is sold as *Hyoaspis miles*). The Rove Beetle *Dalotia coriaria* and its larvae also feed on fungus gnats, and the parasitic nematode *Steinernema felitae* invades fungus gnat larvae, feeding internally, reproducing and eventually killing it. Recently a form of *Bacillus thuringiensis* (known as Bt-i), used to control mosquitoes, has proved to be effective against fungus gnats.

WEEVIL LARVAE

The larvae of several species of weevils feed on roots, either boring tunnels or chewing grooves in the surface. These weevils, including the large, colourful Diamond Weevil, the wattle pig beetles and the Elephant Weevil, are detailed in Chapter 9.

EXAMPLES

Native Apple Root Borer, *Leptopius squalidus*

Description: these are dull-grey weevils, the females about 2cm long, the males much smaller. The females lay eggs in batches on a leaf, then fold and gum the leaf margins together to protect the eggs. The larvae drop to the ground and bore into roots. Mature larvae are fat, legless, white grubs measuring about 2cm long.

Climatic regions: temperate regions, mainly coastal.

Host plants: the weevils and larvae feed on species of *Acacia* and *Eucalyptus*; they have also expanded their range of hosts, attacking the roots of fruit trees including apples, pears, citrus and grape vines.

Feeding habits: the weevils feed by grazing the leaf surface; larvae bore tunnels in roots, especially the very deep roots.

Notes: a dull-coloured, slow-moving, rarely noticed native weevil. Damage is usually of a minor nature although the effects on large anchorage roots can be significant. This species achieved notoriety when its feeding habits were expanded to include orchards and vineyards. Attacks by this pest were very common in orchards planted on recently cleared land and sticky traps on the trunk were used as a control measure. The fact that one orchard at Windsor in New South Wales trapped some 60,000 weevils in 1929 provides an indication of the abundance of this species in that year. The related species *Leptopius tribulus* also attacks the roots of wattles.

Control: control of larvae in roots is impractical, reducing adult numbers through use of sticky traps or spraying may be more practical.

Other Root Feeding Weevils

• Apple Weevil or European Apple Root Borer, *Otiorhynchus cribricollis* – a European weevil that has become established in temperate areas of Australia, feeding on a range of plants including some natives. It is also a significant pest of apple trees. The weevils, 1–1.2cm long, are brown with an elongated snout-like nose and long antennae. The wing-covers, which have rows of small sunken pits, are covered with short white bristles. The larvae, the most damaging stage of this pest, are root borers.

• Fruit Tree Root Weevil, *Leptopius robustus* – a native weevil that occurs in temperate regions. The female weevil often curls and glues the leaf margins around its eggs after laying. The larvae are fleshy white grubs that tunnel in the roots. The pale grey weevils, which range from 1.2–1.5cm long (females are smaller than males), are densely covered with short soft hairs (including legs and antennae). The surface of the body is slightly roughened, the black eyes are prominent and the snout thickened. The weevils feed on the foliage of wattles. The larvae are damaging root borers that attack native plants and fruit trees, particularly apples.

COCKCHAFER GRUBS (SCARAB BEETLE LARVAE)

The larvae of many species of scarab beetle in the family Scarabaeidae, subfamily Melolonthinae, are known as cockchafer grubs. They are mostly found in areas of reliable rainfall that receive 500mm or more annually. There are many species of cockchafers, all living underground and feeding on plant roots. It is not uncommon to find several species of cockchafer feeding together in a patch of soil. Numerous species of cockchafer eat the roots of grasses and can cause serious damage to lawns, bowling greens, golf courses and pastures. Some species damage crops such as sugarcane, maize and sorghum. Others target nursery plants and it is not uncommon to tip out an ailing potplant only to find the potting mix riddled with cockchafer grubs.

Cockchafers can be divided into two groups – those with beetles that do not feed at all, and those with beetles that feed on eucalypts (see also entries for Christmas Beetles and Swarming Scarab Beetles, chapter 9).

Life cycle: female beetles lay their eggs in moist soil among grass in late spring-early summer. The larvae grow steadily, passing through three larval stages and after about 20 months are fully grown. Most damage is done by the third stage larvae which feed over about 14 months before pupating in an oval chamber in the soil, usually in spring. After emergence from the pupae, the newly developed beetles tunnel upwards to wait near the soil surface, emerging during warm moist weather (thunderstorms often trigger a mass emergence). In abundant years the beetles fly in large swarms, some in mornings, others at dusk, depending on the species. The beetles live for up to two months, some feeding on eucalypt foliage (see also Leaf-eating Beetles, chapter 9), others not feeding at all.

Many scarab beetles have larvae that feed on roots. [D. Jones]

The larvae of this common scarab beetle are curl grubs. [D. Jones]

A common scarab beetle, Sericesthis geminata. [D. Jones]

Scarab beetle. [D. Jones]

The larvae of these nectar scarab beetles, Phyllotocus *sp., are curl grubs.* [T. Wood]

The larvae of this dynastine beetle feed on roots. [D. Jones]

EXAMPLES

Curl Grubs, White Grubs or Pasture Grubs

Description: these are the familiar underground larvae of several species of scarab beetles or cockchafer beetles that are dealt with in chapter 9, including those of the commonly encountered and colourful Christmas beetles. The larvae are cream to white fleshy grubs that curl in a characteristic C-shape when resting or disturbed. They have a dark head with prominent jaws and true legs on the thorax. The legless abdomen, which is large and shiny, is often dark grey because of the gut contents showing through the wall. The grubs can be 2–7cm long depending on age and species.

Climatic regions: tropics, subtropics and temperate regions, coastal and inland.

Host plants: the roots of a wide range of plants are eaten, including native and exotic grasses, lilies, bulbs, shrubs and trees. Seedlings of *Acacia* and *Eucalyptus* are very susceptible.

Feeding habits: root tips and the smaller or finer roots are eaten first. The larger grubs sometimes live in a small earthen cavity and frequently all roots in the vicinity are eaten.

Notes: curl grubs all have a basically similar structure and are difficult to identify. They can often be found feeding on grass roots in pasture – studies on some pasture samples have revealed population densities as high as 24,000 grubs per hectare. They are also significant pests of young eucalypt plantations, especially those planted in land that was previously pasture. The grubs congregate on the root system of the seedlings, eating the young roots, which results in stunted growth, loss of anchorage and plant death. The root damage becomes apparent when seedlings are lifted and examined. The larvae of *Heteronyx* beetles are known to feed on the fine roots of young eucalypts causing their death. This includes potted seedlings.

   Curl grubs can also cause significant damage to the roots of potted plants and they are sometimes an economic pest in plant nurseries. The beetles of some species seem to be attracted to modern soil-less potting mixes as a site to lay their eggs. The grubs feed on the plant roots as they develop and often several grubs can be present in a pot. Typically any potting mix impacted by curl grubs appears to be loose as if disturbed by the grub's activities, and the plant wilts or wobbles in the pot due to a reduced root system. In some cases the affected potting mix can appear to be excessively wet because the damaged roots are not taking up water.

Natural control: natural predators include birds (magpies, butcherbirds, ibises, ravens), some parasitoid flies and wasps.

Other control: small numbers of curl grubs cause limited damage but large infestations can be serious. Chemicals can sometimes be used but generally do not penetrate the soil well enough to ensure control.

Curl grub feeding on roots of a potted plant. [D. JONES]

Curl grub showing characteristic C-shape. [D. JONES]

Damage by a single curl grub in a potted plant. [D. JONES]

Other beetles with root-feeding larvae – see chapter 9

MOTH LARVAE

The larvae of some species of moths in the family Hepialidae live in the soil and feed on the roots of shrubs and trees.

EXAMPLES

Rain Moth or Bardi Moth, *Trictena atripalpis*

Description: the large moths have a wingspan of 12–16cm. They are dark grey with prominent white splashes on the forewings and a filigree of thin pale lines. The female moths produce masses of eggs that are released while flying. The slender, pale larvae grow to about 10cm long and are sparsely hairy with a brown head.

Climatic regions: subtropics and temperate regions, coastal and inland.

Host plants: probably a range of trees but certainly species of *Eucalyptus* including *E. blakelyi*, *E. camaldulensis* and *E. tereticornis*; also *Casuarina cristata* and *C. pauper*.

Feeding habits: the larvae tunnel in the soil and feed on roots. The tunnels are lined with a matted webbing material. The size of the tunnel increases as the larvae grows, eventually becoming quite large (2cm or more across).

Notes: the moths, which appear in large numbers during rain, are very short-lived. The large larvae, commonly known as bardi grubs, are used by fishermen as fish bait in inland streams.

Control: impractical as the larvae live underground.

Rain Moth. [D. JONES]

Pindi Moth, *Abantiades latipennis*

Description: the large moths have a wingspan of 8–10cm. They are dark grey with prominent silvery-white bars on the forewings. The female moths produce masses of eggs that are released while flying. The slender greyish or greenish larvae grow to about 9cm long and are sparsely hairy with a brown head.

Climatic regions: temperate regions of south-eastern Australia, coastal and inland.

Host plants: species of *Eucalyptus* growing in wetter forests, including *E. obliqua* and *E. regnans*.

Feeding habits: the larvae tunnel in the soil and feed on roots. The tunnels, which can be 15–60cm deep and 10cm across, are lined with a webbing material and the entrances covered with webbing and litter.

Notes: this species proliferates in areas of forest regrowth. Damage to roots includes girdling and subsequent root death. The pathogenic fungus *Armillariella* also enters through damaged roots. The moths, which appear in large numbers during rain, are very short-lived.

Control: impractical as the larvae live underground.

Pupal case of a root-feeding moth.
[D. JONES]

ROOT COCCOIDS OR GROUND MEALYBUGS

Several groups of specialised sap-sucking coccoids, strictly called root coccoids, live underground and feed on the roots of plants. They are also known as ground mealybugs, root aphids and root mealybugs. Various species are found from tropical to temperate regions, nearly always in relatively dry, well-drained soil. Certain species, which are significant pests of commercial crops, can transmit specific plant diseases. Many species of root coccoid live in a close mutualistic association with ants. The immature stages of some species feed on the above-ground parts of plants before moving underground to feed on roots. Some species form cysts on the roots in which they live and feed.

Description: the adults of commonly encountered species of root coccoids are generally small, 0.1–0.2cm long, globular to ellipsoid in shape and covered with waxy secretions. They can be grey, white, green, yellowish or red (sometimes depending on their food source) and are usually surrounded by white fluffy wax that is produced from pores along the side of the insect. They have short legs and a pair of very short antennae. When squashed they exude bright-red juice. Adults are winged. Reproduction is from eggs. Nymphs are smaller versions of the adults.

Life cycle: adult females lay eggs after mating and the juveniles develop through several moults. Tiny crawlers move about freely to find a feeding site, later stages become fixed and immobile. Root coccoids are found as scattered individuals or in loose groups feeding on the smaller fibrous roots. They usually occur in mixed colonies of adults and nymphs. They are often attended and moved by ants.

Plant damage: root coccoids cause damage to roots and stunt plant growth, sometimes resulting in leaf discolouration and premature leaf fall. Their white waxy exudates are often noticed on the root

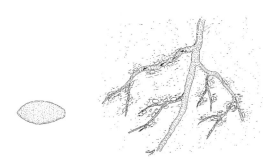

systems of potted plants at repotting time or when planting out (palms and ferns for example). Closer examination of the affected sites shows the coccoids attached to the roots. They attack a wide range of shrubs, trees and herbs, not only those growing in the ground but also in containers. Some species are serious pests of pasture grasses. In gardens these pests can also be found on the roots of herbaceous plants such as bulbs, daisies, sedges, lilies and grasses.

Root coccoids. [D. Jones]

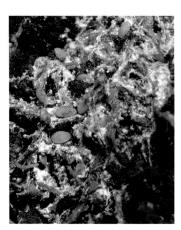

Root coccoids feeding amidst waxy detritis. [D. Beardsell]

Root coccoids on the roots of Brachyscome. [R. Elliot]

Root coccoids up close. [D. Jones]

Notes: root coccoids are widely distributed throughout Australia. Their feeding weakens plants and affected plants become susceptible to disease. These pests are often attended by ants that collect the honeydew produced from the coccoids. The ants move the coccoids to new sites. Infestations of root coccoids in container-grown plants often occur in patches of dry potting mix, perhaps as a result of the water-repellent wax produced by the coccoids.

Natural control: little is known about the natural control of root coccoids.

Other control: control is difficult because root coccoids are out of reach of contact insecticides. Some insecticides such as malathion may give control, but drenching with chemicals can be ineffective as soils can be very difficult to wet satisfactorily. Flooding the soil with water disrupts their feeding. Badly affected potted plants may need complete removal of the potting mix followed by root washing and repotting in a sterilised mix.

NEMATODES

Three groups of microscopic eelworms that burrow into roots and cause stunting, abnormal branching, swellings and gall-like growths are detailed in chapter 6.

OTHER ROOT PROBLEMS

Root Diseases

Several fungus diseases attack the root systems of plants. Common and widespread root pathogens include species of *Fusarium*, *Pythium* and *Rhizoctonia*. Species of *Phytophthora* and *Armillariella* are vigorous root pathogens that commonly cause the death of shrubs and trees. Diseases are dealt with in chapter 19.

This established eucalypt died suddenly following roadworks that affected its root system. [D. Jones]

Root Disturbance

Some native plants are sensitive to root disturbance. Even established plants can suffer problems when significant parts of their root system are disturbed or destroyed. Frequent problems with established trees follow roadworks. Activities such as road widening and the construction of footpaths often result in the death of trees soon after the work is finished. These activities can also cause disease entry through damaged roots which can result in dieback and slow death. Gardening activities around established eucalypts can also cause problems. A common system of establishing gardens on large blocks is to add soil around existing trees and then plant this as a garden. If the soil covers too much of the tree's root system, it suffocates because air movement in the soil and gas exchange with the atmosphere is restricted or curtailed.

Lack of Root Nodules

Specialised nodules on the roots of some plants contain nitrogen-fixing bacteria (commonly referred to as rhizobia) that exist in a symbiotic relationship with the plant. These bacteria have the ability to convert nitrogen gas from the atmosphere into ammonia (termed nitrogen fixation) which can then be used by the plant

This established eucalypt died after a new garden bed was built around its base. [D. Jones]

This Acacia denticulata *is suffering from lack of root nodules.* [D. Jones]

to form amino acids and other cellular components. Most fixation occurs under harsh conditions when nitrogen is limiting plant growth. Rhizobia commonly infect legumes (pod-bearing plants) in the families Fabaceae and Mimosaceae. Often there is a host-specific relationship between the plant and a species of bacteria. Legumes which either have no root nodules or poorly developed nodules are stunted, weak and linger for a time before dying. These plants generally have a pale or yellowish appearance and the older leaves become bright-yellow and fall prematurely. Mulching with some bush litter or leaf mould from under another plant of the same species or a close relative may induce nodule formation. It is a wise procedure to check the nodulated root systems of legumes before planting (before purchase is even better) as nodule formation may be poor when plants are grown in sterilised potting mixes. In some cases lack of nodules is caused by a soil deficiency of molybdenum.

Lack of Mycorrhiza

Some plants have associations with symbiotic fungi in their roots. These are known as mycorrhizal fungi and they aid in the plants' nutrition. Mycorrhizae are common in plants growing on poor sandy soils such as those found in heathlands. The root system of these plants contains short knobby or distorted roots that are the site of mycorrhizal acivity. In the absence of these fungi some plants grow weakly and appear unthrifty. They lack vigour and also may have yellowish leaves. Such plants do not generally respond to the application of fertilisers. The addition of some leaf mould from under established bushes of the same species will generally produce a marked improvement in the plant, although not always.

Mycorrhizae are found in such plant groups as orchids (*Caladenia, Calochilus, Dipodium, Gastrodia, Prasophyllum*), rushes and sedges and also a number of shrubs, especially those that occur naturally on infertile soils such as heaths, and *Ricinocarpos pinifolius*. Certain eucalypts have a mycorrhizal relationship and seedlings grow poorly in their absence, for example *E. dives, E. macrorhyncha, E. marginata, E. obliqua, E. pauciflora* and *E. radiata*. The proteoid roots found in members of the family Proteaceae do not contain mycorrhiza.

Nutrient Deficiences

An imbalance of nutrients in soil or the inability of plants to take up an element can affect the healthy growth of plants (plant nutrition is detailed in chapter 20). Elements that directly affect the growth of plant roots include phosphorus, potassium, sulphur and molybdenum.

Frost

It is not generally realised that heavy frosts can damage or even kill the root systems of plants. The damage is worse if the roots are growing actively at the time of the freeze. This is one reason why heavy frosts occurring early in the season (in late autumn or early winter) often cause so much damage. Frost damage is dealt with in chapter 21.

Soil Salt

High levels of salt in the soil cause damage to the root systems of plants, usually resulting in dieback or death. Salt damage is dealt with in chapter 21.

Waterlogging

Excess water in the soil results in a reduction in soil aeration and affects the health of plants. Some plants tolerate waterlogged soil quite well, whereas others suffer root damage, dieback and death. Waterlogging damage is dealt with in chapter 21.

CHAPTER 15

ANTS AND TERMITES

ANTS

Ants are social insects that are placed in the family Formicidae in the order Hymenoptera. There are thousands of species of native ants, very few of which are garden pests. Vigorous and aggressive colonising ants probably cause most angst with gardeners. Two groups in particular fit these categories. The widespread meat ants of the genus *Iridomyrmex* (sometimes called tyrant ants because of their ability to dominate), form extensive colonies and can occasionally take over a garden. They feed on honeydew secreted by sap-sucking insects, but also kill other small creatures and devour carcases. They can also vigorously attack humans, often disrupting outside activities by their concerted efforts. Green Tree Ants, *Oecophylla smaragdina*, also commonly take over the shrubbery in tropical gardens. Although they are beneficial in controlling other insects, their aggressive behaviour towards any disturbance of the foliage readily alienates them with gardeners.

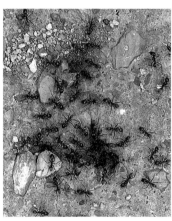

Meat ant nest. [D. JONES] *Part of a meat ant colony.* [T. WOOD] *Green Tree Ants swarming after damage to their nest.* [D. JONES]

More significantly, about 14 species of exotic ants have become naturalised as significant pests in Australia. Some of these ants pose serious threats to gardens and in some cases also to the gardeners themselves.

- **Argentine Ant**, *Linepithema humile*. This ant, which is native to several countries in South America, is now well established over much of Australia. Its dark brown workers, which are only about 0.3cm long, can squeeze into tiny holes and cracks. They have a musty smell when squashed, quite distinct from the usual formic acid smell of most ants. Argentine Ants can form supercolonies consisting of millions of individuals. New colonies establish readily when a queen with some workers separate from the main one. They are also readily spread to new sites by moving infested soil or potplants. Argentine Ants impact in gardens by quickly displacing the native ants from their territory and also by farming colonies of sap-sucking insects. They also have serious ecological impacts in natural communities, even attacking nesting birds.

- **Yellow Crazy Ant**, *Anoplolepis gracilipes*. This exotic species was first found in eastern Arnhem Land in the Northern Territory during the 1970s, and later in Darwin. Since its discovery in Cairns in 2001 it has spread to so many areas in tropical and subtropical Queensland (south to the Gold Coast) that eradication in that state is no longer feasible. The yellow to brownish workers, which are about 0.5cm long, have a dark brown abdomen and long legs. They walk erratically and become frantic when disturbed, spraying formic acid as a defence mechanism. They can disrupt gardens and natural areas, causing significant damage to plants and wildlife. They also have disruptive impacts on plant nurseries and the movement of potted plants between nurseries.

- **Red Imported Fire Ant**, *Solenopsis invicta*. This ant, which is native to South America, has become widespread around Brisbane since it was first detected in 2001 (although it probably arrived much earlier). Recently a new population, probably originating from the southern USA, was discovered in Gladstone. These recently arrived fire ants are believed to have been introduced in construction equipment offloaded from gas industry ships. Fire ant workers, which range from 0.2–0.6cm long, are reddish brown with a dark brown abdomen. New colonies are formed by winged males and females which mate in flight. They build a dirt nest, up to 40cm tall, with subterranean tunnels through which the workers move. These aggressive ants swarm when disturbed and each ant can inflict a painful burning sting followed by a blister. They usually swarm and bite together at once causing an intense burning pain. They eat plant material, including tubers and seeds, as well as insects (especially ground-dwelling bees and wasps) and small animals. They also collect honeydew from sap-sucking insects. These pests not only disrupt gardens and natural areas, but also have significant impacts on plant nurseries and the distribution of plants. They also have the ability to disrupt the way we live.

- **Tropical Fire Ant or Ginger Ant**, *Solenopsis geminata*. An exotic ant that is native to South America and southern states of the USA. It has become well established in the Darwin region and around Katherine in the Northern Territory. This ant, which is extremely aggressive and with similar stinging and feeding habits to the previous species, has apparently been present in northern Australia for more than 80 years. Two small infestations which became established in plant nurseries in Western Australia (one in Perth, one in Port Hedland) have been eradicated.

- **Electric Ant**, *Wasmannia auropunctata*. An exotic species that was first found at Smithfield near Cairns in north-eastern Queensland in 2006, and was later discovered in several nearby areas. It is a tiny, slow-moving, golden-brown ant, only about 0.15cm long, that packs a very painful sting that can itch for several days. These ants do not establish nests but colonise crevices, litter, garden waste and rubbish. They feed in groups, attacking a wide range of other insects, both on the ground and in trees, often completely eradicating native ants from an area within a short period. They also attack many animals, including snakes, lizards and frogs, and farm sap-sucking colonies of aphids, scales and mealybugs.

- **African Big-headed Ant**, *Pheidole megacephala*. An exotic ant from Africa that has been present in northern Australia for more than 100 years and is now also well established in south-west Western Australia. The workers are ginger brown with a darker shiny abdomen. Two different types of worker ants occur in the colony: most workers are 0.2–0.3cm long with normal heads; however, a low proportion of workers are 0.35–0.45cm long with very large heads and powerful jaws. These ants, which are slow moving and do not sting, have no odour when squashed. They form supercolonies which can cover several hectares. They displace all native ants from an area and their feeding activities disrupt the ecology of large areas of bush. Their tunnelling activities and excavations can be so extensive that the root systems of shrubs and trees can be totally disrupted, resulting in weakened, dying and dead plants. These ants pose a major threat to natural ecosystems, especially rainforest.

Control: Accurate identification of an introduced ant is critical because the control measures are often specific to a particular species. Ant colonies can be destroyed by applying a slow-acting poison bait which will be collected and carried back to the nest by the workers. After several days the poison will be spread throughout the colony resulting in its demise. Seek help from local authorities as to the best registered poison to control each species. The simple technique of painting a band of a sticky substance around the trunk of a tree or shrub prevents the movement of ants and aids in the control of the pests they have been farming.

TERMITES

Termites, often incorrectly called white ants, belong to the family Termitidae in the insect order Isoptera. They are very destructive social insects that have the ability to increase in numbers quickly and form large colonies. They are widely distributed throughout Australia, where some 348 species (many yet to be described) are known to occur. Some species build the familiar termite mounds which extend above ground, whereas others are less conspicuous, forming colonies underground, or in the root system of trees (living and dead), logs and stumps. They can also form internal nests within a tree and arboreal nests on the outside of trunks and branches. Humidity within a colony is maintained at very high levels to prevent desiccation (termites are thin-skinned and dehydrate quickly). Temperature levels are also controlled. Termites search for food sources in closed tunnels which can be very long. The colonies themselves can be long-lived – well in excess of 50 years.

Did termites kill this tree? Probably not.
[D. Jones]

Termites are thin-skinned, light brown to cream or white fleshy insects with a darker head. The area between the thorax and the abdomen is poorly differentiated and there is no distinct waist. Termites have a highly developed social structure that resembles those of bees and ants. The queen, which is the egg-producer and is mated by a king termite, is usually larger than the other termites and longer-lived (20–70 years). The workers, which are usually cream or white, build the nest, forage for food and tend the queen. The soldiers, which usually have darker heads than the workers, defend the colony against attack. Finally there is also a group known as the 'reproductives' that have the ability to replace the king and queen if disaster strikes. Members of this caste eventually develop wings (termed alates) and fly away from the colony in warm humid weather, seeking new sites for colonisation.

Most species of termite are harmless to mankind and make important contributions to ecology and soil fertility by recycling grass, fallen bark and other litter; their activities also impact positively on soil aeration, soil structure and water penetration. Some termites obtain protein from certain fungi. There are about 20–30 species of termites that cause problems, especially those species which eat wood and wood products. They can be very destructive of wooden structures and their feeding activities often impact badly on dwellings. The cellulose in the wood is digested by protozoa in the termite's gut. Some termites can be very destructive of living shrubs and trees. Attacks by these termites are often insidious, with the insects entering the tree via the larger roots or directly through the trunk at or below ground level. Once in the plant the sapwood is eaten (the heartwood is not eaten) and the insects work up into the trunk and branches, greatly reducing structural strength. On eucalypts with fibrous bark, particularly bloodwoods, the termite tunnels show up as dark lines or strands in the bark. Trees attacked by termites can linger but eventually collapse and die. Symptoms of termite attack include a general weakening and debilitation of the tree, the loss of larger branches and a reduction in the number of leaves in the canopy. Sudden wilting and yellowing of the leaves on a branch

Termite infestation at the base of Eucalyptus wandoo. [S. Jones]

may indicate that termites are feeding on the living wood of that branch. Sometimes the bark is eaten and the plant dies from ringbarking. Badly damaged trees frequently lose branches and break or fall over in a storm.

Natural control: ants are a major predator of termites, commonly raiding nests and tunnels to carry away the workers as food. Some ants, such as the foaming ants found in tropical parts of Australia, are specialist predators of termites. Colonies and forage lines are also attacked by echidnas, bandicoots, marsupial mice, phascogales and lizards. Termites are very susceptible to attack while on their winged flights. At this stage they are at the mercy of the elements and many predators including lizards, frogs, birds, dragonflies, ants, spiders and robber flies.

Other control: termites are very difficult to contain and should be controlled by a licensed pest controller. Preventive measures are always better than a cure. In tropical and inland regions it is virtually impossible to protect a garden from attack by termites and regular inspection and spot treatment is the only answer. Dead or dying trees should be removed completely (roots and all) and burnt, and any nests destroyed. Feeding lines can range a long way from a termite colony and large areas may need to be examined. Movement of the chemicals through the soil is minimal and it is useless applying chemicals to the surface of the soil in the belief that they will seep down to the pests. Garden trees should be examined regularly and if infested the termite tunnel should be excavated and poisoned before the tree is killed. Nests can be confirmed with temperature probes as the colonies are maintained at controlled temperatures. Healthy trees that are infested by termites can be treated by drilling into the wood near the base and a control agent pumped in by a licensed pest controller, after which the hole is sealed. Chemical control varies with the species and hence accurate identification is essential. Samples of termites should be sent to a specialist for accurate identification. Many garden plants, both native and exotic, are susceptible to termite attack. Those frequently damaged include wattles, eucalypts, palms, umbrella trees and cassias. Resistent trees, such as species of *Callitris,* can be grown in areas of severe infestation.

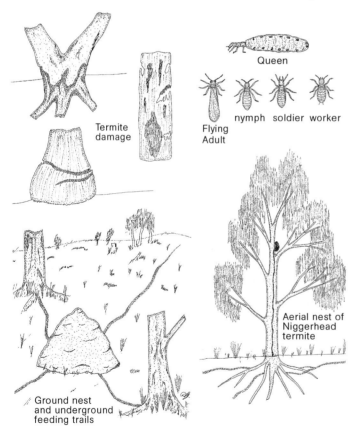

Termite damage and nests. [D. JONES]

EXAMPLES

Giant Termite or Darwin Termite, *Mastotermes darwiniensis*

Description: a distinctive brown termite; workers are about 1cm long, soldiers about 1.2cm long and the winged alates 3–3.5cm long with a wingspan of 4–5cm.

Climatic regions: mainly tropical parts of Queensland, Western Australia and Northern Territory; however, a colony has recently been found in an inland suburb of the Gold Coast.

Host plants: a very wide range of plants including native species occurring naturally in the bush and those grown in gardens; also numerous exotic garden plants, palms, fruit trees and vegetables.

Feeding habits: these termites form underground nests as well as in fallen logs, stumps and dead or dying trees. Feeding termites enter a plant from below ground and damage or kill it by ringbarking. They are voracious feeders and can kill a large tree in a short time. Dead trees commonly become the headquarters of a new colony. Foraging tunnels can extend up to 200m from the colony. When colonies become large enough, they break up and form separate colonies in new sites.

Notes: this remarkable termite, the most primitive form of termite in existence and sharing many structural features with cockroaches, is one of the most destructive species in the world. It not only destroys timber structures, such as buildings, but also eats anything organic (including rubber, plastic and insulation on wire) and kills garden plants. It also significantly impacts on natural vegetation.

Control: in Darwin a commercial baiting technique using a slow-acting poison has been developed and proved to be safe and effective against *Mastotermes darwiniensis.* Other species of termite are not controlled by this technique.

Termite damage to base of palm trunk. [D. JONES]

Termite damage extending up a palm trunk. [D. JONES]

Other Termites

- A number of species of destructive termites attack native plants, especially eucalypts, in parks and gardens throughout Australia. More species are active in tropical regions than in southern Australia.

- **Latex Termite**, *Coptotermes acinaciformis,* is one of the most destructive native termites. It occurs in all Australian states except Tasmania and will kill living trees. It does not build a mound but nests at the base of living and dead trees and stumps and travels underground seeking food. Nests also occur inside the trunks of large trees. A colony can contain up to a million individuals. The aggressive soldiers produce milky latex when disturbed.

Termite nest, probably a species of Coptotermes, *at the base of a large eucalypt.* [D. JONES]

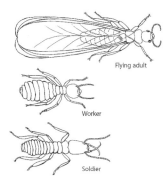

Latex termites. [T. BLAKE]

Flying adult

Worker

Soldier

- **French's Latex Termite**, *Coptotermes frenchi*, is another very destructive native termite. It occurs in coastal areas of the mainland states and will kill living trees. It does not build a mound but nests at the base of living and dead trees, inside large trees and stumps. The workers travel underground seeking food. The soldiers have distinctly curved mandibles and produce milky latex when disturbed.

- **Mounding Latex Termite**, *Coptotermes lacteus*, is widely distributed along the east coast of Australia. It mainly lives in forested areas and builds a mound 1–2m tall. The termites produce milky latex when disturbed.

- **Marri Termite**, *Coptotermes michaelseni*, is the most destructive termite in Western Australia. It attacks living trees and is often found feeding on or near Marri, *Corymbia calophylla*. The soldiers produce milky latex when disturbed.

Mound of Coptotermes lacteus. [D. Jones]

- **Subterranean Termite**, *Schedorhinotermes intermedius*, occurs in coastal areas of the Northern Territory, Queensland and New South Wales. It feeds on the lower trunk and larger roots of living and dead trees.

- *Neotermes insularis,* occurs in coastal areas of the Northern Territory, Queensland, New South Wales and Victoria. It feeds on the trunks and branches, sometimes causing serious damage. It can form small arboreal nests.

- *Nasutitermes exitiosus* is a common mound-builder in eastern and southern Australia. It builds mounds 50–75cm tall. In drier areas it can nest in tree stumps or below ground. The soldiers have a dark brown head and produce milky latex when disturbed.

- Arboreal nests are built by *Nasutitermes walkeri* and *Microtermes turneri*, with foraging tunnels extending over the bark. These species mostly cause minor damage.

Arboreal termite nest. [D. Jones]

Arboreal termite nest in branch junction of rainforest tree. [D. Jones]

Termite strands on trunk of Alphitonia excelsa. [D. Jones]

CHAPTER 16

MISCELLANEOUS FERAL AND NATIVE ANIMAL PESTS

This chapter provides details on a range of miscellaneous animal pests including birds, earthworms, millipedes, native marsupials (possums, kangaroos, wallabies) and rodents (bush rats), feral animals (pigs, camels, deer, goats, horses, donkeys, rabbits, hares, rats, cane toads, mice) and slaters.

FERAL ANIMALS

Numerous exotic animals have become established as pests in Australia. Some are not troublesome to gardeners but the majority have serious impacts on the environment. Many of these feral animals are herbivorous and feed on a range of plants from grasses to shrubs and trees. Feeding is often selective, meaning that preferred palatable plants are browsed at greater levels than others. Complete defoliation and significant physical damage to shrubs and trees is common. Browsing also impinges on flowering, fruit set and recruitment of seedlings, leading to significant changes in the ecological balance of an area. Additionally, each type of animal tends to congregate in habitats suited to their life style, which also produces biased pressure on the natural ecology. Other problems caused by feral animals include hard feet breaking up the surface crust of the soil, chewing and rubbing bark which weakens and kills trees, degradation and fouling of waterholes, competition with native animals and stock, carriers of exotic diseases, and damage to fences, buildings and other structures. All of these damaging factors become accentuated and more critical during periods of drought.

Birds

Some introduced birds damage garden plants while scratching actively through mulches and litter when searching for food. Apart from creating a mess in gardens, it is not uncommon for seedlings and small plants to be either uprooted during this process or covered with litter. The worst culprit in temperate gardens is the introduced Eurasian Blackbird, *Turdus merula*, which can be quite a pest.

Control: protect susceptible plants with fences or enclosures, or cover with branches. Spraying with repellent solutions containing garlic, capsicum or chilli may influence the birds' activities for a short time.

Cane Toads

The Cane Toad, *Bufo marinus*, is a serious naturalised pest that is now very common in most parts of Queensland, northern parts of the Northern Territory and northern New South Wales. It is spreading steadily southwards and westwards. Apart from being of repulsive appearance, highly poisonous and impacting disastrously on Australian wildlife and ecology, it is also a nuisance to gardeners and nurseries. These toads like to burrow and loose, open potting mixes make digging easy. They are also attracted to the humid, warm conditions of structures such as bushhouses, greenhouses and glasshouses, and can be a nuisance to plants such as orchids, bromeliads and ferns. There is no spray that will kill or deter toads, however population levels can be reduced by night-time culls.

Pigs

Feral pigs pose the most serious threat to Australia's natural environment because of their sheer numbers – current estimates suggest that there are more than 23 million feral pigs in Australia – and the fact that they concentrate their feeding in the vicinity of drainage lines and floodplains (pigs must drink water regularly, daily in hot weather). The largest populations of feral pigs occur in eastern Australia, especially in the tropics and higher rainfall areas. They can begin to breed at 7–12 months of age and have a reproduction capacity similar to rabbits. A single animal can roam and feed over a 20–40 kilometre square region. In dry times they gather around waterholes to drink and wallow. Pigs root up the ground when feeding and cause enormous environmental damage. They often concentrate their feeding activities in moist to wet sites, and eat a wide range of animals and plants, including subterranean roots and tubers. They are known to eat the tubers of terrestrial orchids, and in some instances have completely eliminated orchid populations from some areas. Feral pigs impact on gardens in some areas, even large municipal gardens such as botanic gardens (Cooktown Botanic Gardens, for example) have been damaged or trashed by their feeding.

Feral pig damage – note exposed fleshy roots. [T. Wood]

Extensive feral pig damage in a subalpine habitat. [T. Wood]

Camels

Estimates suggest there are between half a million to a million feral camels living in inland Australia and they have the potential to double their numbers every 9–10 years. They have no natural enemies apart from man, and are long-lived (30–50 years). They begin breeding when 3–4 years old. They are inflicting untold damage on desert and dryland ecosystems, pastoral properties and remote communities. Camels eat a wide range of plant food and can graze vegetation 3–3.5m high. Preferred species include *Acacia sessiliceps*, *Codonocarpus cotinifolius*, *Erythrina vespertilio*, *Pittosporum angustifolium* and species of *Santalum*, including the quandong, *S. acuminatum*.

Damage to the bark of young Australian Red Cedar, Toona ciliata, *by feral deer.* [D. Jones]

Deer

Six species of deer have become successfully established in the wild in various parts of Australia, from temperate zones to the tropics. They live as individuals or in herds, feeding on a wide range of plants and damaging the environment. They are prominent in grassy forests where shelter is available, but also inhabit wet forests, rainforest and grassland. They commonly chew bark and kill or deform shrub and tree saplings (including eucalypts), often chewing branches and defoliating palatable plants. They can cause significant damage in moist forests and along stream banks, leading to erosion. In some areas they impact seriously on gardens and can be difficult to keep out, even by fencing. Paradoxically they are protected as a hunting resource in New South Wales, Victoria and Tasmania.

Goats

Feral goats are well established, significant environmental pests in many parts of Australia, especially in arid and semi-arid areas. They also occur on many offshore islands. Estimates suggest that as many as 2.5 million feral goats may be present in Australia. Feral goats occupy a wide range of habitats but their impacts are often severe on rocky hills where they completely eliminate the ground flora, leaving only the trees. They are very adept at stripping bark from trees and also eat regenerating shoots and seedlings, maintaining the bare ground and in severe cases even eating the lichens from the rocks. When hungry they will climb part way up shrubs and small trees, breaking branches to obtain the leaves, and chewing bark. Feral goats cause erosion, land degradation and also pollute waterholes.

Feral goat standing erect to feed on shrubs. [S. JONES]

Damage to eucalypt sapling by feral goats. [D. JONES]

Horses

Estimates place the number of feral horses in Australia at between 300,000 and 400,000. They are found in many parts of Australia in habitats as diverse as semi-arid deserts and well-watered subalpine regions. Horses gather in groups or herds and feed on grass and other vegetation. They also chew tree bark. Horses often concentrate their activities in moist to wet areas and can significantly degrade these habitats as well as waterholes. Their activities in subalpine and alpine areas have impacted badly on some populations of ground orchids that grow in moist to wet sites. Other environmental damage includes trampling fragile alpine vegetation, soil compaction, soil erosion and spreading weeds.

Donkeys

There are about 5 million feral donkeys in Central Australia, the Kimberley region of Western Australia and the Top End of the Northern Territory. They live in small groups and eat a select range of palatable plants, sometimes eliminating these species from an area.

Rabbits and Hares

In areas where these animals are pests, valued shrubs or trees should be protected until they are well established. These animals commonly gnaw young bark and can seriously damage or kill young plants and saplings. Hares, which prefer to live in open grassy habitats, often simply chop off a plant by biting completely through the stem. Rabbits, which find young *Callitris* plants irresistible, often impair natural regeneration by eating any seedling that appears. They will also dig in the soil around recently planted *Callitris* seedlings and gnaw the roots causing death of the trees. Once a hard corky bark is formed, rabbits and hares generally lose interest, although in times of food shortage they may resort to bark gnawing. Rabbits also feed on soft plants such as forbs and ground orchids, sometimes completely eliminating them from an area.

Rabbit diggings. [D. JONES]

Netted enclosure protecting herbaceous plants from rabbits.
[D. JONES]

The following techniques can be useful in reducing damage by these pests.

1. Circle the trunk with a wire netting guard that will prevent access to the plant. This barrier should be about 30 cm from the trunk at any point and supported by strong stakes. It is best to bury the bottom of the wire to a depth of about 15cm to prevent burrowing beneath the guard. Commercial tree guards are available.

2. Wrap bituminous paper around the trunk to about 60 cm high.

3. Newly established plantations should be surrounded by a secure rabbit-proof fence.

4. Bitter-tasting repellents such as the fungicide thiram, garlic spray and sprays made from *Quassia* chips (sometimes available from chemists) can provide a short-term barrier.

5. Reduce rabbit populations by any legal method.

Rats and Mice

These introduced rodents are generally regarded as household pests, however they can also be very destructive in glasshouses, shadehouses and ferneries. Both types of rodents feed on bulbs and tubers, often burrowing in soil to reach them. They also eat the pseudobulbs of some orchids, especially *Cymbidium* and *Dendrobium*. Their feeding is obvious from the longitudinal grooves left by their gnawing teeth. In ferneries, rats have been known to bite off the fern fronds to make nests. They also gnaw through the trunks of young palms, presumably to feed on the apical bud. Rats and mice also eat seeds and seedlings and in native plant nurseries and orchid nurseries they have been known to destroy a season's crop in a single night's feeding. Rats and mice can be controlled by spreading commercially prepared baits or by trapping.

Slaters, Woodlice and Pill Bugs

Slaters (or woodlice) and pill bugs are oval, flat, hard-shelled crustaceans placed in the order Isopoda. They have segmented bodies and six or seven pairs of jointed legs. Slaters, which are flat, dull and oval in shape, lack the ability to roll into a ball, whereas pill bugs are deep-bodied, shiny and can roll into a ball when disturbed. Mostly greyish or blackish, these crustaceans grow 1–1.5cm long. Slaters desiccate in sunshine and therefore congregate in damp, shady places where there is an abundance of decaying organic matter on which they feed. Pill bugs also seek out similar living conditions but are able to cope with some sunshine and dryness. Both groups feed on fungi and decaying plant material, but they can sometimes be a pest.

Control: numbers can be reduced by removing hiding places, clearing up accumulations of organic matter, and spreading lime. Baiting with commercial snail control preparations is an effective means of control. *Eucalyptus* sprays seem to act as a deterrent to slaters.

EXAMPLES

- The slater most commonly encountered in Australian gardens, *Porcellio scaber*, is an introduced species from Europe. As well as feeding on decaying organic matter, this species can also eat seedlings, soft shoots and root tips. Populations establish readily in the warm humid conditions of glasshouses and cause damage by feeding on the young growing root tips of ferns and epiphytic orchids. A whole season's root growth of an epiphytic orchid can be curtailed by this feeding, impacting significantly on the plant's growth and establishment. Slaters also live in the fibrous trunks of tree ferns, eating the growing tips of the aerial roots.

- The Common Pill Bug, *Armadillidium vulgare*, also introduced from Europe, does not appear to damage living plant tissue although many gardeners claim the opposite view. It is readily distinguished from slaters by its darker shiny appearance and the ability to roll into a ball when disturbed.

Slater. [S. JONES]

Pillbugs. [S. JONES]

Millipedes

Millipedes are hard-shelled, segmented, worm-like creatures which have numerous legs (two legs on each segment) and a characteristic habit of coiling when disturbed. There are about 2,000 species of millipedes native to Australia, mostly feeding nocturnally on leaf litter and mosses and none of concern to gardeners. About 14 exotic species also occur in Australia and one, the Portuguese Millipede, *Ommatoiulus moreleti*, is a plant pest. This species, introduced in 1953, is now widespread and abundant in southern Australia. It feeds on a range of plant material including tubers, bulbs and fleshy roots, and it causes problems with orchids, lilies and palms. This pest, which is common in damp situations where there is an abundance of decaying material, is a frequent resident in glasshouses.

Control: numbers can be reduced by collecting at night and squashing (they release a pungent caustic material when handled). Remove accumulated organic material and dust with lime. Commercial snail baits can reduce millipede numbers.

Portuguese millipedes. [D. JONES]

Portuguese millipede in defensive position. [D. JONES]

Native millipede. [T. WOOD]

NATIVE ANIMALS

Some native animals cause problems in gardens, mainly in urban areas and suburbs developed on the fringe of natural bush. Often these visitors are welcomed into gardens but they can also create problems and become a nuisance.

Ants

Australian native ants are present in almost every garden and provide benefits such as helping to control a range of pests such as caterpillars. They are also the main enemy of termites and contribute to ecology by recycling nutrients, mixing and aerating soil and dispersing seeds. Many native ants are a nuisance to gardeners because they farm sap-sucking pests. These ants are attracted to the honeydew excreted by sucking insects such as aphids, lerps, scales, mealybugs and leafhoppers, and frequently encourage the activity of such pests by guarding them against predator attack and moving the young onto fresh growth. In many regions ants construct a byre of litter to cover colonies of scales, leafhoppers and mealybugs. The ants attend the sucking pests within the protection offered by the byre (see also chapter 4).

Birds

Some native birds damage plants while scratching through mulches and litter when searching for food. They can create havoc in gardens by covering small plants with piles of litter, changing garden borders and covering paths with soil and mulch. Apart from creating a mess, it is not uncommon for treasured small plants to be uprooted during this process. Australian Brush-turkeys are powerful scratchers and are undoubtedly the most persistent and damaging of the native birds that visit gardens on the east coast. They commonly cause problems in gardens established close to bushland and are notoriously difficult to deter. Lyrebirds also move large volumes of surface litter, compost and topsoil in their quest for food. In nature they have been observed to expose terrestrial orchids, which are sometimes eaten. White-winged Choughs are also vigorous scratchers that have also been observed to unearth and eat the tubers of terrestrial orchids. Others, including bowerbirds and whipbirds, may be a nuisance in limited areas. Birds can also damage potted plants in shadehouses and ferneries, often pecking young growth or flowers. Bowerbirds eat the leaves of numerous different plants, including some natives. Male Satin Bowerbirds collect blue flowers, such as *Lechenaultia biloba*, to decorate their bowers.

Control: native birds are protected and cannot be killed. Instead shield susceptible plants with fences or enclosures or cover with branches. Spraying with repellent solutions containing garlic, capsicum or chilli may influence the birds' activities for a short time.

Bush Rats

Native rats are a significant garden pest in some coastal areas. They form localised colonies, living in burrows, sheltering in dense tussocks and shrubs, and constructing tunnels at ground level that ramify through the surrounding vegetation. Native rats generally hide during the day and are active at night and on cloudy days. They eat the roots, bulbs and fruit of many native plants. They are especially fond of mat plants and tussocks including species of *Lomandra*, *Dianella*, *Arthropodium*, *Dichopogon* and *Caesia*. They commonly eat the roots of these plants while living in an underground tunnel. Symptoms include dieback of parts of the plant or the whole plant collapsing after the roots have been completely eaten away. Perhaps surprisingly these animals also eat the roots of some shrubs and trees. They are particularly fond of species of *Brachychiton* and have been known to kill a plant of *B. discolor* about 4m tall by eating all of the roots back to the trunk. They also attack palms and have been known to kill small (1–2m tall) but well-established plants of *Archontophoenix cunninghamii* and *Livistona australis*. Control of native rats can be difficult since they are protected fauna.

Bush rats have eaten the apical bud out of this Livistona decora. [D. Jones]

Bush rat hole under Lomandra *clump, note dead* Lomandra *leaves.* [S. Jones]

Closer view of bush rat damage to Livistona *palm.* [D. Jones]

Sudden death of Grevillea *due to root interference by bush rats.* [D. Jones]

The base of this tree was gnawed by bush rats until it fell over. [D. Jones]

Goannas (Lace Monitors)

These stunning and intriguing native reptiles generally cause minor problems in gardens, however they quickly learn where to obtain food, and can become addicted to petfood. They also like to explore freshly laid mulch looking for food with the result that mulch can be strewn over a large area and recent plantings may be damaged. The nuisance value they create is greatly offset by the excitement and enjoyment gained from the presence of these protected animals in gardens.

Brush-tailed possum. [S. Jones]

Possums

Possums, which feed on a variety of plants, can be destructive in parks and gardens. Two types are commonly encountered in urban gardens, brush-tailed possums and ring-tailed possums. Both types have their favourite food plants and can be pests. They often reside locally but can also travel long distances in search of food. Possums are especially fond of young tender leaves, developing shoots, flower buds and fruit. They can ruin a year's flowering and fruiting in a couple of nights' feeding and regular browsing can stunt growth and even result in stem dieback. Possums can also severely scratch the bark of smooth-barked gums and other plants causing sap exudation and perhaps creating entry points for diseases, and pests such as borers.

The use of metal, plastic or fibreglass bands around the trunks of isolated trees prevents access by possums. These bands should be loosened at intervals and checked to see that insect pests are not exploiting the covered area of bark. Spraying affected growth with bitter-tasting repellents such as the fungicide thiram, chilli spray, garlic spray and sprays made from *Quassia* chips (see deterrents, chapter 2) can be effective for short periods.

Kangaroos and Wallabies

These highly mobile animals are frequent visitors to gardens in some parts of Australia. They cause significant physical damage to plants when hopping and also feed on suitable shoots and leaves (especially wallabies). They also have the habit of clawing and tearing branches with their arms.

These animals are also naturally curious and have a habit of checking out recently disturbed areas such as new plantings. Most people who have garden visits from these animals agree that wallabies cause more damage than kangaroos. High external fences will keep these animals out of a garden and wire mesh barriers are effective at protecting precious plants. Spraying with bitter-tasting repellents such as the fungicide thiram, chilli spray, garlic spray and sprays made from *Quassia* chips (see deterrents, chapter 2) can also be effective for short periods.

Wallaby damage to kangaroo paw leaves. [D. JONES]

Earthworms

Earthworms are highly beneficial in most environments, however worms in potted plants can be a nuisance as their feeding activities cause a breakdown of the potting compost and subsequent blockage of the drainage holes. The situation is worse when the pot is placed on the ground because the worms can enter through the drainage holes. They can also build castes around the base of the pot, effectively blocking the drainage. Worms can even disrupt the compost of plants held on benches if conditions are moist enough for them to live and move about. Raising pots and other containers off the ground with a ceramic tile or similar support, as well as placing them on strong wire benches can alleviate the problem. Solutions of metallic salts such as ferric sulphate or aluminium sulphate reportedly discourage worm activity.

CHAPTER 17

PARASITIC PLANTS

Some plants are parasitic on others, directly tapping into the host's sap stream, causing growth reduction and, in extreme cases, plant death.

MISTLETOES

Mistletoes are parasitic plants that are entirely dependent on another plant for their survival. Mistletoe seeds, which are usually deposited after being eaten by a bird, germinate on the trunk or branch of the host plant and quickly develop a specialised growth that penetrates into the sapwood of the tree, forming a haustorial attachment that taps into the host's sap stream. The mistletoe grows to form a substantial clump, photosynthesising from its leaves, flowering within a couple of years and producing its own fruit and seeds.

Australia has about 70 species of mistletoe in 12 genera. Mistletoes are widely distributed over the Australian mainland, but do not occur in Tasmania. They occur in most habitats, including desert floras and rainforest. Some mistletoes are highly specific, being confined to a single host species or host genus, whereas others are known to parasitise a wide spectrum of plants, including exotics. Some mistletoes have foliage similar to the host plant and are therefore somewhat difficult to discern from the host, others are quite dissimilar to the host and easy to spot. Research suggests that mistletoes are more abundant now than at any other time in recent history.

Mistletoes are an important plant group that makes a tremendous contribution to the ecology of an area. They can also add greatly to a garden. Most native mistletoes produce interesting and often colourful flowers that are rich in nectar and attract honey-eating birds. They flower consistently, even during dry times and droughts, and can produce impressive floral displays. They also produce fleshy fruit that consists of a seed surrounded by a sticky nutritious flesh. These fruits are eaten by birds, especially the small and beautiful Mistletoebird, which is the main dispersal agent. Many birds choose mistletoe clumps as a nesting site because they provide security and shelter. The larvae of some butterflies, particularly species in the family Pieridae such as the black jezebel, *Delias nigrina*, feed on mistletoe leaves. The caterpillars are in turn eaten by birds such as bronze-cuckoos, cuckooshrikes and shrike-thrushes. Arboreal marsupials such as possums and gliders also feed on mistletoes.

Caterpillars of the Black Jezebel, Delias nigrina, *on the mistletoe* Dendropthoe curvata. [D. JONES]

Mistletoes are often absent or remain in low numbers on trees in densely planted sites and in shade. A few plants of mistletoe in a park or garden add interest and do very little harm. Heavy infestations, however, severely weaken a plant and can cause its death. Large numbers of strong vigorous clumps can even kill large eucalypts. Mistletoes die after the death of the host plant. Severe mistletoe infestations are mostly seen on trees in open situations, especially isolated specimens left in paddocks after clearing and trees growing alongside roads and highways. Plants stressed by drought or suffering from dieback are more susceptible to attack and it is not uncommon to see branches of eucalypts heavily laden with several huge clumps of mistletoe. The clumps, which are usually conspicuous because of their dense nature and different colouration to the host (they are often reddish), become even more conspicuous as the host weakens and loses foliage.

Control: mistletoe clumps that are accessible can be cut off and destroyed. The traditional method of control involves spraying mistletoe clumps with herbicide. This is impractical for clumps high up in the tree canopy. Trunk injections have also been tested but have been found to be unreliable. Modern research suggests that changes in management may be more successful. Controlled burns can impact on clumps situated low in the canopy since mistletoe is readily killed by fire. The encouragement of arboreal marsupials by installing nesting boxes can also produce positive effects. Otherwise enjoy the benefits they provide.

Heavy infestation of mistletoe. [D. Jones]

Infestation of mistletoe. [D. Jones]

Large clumps of mistleoe, Amyema muelleriana, *on a relatively small tree of* Eucalyptus scoparia. [D. Jones]

DODDERS AND DODDER LAURELS

Two types of leafless parasitic plants, both commonly known as dodders, are occasionally problematic to native plant-growers. Although displaying amazing similarity due to convergent evolution, they belong to distinctly different plant groups and are dealt with separately.

Dodders

Species of *Cuscuta,* in the family Convolvulaceae, are annual stem parasites that lack roots and leaves. They germinate from a seed and quickly become attached to a nearby plant, invading its stems and forming haustorial attachments that tap into the host's sap stream. *Cuscuta* stems are straight but they spread by long thin tendrils that twine and form coils on the host plant. These tendrils also develop haustoria to draw on the host sap stream. Clusters of small whitish bell-shaped flowers are produced in abundance as the plant develops (within three weeks of germination), followed by pale brown fruit, each containing 1–4 seeds.

About 14 species of *Cuscuta* are found in Australia, including three native species. All 11 exotics are declared noxious weeds. The Golden Dodder, *Cuscuta campestris*, is a frequent exotic pest of nurseries and gardens in subtropical and temperate parts of Australia. It has golden-yellow stems and its attacks can be severe on soft plants such as species of *Brachyscome, Dampiera, Goodenia, Scaevola* and *Xerochrysum*. It can also damage woody plants including shrubs and small trees. Its spread from plant to plant in nurseries is rapid and control should be initiated when first noticed. It also impacts on herbaceous crops such as lucerne and clover. The strong wiry stems entangle the host, causing strangulation and smothering growth. Affected stems break readily at points of haustorial attachment. **Control:** spread of Golden Dodder is mainly by water but fragments moved by birds, animals or humans can become attached to another plant. Seedling plants must become attached to a host plant within a few days of germination otherwise they shrivel. Small plants impacted by dodder are best removed and burnt. Control in larger plants is often impractical but sometimes removing infested branches can be effective.

Dodder Laurels

Species of *Cassytha* (sometimes also called devil's twine or strangle vines), in the family Lauraceae, are large, perennial parasitic climbers that form tangled smothering clumps. About 14 species of *Cassytha* are native to Australia. Although commonly viewed as pests because they are untidy and smother other plants, they are an important part of the ecology in many open habitats. *Cassytha* seedlings can only survive if they become attached to a host plant, usually a woody shrub or tree, before they die. After attachment their root system dies and they rely on water and nourishment from their host. Young plants of many species are initially green but after they have parasitised a host plant they lose chlorophyll and become yellow, orange, reddish or brown. They do however retain the ability to re-green and again become photosynthetic if the host drastically slows growth or becomes dormant. Some other species may retain their chlorophyll throughout life. Clumps of *Cassytha* often spread to adjacent plants, sometimes covering patches of vegetation with a tangled mess. Large clumps are heavy and can break branches and smother the host plants. The dodder cannot survive if all the host plants die.

Examples: *Cassytha filiformis*, a pantropical species, commonly smothers coastal vegetation in parts of tropical Australia and on many Pacific islands. It is also a frequent pest of coastal gardens in the tropics. The Mallee Strangle Vine, *C. melantha*, forms massive clumps of crowded stems over vegetation in drier parts of the ranges and inland areas of many states. The sheer weight of a large clump can cause the supporting shrubs and trees to collapse and die. Another native species, *C. pubescens*, has been observed to strongly parasitise populations of the introduced Scotch Broom, *Cytisus scoparius*, and it has been suggested that it could be a useful control agent of this serious invasive weed.

Control: species of *Cassytha* are sometimes a problem in parks or large gardens. Plants are best removed when small to prevent their development and spread. Large clumps can only be controlled by removing all the host plants.

The Mallee Strangle Vine, Cassytha melantha, *can smother patches of vegetation.*
[D. Jones]

Cassytha filiformis *is an annoying strangler in coastal gardens of the tropics.* [D. Jones]

CHAPTER 18

IMPACTS OF DISEASES AND THEIR CONTROL

CAUSATIVE AGENTS, EFFECTS OF DISEASE, DISEASE CONTROL, SANITATION, FUNGICIDAL SPRAYS

Australian plants are afflicted by a range of diseases, with the severity of the attack depending on many factors – chiefly the health of the plant. Some diseases are extremely virulent and can cause the death of large, well-established plants and may spread to ravage whole areas. Most diseases, however, only cause limited damage and occur sporadically when a particular set of conditions supports an outbreak. A few diseases are insidious and cause stunted growth or reduced flowering over many years without ever really becoming obvious.

CAUSATIVE AGENTS

Diseases are primarily caused by four groups of pathogens – fungi, bacteria, protists and viruses. Phytoplasmas can also cause some specific diseases.

Fungi

These are the most common pathogens attacking native plants. All parts of the plants may be attacked but the leaves, roots and timber are particularly susceptible. Fungi spread by spores (mostly air-borne) and some types are very troublesome during warm, humid, still weather. They can spread rapidly under ideal conditions and control steps should be taken immediately damage is noticed.

Bacteria

In fruit crops, pathogenic bacteria cause wounds associated with cankers, gumming and blight of blossoms and young leaves. The only important bacterial pathogen known to significantly damage native plants is the crown gall organism, *Agrobacterium*, but there are probably others of importance. Cultivated orchids, exotic and native, also suffer from bacterial diseases. In tropical and subtropical regions the leaves of Bleeding Heart, *Omalanthus populifolius*, are often disfigured by a disease that may be bacterial.

Control of bacterial diseases is best achieved through preventing outbreaks, as bacterial infections spread through water splash, insects or on equipment. Good hygiene practices as detailed under the section below should be followed to prevent the introduction of bacterial diseases and their spread. Spraying with a copper-based fungicide can be used for control, although copper-resistant bacterial strains are now present in the environment.

Protists

Protists are a group of unicellular or multicellular organisms that do not have any specialised tissues. They are placed in the kingdom Protista. They include the water moulds, which have traditionally been treated as oomycetes and included with the fungi. Several protists cause serious plant diseases including species of *Phythophthora* and *Pythium*. More details, including control measures, are included in chapter 19.

Viruses

Viruses are minute particles which live within plant cells. If present in sufficient numbers or in combinations of different virus species, plant damage can occur. Typical virus symptoms include irregular mosaic patterns on leaves, distortion of young shoots and leaves, colour breaks in flowers, and reduced size and rosetting of leaves. Virus attacks stunt plant growth and severe attacks can kill. Sucking insects such as aphids, leafhoppers and jassids transmit viruses. Plants raised from seed are generally free of viruses although certain viruses are known to be seed transmitted.

The effects of a virus are promulgated by vegetative propagation. Cuttings, divisions or scions for grafting should only be taken from healthy vigorous plants and never from unthrifty plants with distorted or mosaic-patterned leaves. Plants suspected of being virus-infected should be destroyed. Aphids and other possible vectors of virus transmission should be controlled when noticed. Secateurs and knives should be sterilised if they have been used to cut plants suspected of being virus infected.

Viruses are known in a few native plants and are suspected in many others because of symptoms such as mosaic patterns which appear in the leaves of species such as *Solanum aviculare* and *S. laciniatum*. More details are included in chapter 19.

Phytoplasmas

These are specialised bacterial parasites found in plants. They were originally termed 'mycoplasma-like organisms' but research has thrown light on their true identity. Phytosplasmas are transmitted by sucking insects such as leafhoppers and they impact on plant phloem tissue. Disease symptoms range from mild impacts to plant death. They are also thought to be implicated with growth features such as the replacement of flowers by leaf-like structures (termed phyllody) and witches' brooms. Control is mainly by selection and propagation of clean stock.

SYMPTOMS OF DISEASE

Spots

Small to large spots on leaves and stems can be the result of fungal attack or unfavourable environmental conditions. Spots can have different coloured zones and are often surrounded by a zone (halo) of light green or yellow tissue. The dead tissue in a spot may fall out, leaving a hole. The condition of 'shot-hole' arises when numerous small spots fall out in this way.

Blotches

When spots coalesce or are large and irregular in shape they are usually called blotches.

Blights

These are diseases that kill young shoots, stems and flowers. They are often associated with soft stem dieback.

Bleaching and Scorching

Yellowing or whitening of tissue which eventually turns brown and shrivels.

Necrosis

Death or decay of tissue. This condition is common following attacks on young tissues.

Lesions and Cankers

Growths on leaves, branches or trunks which are usually black and hardened by the exudation of gum. They are often irregular in shape.

Wilting followed by dieback

Young shoots wilt followed by dieback which extends down the branches. These symptoms indicate interruption to the water supply system of the plant such as a vascular blockage or damage to the roots.

Stunting

Reduced growth and stunting is characteristic of many diseases.

Rots

Rots occur when a disease causes the disintegration of plant cells and the cell-contents exude as a slimy material that can emit an odour.

Gummosis

A term used to designate the exudate of gum from the trunk or branches. This may be caused by diseases or result from the activities of borers.

Moulds and Mildews

Fungal threads or fruiting bodies are usually obvious on the surface of affected tissue as a grey fluffy layer.

DISEASE CONTROL

Methods of disease control vary with such factors as plant health, stage of advancement of the disease and the type of disease involved. The best disease control relies on prevention rather than cure.

Practices to Reduce Disease Incidence

1. Vigorous healthy plants are better able to resist disease than are weak or unthrifty plants. Only select those plants which are healthy and vigorous. Attend to all of the plants' needs, such as drainage, fertilisation and watering, and remember that any setback can contribute to the entry of disease.

2. Avoid sowing seed thickly or planting gardens too densely. Air movement is an excellent control of some fungal diseases such as moulds. Dense planting encourages rapid spread of disease. Dense planting can be useful to give a rapid effect in a garden but thinning may be necessary after a couple of years.

3. Avoid large areas of a single species, as this encourages the rapid spread of any disease which attacks that species.

4. Trim ragged edges of large wounds with a sharp instrument, and dress with Bordeaux mix to prevent the entry of disease. Traditionally mastic sealants have been used to treat large wounds but some arborists believe these materials may inhibit healing.

5. Remove weak or dying trees and check plants at regular intervals.

6. Initiate control measures as soon as diseases are first noticed. Use fungicidal treatments as a last resort and then in strict accordance with safety procedures and the manufacturer's directions.

7. Choose the correct fungicide for the job and apply in accordance with directions, ensuring that treated foliage is thoroughly covered.

Hygiene Practices

The causative agents of many diseases can be transmitted on seeds or on leaves, stems and flower buds of material collected for cuttings. Infectious material of soil-borne fungi, bacteria and also insect eggs can also be carried on shoes, clothing and garden tools. A simple set of hygiene practices can be used to protect an area, be it crops or gardens, from the spread and introduction of plant diseases or pests.

It is a good practice to wash hands with soapy water before and after handling plants and seeds, especially if you have been gardening around unthrifty plants, or visiting another garden.

Consider using disposable gloves when working with vulnerable plant practices such as taking cuttings, sowing seed or handling seedlings. Cutting material should only be taken from healthy plants. If a disease is obviously present, taking cuttings from that plant will only promulgate the disease. Hot water dips can be useful for ridding plants or cuttings of mites, nematodes and some diseases. Temperatures of about 40°C for 15–30 minutes can be satisfactory, however the temperature and immersion time needed varies with the species of plant and the pest to be controlled.

Keep garden equipment such as secateurs clean, remembering to sterilise them after working on plants you think may be suffering from a disease. It is a good practice to clean tools by brushing off all dirt or plant sap and scrubbing with steel wool dipped in rubbing alcohol. Dry tools, preferably in the hot sun, and apply a product such as WD40 to prevent rusting. Alternatively tools can be brushed clean and dipped in a mixture of four parts water to one part household detergent such as lysol and then dried as above. Soaking in a saturated solution of Trisodium phosphate is very effective.

Always dispose of prunings of sick plants carefully. Do not compost, use as mulch or leave in the garden, but seal in a plastic bag and leave in the sun (solarise) or burn to remove the risk of disease spread. When choosing new plants, make sure they are healthy and are not exhibiting any symptoms of disease such as dead shoot tips, spots on leaves, curling leaves or unusual markings on leaves. This is particularly important when sourcing plants from markets or garage sales. Don't be afraid to tip the plant out to check the health of the roots and make sure snails, earwigs and other pests aren't included in the purchase.

Use certified 'free from pests' seed or propagation material and use trusted suppliers.

Make your own mulch or source it carefully, making sure it is well aged, or certified disease free.

Fungicidal Sprays

Fungicidal sprays, pastes or dusts can be extremely useful for controlling diseases, but they should be handled carefully as outlined in chapter 2. Not all fungicides are compatible with each other, or with pesticides, and may cause problems if mixed together. Information on compatibility can be obtained from the product label or Material Safety Data Sheet. Fungicides can be systemic, have curative properties or act to prevent infections in plants. Commercially prepared products often contain combinations of systemic and curative active ingredients with preventative elements such as copper.

Bicarbonate of Soda

A weekly spray of a mixture comprising the components below has been found as a suitable preventative treatment to minimise black spot and mildew. Mix one tablespoon of bicarbonate of soda with 4.5 litres of water and one tablespoon of horticultural oil such as white oil. Shake to mix thoroughly and spray lightly onto the infected leaves.

Bicarbonate of Potash

A spray with bicarbonate of potash replacing bicarbonate of soda in the above recipe is also very effective for treating powdery mildew. Commercial preparations are also available.

Bitertanol

Bitertanol is a broad spectrum, systemic fungicide that has both protective and systemic properties. It can be used on rust in *Boronia* and to treat powdery mildew in eucalypt seedlings. Multiple applications are often required.

Bordeaux mixture

This mixture is one of the oldest fungicides known but it is still very effective. It is not very soluble in water and must be kept agitated while spraying. It is effective against a range of diseases and is relatively safe to use but can damage foliage and should only be applied during cool weather. Bordeaux mixture can also deteriorate and only fresh mixes should used. It can be prepared by dissolving 40g of copper sulphate in 5 litres of water in one plastic bucket and 40g of hydrated lime in 5 litres of water in another plastic bucket. The two can then be slowly mixed together with constant stirring. Bordeaux paste is a useful protectant to cover wounds.

Captan

Captan is a fungicide used as a protectant fungicide or drench. It is very useful against *Botrytis* (grey mould), black spot and downy mildew and for the prevention of damping-off of seedlings caused by *Pythium*. It is often recommended in conjunction with a copper spray. It can also be used as a dust to prevent fungi from damaging stored seed.

Chamomile

A preparation of this medicinal herb is reputed to control fungal diseases, especially mildews, but effectiveness may vary from season to season. Approximately two handfuls of foliage is covered with 1 litre of boiling water and allowed to soak for 30 minutes. Then strain the contents and mix with equal parts of water. The addition of a wetting agent, for example soft soap or 'Clensel', may make the applications more effective.

Chlorothalonil

This broad-spectrum, organic, non-systemic fungicide is used mainly for treating fungal diseases of fruit and root crops, such as *Botrytis* (grey mould), rust diseases and *Alternaria* leaf spot.

Copper Oxychloride and Cupric Oxide

Used as a protective treatment, copper fungicides work to prevent disease by placing a thin barrier between the plant and fungal spores. Copper fungicides can be used in place of Bordeaux mixture and have smaller copper particles that provide good leaf coverage. They have proved very useful for controlling the various leaf spots that infect a variety of plants and are relatively safe to use. A commercial preparation is available in combination with Metalaxyl for use on trunk canker and *Phytophthora* root rot in macadamias.

Cow's Milk

In recent times diluted cow's milk has become popular for controlling powdery mildew on a large range of plants. The recommended mixture is one part milk (any type of milk) to nine parts water. Milk reduces the acidity on leaf and stem surfaces, which provides a deterrent to development of powdery mildew. It is best to begin spraying as soon as the first outbreak of powdery mildew is observed. Best results are usually gained by spraying weekly, in the morning hours and on non-cloudy days, making sure to thoroughly wet all of the leaves (including undersurfaces) and stems. Do not water the leaves for at least 12 hours.

Fosetyl

Fosetyl is included as an aluminium salt and is thought to be effective for control of *Phytophthora* and *Pythium* root rot. It can be applied as a spray or in a drench.

Iprodione

Iprodione is a member of the dicarboxamide group of fungicides and is a contact fungicide. It inhibits the germination of fungal spores and blocks the growth of fungal mycelium. It is useful against *Botrytis*, *Sclerotinia* rot and moulds.

Lime Sulphur

A very old and smelly mixture which works on a variety of pests and diseases such as powdery mildews. It should not be sprayed when the temperature is above 29°C and should be kept away from walls, vehicles, etc, as it can stain. On ornamentals it is used at a dilution of one part to 50 parts of water.

Mancozeb

Mancozeb is a protectant fungicide that is often used in combination with Iprodione. This chemical is effective against certain leaf spots and for the control of downy mildew and certain foliar diseases on a number of crops including poppies and grape vines. It has also been used against myrtle rust.

Metalaxyl

Metalaxyl is a systemic fungicide that also has curative properties. It is often used in combination with Mancozeb or a copper-based spray. It can be used for *Phytophthora* root rot, *Pythium* root rots and smuts.

Neem Oil

Organic Neem oil is considered to be a broad-spectrum fungicide. A home mix can be prepared by mixing 2 tablespoons and 1 1/2 teaspoons dish soap with 4 litres of water. Spray all plant surfaces, including the undersides of leaves, until wet.

Peroxy Acetic Acid (Peracetic Acid)

Peroxy Acetic Acid is a non-selective surface sterilant, acting as a contact fungicide. It has corrosive properties and is most commonly used on grapes. It should not be used on new growth and is reported to control downy mildew, powdery mildew and white blister.

Phosphorous Acid (also known as Phosphonic Acid or Potassium phosphite)

The salt of phosphorous acid, potassium phosphite (KH_2PO_3), also known as phosphite or phosphonate, has fungicidal properties [NOTE - phosphorous acid should not be confused with either phosphoric acid, which is a strong inorganic acid used for cleaning, or phosphorus (P), which is a component of fertiliser]. A number of new fungicides are available that have phosphorous acid as the active ingredient. They are suitable for use on root rot caused by *Phytophthora* spp., *Pythium* spp., collar rot, crown rot and downy mildew. These fungicides are systemic and move in both the xylem and phloem of the plant, travelling from leaf tissues to the crowns and roots. They can be applied through a stem injection or by spray application. [NOTE-phosphorous acid does not supply the element phosphorus (P) for a plant's nutrition; conversely phosphate supplies P for plant growth but does not control disease beyond improving the general health of a plant].

Potassium Salts

Salts of potassium such as potassium sulphate (potash), potassium carbonate, potassium bicarbonate and potassium permanganate (Condy's crystals) are extremely water soluble and useful tonics to raise soil pH and provide potassium to plants. All have been described as having some fungicidal properties and can be used to boost plant health.

A solution of Condy's crystals can be used as a general fungicide and is reported to be useful for black spot and mildews. Mix 5g of Condy's crystals with 5 litres of water, shake well and spray onto leaves. Another general fungicide recipe can be made by mixing one level teaspoon of bicarbonate of soda into one litre of water. Add one litre of skim milk and a pinch of Condy's. Shake thoroughly and spray onto leaves.

Propiconazole

This systemic fungicide is suitable for use on rusts, including Boronia Rust, *Puccinia boroniae*.

Propomacarb

Propomacarb is a systemic fungicide that can be used on native plants as a dip for cuttings, or when planting out, or applied directly to soil to prevent seedling damping-off as a result of infection with *Pythium*. It is also used in forestry production nurseries to prevent damping-off of eucalypt seedlings caused by *Pythium*.

Sulphur

Sulphur is available as a wettable powder that is effective as both a fungicide and an insecticide. It is safe to use and is effective for control of powdery mildew and leaf blights. It should not be sprayed in hot weather or in conjunction with an oil spray. It is recommended that a wetting agent is added to the solution to give greater coverage of the mixture.

Tebuconazole

Tebuconazole is a systemic fungicide that is absorbed into the sap stream of a plant and moves around the plant and destroys disease. It acts both as a preventative and a curative, destroying diseases already in the plant and preventing disease infestations. It is suitable for use on powdery mildew, rusts and leaf spots.

Thiram

A very useful fungicide for control of leaf spots and rusts. It is also a very good repellent for hares and rabbits and a useful dressing for stored seed. Mainly used as a fungicide for treating turf and as a seed treatment.

Triforine

A systemic fungicide which is absorbed into the sap stream of the plant, acting as both a preventative and curative agent. It is used by spraying onto foliage and is recommended for control of powdery mildew and rusts, including myrtle rust.

Triadimenol

Triadimenol is a broad-spectrum systemic fungicide that is applied to leaves through a spray. It is quickly absorbed by the plant and transported within the plant tissues. It is suitable for use on powdery mildew and myrtle rust.

Zineb

Zineb is a broad-spectrum protectant that prevents infection of new foliage and fruit. It is chemically similar to Mancozeb and can be alternated to assist in managing disease resistance. It is commonly used to treat mildews and rusts.

CHAPTER 19

DISEASES OF NATIVE PLANTS

Cautionary note on disease identification. Accurate identification of the pathogen causing a disease is essential if the correct control methods are to be applied. Many plant diseases can be very difficult to identify and the notes provided in this chapter should only be used as a general guide. If in doubt consult an expert such as a plant pathologist, especially for soil-borne pathogens which can be very complex and difficult to identify accurately. Also be aware that it can be difficult to separate some of the symptoms exhibited by stressed plants from those caused by a plant pathogen. Plant stress can result from a whole range of factors, including structural damage to stems and roots, repotting, inadequate soil drainage, unsuitable pH, incorrect nutrition, climatic extremes and watering (too much or too little). Additionally, stressed plants can suffer severe damage from relatively weak pathogens that would be innocuous on healthy specimens.

FUNGAL DISEASES

Fungi are the commonest organism to produce disease in plants. They are an incredibly complex group and most species should only be identified by experts.

A note on fungal names. Fungi are mainly classified by their sexual structures. Some fungi that cause plant diseases are difficult to classify because they only reproduce by asexual methods. Even those that produce both asexual forms (termed anamorph) and sexual (termed teleomorph) can be difficult to identify and it is not uncommon for a separate scientific name to be applied to the asexual and sexual states of a single species. Both names are given for a couple of the disease-forming fungi included in this chapter (the teleomorph name appears first).

Kino discharge as a result of vascular damage caused by a fungal disease in a Corymbia. [D. JONES]

Bark stain in Alphitonia excelsa *caused by a pathogenic fungus.* [D. JONES]

Fungal infection in trunk of Banksia integrifolia. [D. JONES]

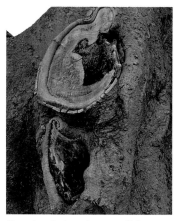

Fungal rot in branch stump of Moreton Bay Fig. [D. Jones]

Armillaria Root Rot, *Armillariella luteobubalina*

This vigorous native fungal pathogen attacks the root systems of established plants causing loss of vigour, dieback and/or death. It occurs naturally in many forested areas and causes economic losses in forest plantations, orchards, vineyards and gardens. A wide range of plant types is susceptible to attack by this fungus, both natives and exotics, including ferns and other understorey plants, tree ferns, shrubs, vines and trees. Because of the colour of its fruiting bodies (mushrooms) it is commonly known as the honey fungus. It is also sometimes called shoe-string fungus because of its specialised hyphae.

This fungus becomes established in old stumps or pieces of large roots or other dead wood. It often becomes prominent after an area is cleared prior to cropping and can be spread further by operations such as ploughing and raking. It can also be spread in infected mulch. The fungus itself grows through the soil by means of thick, flat, black fungal strands which resemble shoe laces and are known as rhizomorphs. These rhizomorphs, which can enter plant roots, are one means of infection by the fungus, the other is direct transfer of the fungus from infected roots which come into contact with healthy roots in the soil. After entering a root, the fungus spreads through the root system and invades the base of the trunk, developing beneath the bark. Within infected roots and stems the fungus can be seen as a creamy layer beneath the bark.

Symptoms: plants infected by armillaria root rot may linger for many years, dying slowly. Infected plants lose vigour and become unthrifty. Branches die back from the tips and leaves become pale and chlorotic. Splits often develop in the trunk of infected trees and patches of gum may also exude from the bark. Lifting the bark may reveal patches of spreading cream to white fungal threads that have the typical smell of mushrooms. Eventually the plant dies when the fungus encircles the bark of the stem or trunk. Clusters of honey-coloured toadstools are produced in early winter from the base of infected plants and also from any decaying wood in the vicinity.

Control: avoiding injuries to plants and keeping them healthy by fertilising and watering may limit the damage caused by this disease. Control of this pathogen is impractical and infected plants are best removed and burnt. If noticed early, infected plants may benefit from having a trench of soil dug around the drip line to locate and destroy any rhizomorphs. The use of root barriers 60–80cm deep can isolate and protect valuable plants from attack. Excavating soil around the stem base exposes infected tissue, causing it to dry out and slowing the progress of the pathogen. Prevention is much better than

Dieback symptoms in fern-leaved Acacia *caused by Armillaria Root Rot.* [D. Jones]

Dieback symptoms in phyllodinous Acacia *caused by Armillaria Root Rot, D. Jones]*

Fruiting bodies of Armillaria Root Rot at base of dead shrub. [D. Jones]

Fruiting bodies of Armillaria Root Rot. [D. JONES]

Rhizomorphs of Armillaria Root Rot. [D. JONES]

cure and an important control measure is the complete removal and destruction of all stumps, roots and pieces of decaying wood from a site prior to planting. The area should be left unplanted for some years (the fungus can survive in the soil for at least 20 years). If complete removal of tree material or leaving the area unplanted is impractical, infected areas can be fumigated before replanting.

CANKER DISEASES

Canker is the term applied to a localised patch of damage in the stems of woody plants. Stem death and branch dieback are often the first noticeable signs of an infection. The damage, which is sometimes hidden beneath the bark, consists of splits and cracks in the bark, discoloured sunken areas, discrete lesions and callus growth, sometimes associated with abnormal growth and also the exudation of gum. Frequently, affected branches die off, crack or break prematurely at the point of infection. Symptoms of this type are caused by a wide range of fungal pathogens (also bacteria). Sometimes the impacts are quite localised, affecting only one or a few branches in the canopy, but serious infections can result in canopy loss and plant death. There are instances in Western Australia of devastating impacts affecting entire communities of *Banksia* (see below), including threatened species. Spread is by water splash. Weak or damaged plants are most susceptible.

Control: cankers are generally difficult to control. Localised infections should be removed by cutting the stem back to healthy tissue (burn infected tissue). The cuts should be treated with Bordeaux mix paste or a copper-based fungicide to limit reinfection. Spraying with copper sprays may give some control but complete removal and destruction of the infected plant is often the only way to prevent spread.

Large old canker on a wattle trunk.
[D. JONES]

Canker on a eucalypt stem. [S. JONES]

Small canker on a eucalypt stem.
[D. JONES]

EXAMPLES

Aerial Canker of *Banksia, Zythiostroma* sp.

This fungus is a serious disease of banksias in Western Australia. It is known from several areas in the south-west of the state and has been found in different plant communities, including Jarrah forest, sandplain vegetation and heath. It affects a number of species of *Banksia* and poses a serious threat to

the survival of rare species such as the vulnerable *B. verticillata* and the endangered *B. brownii*. Dead branches are often the first noticeable symptom and close examination will show the presence of cracked bark and cankers. This disease is easily capable of killing individual plants and can devastate whole *Banksia* populations.

Control: no effective sprays are currently known to control this disease.

Dead Banksia *in natural habitat, Western Australia.* [C. FRENCH]

Dead Banksia *in sandplain vegetation, Western Australia.* [S. JONES]

Banksia coccinea Canker or *Diplodina* Canker, *Cryptodiaporthe melanocraspeda / Dipladina melanocraspeda*

Whole stands of mature plants of *Banksia coccinea* have been killed in south-western Western Australia by this fungus which causes cankers that encircle a stem, resulting in stem death and dieback. The upper branches die first and the disease progresses downwards killing the plant within a few years of infection. This canker disease has had serious impacts on populations of this unique species in areas free of infection by *Phythophthora cinnamomi*.

Control: no effective sprays are currently known to control this disease.

East Australian *Banksia* Canker, *Pestalotia* sp.

Some east Australian species of *Banksia* such as *B. spinulosa* and *B. marginata* are subject to attack by this fungus which causes black areas, wet patches and sunken or flattened areas on the stems. This disease is noticeable in subtropical regions and stems may die above the lesions or break off prematurely. The tissue inside the lesion is usually discoloured or dead. Plants in shady situations seem most noticeably affected. This disease is usually only of a minor nature, although plants can be stunted.

Waratah Seedling Canker, *Cylindrocarpon destructans*

Gum and sap oozing from site of canker fungus infection in Banksia spinulosa. [D. JONES]

Seedlings of the New South Wales Waratah, *Telopea speciosissima*, can be attacked by this soil-borne pathogen which causes black patches (cankers) in lignotubers and stems, resulting in wilting and plant death. The disease shows up in stressed plants particularly those that have been recently repotted; the fungus also causes dieback and eventual death of older plants.

Control: by ensuring hygienic conditions and potting into pasterurised mixes (see control of Damping-off below).

Stem Canker of Red Flowering Gum, *Quambalaria coyrecup*

Urban plantings in Western Australial of the spectacular flowering gum, *Corymbia ficifolia,* are severely attacked by this debilitating fungus. Strangely enough stem canker is only found on *C. ficifolia* that have been planted and not those growing in natural communities. The disease enters through injuries on the bark and causes large cankers on stems and trunks. These cankers swell and split to reveal the wood beneath. Large amounts of gum are exuded from the infected area, staining the bark of the tree dark red. Large cankers can girdle the infected branch or trunk, resulting in ringbarking and death of growth above the infected area. Eventually the infected sites become coated with a powdery mass of white spores. The leaves and small branches are the first to wither and die, followed by larger branches and finally the death of the tree.

This same stem canker disease was also noticed on the Marri, *Corymbia calophylla,* in the 1970s, and by the 1990s was observed to be causing severe decline and death right across the natural range of this important species. About one quarter of surveyed trees had significant cankers present with the disease especially prominent on remnant trees left along roadsides and in paddocks. The same disease also attacks *C. haematoxylon* in Western Australia, although causing less damage than on *C. ficifolia* and *C. calophylla.*

Control: extremely difficult and to date no fungicides have been found to be effective, although tests with recently developed fungicides are continuing. Wounds should be trimmed and dressed to prevent the entry of the fungus and infected branches should be removed immediately when noticed. Trees should be inspected regularly to minimise the chance of infection. Some individual trees are known to be resistant to stem canker and these provide hope for future control. Grafting onto resistant rootstock may be successful.

DAMPING-OFF OF SEEDLINGS

Damping-off of seedlings is commonly caused by soil-borne fungi such as *Fusarium* and *Rhizoctonia* (especially *R. solani*). Protists including species of *Pythium* and *Phytophthora* can also cause damping-off. These organisms, which can survive indefinitely in soil, cause damage when conditions become favourable for infection. The first obvious symptoms of damping-off occur when seedlings begin toppling over. This usually occurs suddenly and without warning.

The major fungal groups that cause damping-off are active under different conditions and can produce different symptoms. *Rhizoctonia*, which is worst in hot humid conditions, causes sunken lesions on the stem at or just below soil level. *Fusarium* causes lesions and stem contraction (wire stem) at ground level. *Pythium* is worst in cool conditions, especially in potting mixes that are too wet. It causes wet, slimy rots of stems (the outer tissue of the stem is easily separated from the core) and roots. *Pythium* is readily spread on the feet of fungus gnats. *Phytophthora* also causes similar rots.

Damping-off spreads very rapidly and should be controlled when first noticed. It is worst when seedlings are sown too thickly, in too much shade or with poor ventilation. Damping-off can spread rapidly through batches of seedlings, especially in poorly aerated conditions. Occasional seedlings will survive an attack but the stem is usually withered and constricted at soil level.

Control: complete control of these diseases can be difficult but its spread can be limited by drenching the soil with captan (ineffective against *Rhizoctonia*), metalaxyl or copper oxychloride. Seed can be treated with a fungicide before sowing or presoaked in a weak solution of bleach (one teaspoon of bleach per litre of water).

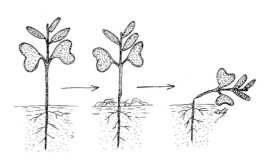

Damping-off of seedling. [D. JONES]

Containers should be sterilised with a weak hydrogen peroxide solution (soak for 10 minutes in 10ml of hydrogen peroxide in 1 litre of water). It is always best to sow into sterilised or pasteurised potting mix (the impact of these pathogens can be reduced or controlled by pretreating the mix in a microwave for 10–12 minutes or heating in an oven at 98°C for 30 minutes). Small amounts of mix can be sterilised by 'solarisation', whereby a quantity of potting mix in a sealed clear plastic bag is placed flat in a hot sunny spot for six to eight hours. Care must always be taken to prevent recontamination of sterilised mix as any pathogen is much more destructive with reduced competition. Research is underway testing the effects of biocontrol agents (yeasts, fungi and bacteria) that are antagonistic to damping-off diseases. Early results suggest there are some benefits from using these in conjunction with fungicides.

EREMOPHILA WILT

Eremophila nivea *suffering from eremophila wilt.* [D. JONES]

Certain species of *Eremophila*, especially those with hairy leaves such as *E. nivea*, suffer from a disease that causes wilting in the young shoots. Sometimes the disease seems to be confined to isolated shoots on the infected plant, on other occasions it forms patches within the canopy. Affected shoots become dry and contrast with healthy growth. Severely affected plants drop numerous leaves, develop an open unhealthy appearance and eventually die. This disease affects seedlings as well as cutting-grown plants and grafted plants. The cause of this wilting disease is unclear but it is possibly the result of attack by the fungus that causes fusarium wilt in a wide range of plants (including cultivated *Eremophila* plants in Italy). This fungus, *Fusarium oxysporum*, is a cosmopolitan soil pathogen that survives on decaying organic matter in the soil and infects plants through wounds in the roots (see also the entry for fusarium wilt below). It is also spread by water splash.

Control: there is no current method of control.

Fern Crown Rot or Hard Crown, *Rhizoctonia* species

Fern crown rot, a common disease of many fern species, usually arises after long periods of heavy rain or still humid weather. It is also often prevalent in potted ferns grown in poorly ventilated conditions. The causative agent is a fungus in the genus *Rhizoctonia* and the first noticeable symptoms are distorted fronds. Commonly the affected frond has a lopsided appearance with patches of normal growth mixed

Impact of hard crown disease on Asplenium. [D. JONES]

with stunted areas and black, wet-looking patches on the stipes. Affected fronds are very brittle. As the disease progresses through the plant, new fronds abort or expand incompletely and the crown develops a thickened, unhealthy appearance. Some infected plants can linger but most die. This disease can have serious impacts on species of *Asplenium*, *Blechnum* and tree ferns, especially *Cyathea cooperi*, *C. baileyana* and *C. rebeccae*, less commonly *C. australis* and *Dicksonia antarctica*. Tree ferns often have a sunken area around the crown where water can collect and crown rot can be started or exacerbated by too frequent watering of this area.

Control: once symptoms appear the disease is usually well established and impossible to control. Affected plants are best destroyed. Careful watering and good aeration help avoid the disease.

Fungus Galls – see chapter 7

Fusarium Wilt, *Fusarium oxysporum*

Species and cultivars of some Australian native daisies are susceptible to this soil-borne pathogen. Early symptoms include wilting of any soft growth followed by plant collapse and death. Close examination reveals dead or dying roots and an upward progression of dead tissue inside the stem (this can be seen by slicing the stem vertically with a sharp knife). Most native daisies are probably susceptible to this pathogen. This disease can be severe on seedlings of annual species, such as *Xerochrysum bracteatum* and *Rhodanthe chlorocephala*, particularly when planted in mass for annual displays. It also affects perennial cultivars of *Xerochrysum bracteatum* and can kill other daisies including *Brachyscome multifida* and some *Brachyscome* cultivars, *Chrysocephalum apiculatum. C. baxteri, Coronidium elatum, Rhodanthe anthemoides* and *Xerochrysum acuminatum.*

Control: this disease is difficult to control. Antagonistic bacteria and fungi offer some hope of reducing its impact. A weak solution of baking powder (sodium bicarbonate $NaHCO3$) at about 20 mg per 100ml of water is reputed to reduce growth of the fungus mycelium.

Ganoderma Butt Rot, *Ganoderma applanata*

Although it is an important agent in the decay of fallen trees and logs, the bracket fungus *Ganoderma applanata* (formerly included in the genus *Fomes*) can also attack the roots and trunks of living trees such as wattles and eucalypts. The fungus enters through tissue that has been damaged in storms or from borer holes, termite holes or other injuries, and then spreads through the wood around the entry point. It decays both sapwood and heartwood, eventually killing the tree. Early symptoms include an unthrifty appearance and leaves that are smaller than normal and often pale. Eventually the fungus forms a large woody fruiting body on the trunk, usually near the base of the infected tree. This structure, which is stemless, projects outwards like a shelf. It lasts many years, increasing in size each year, and can eventually reach 30–50cm across and 5–10cm thick. The dark brown to blackish upper surface, which often appears shiny, is roughened with a whitish rim. The underside is cream to white but bruises easily.

Bracket-like fruiting body of Ganoderma Butt Rot. [S. Jones]

Control: there is no control available for this disease. Prevention is better than cure and all wounds should be neatly trimmed and treated with Bordeaux mix paste [note – wound dressings or sealants which have been generally recommended for treating large wounds may impact on development of the cambium layer and inhibit healing].

Ganoderma *fruiting bodies on* Corymbia maculata. [D. Jones]

Fungal fruiting body, possibly Ganoderma, *on* Agonis flexuosa. [S. Jones]

Fruiting bodies of bracket fungus, possibly Ganoderma, *on dying* Acacia. [D. Jones]

Geraldton Wax Petal Blight, *Alternaria alternata*

The flowers of Geraldton Wax, *Chamaelaucium uncinatum*, a popular cut flower that is grown in gardens and commercial plantations in several Australian states, can suffer from a disease caused by this fungus. It causes brown patches on the petals, usually starting on the edges and progressing inwards to the centre of the flower. Severely affected flowers shrivel and drop prematurely. The disease can infect plants in the ground, but also develops in cut flower stems packaged for export, spoiling the consignment.

It should be noted that Grey Mould, *Botrytis cinerea* (see below) can also infect the flowers of Geraldton Wax, but in this case the disease appears in the centre of a flower and progresses outwards. **Control:** avoid overcrowding and prune or stake plants to improve air movement and circulation. Spraying infected plants with captan or iprodione may give some control.

Grevillea Leaf Scorch, *Verrucispora proteacearum*

A disease mainly of the wet season in tropical and subtropical regions, but also occurring in cooler climates after periods of hot humid weather. Affected leaves have large grey areas which eventually

Grevillea Leaf Scorch. [D. Jones]

become brown as if they had been scorched in a fire. In severe cases, usually in very wet seasons, a high proportion of the leaves may be affected. Usually only the recently matured leaves near the end of the branches are affected and older leaves remain normal. Affected leaves are shed prematurely and if these are examined closely, clusters of small, black fruiting bodies can be seen. The disease is most noticeable on grevilleas, particularly Silky Oak, *Grevillea robusta*, but also *G. banksii* and its hybrids (including 'Mason's Hybrid', 'Robyn Gordon' and 'Superb'), *G. hilliana* and *G. venusta*. It also infects species of *Hakea*.

Control: controlling this disease on large trees is impractical, especially on Silky Oak, which seem to be the species most affected. Small plants or prized garden subjects can be sprayed with copper oxychloride at the first sign of symptoms.

Grey Mould, *Botrytis cinerea*

Grey Mould often grows on the foliage and ageing or decaying flowers of many native plants when the conditions are still and very humid, or in drizzly weather. Attacks are worse on species that have dense fine foliage, those with soft hairy foliage, fleshy foliage or species that have flowers with fleshy petals or soft-petals. It can also be bad on plants with a prostrate growth habit. Most attacks of grey mould are of a transient nature and cause little damage, but severe attacks resulting in death of stems and leaves can occur during long periods of weather suitable for the disease. Frequently the mould starts on decaying tissue such as spent flowers, rotting fruit or frost-damaged organs. It can also start in dense foliage, especially if wet for several days. Seedlings of hairy plants can also be very badly affected.

Grey Mould is easily recognised as it covers affected plant tissue with a mass of fine grey threads. When the fungus is sporing the tiny, pinhead-like, grey fruiting bodies are especially obvious. The disease becomes conspicuous in misty or wet weather as the fungal threads trap droplets of water.

Grey Mould can be a very significant disease in enclosed structures such as greenhouses and glasshouses. If ventilation is inadequate or if the plants are too crowded, the disease can spread rapidly. Decaying plant material, such as fallen leaves and ageing flowers (especially those flowers that are fleshy or with a high nectar content such as *Hoya* flowers), are common sources of infection in a greenhouse, although rotting snail pellets often also provide a significant starting point. Greenhouse-grown plants that are susceptible to Grey Mould include native orchids, gesneriads such as *Boea* and *Lenbrassia* and many ferns (developing fronds are especially susceptible). Grey Mould can also be very damaging to seedlings and cuttings in propagation structures.

In the garden this disease can damage fine-leaved species such as *Acrotriche serrulata, Astroloma ciliatum, Epacris* species, *Grevillea pilulifera* and *Thryptomene baeckacea*. It can also be bad on plants with soft or hairy leaves, such as Sturt's Desert pea (*Clianthus formosus*), *Eremophila densifolia, E. glabra* (especially forms with woolly-leaves), *E. hillii, E. nivea, E. subfloccosa, E.* 'Kalbarri Carpet' and species of *Ptilotus*. Damage is more severe if the plants are overcrowded or grown in an unsuitable environment such as frequently occurs when *Eremophila* species from arid and semi-arid regions are planted in high rainfall areas. The flowers of *Eremophila, Scaevola* and *Lechenaultia* are especially susceptible to Grey Mould and the damage can spread into the leaves and stems.

Grey Mould can also cause commercial damage to the flowers of Geraldton Wax, *Chamaelaucium uncinatum*, infecting plants in the field, resulting in premature flower drop and also spoiling harvested material while in transit to overseas destinations (also see above entry for Geraldton Wax Petal Blight). As a general rule plants grown in sunny, well-aerated situations do not suffer as badly from the impact of Grey Mould as plants growing in shady or protected situations.

Control: avoid overcrowding and prune or stake plants to improve air movement and circulation. Regularly remove fallen plant material and decaying snail baits in glasshouses. Spray infected plants with captan or iprodione. Sulphur-based sprays may protect plants during humid or drizzly weather.

Grey Mould infestation of Scaevola *clump – the plant died soon after this infection was noticed.* [D. JONES]

Powdery grey spores of Grey Mould infestation. [D. JONES]

Grey Mould infesting Sturt's Desert Pea. [D. BEARDSELL]

Foliage damage in Xerochrysum *caused by Grey Mould.* [D. JONES]

Blighted flowers of Melaleuca huegelii *damaged by Grey Mould which developed in humid conditions.* [D. JONES]

INK DISEASE OF KANGAROO PAWS

Kangaroo paws (*Anigozanthos* species and cultivars) are subject to attack by a fungus commonly known as ink disease or ink spot. It is caused by the fungus *Alternaria alternata*, which is a widespread pathogen that attacks numerous plants (see also Geraldton Wax Petal Blight). This fungus is not uniform but occurs naturally in a range of distinct races which differ in their virulence or ability to cause infection. Secondary or opportunistic fungi (such as *Botrytis*) may also be present in infected tissue. Ink disease of kangaroo paws is readily identified because it causes a general blackening of leaves and flower stems, although closer examination shows symptoms ranging from brown to black sunken lesions, often with a black rim. The symptoms usually show up first on the apical parts the leaves and then the disease spreads downwards and into the rhizomes. Susceptible species (such as *Anigozanthos manglesii*, *A. gabrielae* and *A. humilis*) are commonly killed by the disease, whereas more vigorous species (*A. flavidus*, *A. pulcherrimus* and *A. rufus*) are better able to cope with infection. A range of susceptibility is also apparent in the large number of *Anigozanthos* hybrids currently available. Those with *A. flavidus* in their parentage are less affected than others. Vigorous plants of any species or hybrid are more resistant to ink disease than weak ones.

Control: ink disease can be difficult to control as a range of factors influence the results. Plants growing in cool moist climates are much more susceptible than plants growing in hot dry climates, and there is some indication that nutrient deficiencies in soils (particularly potassium and calcium) may increase the plant's susceptibility. The best means of control is to keep the plants vigorous and healthy. If they are still attacked, then spraying with a fungicide such as mancozeb is recommended. Applications of potash and calcium are also worth a try. On badly infected plants it is best to remove and burn affected leaves.

Some success in controlling ink disease may also be obtained by fire-torching infected plants (not a slow hot burn that can kill plants) in late autumn. Slashing the foliage first followed by a quick burn (known as the 'slash and burn' technique) is another useful method. Treated plants respond to the reduced fungal load by producing clean new growth in late winter-spring. In cool moist climates, susceptible species, such as *A. humilis*, *A, gabrielae*, and *manglesii,* are best treated as annuals, raising a fresh batch from seed each year.

Selection and propagation of disease-resistant strains of kangaroo paws, both species and hybrids, has proved successful against some isolates of ink disease. However, sometimes new selections are released without thorough testing and they later prove to be susceptible to untested or virulent isolates of the disease.

Ink Disease on the leaves of kangaroo paw.
[D. Jones]

LEAF SPOTS, BLOTCHES AND BLIGHTS

The leaves of a wide range of native plants become discoloured, disfigured or distorted following attacks by various fungi. Most attacks are sporadic and do little more than render the leaves unsightly, but severe infestations occasionally occur and these can result in significant defoliation, stem dieback and plant death. Mature healthy plants often recover from minor infestations with little more than a setback, but seedlings and young trees can become completely defoliated and die. Leaf diseases occur in most areas of Australia and they are generally more prevalent during warm humid conditions and after periods of heavy rain. Some diseases mainly infect adult leaves, others attack young expanding juvenile leaves. Leaf diseases often start off as a discrete pale spot that becomes grey or brown. The size, shape and colour of the spot can be useful for identification of the fungus. Often the brown areas have a surrounding halo of paler tissue, or coloured margins. The brown lesions sometimes develop a corky texture with age. Black spots, pinhead structures and raised areas within the lesion can be the sites of spore production. A few examples of the numerous leaf diseases affecting native plants are given below.

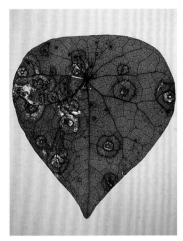

Macaranga *leaf damaged by leaf-spotting fungus disease; note halo spots around infected sites.* [D. Jones]

Leaf-spotting fungus disease on Banksia integrifolia. [D. Jones]

A leaf-spotting fungus disease on the rainforest tree Dillenia alata. [D. Jones]

Symptoms of leaf-spotting fungus disease on phyllodes of Acacia pycnantha. [D. Jones]

Unknown fungus disease on the leaf tips of Cymbidium suave *in the wild.* [D. Jones]

Alternaria Leaf Spots, several *Alternaria* species

The genus *Alternaria* contains many species of pathogenic fungi that cause necrotic spots on the leaves of a wide range of plants. Although species of *Alternaria* cause major disease problems in vegetables such as cucurbits, tomatoes or brassicas, they are usually of a fairly minor nature in native plants (but see also ink disease above). Common symptoms are brown necrotic areas (often angular) surrounded by concentric rings of paler tissue that can coalesce to form a halo. Often the damage is most prominent on older leaves. Leaf spots spread during warm, humid or rainy weather and are worse when plants are overcrowded. Frequently *Alternaria* species also invade damaged plant tissue as a secondary or opportunistic pathogen.

A species of *Alternaria* has been recorded as causing leaf spots on the native species of *Passiflora* and some species of *Eremophila*. Species of *Eremophila* can be badly affected by this disease when they are grown in areas of high rainfall, humid coastal districts or following periods of warm humid weather. Geraldton Wax, *Chamaelaucium uncinatum*, can also suffer badly from attacks by a species of *Alternaria* which causes leaf spotting and premature leaf fall, resulting in sparse unthrifty plants.
Control: the impacts of minor infestations can minimised by pruning infected stems and leaves. Severe infestations may need to be sprayed with mancozeb.

Anthracnose Spot of Wattles, *Glomerella cingulata/Colletotrichum gloeosporioides*

A widespread disease that attacks many species of wattle in various parts of Australia and also in several other countries. It damages wattles in the wild as well as plants grown in gardens, parks and windbreaks.

Anthracnose damage on stems and phyllodes of Acacia saligna. [D. Jones]

A leaf-spotting fungus disease on Eucalyptus camaldulensis. [D. Jones]

Mycosphaerella Leaf Disease on juvenile leaf of Eucalyptus globulus. [D. Jones]

It can also cause substantial economic damage to nurseries and plantations, especially in tropical regions. This fungus attacks wattles with phyllodes and those with bipinnate leaves. *Acacia mangium*, *A. auriculiformis* and *A. aulacocarpa* are three significant commercial timber species that are susceptible. Symptoms of the disease include small dark brown to black spots on the phyllodes and stems that develop into larger blotches which often develop cracks. Frequently the infected area includes dead tissue with darker margins. Sometimes the diseased area is concentrated near the phyllode tips (tip necrosis). Infected phyllodes fall prematurely, leaving the plants with a sparse appearance. Often the older phyllodes are affected first and eventually black lesions become prominent on the stems. Badly affected plants frequently die and those that survive become weakened and unhealthy.

This disease is most noticeable during the wet season in the tropics. It is probably accentuated by the plants being too wet and not being able to dry out in the still, humid, warm conditions. The disease can be particularly prominent in nurseries and has been known to cause very high mortality rates in *Acacia* seedlings.

Control: spraying large trees to control Anthracnose is impractical because of the difficulty of obtaining good coverage. Often by the time damage is noticed the disease has become well established and difficult to control. Regular spraying of seedlings with copper oxychloride or mancozeb is recommended for control.

Leaf Diseases infecting *Eucalyptus* and *Corymbia*

More than 500 species of fungi are known to infect the leaves of *Eucalyptus* and *Corymbia* species. Some of these fungi cause serious infestations affecting the plants' health, others are more cosmetic and have a very limited impact on growth. The impacts of some diseases become much worse on cultivated plants grown in conditions different from those experienced in their natural habitat. Leaf diseases not only impact on eucalypts grown in parks and gardens, but also affect plantation-grown trees, forestry plots, windbreaks and those grown for ornamental cut foliage.

• **Mycosphaerella Leaf Disease**, *Mycosphaerella cryptica* and *M. molleriana*, sometimes also called Crinkle Leaf Disease, is a serious disease of plantation-grown eucalypts, especially *E. globulus and E. nitens*. It also has impacts on trees in parks, gardens and windbreaks. Attacks are severe on young expanding leaves resulting in necrotic patches that dry out causing the affected leaf margins to distort and crinkle. Adult leaves can also be impacted. Often both species of *Mycosphaerella* can be present in infected tissue. Plantations of *E. globulus* in Western Australia are badly affected by two other species of *Mycosphaerella*, *M. aurantia* and *M. ellipsoidea*. Juvenile foliage is the main site of attack.

- **Tuart Leaf Disease.** This is a disease also caused by *Mycosphaerella cryptica*, which impacts badly on many eucalypts but in particular Tuart, *Eucalyptus gomphocephala*, an important woodland tree in Western Australia. Studies on natural seedling recruitment and planted Tuart seedlings showed significant mortality caused by the pathogen.

- **Kirramyces Leaf Blight.** Several species in the fungal genus *Teratosphaeria* (formerly *Kirramyces*) infect both mature and young leaves of some eucalypts and corymbias. These leaf pathogens are widespread and sometimes common in the tropics and subtropics. One species, *T. corymbiae*, causes severe damage to cultivars of popular flowering gums, particularly *Corymbia ficifolia*, *C. ptychocarpa* and their hybrids. Infection by this pathogen impedes uniform growth and results in reddish spots and blackish lesions on the leaves and stems. Older lesions often crack or split and develop black fruiting bodies. Badly infected trees develop stem dieback and become severely stunted. These flowering gums are propagated by grafting and care must be taken to select clean material for propagation as this disease is extremely difficult, if not impossible, to control. This fungus can also severely damage mature leaves on some West Australian eucalypts including *E. conferruminata* and *E. lehmannii*.

Kirramyces Leaf Blight on the stems and leaves of Corymbia ptychocarpa *hybrid.* [D. JONES]

- **Corky Leaf Spot or Target Spot**, *Aulographina eucalypti*, is a common fungal pathogen that infects the adult leaves of many species of eucalypt causing small round brown lesions with a corky central area.

- **Blue Gum Leaf Spot.** Several fungi cause leaf diseases in the Tasmanian Blue Gum, *Eucalyptus globulus*. Blotches caused by the fungus *Phaeoseptoria eucalypti* often disfigure juvenile leaves. This fungus causes little damage to mature trees but can have disastrous effects on seedlings and should be controlled as soon as noticed. Affected leaves are shed prematurely and new growth collapses. Seedlings of *E. globulus* and *E. regnans* can become infected by the fungus *Haynesia lythri*, resulting in leaf spotting and premature shedding. *Haynesia lythri* also infects a number of eucalypts from drier inland areas of Western Australia and can cause serious damage to these plants when grown in the eastern states.

Damage caused by the leaf-spotting fungus Haynesia lythri *to a juvenile leaf of* Eucalyptus conferruminata. [D. BEARDSELL]

Leaf Diseases infecting other genera

- **Angophora Leaf Spot.** The leaves of *Angophora bakeri*, *A. costata* and *A. hispida* can be severely marked following attacks by the fungus *Mycosphaerella angophorae*, which causes dark brown to red-brown lesions in adult leaves.

- **Syncarpia Leaf Spot.** The leaves of Turpentine, *Syncarpia glomulifera*, are sometimes infected by the fungus *Mycosphaerella syncarpiae* and/or *M. marksii*. Attacks are mostly of a sporadic nature with the fungus causing red-brown spots, cracking and leaf distortion.

A leaf-spotting fungus disease on Angophora costata *(upper and lower leaf surface shown).* [D. JONES]

- **Callistemon and Melaleuca Leaf Spots.** A leaf-spotting fungus known as tar spot is found on some bottlebrushes in subtropical areas. It is caused by fungus in the genus *Phyllachora* and develops hard black raised lesions that resemble blobs of tar. Attacks are usually of a minor nature. Another potentially more serious leaf spot found on some of the larger-leaved melaleucas (such as *M. viridiflora* and *M. leucadendra*) and *Callistemon polandii* is caused by a species of the widespread fungus *Cylindrocladium*. This produces uneven spots and blotches which frequently have a purple halo or an overall purplish appearance. After a short time the blotches become brown and sunken. This disease, which is worst during still humid conditions and/or wet weather, causes affected leaves to fall prematurely.

- **Hakea and Grevillea Leaf Spots.** Dull black velvety blotches on the leaves and fruits of some *Hakea* species (*H. elliptica*, *H. laurina* and *H. conchifolia* for example) are caused by a fungus in the genus *Cladosporium*. Damage to healthy plants is generally minor although weakened plants, such as those growing in poorly drained areas, can become severely disfigured. *Hakea myrtoides*, an attractive and rare species from Western Australia, can suffer debilitating damage from this fungus, especially weakened plants growing on sites that have not been burnt for many years. Trials have shown that healthy regrowth is produced after controlled burning. In subtropical and temperate areas the fungus commonly known as sooty spot (*Placoasterella* species) causes sooty black spots and blotches on the leaves of *Grevillea* and *Hakea*. Affected species include *Grevillea asplenifolia*, *G. barklyana*, *G. longifolia*, *G. venusta*, *Hakea plurinervia* and *H. salicifolia*. This fungus does not penetrate deeply into the leaf tissue and can sometimes be wiped off.

Cladosporium Leaf Spot in Hakea victoriae. [D. JONES]

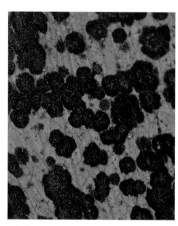

Cladosporium Leaf Spot on Hakea elliptica. [D. BEARDSELL]

Placoasterella Leaf Spot on Grevillea barklyana. [D. JONES]

Placoasterella Leaf Spot in Hakea corymbosa. [D. JONES]

Tar Spot disease on Ficus fraseri. [D. JONES]

Tar Spot disease on Ficus rubiginosa *leaf.* [D. JONES]

- **Ficus Leaf Spot or Tar Spot.** The leaves of some native figs (for example *Ficus fraseri* and *F. opposita*) can be disfigured by the activities of the fungus *Phyllachora rhytismoides*. This fungus forms irregular-shaped, flat, shiny black lesions on the leaf surface. These lesions can resemble blobs of tar.

- **Alpinia Leaf Spot.** Cruciform or diamond-shaped black lesions on the leaves of native gingers (*Alpinia arctiflora* and *A. caerulea*) are caused by the fungus *Phyllachora alpiniae*.

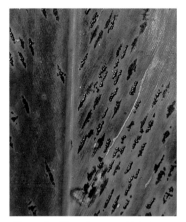

Alpinia Leaf Spot disease on Alpinia caerulea. [D. JONES]

Rhagadolobium *leaf spot on the underside of* Cyathea australis *fronds.* [D. JONES]

Rhizoctonia *rot in* Platycerium superbum. [D. JONES]

- **Fern Leaf Spots.** The fronds of tree ferns (*Cyathea australis* and *C. cooperi*) in Queensland and New South Wales can become infected by the fungus *Rhagadolobium bakerianum*. This causes black blotches and spots on the underside of the fronds. The nest leaves of Elkhorn Ferns (*Platycerium* species) develop black watery spots if overwatered or placed in a poorly ventilated position. This disease, which is caused by a species of *Rhizoctonia*, can spread and kill the plant. The best treatment is to dry the plant out for several weeks before resuming careful watering. Another fungus, *Pseudomonas cichorii*, causes brown spots and blotches 1–3cm across on the edges of the aerial fronds of *Platycerium* species. These lesions are often circled by a halo of pale tissue. Badly infected fronds fall prematurely.

- **Palm Ring Spot**, *Bipolaris incurvata*, is a serious leaf spot disease of palms that flares up periodically in tropical and subtropical regions. Affected leaflets develop small brown spots or blotches that are generally surrounded by a distinct halo of pale green or yellowish tissue. The spots expand and coalesce, forming necrotic areas, often with a darker border. Affected areas can drop out imparting a tatty appearance. Badly affected leaves wither and die. This disease is particularly active during long periods of wet weather and is worse in palms that are crowded and with insufficient air movement. The spores are spread by wind, rain and handling. Susceptible native species include *Caryota albertii*, *Howea belmoreana*, *H. forsteriana* and *Ptychosperma elegans*. Spraying large palms is impractical, but infected leaves should be removed and burnt. Spraying young palms with mancozeb can give some control of the disease.

Palm leaflets affected by Palm Ring Spot. [D. JONES]

- **Palm Leaf Spot**, *Pestalotiopsis* sp. This is a minor fungal disease that causes spots on the leaflets of some native palms such as *Caryota albertii* and *Ptychosperma elegans*. The disease is most obvious on palms growing in shady situations and is usually absent from hardened palms growing in the sun. Control is rarely necessary.

- **Palm Leaf Blight**, *Gloeosporium palmarum*. Widespread in the tropics, this disease causes brown spots and blotches on the leaflets of native and exotic palms. Attacks are sporadic and are usually of a minor nature. This disease has been observed to attack *Archontophoenix alexandrae, A. cunninghamiana* and *Ptychosperma elegans*. Spraying young palms with mancozeb can help control the disease.

- **Control of Leaf Diseases:** spraying large trees to control leaf diseases is impractical because of the difficulty of obtaining good coverage. Often by the time damage is noticed the disease has become well established and difficult to control. Regular spraying of seedlings and young plants with copper oxychloride and mancozeb may provide effective control. Similar methods for disease control apply to other types of plants, such as ferns and gingers that are suffering damage caused by leaf infecting fungi, however it should be noted that copper sprays can severely damage some types of plant – *Dendrobium* orchids, for example.

Macadamia Husk Spot, *Pseudocercospora macadamiae*

Husk spot is a common disease of *Macadamia* plantations in eastern Australia, causing up to 40 per cent loss of production. The symptoms begin as yellowish spots on the green husk of the developing fruit. These later develop into a tan to dark brown hard spot and the fruit fall prematurely. Most commercial cultivars are susceptible but a few show some resistance.

Control: routine spraying with fungicides is necessary in conjunction with the planting of resistant cultivars.

Myrtaceae Tip Blight, *Pestalotiopsis* species

In tropical, subtropical and warm temperate regions the young shoots of some plant species of the family Myrtaceae can die quite suddenly without any apparent cause. Usually only small branches are affected with the tips withering and the leaves browning-off and remaining curled on the plant. The condition is most prominent after the humid, warm weather of the wet season and is aggravated by lack of air movement. Affected species include *Babbingtonia densifolia, Melaleuca bracteata, M. erubescens, M. thymifolia, Neofabricia myrtifolia, Thryptomene oligandra* and *T. saxicola*.

Although this disease is usually of a minor nature it can be persistent and its effects can be very annoying. Tip dieback makes the plants unsightly and severe attacks can kill the plant.

Control: some control has been achieved by spraying with copper oxychloride, however usually by the time the symptoms are noticed the damage is done and spraying is ineffective. Pruning infected tissue improves the appearance of infected plants.

Myrtle Beech Wilt, *Chalara australis*

This disease, which occurs in both pristine and disturbed forests in Victoria and Tasmania, kills established trees of Myrtle Beech, *Nothofagus cunninghamii*. It enters through wounds in the roots and bark, attacking the living wood of the tree causing wilting and death. Initially the leaves of infected plants become pale green to yellow before dying and turning brown. Dead leaves can hang on the tree for more than 12 months. The most notable sign of the disease is the wilting of the tree crown, with death of the whole tree occurring six months to three years after infection. The disease can spread to adjacent trees via air- or water-borne spores (sometimes also via natural root grafts), killing groups of trees within Myrtle Beech communities. Infected trees are subject to attack by the Mountain Pinhole Borer, *Platypus subgranosus*, a small beetle that bores pin-sized holes in the bark. This beetle may help to spread the disease. The appearance of this disease can have dramatic impacts on natural stands of Myrtle Beech. It is listed on the Victorian Flora and Fauna Guarantee Act as a key threatening process for the survival of Myrtle Beech.

Control: control methods are impractical.

Powdery Mildew, *Oidium* species

There are about 100 species of fungi that cause the distinctive and easily identifiable disease that is commonly known as powdery mildew (the symptoms caused by each fungus are basically similar). These fungi, which can only survive on living tissue, grow on the surface of the epidermal cells of leaves, young shoots, flowers and fruits of a wide range of plants. They do not invade internal parts of the plant but penetrate the epidermal cells with specialised haustoria. Symptoms begin as white to grey spots which quickly spread, the afflicted part becoming covered with white to greyish fungal growth and masses of white powdery spores. Ripe spores are readily dispersed in the wind or rain, spreading the infection to other parts of the plant or onto new plants. New growth is infected as it is produced, often withering quickly. Badly infected leaves become distorted and folded or cupped. Older leaves become hard and brittle before being shed prematurely. In autumn, the fungus produces a different type of spore to that produced in the warmer months. These resting spores lie dormant over winter, germinating in spring to renew the life cycle.

Powdery mildew diseases, which are most prevalent from spring to autumn and after the wet season in the tropics, can spread very rapidly. They favour still weather, particularly warm, dry, humid conditions with dewy nights. They are worst when plants are overcrowded or in shady sites and where there is poor air movement and ventilation. Powdery mildews are common diseases of nurseries and glasshouses, but can also infect native garden plants during periods of still humid weather. Susceptible natives include species of *Acacia* (particularly *A. holosericea*), species and cultivars of *Brachyscome*, *Eremophila*, *Hypocalymma*, *Chamaelaucium*, *Thryptomene* and *Verticordia*. Young growth of many eucalypts (especially inland mallees and those with silvery foliage) can also be susceptible to attack by these pathogens. Fruiting cultivars of passionfruit and at least two of the native *Passiflora* species are subject to infection by the Passionfruit Powdery Mildew fungus, *Oidium passiflorae*, especially in tropical regions. Powdery mildew can spread extremely rapidly and should be controlled when first noticed. Seedlings and young plants are particularly susceptible to these diseases.

Control: the impact of powdery mildew can be reduced by ensuring good ventilation. Avoid overcrowding plants in the garden and nursery. Prune regularly to maintain good air movement. Avoid planting susceptible species in shady sites. Improve ventilation of glasshouses and propagation areas. Avoid watering late in the day as this encourages spore germination. Spraying milk on the foliage of susceptible plants acts as a preventative (one part milk to 5–10 parts water). Spraying with wettable sulphur when symptoms are first noticed can give effective control but several applications may be needed.

Powdery Mildew damage on eucalypt seedling. [D. Beardsell]

Powdery Mildew damage on Brachyscome. [R. Elliot]

Passionfruit Powdery Mildew on Passiflora herbertiana. [D. Jones]

Quambalaria Shoot Blight, *Quambalaria pitereka*

This fungus disease, also known as *Ramularia* blight, attacks *Angophora costata* and some species of *Corymbia*. It affects new shoots and young leaves and is often severe in wet weather during summer. It can have severe impacts on growth in timber plantations and forestry plots and can be a serious disease in nurseries, causing growth malformation of seedlings and young plants. New shoots and leaves become distorted (curled leaves and twisted stems) and appear as if they have been splashed with bird droppings or white paint. This material is in fact a mass of tiny white fruiting bodies of the fungus. As the disease progresses stems and petioles split and develop necrotic brown lesions, leaves fall, shoots die back, and the plants take on a sparse appearance. Once a plant becomes infected the disease rapidly spreads to adjacent plants by spores produced from the white pustules.

Quambalaria Shoot Blight on young Corymbia *shoot.* [D. Jones]

Attacks of this disease are persistent and stunt or even kill trees. Cutting back young plants serves no purpose as new shoots are rapidly infected. The disease is favoured by warm, still conditions and is a serious problem during the wet season of tropical and subtropical regions.

Several species of *Corymbia* are badly affected by this disease including *C. citriodora*, *C. eximia*, *C. henryi*, *C maculata* and *C. torelliana*. A strain of *Corymbia citriodora* from the Atherton Tableland appears to be resistant to this fungus.

Control: control is only practical with young plants. Two or three sprays of copper oxychloride at five day intervals provides some control.

Brown Root Rot, *Phellinus noxius*

This disease is found naturally in many rainforests around the world, including those in eastern Australia from Cape York, Queensland, to northern New South Wales. It attacks more than 200 species of native and exotic trees, including palms. With time the fungus has spread to plantations, roadsides and orchards, extending even into urban environments of the east coast where it has become a significant problem in parks, street trees and gardens. It has severe impacts on avocado, hoop pine and figs. The disease infects the roots and spreads upwards, causing decay and tree death. A characteristic early symptom is the development of a stocking of dead or dying wood at ground level, often with white to grey patches of fungus on the upper margin (the upper margin of the fungus becomes brown with age). Soil and plant litter often stick to the fungal growth on the stocking. Young trees can be affected quickly by this disease, first wilting and then declining rapidly and dying. Older trees decline more slowly with yellowish leaves becoming prominent and then falling prematurely, leaving a sparse crown. Large trees can take many years to die but in the meantime the wood of the trunk is decaying internally and affected trees can often fall dramatically and without warning. The fungus fruits after periods of heavy rain, either producing hard, dark brown, bracket-like fruiting bodies that are charcoal grey beneath, or smooth flat fungal masses of a greyish colour on the lower trunk and roots. This fungus is very long-lived and can remain in infected stumps or soil for more than 50 years.

Bracket-like fruiting body of Brown Root Rot, a wood-rotting fungus that affects rainforest plants. [D. Jones]

Control: control is not practical. Avoid planting into infected soil.

RUST DISEASES

The plant diseases which are commonly known as rusts are the result of infection by a range of pathogenic fungi that can only survive in living plant tissue. Symptoms usually consist of small yellow to orange spots (grey to black in *Acacia* rust) on the leaf surface, these spots developing into powdery pustules as the fungus grows and forms spores. Some rusts can induce galls. Ripe spores of rust fungi are readily dispersed in windy weather and during gardening operations. Many rust fungi seem to be fairly specific to the plants that they attack; myrtle rust is a significant exception to this rule. Control of rust fungi can be difficult. Ruthless culling is recommended followed by replacement using plants that show immunity. Chemicals such as bitertanol and oxycarboxin offer some degree of control if applied early in the infection process.

Rust pustules on leaf underside of Notelea. [D. Jones]

Acacia Rust or Wattle Rust, *Racospermyces digitatus*

This is a native species of rust fungus that attacks wattles in temperate and tropical regions. It not only damages wattles in the wild but also plants grown as ornamentals. It also has the potential to cause substantial economic damage to nurseries and plantations, especially in the tropics – *Acacia mangium* and *A. aulacocarpa* are two significant commercial timber species that are susceptible. Wattles attacked by this fungus include those with phyllodes (*Acacia aulacocarpa, A. concurrens, A. mangium, A. notabilis* and *A. podalyriifolia*) and those with bipinnate leaves (*Acacia deanei* and *A. irrorata*). Infected plants develop

irregular crusty or warty grey to black swellings on phyllodes and pods (swellings on the leaflets of bipinnate wattles are smaller than those found on phyllodes). The swellings enlarge as the season progresses, producing masses of spores. Infected plant parts become distorted and may fall early.

Control: attacks are often sporadic and no control is known. Spraying young plants with bitertanol may be worth trying.

Wattle Rust on phyllodes of Acacia podalyriifolia. [D. Jones]

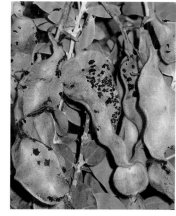

Wattle Rust on seedpods of Acacia podalyriifolia. [D. Jones]

Myrtle Rust, Guava Rust or Eucalypt Rust, *Puccinia psidii s.l. / Uredo rangelii*

This disease, a recent introduction (2010) into Australia from South America, has the potential to cause major ecological changes in the Australian flora and greatly affect the choice of garden plants we grow. As its name suggests, this disease attacks members of the myrtle family (Myrtaceae), of which there are more than 2,200 Australian native species (or nearly 10 per cent of the Australian flora). The potential impact is obviously staggering. Not only is the Myrtaceae very diverse in the Australian flora, but it is also a dominant element in the landscape. Consider for example the dominance of *Eucalyptus* and *Corymbia* species in the open forests and woodlands that cover much of the continent, or the significance of *Syzygium, Lophostemon* and *Syncarpia* in wetter forests, especially rainforest; also the importance of shrubby myrtles as understory plants in numerous habitats from well watered to drier zones. The disease, which is already having significant ecological impacts, could disrupt the ecological

Myrtle Rust on Rhodamnia rubescens *seedling.* [A. CARNEGIE]

Myrtle Rust on the leaf undersides of Gossia inophloia *'Blushing Beauty', a plant which is very susceptible to this disease.* [D. JONES]

Myrtle Rust on leaf of Melaleuca quinquenervia. [A. CARNEGIE]

balance and result in large-scale changes to the composition of plant communities over huge areas.

Consequences: not only does Myrtle Rust have the potential to change the flora, but the ecological impacts could be far-reaching, impinging on a whole range of factors in the ecology of natural communities (such as the timing and degree of flowering, nectar production, pollen flow and seed production). The disease also has consequences for commercial activities, such as forestry, plantation timbers, the nursery industry, cut flower growers, honey production and bush foods. It will also impact heavily on gardeners and landscape designers, greatly influencing the choice of suitable plants in home gardens and urban landscapes for decades to come. Immediate impacts are already obvious in eastern Australia. Replacing infected garden plants with non-Myrtaceous shrubs is recommended by the authorities in infected areas, as is also the use of tolerant species in regeneration sites.

The disease: myrtle rust is part of a complex group of rust fungi native to South America that have spread into Mexico, Florida and Hawaii. Myrtle rust in Australia is caused by the rust fungus *Puccinia psidii* (a variant within the myrtle rust complex that produces distinctive spores is called *Uredo rangelii*).

Symptoms: in the early stages of infection, myrtle rust produces grey to brown spots (often with a reddish halo) on the stems and leaves. These quickly develop into yellow pustules which produce masses of bright yellow powdery spores. The disease impacts mainly on new growth resulting in leaf curling, growth distortion and stem dieback. Later flushes of new growth also become infected. Pustules can also develop on flowers (e.g. *Chamaelaucium uncinatum*) and soft fruit, such as those of *Eugenia reinwardtiana* and some species of *Syzygium*. Seedlings of many species can also be attacked and killed, the yellow pustules even developing on the cotyledons.

Spread: myrtle rust, which now cannot be eradicated in Australia, has become a fact of life in many areas of the east coast and is spreading steadily. The initial spread of myrtle rust was via infected nursery plants. This resulted in outbreaks in eastern Australia in areas as widely separated as Cairns in north-eastern Queensland and many sites in Victoria, with numerous occurrences in between. The disease is well established in south-eastern Queensland (also western parts of the Darling Downs and on coastal islands in Moreton Bay) and many areas of New South Wales, including the heavily populated Central Coast and the less populated South Coast). It has also been found in Townsville, Rockhampton and the Whitsunday Islands. Myrtle rust has not been recorded as yet from Tasmania, South Australia and Western Australia.

Myrtle rust has not only infected gardens but has also spread into bushland, particularly wetter forests, including rainforest. The potential changes could be far reaching and quite devastating. The small spores are readily distributed by wind, but can also be carried by insects, birds and other animals, as well as being spread by human activities including contaminated vehicles and clothing. This disease, which becomes very active in hot, humid weather, has the potential to spread further into the tropics. It is also tolerant of cold and may spread further into temperate Australia, infecting gardens and bushland.

Plant of Agonis flexuosa *badly affected by Myrtle Rust.* [D. JONES]

Myrtle Rust on Agonis flexuosa. [D. JONES]

Severe damage of Myrtle Rust on Agonis flexuosa. [D. JONES]

Hosts: to date myrtle rust has been found to infect more than 185 species of native and introduced Myrtaceae. Much of this knowledge has come from glasshouse trials. Although these plants are known to be susceptible to attack, the impact of the fungus varies significantly between species and even between selections and cultivars. Susceptibility is an ongoing learning process because of the novelty of the disease in Australia, but much has been learnt in a short time. New hosts are being discovered regularly. Even within a species, only some of the individuals became infected.

Extremely susceptible native species include *Agonis flexuosa* (especially the dwarf and dark-leaved cultivars), *Chamaelaucium uncinatum*, *Decaspermum humile*, *Eugenia reinwardtiana*, *Gossia inophloia*, *Melaleuca quinquenervia*, *Rhodamnia angustifolia*, *R. maideniana*, *R. rubescens* and the exotic Rose Apple, *Syzygium jambos*.

Highly susceptible species include *Anetholea anisata*, *Austromyrtus dulcis*, *Backhousia citriodora*, *Callistemon polandii*, *Choricarpia leptopetala*, *Gossia acmenioides*, *G. gonoclada*, *Lenwebbia prominens*, *Melaleuca fluviatlilis*, *M. leucadendra*, *M. nodosa*, *M. viridiflora*, *Rhodamnia costata*, *R. dumicola*, *R. sessiliflora*, *Rhodomyrtus psidioides*, *R. tomentosa*, *Syzygium oleosum*, *Tristania neriifolia* and *Xanthostemon oppositifolius*. Laboratory tests also show that various species and hybrids of *Angophora*, *Corymbia* and *Eucalyptus* are susceptible but have not been affected badly in the wild (to date).

Fortunately some commonly cultivated plants display some tolerance of the disease. These include *Austromyrtus tenuifolia*, *Callistemon formosus*, *Corymbia torelliana*, *Melaleuca nesophila*, *Pilidiostigma glabrum*, *Syzygium australe*, *S. hemilampra, S. floribundum, S. leuhmannii, S. moorei, S. smithii, S. tierneyanum, S. wilsonii* and *Tristaniopsis laurina*.

Control: although it is not technically feasible to eradicate myrtle rust in Australia, there are some controls that can be implemented. A number of fungicides have been registered for the control of myrtle rust, only three of which are available for home gardeners. These are copper oxychloride, mancozeb and triflorine, all of which seem to slow down the disease rather than giving effective control. Triflorine has some systemic properties whereas the other two act mainly as protective agents. Commercial producers can use other specialised chemicals. Check with local Government departments for more details.

It is emphasised that plants infected with myrtle rust should not be handled until after spraying as any movement or disturbance readily disperses the spores. Infected plants can also be sprayed with methylated spirits before removal. After spraying, it is strongly recommended that infected and susceptible plants should be removed from infested gardens, because reinfection is likely even after fungicide application. Infected material should be disposed of correctly. It is best to wear disposable gloves. Seal small quantities of infected material in a plastic bag; wrap larger quantities in black plastic and use packaging tape to seal any gaps. Leave the plastic-wrapped material in the sun for 3–4 weeks to kill off any fungal growth, including spores.

Rust disease on kangaroo paw. [D. JONES]

Rust pustules on the leaves of a kangaroo paw. [S. JONES]

Rust Disease of Kangaroo Paws *Puccinia haemodora*

Kangaroo paws (species and hybrids of *Anigozanthos* and the Black Kangaroo Paw, *Macropidia fuliginosa*) are subject to attack by this parasitic rust fungus which can only survive in living plant tissue. It causes small reddish-brown spots or pustules on the leaves of infected plants, sometimes surrounded by a circle of black tissue. These spread to form dense patches which weaken the plant and cause early death. This disease, which can spread readily from plant to plant, is apparently unknown in natural populations but it causes serious damage to cultivated kangaroo paws in gardens, nurseries and commercial cut flower plantations. The fungus occurs in a range of distinct races, each race more or less specific to a species of *Anigozanthos*. The Tall Kangaroo Paw, *Anigozanthos flavidus*, is mostly resistant to the disease and this immunity is transferred to some of its hybrid progeny. The smaller kangaroo paws, particularly *A. humilis*, *A. gabrielae*, *A. onycis* and their hybrids, as well as *Macropidia fuliginosa*, are particularly susceptible. The impacts of the fungus variants are variable however. For example, research has shown that *Macropidia fuliginosa* is resistant to the strain of the fungus that attacks *A. manglesii* but is susceptible to another strain of the disease.

Control: choose immune species and hybrids. Spray infected plants with bitertanol.

Rust Disease of Boronia *Puccinia boroniae*

Several species of *Boronia* can become infected by strains of the rust fungus *Puccinia boroniae*. It causes yellow-brown pustules on stems and leaves resulting in premature leaf fall and stem death. Infected plants, although appearing sparse and unthrifty, can linger for some years. Plant death is not a common result, but severe damage in commercial cutflower plantations results in loss of suitable stems for cutting and downgrading or rejection of infected material. The impacts of this fungus, which vary from season to season, range from fairly minor damage to serious loss. Economic losses commonly occur with the widely grown *Boronia heterophylla* and *B. megastigma* but other species affected include *B. clavata*, *B. elatior* and *B. coerulescens*. Garden grown plants and those in commercial cut flower plantations are affected by this disease.

Control: choose immune species and hybrids. Destroy infected plants. Spray infected plants with bitertanol.

Rust Diseases of Terrestrial Orchids *Uromyces* species

Three different species of rust fungi in the genus *Uromyces* are known to infect naturally occurring plants of a number of terrestrial orchid species. The rust fungus *U. thelymitra* has been recorded from species of *Thelymitra* and *Lyperanthus nigricans*; *U. microtidis* from species of *Microtis* and *Prasophyllum,* and *U. orchidearum* from species of *Chiloglottis* and *Diuris*. Another rust fungus, which may be a species of *Puccinia*, has also been found on *Caladenia radialis* at Yealering in Western Australia. Orchid plants infected by rust fungi develop yellow to orange pustules on the leaves. In species of *Chiloglottis* the pustules darken quickly after emergence and become small brown lesions. Some growth deformations may occur in infected leaves including lopsided development and twisting or curling. Infected plants of *Chiloglottis* species produce leaves that are quite distinct from those on healthy plants, being narrower (the blade does not expand fully) and held partially erect rather than flat on the ground. They are often also noticeably paler in colour than leaves on healthy plants. Rust fungus infection also appears to reduce or inhibit orchid flowering and can cause premature dieback.

Terrestrial orchids infected by rust fungi have been observed in natural communities in many parts of southern Australia. The disease persists within an infected plant through successive generations, the symptoms appearing soon after leaf emergence. In species of *Chiloglottis*, which increase vegetatively to form clonal colonies, the fungus is transferred annually to daughter tubers, resulting in whole colonies of infected plants.

Control: no control is known. Spraying infected plants with bitertanol may be worth trying.

Colony of Chiloglottis *infected by rust fungus.* [D. JONES]

Pustules of rust fungus on Diuris *leaf.* [D. JONES]

Other Rust Diseases

- **Dampiera Rust.** Several species of *Dampiera* can become infected by the rust fungus *Puccinia dampierae*. This fungus causes yellow pustules on the leaves and stems causing premature leaf fall and growth distortion in some species. This disease infects *Dampiera* in the eastern states and Western Australia. Infected plants lack vigour but can last for several seasons before succumbing.

- **Fern Rust.** The native fern, *Lunathyrium japonicum,* is sometimes subject to attack by a rust fungus. Affected fronds become misshapen and the underside covered with numerous orange pustules. As these mature they release rust-coloured spores and eventually the whole lower surface is an orange-brown powdery mass. Although the disease does not appear to spread to other species of ferns, affected plants should be removed or destroyed to reduce the possibility of infection.

- **Ormosia Rust.** The Yellow Bean, *Ormosia ormondii*, a native tree found in lowland rainforests of north-eastern Queensland, is sometimes attacked by the rust fungus *Atelocauda shivasii*. Grey to brown lesions surrounded by a yellow halo develop on the leaflets, these patches darkening with age and producing masses of bright yellow to yellowish-brown or reddish-brown spores. This is a highly specialised disease of rare and sporadic occurrence.

- **Tricoryne Rust.** In parts of in south-eastern New South Wales the native lily *Tricoryne elatior* is attacked by a rust fungus that causes yellow to brown lesions on the leaves and stems. The plants become weakened by this disease, with whole clumps sometimes collapsing completely. Badly infected leaves wither or are shed prematurely.

Plant of Tricoryne elatior *badly affected by rust fungus.* [D. JONES]

Pustules of rust fungus on Tricoryne elatior *leaves.* [D. JONES]

Soft Rot or Crown Rot *Sclerotium rolfsii*

Also known as Southern Blight, this crown-rotting soil-borne pathogen attacks soft herbaceous plants and small shrubs that have a dense mat of foliage at or close to ground level. It can be prominent in tropical and subtropical areas during the wet season, but is also widespread in temperate regions. It is worst during warm, still, humid conditions and following rainy periods when the foliage remains wet for long periods, but can also appear when plants are overcrowded. Susceptible species include *Actinotus helianthi, Ajuga australis, Ammobium alatum, Boronia deanei, Brachyscome multifida, Chrysocephalum apiculatum, C. baxteri, Dampiera linearis, D. teres, Hibbertia obtusifolia, H. sericea, Laurentia axillaris, Lechenaultia formosa, Rhodanthe anthemoides, Thomasia pygmaea, Xerochrysum bracteatum* (especially prostrate cultivars such as 'Diamond Head'). Some tussock grasses, native lilies and terrestrial orchids are also susceptible.

Susceptible species, particularly soft herbaceous plants such as *Ammobium alatum*, can collapse quickly. The first sign of attack is often wilting followed by yellowing of the lower leaves. These become brown and the whole plant collapses within a few days and may become slimy. White mycelial threads and round white resting bodies (sclerotia) are evident on close inspection of the basal parts of dead plants. The resting bodies eventually become brown and hard. Infected plants of *Ajuga australis* develop bright red colours in the older leaves and new leaves are much smaller than usual. This species can linger for some months before succumbing. Twiggy plants generally react more slowly, becoming unthrifty with the lower leaves becoming yellowish and falling, giving the plants a sparse appearance with green leaves only remaining on the stem tips. Pot grown plants can also suffer from this disease, including King Fern, *Angiopteris evecta*. Laboratory tests also showed that the related Potato Fern, *Marattia salicina*, is also susceptible.

Fungi in the genus *Sclerotinia* can also cause similar symptoms but can be differentiated from *Sclerotium* because they develop black resting bodies resembling small seeds or mouse droppings.

Control: the white resting bodies of this fungus can persist in the soil for many years, infecting later plantings of susceptible species. Infected plants should be removed and burnt and infected soil should be sterilised. Resistant plants are best grown in areas that have been infected with this fungus. Herbaceous plants should only be grown in well-drained soil in a sunny position open to good air movement. A mulch of gravel or coarse sand may help avoid the disease. Spraying or drenching with dichloran may help control this disease.

Soft Rot affecting Ajuga australis.
[D. Jones]

A dry appearance in this Dampiera teres *is an early indication of root disease.*
[D. Jones]

Plant of Dampiera teres *killed by Soft Rot; note white fungal threads.* [D. Jones]

Sooty Mould *Fumago vagans*

This is a common fungus disease that covers plant parts with an unsightly sooty black material. The fungus grows on sugary secretions (honeydew), especially that produced by sucking insects such as aphids, jassids, lerps, mealybugs and scales, less commonly on plant secretions exuded by glands on leaves, phyllodes and the stems and of fast growing shoots. It can also develop on the foliage of shrubs growing under trees infested with sucking pests because their exudates rain down onto the plants below.

Sooty Mould on Notelea *leaves.* [D. Jones]

Sooty Mould has blackened the stems of this Calothamnus. [S. Jones]

Control: Sooty Mould control involves removing the source of honeydew on which it grows. Mostly this disease follows infestations of sucking insects. If these pests are controlled, the Sooty Mould usually disappears within a couple of weeks. Spraying with warm soapy water can assist in cleaning affected stems and leaves.

PROTISTA DISEASES (WATER MOULDS)

Protists, which belong to the kingdom Protista, are a group of unicellular organisms that are gathered together basically because they do not fit comfortably elsewhere. One group, the water moulds (previously treated as fungi), includes a number of very important plant pathogens.

Phytophthora Root Rot *Phytophthora cinnamomi*

This disease is also called cinnamon fungus (although it is not a fungus), jarrah dieback, wildflower dieback, or just dieback. It is a vigorous soil-borne pathogen that kills a wide range of native plants by attacking their root systems. An exotic disease that was probably introduced via European settlement, it is now widespread and well established in forested areas in the wetter parts of most Australian states. It invades natural ecosystems killing susceptible species and causing irreversible changes to the composition of natural floras. Some 202,500 hectares of the magnificent Jarrah forests of WA have been destroyed since about 1927, and it is estimated that much more is doomed. In Victoria the important coastal forests of east and south Gippsland are severely infected as well as the Brisbane Ranges and parts of the Grampians in the west of the state. In South Australia the disease is well established in the Mt Lofty Ranges, Fleurieu Peninsula and over much of Kangaroo Island. Perhaps surprisingly the disease is also very widespread in Tasmania with tens of thousands of hectares of dry sclerophyll forest, heathland and moorland communities being infected. Occurrences in New South Wales are mainly isolated and well separated with significant local impacts in some areas (around Eden, Sydney and the Barrington Tops, for example). In Queensland this pathogen causes severe damage in coastal areas and has become well established in wallum communities (although not always causing problems here). The situation in north-eastern Queensland forests is more complex with the pathogen mainly causing patch deaths around areas of human and pig disturbance.

Phytophthora Root Rot has had major impacts in conserved sites such as National Parks and gazetted reserves, in some cases resulting in restricted access to important sites. Visitors to Sydney can see the impact of this disease with startling changes in the flora of the Sydney Harbour National Park highlighted by the numerous skeletons of dead trees. Populations of the Austral Grass Tree, *Xanthorrhoea australis*, a distinctive but susceptible component of the Victorian flora, have been decimated in some areas in the state. The disease has even impacted the Wollemi Pine, an iconic species known only from small isolated populations within the Wollemi National Park in New South Wales. This impact, although minor (only three trees at this stage), clearly illustrates how easily this disease is spread and the dangers of unauthorised visitation to sensitive sites. *Phytophthora cinnamomi* is also versatile, even killing stressed mangroves suffering from pollution in several localities of coastal Queensland. It also threatens small and isolated natural populations of endangered plants. In Western Australia for example, 10 of the 11 known wild populations of *Banksia brownii* have become extinct following infection by phytophthora root rot. Similarly two of the three known populations of the critically endangered plant *Lambertia orbifolia* are affected by the disease.

Phytophthora Root Rot has had a huge impact on horticultural activities in Australia. It is a common disease of home gardens, orchards, parks and nurseries and causes much anguish to plant lovers. It has devastated some public gardens including the National Botanic Gardens in Canberra where large areas became infected within two years of its introduction.

Life cycle: Phytophthora Root Rot is caused by a water mould that invades the roots of plants, crippling or destroying the root system and killing susceptible species. This pathogen spreads by colourless thread-like growths (mycelia) and survives lean times by the means of resistant (or resting) spores (termed chlamydospores) that survive in soil and decaying plant roots. At temperatures above about 15°C the resting spores germinate and produce a short tube with a terminal sac containing 20 to 40 motile zoospores. These spores, which are chemically attracted to plant roots, swim for short distances in water to the root tips of any plants in the vicinity (they can be dispersed over larger distances in flowing water). Infection of the roots takes place and the mycelial threads grow into the roots invading vascular tissue and impeding

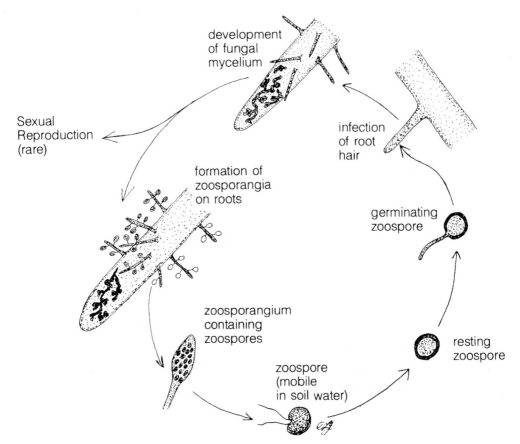

Life cycle of Phytophthora cinnamomi. [D. JONES]

Advanced stage of Jarrah dieback. [C. French]

Warning sign in Western Australia. [S. Jones]

Forest affected by Jarrah dieback. [C. French]

water uptake. More resting spores are produced which are released into the soil as the infected roots rot. Thus the cycle turns over quickly and in suitable conditions the disease progresses rapidly. It has been estimated that in warm wet conditions (such as occur during heavy rainfall in summer) the fungus can complete its life cycle within 24 hours. Sexual reproduction also occurs in this pathogen, resulting in the production of a third type of spore, known as an oospore. Both oospores and chlamydospores can survive for a long time in soil and decaying roots, reinfecting plants when conditions become favourable. Little infection takes place below 15°C.

Symptoms: Phytophthora Root Rot causes a variety of symptoms in plants, depending on the species attacked and the soil conditions where it is growing. The most startling appearance of the disease is the sudden and dramatic collapse within a couple of days of a plant that was apparently healthy. The first noticeable symptom is a dryish look in the leaves followed by rapid decline in plant health. This type of collapse is common in susceptible species such as those originating in areas with a dry climate, including numerous species of *Banksia, Dryandra, Isopogon* and *Adenanthos* and some species of *Hakea, Grevillea* and *Eucalyptus* (particularly those from Western Australia). If the species is very sensitive to

Macrozamia communis *killed by* Phytophthora cinnamomi, *with ailing plants in background.* [D. JONES]

Grass trees are very sensitive to root diseases, especially Phytophthora. [D. JONES]

Dying growth in Isopogon *caused by root problems.* [D. JONES]

Dryandra *killed by* Phytophthora cinnamomi. [D. JONES]

phytophthora root rot, sudden plant collapse can occur irrespective of whether the soil is well-drained or waterlogged. The collapse usually occurs during hot weather and is a direct result of the decay of the plant's root system. Often plants that collapse in this way have been able to survive because the remaining roots have gathered sufficient water for the plant's needs. This becomes impossible under conditions of severe stress and affected plants collapse rapidly. Wilting of young shoots is another common early symptom of the presence of this pathogen, followed by progressive dieback of the stems. Infected plants often appear to be loose in the soil due to loss of the root system. Typically when the root system is examined few or no healthy small roots will be present.

Phytophthora Root Rot is most active in acid soils and has minimal impact in calcareous soils. It becomes much more virulent in heavy or waterlogged soils and virtually only resistant plants can survive such a combination. In well-drained soils, the root rot may not kill the plants but can cause stunting, slow growth and dieback of the growing tips. The small feeding roots are attacked and the leaves of affected plants are often yellow with brown margins and tips. Plants in such situations may linger for quite a long time but look unthrifty and often succumb when conditions favour the fungus (warm wet periods of 24 hours or longer).

Control: the low toxicity chemical compound phosphonate, which contains phosphorous acid (not phosphoric acid) has proved to be effective against Phytophthora Root Rot. This compound, which is available as commercial products that are safe to use on native plants, can be applied as a foliar spray, soil drench or trunk injection. It acts by inhibiting sporulation of the pathogen but has only limited effect on the growth of the mycelium. Research has also shown that it has minimal impacts on mycorrhizal fungi. Two other chemicals have been used with success against Phytophthora Root Rot. These are metalaxyl and fosetyl. Metalaxyl is taken up by the roots and spreads up through the trunk and stems. Infected plants should be drenched with 2–4 litres of metalaxyl mixed at the rate of 0.5–1g per litre. Fosetyl has been applied as a protective spray at the rate of 4g per litre per plant.

Improvement of soil drainage by installing drains, raising garden beds and/or using sunken paths greatly increases the chances of plants surviving attack. The addition of mulches (material rich in fibre such as straw or chunky bark works best), compost and other organic material is extremely beneficial in suppressing the disease. Such organic materials encourage a wide spectrum of other organisms that provide a vigorous soil rhizosphere which competes with or is antagonistic to the pathogen. Only tolerant or immune plants should be grown in areas known to be infected with phytophthora root rot (see list in Appendix 1).

Contaminated soil is the usual source of infection in gardens. Infected soil should be treated before replanting unless resistant species are being used. Infected soils can also be planted with grasses and sedges in the hope of eliminating resistant spores of the pathogen. The pathogen is readily transported by water flow as well as on contaminated tools, shoes and vehicles.

Macrozamia communis *dying from* Phytophthora cinnamomi *infection.* [D. Jones]

Dying Adenanthos sericea *in natural habitat.* [S. Jones]

Phytophthora cinnamomi *has killed one plant in this clump of* Banksia repens. [D. Jones]

Note the absence of fibrous roots in this plant of Grevillea asparagoides *killed by a root-rotting fungus, probably* Phytophthora cinnamomi. [D. Jones]

Collar Rot several *Phytophthora* species

As the name of this disease implies, the primary site of attack by this pathogen is the collar or lower part of the stem or trunk at soil level. Collar rot can be caused by *Phytophthora cinnamomi* but is more usually the result of attacks by other species of *Phytophthora* (such as *P. cactorum* and *P. multivora)*. The trunk is usually infected within 15 cm of soil level, with the infected part turning brown and decaying. Infected plants appear unthrifty and the leaves become pale to yellow. The bark above the rot may crack and exude gum. The wood beneath infected bark is brown and often has a sour smell. Usually there is a distinct margin between healthy and diseased tissue. Infected plants can linger but generally die when stressed or the decay completely encircles the stem or trunk.

Control: collar rot can be controlled by products containing potassium phosphonate applied as a soil drench, foliar spray or trunk injection (see also previous entry). The dead bark of afflicted plants can be cut away to expose healthy tissue and the area painted with a potassium phosphonate compound, Bordeaux mix paste, or mastic containing a fungicide. If the fungus is killed and the wound is not too extensive, the plant will often recover from such treatment.

Collar rot is favoured by saturated soil and exacerbated by cultural practices such as mounding soil or crowding mulches around the plant stem or trunk. Allowing dense weed growth to smother this region can also create conditions for infection. Every effort should be made to reduce humidity in the region of the plant collar at soil level. Soil drainage should be improved if this is seen as a cause of the disease. A heavy dressing of lime on the soil surface around the trunk is claimed to reduce the incidence of collar rot. Budded or grafted plants should be planted with the union well above soil level.

Impact of Collar Rot on Crowea. [D. Jones]

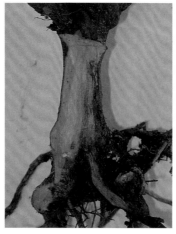

Collar Rot-affected tissue on the stem of Crowea – *the fungus attacked from the right.* [D. Jones]

Collar Rot damage to Adenanthos sericea. [M. Wood]

Collar Rot damage to stem Adenanthos sericea. [M. Wood]

Hakea leuhmanniana *killed by Collar Rot – note heavy growth of groundcover around the base.* [D. Jones]

Banksia spinulosa *killed by Collar Rot.* [D. Jones]

Dying growth in Isopogon *caused by root problems.* [D. Jones]

Other Phytophthora Diseases

The genus *Phytophthora*, which contains more than 100 species, includes some of the most devastating plant pathogens in the world. These pathogens have caused major dieback diseases in many countries, as exemplified following the introduction of *P. cinnamomi* into Australia. Recent studies using molecular techniques have identified many new species of *Phytophthora*, including several isolated from natural ecosystems in Western Australia. Some of these new species may well be endemic in Australia and early studies suggest they may have co-evolved with the local vegetation where they occur. One species, *P. multivora*, is widely distributed in south-western WA and has been associated with decline and death of significant tree species including Tuart (*Eucalyptus gomphocephala*), Jarrah (*E. marginata*), Peppermint (*Agonis flexuosa*), several banksias including *B. grandis*, *B. menziesii* and *B. prionotes*, and understory plants such as species of *Bossiaea, Patersonia, Podocarpus* and *Xanthorrhoea*. As with *Phytophthora cinnamomi*, this pathogen infects the fine roots but has also been associated with stem girdling and collar rot in *Banksia littoralis*. Unlike *P. cinnamomi* however, this species is also active in calcareous soils.

Loss of leaf sheen and a dry appearance are early symptoms of impending death in Banksia, *most often caused by a species of* Phytophthora. [D. Jones]

Two other recently described species, *P. arenaria* and *P. constricta,* have both been isolated from Kwongan (sandplain) vegetation, the former from warm dry northern areas in Western Australia, the latter from cooler southern areas around Perth. Both species have been associated with plant deaths, but their impact on natural ecosystems is largely unknown. It is suspected, however, that *P. constricta* may have caused significant *Banksia* deaths following periods of heavy summer rain. The impacts of these pathogens may be exacerbated if infested material is moved to sites out of their normal range. Also the impact of climate change, with predicted higher summer rainfall, may significantly change their impact.

Control: largely unknown. Chemicals used against *P. cinnamomi* may offer some control but further research is needed.

Pythium Root Rot *Pythium* species

Pythium root rots are common soil-borne diseases that attack a wide range of plants, including commercial crops and garden plants. Species of *Pythium* are widespread and long-lived in soils, particularly acid soils of high fertility. *Pythium* root rot can be prominent during periods of heavy rain but it can also be active during drought. *Pythium* can also interact with soil-borne fungi such as *Fusarium* and *Rhizoctonia*, increasing the severity of infection.

The Goodenia *on the left has been killed by Pythium Root Rot.* [D. JONES]

  Pythium root rot is well known for causing seeds to rot in the ground and killing turf. Another strain (or species) causes damping-off of seedlings in plant nurseries. *Pythium* can also infect plant roots and can weaken or kill stressed plants. Native groundcovers, including species of *Goodenia*, *Scaevola* and *Myoporum parvifolium* can be susceptible. Infections of *Pythium* have caused serious losses in commercial ginger plantations in Queensland. *Pythium* can also be devastating in hydroponic production systems and vegetable crops grown in greenhouses.

Control: *Trichoderma*, a parasitic fungus, is antagonistic to *Pythium* and has been used successfully as a biocontrol in soil trials. A strain of the bacterium *Pseudomonas* also shows potential as a biocontrol agent. The chemical metalaxyl has also been used against *Pythium*.

BACTERIAL DISEASES

Bacteria are tiny single-celled organisms that cannot be seen with the naked eye. Bacteria pathogenic to plants, cause a range of symptoms including spots, blotches, blights, pustules and smelly rots. A few serious bacterial diseases cause economic damage to crop plants. Bacteria infect plants through wounds or natural openings such as stomata. They are readily spread by water, insects and careless handling of infected material.

Crown Gall

Crown gall is a widely distributed bacterial disease that is known to attack over 600 species of plants around the world. Until recently the bacterium causing the disease was well known as *Agrobacterium tumefaciens*, however, genetic studies published in 2001 resulted in a new name, *Rhizobium radiobacter*. This bacterium attacks plant stems at ground level and also the underground stems of plants, causing swollen lesions, galls and cankers. These are usually roughened and misshapen and constrict the flow of sap through the area attacked. As a result infected plants are stunted and severe attacks can kill even healthy plants. The galls also weaken the attacked area and badly affected plants can blow over in windy weather.

  Crown gall is mainly a disease of cultivated soils and appears to be unknown in undisturbed sites. It is a serious pest of deciduous trees, especially fruit crops, but several native plants are known to be susceptible including *Abutilon theophrastii*, *Araucaria bidwillii*, *Brachychiton acerifolius*, *B. populneus*, *Cordyline stricta*, *Eucalyptus tereticornis*, *Ficus benjamina* and *Lythrum salicaria*. Nurseries are not allowed to sell plants infected with crown gall.

Control: crown gall is extremely difficult to control once established in a plant. Vigorous healthy plants will withstand infection better than weak plants and may not even show symptoms. Soils known to be

Base of trunk and roots affected by Crown gall. [D. JONES]

infected with crown gall should be fumigated with a commercial preparation before planting. A genetically modified benign strain of the crown gall bacterium which prevents infection by competing actively with the disease-producing type, is available commercially under a trade name. The usual method of application is to dip the base of cuttings or the root system of plants in a solution of this benign strain before planting.

Bacterial Leaf Spots of Orchids

Bacterial pathogens in the genera *Acidovorax* (previously *Pseudomonas*) and *Erwinia* cause significant economic losses in orchid nurseries and cut-flower enterprises in many tropical countries. They can also cause significant damage in the collections of hobbyists. Leaf symptoms usually start off as faint discoloured sites that spread into dirty green or brown water-soaked blotches. Often yellow margins develop around the infected area. Sometimes the spots dry out and become isolated from the rest of

the leaf as a brown sunken area but they can also spread, eventually rotting the whole leaf or causing premature leaf fall. The diseases can also spread into canes causing a mushy smelly rot. Development of these diseases can be very rapid, whole plants being killed within a few days in warm to hot humid conditions. Infected sites ooze liquid laden with bacteria and these infections can be spread by water splash, contact between plants, insects and contaminated clothing. The bacteria also readily enter plants through wounds or sites damaged by slugs, snails or insects.

Control: avoid overhead watering if bacterial infections are present. Spread plants apart to avoid contact and improve aeration. Spray affected plants with copper oxychloride (note that many dendrobiums can be damaged or even killed by copper sprays).

Bacterial brown spot on leaf of Dendrobium kingianum. [D. JONES]

VIRUS DISEASES

Parasitic viruses that live within plant cells are the most destructive type of plant pathogen known because the plants have no defence against viral attack. Composed of a nucleus surrounded by a protein sheath, they are minute particles that can only be seen with the aid of an electron microscope. Once inside a plant cell the virus takes over the cell nucleus to divert the host's metabolism into producing more virus particles that invade surrounding cells. Substantial plant damage can occur if the virus particles are present in sufficient numbers or in combinations of different virus species. Many of the bad plant viruses are of exotic origin but modern studies are revealing a suite of newly identified viruses in natural populations of native plants.

Symptoms: typical virus symptoms include irregular mosaic patterns and black streaks on leaves, distortion of young shoots and leaves, colour breaks in flowers, and reduced size and rosetting of leaves. Virus attacks stunt plant growth and severe attacks can kill the plant. Sucking insects such as aphids, leafhoppers, jassids, mites, etc, commonly transmit viruses. Even some cockroaches can transmit viruses in the course of normal feeding. They can also be transferred in infected sap on implements such as knives and secateurs. Some viruses can even be transmitted via plant to plant contact. Plants raised from seed are generally free of viruses, although a few viruses are known to be seed-transmitted.

EXAMPLES

- **Hardenbergia Mosaic Virus.** Plants of *Hardenbergia comptoniana* infected by this virus have been found in natural populations in coastal areas of south-western Western Australia. The eastern species, *H. violacea*, is also susceptible to the virus which first induces mosaic symptoms in young leaves. Symptoms of severe damage include distorted, puckered and blistered leaves.

Pterostylis obtusa *infected by Tospovirus.* [M. Clements]   *Symptoms of Potyvirus in* Pterostylis. [D. Jones]

- **Pterostylis Blotch Tospovirus.** Wild collected plants of several species of *Pterostylis* showed irregular leaf blotching consisting of dark green and lighter green patches. Six species of *Pterostylis* were tested from locations in New South Wales, Australian Capital Territory and Victoria.

- **Pterostylis Potyvirus.** This virus has been isolated from five genera of Australian terrestrial orchids – *Chiloglottis, Corysanthes, Diuris, Eriochilus* and *Pterostylis.* Symptoms are mainly light and dark green blotching in the leaves.

- **Yellow Tailflower Mild Mottle Virus.** Some populations of a common Western Australian coastal plant, the Yellow Tailflower, *Anthocercis littorea*, have been found to be infected by this virus. It probably originated in the Australian flora, and was first identified in natural populations north of Perth. Symptoms include stem dieback and mottled leaves. This virus is readily spread by plant to plant contact, touching and on tools.

- **Caladenia Virus.** Cultivated plants of three species of terrestrial orchid native to Western Australia were found to be infected with this potyvirus. The orchids are *Caladenia latifolia, C. arenicola* and *Drakaea elastica*. Symptoms were not recorded but mites are a possible vector.

- **Donkey Orchid Symptomless Virus.** A few plants in natural populations of *Diuris longifolia* and *Caladenia latifolia* in Western Australia were found to be infected with this virus which did not produce symptoms in the plants.

Other virus examples

- Routine studies have located viruses in several native plants but the majority of species have never been tested. Some native grasses, including Windmill Grass (*Chloris truncata*), Wallaby Grass (*Danthonia tenuior)*, *Eragrostis benthamii*, and Kangaroo Grass (*Themeda triandra*) have been found to be infected by Barley Yellow Dwarf Virus which causes stunting and leaf yellowing.

- At least four species of virus are known to infect epiphytic native orchids including *Cymbidium madidum* (infected by Cymbidium Mosaic Virus), *Dendrobium bigibbum* (Cucumber Mosaic Virus) and *D. kingianum* (Orchid Fleck Rhabdovirus).

- Among legumes, three species of *Glycine* (*G. clandestina, G. tabacina* and *G. tomentella*), *Indigofera australis, Kennedia coccinea* and *K. rubicunda* have been found infected with various mottle and mosaic viruses.

- Bean Yellow Mosaic Virus has been isolated from several species of *Diuris* and *Pterostylis* in eastern Australia.

Symptoms of Cymbidium Mosaic Virus in Cymbidium madidum *leaf.* [J. Fanning]

- A potyvirus has been isolated from plants of *Diuris orientis* in a natural habitat in western Victoria.

- Plants of Sturt's Desert Pea, *Clianthus formosus*, have been found infected with both Bean Yellow Mosaic Virus and Tomato Spotted Wilt Virus.

- The native banana *Musa banksii* has been found infected with Bunchytop Virus.

- *Passiflora aurantia* is commonly infected with Woodiness Virus and similar symptoms have been observed in another native passionfruit, *P. cinnabarina*.

- Other native plants known to be infected by various viruses include *Alocasia brisbanensis*, *Atriplex suberecta*, *Ipomoea plebeia* and *Nicotiana velutina*.

Control: plants suspected of being virus-infected should be destroyed by burning as there is no successful treatment available. The effects of a virus are promulgated by vegetative propagation. Cuttings, divisions or scions for grafting should only be taken from healthy vigorous plants and never from unthrifty plants with distorted or mosaic-patterned leaves. Aphids, jassids, leafhoppers, thrips, mites and other possible vectors of virus transmission should be controlled when noticed. Secateurs and knives should be sterilised if they have been used to cut plants suspected of being virus infected. Precleaning the knives or secateurs and soaking for 10 minutes in a saturated solution of Trisodium sulphate is effective (add the compound to warm water until undissolved crystals remain). Soaking for 10 minutes in 70 per cent alcohol is also effective. Wash hands with soap and water after handling plants.

WITCHES' BROOMS

Witches' brooms, although included here with diseases, are unusual growth abnormalities or deformities in a woody shrub or tree. Basically the natural branching is changed such that a dense patch of shoots arise from a single point. This creates a mass of growth that is startlingly different from the rest of the plant and resembles a broom, bird's nest or possum's nest. These growths are often prominent near the end of a branch and sometimes develop into unusual elongated hanging structures. They can also be found lower down on the main branches and on the trunk.

Witches' brooms can be caused by range of organisms including fungi (*Uromycladium*, for example), phytoplasmas (bacteria-like organisms), eriophyid mites, armoured scales, nematodes and aphids. They are relatively common but little appears to be known about the cause and development of witches' brooms, especially those associated with native plants. Genera prone to witches' broom development

Witches' broom on the trunk of Acacia terminalis. [D. JONES]

Witches' broom on Banksia. [S. JONES]

Small witches' broom on Acacia melanoxylon. [R. ELLIOT]

include *Acacia* (*Acacia saligna* caused by phytoplasma, *Acacia terminalis* cause unknown), *Allocasuarina* (*A. fraseriana* caused by phytoplasma, *A. littoralis* possibly caused by mites), *Banksia* (possibly caused by mites), *Eucalyptus* (especially *E. blakelyi*, *E. bosistoana*, *E. microcarpa*, *E. obliqua*, all caused by armoured scales), *Exocarpos cupressiformis* (possibly caused by mites) and some species of *Leptospermum* (*L. laevigatum*, *L. myrsinoides* and *L. myrtifolium*, cause unknown).

Control: If the witches' brooms are seen to be a problem they are best removed and destroyed.

Witches' broom on the end of a stem of Allocasuarina littoralis.
[D. Jones]

Witches' broom on Kunzea. [S. Jones]

Stem bunching and little-leaf on Acacia melanoxylon, *possibly caused by a phytoplasma.* [D. Jones]

Stem bunching and little-leaf on eucalypt, possibly caused by a phytoplasma. [D. Jones]

INTERACTION OF DISEASE WITH THE WEATHER

The damage incurred by plants during storms can be much worse in diseased plants compared to healthy plants. Fungus diseases enter through wounds and other damaged sites on the branches and trunk causing wood decay and weakening the structure of the tree. Water and plant debris also collects in crotches on the trunk and sunken areas where large branches join, eventually leading to the entry of fungi and wood decay. Strong winds impact on these weakened areas, often dramatically. In a similar way pathogenic soil fungi can invade the plant's roots system affecting its anchorage in the soil.

This steep crotch angle is a weak point in this eucalypt. [D. JONES]

One branch has been lost to a northerly wind; note the decayed wood. [D. JONES]

The second branch has been lost to a southerly wind. [D. JONES]

Storm damage in this Banksia integrifolia *was accentuated by fungus infection low down on the trunk.* [D. JONES]

Wind-damaged Grevillea. *Note the contrast between the area of fungus-infected wood and the healthy green wood.* [D. JONES]

CHAPTER 20

NUTRITIONAL DISORDERS OF NATIVE PLANTS

PLANT NUTRITION

Plants extract different elements from the soil so that they can grow, flower and produce fruit. Most of these elements are essential for normal plant growth. Some elements are required in large quantities and are known as the major elements, while others are required in small quantities and are known as the minor or trace elements. Both types of elements are listed in the accompanying table.

MAJOR ELEMENTS	MINOR ELEMENTS
Nitrogen	Iron
Phosphorus	Manganese
Potassium	Boron
Magnesium	Zinc
Calcium	Copper
Sulphur	Molybdenum
	Chlorine
	Cobalt

Plants take up nutrients depending on their availability in the soil. These elements are generally available from soil in sufficient quantities for normal plant growth. Soil reserves, however, are depleted by cropping or leaching following heavy rains and it is necessary to build up levels again by adding organic material and fertilisers. Some soils may be deficient in a particular element due to the chemistry of the rock from which the soil originated. The absence of an element from soil, its immobilisation, or its excessive release by chemical factors within the soil, affects the growth of plants and causes symptoms of deficiency. Plants grow better if a balance of nutrients is available to them in the soil rather than if there is an excess of one particular element. The nutrients interact within the plant and an imbalance can upset the plant's growth.

SOIL PH

The pH unit is a logarithmic function of the concentration of hydrogen ions in the soil. Put more simply it is a measure of the acidity or alkalinity of a soil. The pH scale ranges from 0 to 14 with 7 being neutral, 14 very alkaline and 1 very acid. Soils commonly range from pH 4 to 9. The figure below illustrates the pH scale.

The measure of pH in a soil is of critical importance to plant growers for three reasons.

1. Plants have pH preferences, some being suited to grow in acid soils, others preferring alkaline soils. Some plants have a very narrow pH tolerance whilst others grow in a wider range.

2. Extremes of pH can indicate a deficiency or excess of some of the major and minor elements. In very acid soils phosphorus and molybdenum can become deficient whereas in alkaline soils sodium or calcium is often in excess.

3. Interactions occur between all elements and soil pH. Thus under very acid conditions aluminium and manganese may reach high levels which are poisonous to plants. Similarly under very alkaline conditions trace elements such as iron and zinc may be immobilised to the extent that plants cannot obtain sufficient supplies for normal healthy growth.

The optimum soil pH at which most Australian plants like to grow is in the range 6 to 6.5.

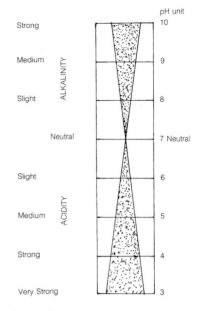

Soil pH Scale. [T. Blake]

NUTRIENT DEFICIENCIES

Nutritional disorders in plants are not uncommon. Each plant species produces a characteristic set of symptoms when growing in soils low or deficient in an element. These symptoms are related to the use of the element within the plant's structure and chemistry. Such deficiencies have been studied for many years and the basic response of plants to low levels of each element is known, although the response varies to some extent with each species. Excessive levels of some elements also induce characteristic growth responses (see below).

SYMPTOMS OF NUTRIENT DEFICIENCY

Young leaves affected

Distortion, blackening, early shedding.. *calcium*
Wilting, leaf thickening and distortion, early shedding, dieback.. *boron*
Pale-green to yellow leaves, reduced size, stunting... *sulphur*
Inter-veinal yellowing, veins remain green, necrosis... *iron*
Tips wilt and die, leaves grey to bluish-green... *copper*
Reduced leaf size, shortened internodes, yellow mottling between veins............................. *zinc*
Chlorotic patches on leaves, veins green, leaf cupping, necrotic spots................................ *manganese*

Old leaves affected

Leaves uniform pale-yellow, growth retarded... *nitrogen*
Leaf mottling and curling, lack of nodules (legumes)... *molybdenum*
Deep-green to purple-green leaves, stunting... *phosphorus*
Yellow to brown margins on leaves, some spotting... *potassium*
Interveinal yellowing with dark-green veins, often very green around midrib, sometimes with reddish tinges........
.... *magnesium*

CORRECTION OF DEFICIENCIES

Most deficiencies are readily corrected by adding the deficient element to the soil in the correct quantity. Speed of response by the plant will be variable depending on factors such as the element involved, the temperature and time of the year, and how deficient and stunted the plant became before the missing element was added.

Occasionally adding the element to the soil may not give the desired response. This is common in alkaline soils where elements such as zinc are tied up by the high calcium levels and are therefore unavailable for plant growth. Any added zinc is also tied up in the same way. To overcome this problem zinc is best applied as a liquid solution to the foliage. Similar problems may occur when phosphorus and molybdenum are added to very acid soils.

A single deficient element can completely stop plant growth or cause stunted growth or distortion of stems and leaves. **The only way to restore normal growth is to add that missing element. Increasing the levels of other elements will bring no response until the deficit of the missing element is corrected.**

NUTRIENT TOXICITIES

Some elements are toxic to plants when present in the soil in excess quantities. Usually an excess is due to an imbalance such as can occur at extremely low soil pH (very acid) or high soil pH (very alkaline). Interaction between some elements may also induce a toxicity. Nutrient toxicities produce characteristic growth symptoms and these are outlined where known under each of the elements in the following pages.

Correction of Toxicities

Toxicities are more difficult to correct than are deficiencies. Often it is necessary to change the pH of the soil or else increase the level of some other element with which the toxic element interacts. Toxicities are fortunately rare in soils and are usually caused by overuse of fertilisers or lime.

NATIVE PLANT NUTRITION

The nutrition of native plants is a relatively new subject and research is revealing how some species respond to different nutrient levels. As a general rule most plants have basic similarities in their behaviour to nutrient disorders and thus symptoms for crop plants can be used as a guide for other plants. The following deficiency symptoms are based on those for food crops but have been modified where the authors have had experience with a deficiency in native plants. For convenience the elements are dealt with in alphabetical order.

DETAILS OF EACH ELEMENT

Boron

Boron is important in actively growing areas such as meristems and root tips and is vital for cell wall formation, protein synthesis, sugar transport and the development of reproductive structures. It is related to the use of calcium and potassium. Boron occurs in the plant as both an insoluble and a water-soluble form. Boron is immobile within plants.
Deficiency symptoms: wilting and defoliation of the upper parts of shoots followed by death of the terminal bud and dieback of the shoots. Leaves may become thickened and

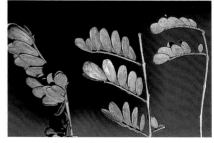

Boron deficiency in Acacia spectabilis. [D. Beardsell]

lateral buds often develop.

Correction: boron compounds such as borax or boric acid applied at low concentrations to the soil (1–4g per 10 square metres) or as a foliar spray (boric acid at 1g per litre of water).

Remarks: deficiency symptoms of boron are extremely variable depending on the species. Wattles (*Acacia* spp.) may be badly affected, particularly *A. adunca* and *A. spectabilis*. Only small amounts are required by plants and excessive quantities are very toxic.

Calcium

This element is important for the strength and structure of cell walls and proteins, in the formation of flowers and in a number of chemical reactions. It is also very important for healthy root growth. Calcium interacts with many other elements within the plant.

Deficiency symptoms: Partial or complete failure of terminal buds. Young leaves, if they develop, are distorted with necrotic areas and sometimes with hooked tips. Dieback of twigs and stunting occurs. The root system is greatly reduced due to death of the growing points.

Correction: by the addition of lime, dolomite or gypsum to the soil at 1–3 kg per 10 square metres. The amount needed will vary according to the soil pH level. Lime is the most effective agent on very acid soil since it raises the pH more significantly than dolomite. Gypsum supplies calcium but does not alter the soil pH.

Remarks: calcium deficiency is most common in very acid soils. Calcium uptake is also related to transpiration and factors such as water shortage or excessive cold spells reduce transpiration and calcium uptake.

Chlorine

Chlorine is needed for photosynthesis and iron balance within the plant. There has been very little research on impact of chlorine deficiency on native plants, although it is known that excess of the chloride ion in soils is toxic.

Copper

Copper is required in small quantities for cell wall metabolism, photosynthesis, respiration and the function of important enzyme systems. It is concentrated in green leaves.

Deficiency symptoms: stunted growth associated with wilting and dieback of shoot tips. Leaves of legumes show a greyish or bluish colouration. Leaves are often pale green and collapse easily. Gum may be exuded from the bark.

Correction: copper sulphate ($CuSO_4$) applied to the soil (12g per 10 square metres) or as a foliar spray of copper oxychloride (0.5g per litre).

Remarks: copper can be very toxic if available to the plant in appreciable quantities. Care must be taken not to over-correct this deficiency. Also the overuse of copper-based fungicides can cause localised problems.

Iron

Iron is needed in continuous small quantities for the proteins used in chlorophyll synthesis. Iron is very mobile within plants.

Deficiency symptoms: earliest symptoms are pale-green new growth. This is followed by chlorosis of the leaf areas between the veins which remain green, giving the leaves a mottled appearance. In severe cases all of the green may disappear from the leaves which become yellow or white and die and this is followed by dieback of the stems. In some species, such as *Hypocalymma cordifolium,* the deficient new growth may be pinkish instead of pale-green. Iron deficiency can be confused with manganese deficiency.

Correction: by applying iron salts to soils or by foliar applications. Research has shown that soil drenches are more effective than foliar applications. The fertiliser used depends on the soil pH. Ferric sulphate is used only for soil applications (0.5kg per 10 square metres). It is only useful in the pH range 5–7 and acidifies the soil. Iron chelates are effective in the pH range 6.5–8.5 and are useful for correcting iron deficiency in calcareous soils. Soil drench applications are usually between 1–50 litres of 20g per litre concentration per plant depending on the plant's size. Foliar sprays at the same concentration can be effective, but up to three sprays may be necessary to completely correct the deficiency.

Remarks: iron is rarely at deficient levels in soils but is frequently held in a form unavailable to plants. Iron is commonly unavailable to plants in soils with a pH above 6.5. The situation is even worse in calcareous soils where the deficiency known as 'lime-induced iron chlorosis' is very common.

Iron deficiency in Pomaderris. [D. Jones]

Early stages of iron deficiency in Calothamnus quadrifidus. [D. Jones]

Iron deficiency in Banksia ericifolia. [D. Jones]

Severe iron deficiency in Banksia ericifolia. [D. Jones]

Iron deficiency in Banksia spinulosa. [D. Jones]

Severe iron deficiency in Banksia spinulosa. [D. Jones]

Iron deficiency in Banksia integrifolia.
[D. BEARDSELL]

Iron deficiency in Banksia robur.
[D. BEARDSELL]

Iron deficiency in Eucalyptus gunnii.
[D. BEARDSELL]

Magnesium

Magnesium is an important constituent of chlorophyll and is used in photosynthesis. It is also related to the use of phosphorus in the plant. Magnesium is highly mobile within plants.

Deficiency symptoms: usually show up first on older leaves because the magnesium is transferred into new leaves and developing fruit. Parts of the leaves become chlorotic with the veins, especially the midrib, remaining green. A green band or V-shape is often conspicuous along the midrib. In severe cases the whole leaf becomes pale-yellow and eventually necrotic. Old leaves may become reddish and are shed. Root growth is poor and flowering is adversely affected.

Correction: magnesium sulphate ($MgSO_4$ – Epsom Salts) applied to the soil (50g per 10 square metres) or as a foliar spray (15–20g per litre). Foliar sprays work rapidly but are not long lasting and both soil and foliage treatments may be necessary. Magnesium carbonate and dolomite can also be applied to the soil. Both however will raise soil pH (magnesium sulphate has a negligible effect on pH).

Remarks: magnesium is most commonly deficient in very acid sandy soils. High levels of potassium can also aggravate its deficiency. Magnesium is also deficient in some clay soils in the tropics, for example around Darwin.

Magnesium deficiency in
Hymenosporum flavum. [D. JONES]

Magnesium deficiency in Banksia
spinulosa. [D. JONES]

Magnesium deficiency in Banksia
spinulosa – *note green around leaf
midribs.* [D. JONES]

Magnesium or iron deficiency in Corymbia. [D. Jones]

Magnesium deficiency in Garcinia dulcis. [D. Beardsell]

Magnesium deficiency in Hypocalymma. [D. Beardsell]

Manganese

Manganese is necessary in photosynthesis, nitrogen metabolism and the function of important enzyme systems concerned with oxidation and the formation of sugars and chlorophyll. It also has an important interaction on the solubility of iron. It is concentrated in the green tissue of plants. Manganese is immobile in plants.

Deficiency symptoms: usually first noticeable on recently mature leaves. Chlorotic patches occur on the leaves with the veins remaining green. The symptoms are very similar to iron deficiency but often associated with cupping of the leaves and with brown flecks or patches on the chlorotic areas. In severe cases the leaves become entirely yellow with necrotic patches. Premature leaf-fall is common.

Correction: manganese sulphate ($MnSo_4$) applied to the soil (50–150g per 10 square metres) or as a foliar spray (2g per litre).

Remarks: manganese deficiency is common in calcareous soils rich in organic matter. It also occurs in poorly drained soils. The excessive use of lime is a common cause of manganese deficiency.

Manganese excess: manganese becomes available in large quantities in very acid soils and can be toxic to plant growth.

Correction: applications of lime to reduce acidity. A pH of 6–6.5 is optimum for manganese availability.

Manganese deficiency in Hymenosporum flavum. [D. Beardsell]

Molybdenum

Molybdenum is needed for the formation of nodules in legumes and in enzymes used in nitrogen nutrition. Molybdenum is also important in sulphur metabolism within the plant. Molybdenum is mobile in plants.

Deficiency symptoms: lack of nodules on the roots of legumes. A slight mottling of the leaves associated with yellowing and curling of the leaf margins. Sometimes the leaves of affected plants develop abnormally.

Correction: sodium molybdate (Na_2MoO_4) applied lightly to the soil (1g per 10 square metres) or as a foliar spray (0.05g per litre).

Remarks: deficiency is most common in very acid soils and liming of such soils will be necessary to prevent added molybdenum becoming unavailable to plants. It is only required in minute amounts.

Nitrogen

This element is important in the formation of chlorophyll which is essential for photosynthesis. It is also used in the formation of amino acids which are the building blocks of proteins. Nitrogen encourages vegetative growth and produces a lush greenness in foliage.

Deficiency symptoms: usually appear on the older leaves first. The first sign is often the development of purplish margins, particularly in winter. Leaves then become a uniform pale-yellow and the plants are stunted. Younger leaves may be green. In severe cases the leaves become severely bleached (often whitish) and develop necrotic patches which spread until the whole leaf dies.

Correction: soil applications of either inorganic fertilisers or organic fertilisers. Inorganic fertilisers dissolve readily and are fast acting. They include ammonium nitrate, ammonium sulphate, urea, potassium nitrate, calcium nitrate and sodium nitrate. Organic fertilisers such as blood and bone and animal manures are also rich in nitrogen but are generally slow-acting. For rates of nitrogen fertiliser see the accompanying table. Foliar sprays of calcium nitrate (2g per litre) or urea (1.5g per litre) are also effective. Liquid fertilisers, such as ammonium nitrate (10g per litre) or calcium nitrate (12g per litre), can also be applied to plants in the ground. For plants in containers ammonium nitrate applied at 1g per litre is adequate.

Remarks: nitrogen is extremely important for plant growth and adequate supplies are characterised by deep-green leaves and vigorous healthy growth. Nitrogen is not commonly deficient in native plants.

Nitrogen excess: nitrogen is readily supplied in excess, the symptoms being lush soft growth and a retardation or inhibition of flowering. Excess nitrogen increases a plant's susceptibility to disease, wilting, death of roots and frost damage. The stems of young plants may become thin, sappy and weak. Excess ammonium can build up in the soil during cold or wet weather and cause plant damage including leaf curling and patches of dead tissue. Plants can die readily from excess ammonium.

Correction: nitrogen is readily leached from the soil by heavy watering. An application of phosphorus can help counteract the effects of excess nitrogen.

Foxtail Palms, Wodyetia bifurcata, *severely deficient in nitrogen.* [D. JONES]

Severe nitrogen deficiency in Brachyscome multifida. [R. ELLIOT]

Nitrogen deficiency in Lagunaria patersonii. [D. BEARDSELL]

NITROGEN FERTILISERS AND THEIR RATES OF APPLICATION

FERTILISER	NITROGEN %	AVAILABILITY	APPLICATION RATE kg/10m²	COMMENTS
Blood and bone	7	slow	1–1.5	slow release
Hoof and horn	12–14	slow	1–1.5	slow release
Dried blood	12–13	slow	1–1.5	slow release
Ammonium nitrate	35 15	rapid	0.25	rapidly available
Calcium nitrate	20 13	rapid	0.5	rapidly available
Ammonium sulphate		rapid	0.5	acidifies soil
Potassium nitrate		rapid	0.5	also supplies potassium
Sodium nitrate	15	rapid	0.5	useful in winter
Urea	46	rapid	0.15	use carefully

Phosphorus

Phosphorus plays an important role in the storage and use of energy in chemical reactions involved in respiration and photosynthesis. It is also particularly important for root growth, stem strength and the development of flowers, fruits and seeds. Phosphorus is very mobile within plants.

Deficiency symptoms: stunting and development of deep-green leaves sometimes associated with purplish or bronze tinges in the leaves and stems. The root system is also greatly reduced.

Correction: soil applications of superphosphate, triple superphosphate, or soluble orthophosphate compounds. Superphosphate is the most commonly available form of phosphorus fertiliser. It contains 20 per cent phosphorus and is used at the rate of 0.3–1kg per 10 square metres. Triple superphosphate is a concentrated form which contains 50–60 per cent phosphorus. It is applied at about one third the rate for superphosphate. Monocalcium phosphate is expensive but is very soluble and useful where rapid uptake is needed. Organic manures such as hoof and horn, bone meal or blood and bone are rich in phosphorus. Phosphate fertilisers work best if very finely ground.

Remarks: heathland communities have developed on phosphorus-deficient soils and have evolved a very tight phosphorus cycle so that the element is not wasted. The element is usually withdrawn from all old leaves before they are shed and specialised root systems (such as proteoid roots) play a significant role in the uptake of phosphorus. Many commonly grown native plants respond well to the use of phosphorus fertilisers, but some can be sensitive to their use.

Phosphorus excess: phosphorus toxicity arises from excessive use of fertilisers containing phosphorus. Superphosphate and especially the concentrated phosphates must be used with care. Blood and bone is high in phosphorus and can cause phosphorus toxicity if used in excess. Phosphorus toxicity is mainly a problem of plants grown in containers and is not a common problem in garden-grown plants because the phosphorus is fixed and rendered immobile by soil. The proteoid roots found in species of Proteaceae are however, very efficient at extracting phosphorus from the soil and toxicity can occur in gardens established in vegetable growing areas where high levels of fertilisers and animal manure have been used over the years.

Excess quantities of phosphorus can be toxic to some native plants. Many of these phosphorus-sensitive plants originate in habitats where levels of soil phosphorus are extremely low and the plants

Phosphorus toxicity in Brachyscome. [R. ELLIOT]   *Phosphorus toxicity in* Macadamia integrifolia. [D. BEARDSELL]

have developed efficient mechanisms of phosphorus uptake. Plants sensitive to high levels of phosphorus are mainly found in the families Proteaceae, Rutaceae and Haemodoraceae. Great variability is shown even within these families however, and there are many species that respond with normal growth following applications of phosphorus. Some members of the Fabaceae, Mimosaceae and Myrtaceae are also sensitive to high levels of phosphorus.

Toxicity symptoms include poor growth, chlorosis with green veins in young leaves, red or purple colouration in older leaves which often fall prematurely, marginal leaf burn and tip dieback on older leaves. Tip dieback spreads towards the leaf base and is followed by premature defoliation giving the plants a sparse appearance. The toxic effects of high levels of phosphorus are made worse by high levels of calcium and can be offset by high levels of nitrogen and/or iron.

Correction: phosphorus toxicity is difficult to correct and often results in the death of plants. It can be offset to some extent by increasing the levels of nitrogen and iron available to plants. Do not increase calcium as this element makes the toxicity worse.

Potassium

Potassium is essential for the production and transport of carbohydrates (especially starch) in plants. It strengthens stems and roots and builds up resistance to disease. It is also important in chlorophyll formation, respiration and root development. Potassium is mobile within plants.

Deficiency symptoms: these usually appear first on older leaves which develop a marginal necrosis or burn that spreads towards the tip. Young leaves can become reddish and red patches can occur on the underside of leaves. Eventually the whole leaf can become necrotic. Root growth is also impaired.

Correction: soil applications of potassium sulphate (K_2SO_4) or potassium chloride (KC1). Both can be applied at 50–100g per 10 square metres. Potassium sulphate is preferable for soils with high chloride levels. Potassium is also found in seaweed (25 per cent) and ash (3–10 per cent). Seaweed contains high levels of sodium chloride and may need leaching before use.

Remarks: potassium is required in large quantities by plants. It is usually present in soils in sufficient quantities but is not always readily available to plants. It can be deficient in sandy soils. Excessive levels of potassium can induce magnesium deficiency.

Potassium deficiency in Banksia integrifolia. [D. Beardsell] *Potassium deficiency in palm.* [D. Jones]

Sulphur

Sulphur is a constituent of many important chemicals such as amino acids. It is also important for chlorophyll synthesis, nitrogen metabolism and root formation. Sulphur is not mobile within plants.

Deficiency symptoms: retardation and stunting of growth. Uniform pale green chlorosis of all leaves except those which are very young. Sometimes the veins show up pale in an otherwise green leaf. Leaf margins can also be rolled under.

Correction: soil applications of elemental sulphur or gypsum (1–3kg per 10 square metres). Elemental sulphur increases soil acidity which can be useful in calcareous soils whereas gypsum does not affect pH.

Remarks: sulphur is not commonly deficient or toxic. On very rare occasions an excess of sulphur can lead to symptoms similar to those of nitrogen deficiency.

Zinc

Zinc acts as a catalyst for many chemical reactions in plants, particularly those leading to the production of auxin, the major plant hormone responsible for normal stem and leaf growth. Zinc is immobile in plants.

Deficiency symptoms: appear on young leaves first. Irregular yellowish or chlorotic areas between the veins, stunting of growth and shortened internodes giving a rosetted appearance. Leaves are often greatly reduced in size (termed 'little-leaf') and more crowded imparting a bunched appearance. Leaves become pale and the older leaves brown and drop off. Affected stems die back from the tips.

Correction: zinc sulphate ($ZnSO_4$) applied to the soil (12g per 10 square metres) or as a foliar spray (0.5g per litre).

Remarks: zinc is a complex element and deficiency can occur under a wide variety of conditions. It is required in minute quantities and excess levels are toxic to plants. Zinc deficiency is common in calcareous soils. An excess of zinc can lead to stunted growth and may also cause a deficiency in iron.

CHAPTER 21

OTHER AILMENTS AFFECTING NATIVE PLANTS

Some factors in the plant's environment, other than those dealt with in the previous chapters, may adversely affect the growth of native plants. These factors can range from physical injuries induced by adverse climatic conditions to chemical poisoning, environmental changes and unusual growth features. The symptoms vary correspondingly.

CLIMATE-INDUCED AILMENTS

Cold

Leaf of Tecomanthe *responding to cold weather by changing colour.* [D. Jones]

Many plants show a reaction to cold which is not necessarily associated with frost damage. These plants often originate from subtropical or mild coastal districts where they are not usually exposed to much cold weather. Examples are *Grevillea* x *gaudichaudii*, *Buckinghamia celsissima* and *Omalanthus populifolius*. Typical symptoms are a darkening of the leaves, frequently with black patches appearing on the margins. New growth during the warm weather is quite normal. On some species the leaves redden or have bluish tones following periods of cold.

Dryness

Wilting is the most common symptom of dryness and is an indication that the roots cannot supply water fast enough to the above-ground parts of the plant to keep up with transpiration losses from the leaves. Wilting is most common in dry ground and is obvious on plants with soft, young growth. Temporary wilting however, can also occur, even in moist or wet soil. Temporary wilting is usually caused when there is a sudden increase in the demand for water which the plant cannot cope with in the short term. It is most often seen in vigorously growing plants on days of intermittent cloud and hot sun. It is also common on potted plants that have just been moved from a shady or protected position into the sun, and it also happens after the root system has been damaged.

Sudden wilting often occurs on very hot days. [D. Jones]

Plants which are not in vigorous growth react differently to dryness. The leaves on plants with tough foliage, such as species of *Dryandra*, *Hakea*, *Grevillea* and *Banksia*, lose their lustre, take on a dry appearance and often curl inwards towards the stem. The older leaves often turn yellow and fall off prematurely following a dry spell. These plants frequently do not recover from dry periods, especially if a significant portion of the root system dries out. In shallow rooted plants, such as *Brachyscome*, *Lechenaultia*, *Goodenia* and some *Scaevola*, the leaf margins curl inwards, the leaves appear dull and dry, and fall prematurely. Affected plants look sparser than normal. In large-leafed plants it is not uncommon for the leaves to droop downwards against the stems, followed by the older leaves falling prematurely after becoming yellow.

Fog

Fogs, especially those lasting more than a day, cause plants to sweat because the stems and leaves are continually moist and transpiration is restricted. Sweating is worst in plants with dense foliage, compact growth habit and hairy leaves. Groundcovering plants are commonly affected and grey-foliaged plants can also suffer badly. Leaf drop is common after the leaves turn brown or black and attacks by grey mould can be severe on affected plants. Species of *Eremophila* and *Verticordia* are very prone to damage from sweating. Sweating is discussed further below.

Frost

Frost is a common plant killer. During frosts plant tissue is frozen and as the tissue in plant cells freezes it contracts and expels water into spaces between the cells. This water freezes to form ice crystals. When thawing begins, the ice melts and the water is absorbed back into the cells by osmosis. If thawing occurs quickly there is little or no damage to the leaves, but if the ice melts slowly the cells collapse and the leaf tissue 'burns' from desiccation. This damage can be aggravated by the fact that the atmosphere at or near ground level is dry, due to all moisture being frozen. Frost damage thus occurs when plants are thawing, not when freezing is taking place. Frost damage can be worse if rapid thawing takes place which can occur when the sun strikes frozen plant tissue. This damage can be lessened by spraying plants with water, near sunrise (a technique used successfully in commercial cut flower plantations and fruit orchards). Plants can also suffer damage to their roots when the soil becomes frozen during heavy frosts. Damaged roots die and the plants suffer from a reduced ability to take up water. Frosts can also damage buds, impacting on flowering and fruit set.

Frost damage is more severe on young soft growth than on mature hardened growth. Frost hardiness is gained in plants as their growth slows down in autumn, due to colder weather and shorter hours of daylight. Early light frosts also harden foliage off which explains why heavy frosts in late autumn/ early winter cause so much damage. Hardened plants may suffer little or no frost damage at -1 to -2°C, whereas damage can be severe at temperatures of -5°C or below. Unseasonal frosts can also cause significant plant damage. Frost damage mainly occurs at or near ground level, decreasing in severity with height depending on the temperature attained. Temperatures at ground level can be up to 5°C lower than 1 or 2m above. Trees that are frost tender when young, do not become frost tolerant as they mature, but their increased size enables them to cope without suffering extreme damage, unless exposed to very heavy frosts.

Frost damage commonly shows up as wilted growth and which becomes watery and then blackens. New growth can be distorted.

Frost-damaged fronds of Cyathea cooperi. [D. JONES]

Badly frost-damaged tree fern. [D. JONES]

Frost-damaged young growth of
Eleaocarpus eumundi. [D. Jones]

Frost damage can be reduced or minimised by the following actions:

- Choose frost-tolerant plants. There has been limited study into the effect of frost on Australian native plants. Species from alpine, mountainous and inland regions are usually quite frost hardy and those from high elevations in tropical and subtropical regions have also proved to be relatively frost tolerant. Frost tolerance can also vary within a species, especially those with a wide geographical distribution. For example, seedlings from coastal populations may be prone to frost damage, whereas those propagated from inland provenances are more tolerant.

- Look for microclimates where frost damage is lessened. Avoid low-lying areas which can trap frost, plant on slopes where there is good air drainage and make use of the protection afforded by larger shrubs and trees. Frost tender plants can be grown against buildings, with eaves offering protection, and the walls, especially if brick or some other solid construction, will radiate warmth when the temperatures drop. Solid fences or walls restrict air flow and can create frost pockets.

- Do not use nitrogenous fertilisers later than in mid-summer. This avoids late growth flushes and allows plant tissues to harden. Plants with young growth during autumn and winter are prone to frost damage.

- Keep the ground around plants clear of weeds.

- Keep the ground moist.

- Do not cultivate around plants. It is best to have a hard, flat surface.

- Do not mulch with organic material. The best mulch to use in frost prone regions is coarse sand or screenings.

- Covering plants with material such as hessian or frost cloth can help to retain the radiated ground heat around plants. Do not cover plants with plastic bags, as they do not always exclude frost, also when the sun reaches the plant, hot air will be trapped within the bag which will promote rapid thawing.

- Avoid pruning frost-damaged plants until the danger of frosts has passed. Premature pruning can stimulate new growth, which is soft and susceptible to late frosts.

Hail

The impact of hail causes obvious physical damage to plants, but accumulations of hail sitting on plants also result in damage from chilling or freezing. Physical damage caused by hail mostly shows up

Hail damage to Scaevola *caused by both impact and freezing; the plant died about a week later.* [D. Jones]

soon after the storm has passed, although damage to growth buds, flower buds and fruit may not show up for weeks or months when malformed or distorted growth appears. Fruit damage may not be obvious until maturity.

Sensitive plants, such as ferns, palms, orchids and gingers, can suffer 'freezer burn' if hail collects around the stems of young plants. Similar impacts occur with prostrate plants and groundcovers. It is not uncommon for such 'burnt' plants to die following a severe hail storm. The impact of hail is usually more severe during spring and summer, because the hailstones are commonly much larger than in cooler weather. Irregularly shaped hailstones and those with sharp edges cause severe damage to plants. Young plants or valued specimens can be protected by a temporary overhead covering of hessian or shade cloth; alternatively plants can be grown in the protection of a shadehouse. Hail-damage to the foliage of kangaroo paws, *Anigozanthos* species and cultivars, may lead to infection by ink-spot disease (see chapter 19).

Humidity and Drizzly Weather

Long periods of high humidity can cause damage to susceptible plants because they are unable to transpire efficiently. Long periods of misty or drizzly weather also cause similar problems. Plants affected worst by high humidity are usually species from dry inland areas where humidity is low for much of the year. Plants with hairy leaves, such as *Adenanthos sericeus*, *Banksia ornate* and some *Eremophila* species are often very badly affected. Symptoms include sweating, leaf-bronzing and premature leaf fall. The impacts of leaf-spotting diseases and grey mould are also exacerbated. By contrast long hot spells with very low humidity can cause desiccation.

Lightning

Hundreds of trees each year are struck by lightning in Australia. Very tall trees or isolated trees in open areas are most likely to be struck, but even single trees within a forest can be affected. Trees retaining large volumes of water are susceptible. The damage from a lightning strike includes minor bark burns, scorched areas, exploded patches of bark, expanded strips of bark, splits in the wood (including spiralling splits down the trunk), and complete shattering of the tree (or parts of the tree) with debris thrown out from the canopy. It is common for patches of leaves to turn brown and die within a few days of a strike. Most trees (especially eucalypts), die after a lightning strike. Even trees showing limited physical impacts commonly die, usually from unseen damage to the internal vascular system of the plant, and also from damage to the roots. Valuable trees can be protected by the installation of conducting rods. For information contact your local electricity authority.

Damage to eucalypt caused by lightning strike. [D. Jones]

Smog

Plant damage caused by photochemical smog is relatively common close to the larger capital cities. Smog contains several phytotoxic components, such as ozone, nitrogen dioxide, peroxyacetyl nitrate, sulphur dioxide and other oxidant gases. Sulphur dioxide and ozone, when combined, are more damaging to plant tissues than when acting separately. Most plant damage occurs to the mature leaves, whereas young leaves and very old leaves only show minimal damage. Visible features of smog damage are similar to that which is suffered from leaf parasites or in some cases mineral deficiencies. The symptoms vary from patches that appear to be wet (termed water-soaked) to dark necrotic spots or bleached areas in the leaves. Banding, silvering, flecking and stippling of older leaves can occur. It is usual for the upper leaf surface to show more damage than the undersurface. Affected leaves are shed prematurely. It is thought that smog-affected plants are more susceptible to attacks from insect pests and diseases such as root fungi.

The air-borne chemicals found in smog are toxic to some species of Australian plants. Generally, soft-leaved species are more sensitive than hard-leaved species. Three natives that have proved to be very sensitive are *Banksia marginata*, *B. spinulosa* and *Eucalyptus cernua*.

Snow

Snow actually protects plants from excessive cold because it acts as an insulator. However, it is also heavy and accumulations of snow cause breakage and other physical damage from its sheer weight. It can also influence the shape of plants since snow lying for long periods on a branch can change the angle on which it grows after the snow melts. Snow can also cause sweating on some plants if it lies on them for long periods.

Sun Scorch and Excessive Heat

Some plants are sensitive to exposure to hot sun, suffering sunburn and scorching if planted in an exposed situation. Affected leaves develop damaged areas which become papery and turn white. Young growths often wilt or even collapse completely. Shade-loving plants, which usually have large thin-textured leaves, are readily damaged by excessive hot sun. Sensitive species include aroids, gingers, orchids, lilies and ferns; also trees and shrubs that occur naturally in shady forests, especially rainforest. The damaging effects of sun are exaggerated by hot winds.

Sun damage can also occur when protection is suddenly removed. This is common following storms and sheltered plants exposed this way can scorch badly. Sun damage is also common when plants are removed from a sheltered verandah, glasshouse or greenhouse during hot weather without being

The orchid Corunastylis morina *damaged by very hot weather in natural habitat.* [D. Jones]

hardened-off by gradually increased exposure to sun over a few days. Nursery plants with soft foliage can also suffer damage when newly planted. Foliage damage to plants growing on verandahs, on or beneath decks and balconies can also occur as sun exposure changes with the season.

Sun damage can also occur to bark if it is suddenly exposed to sun after defoliation by insects or following storm damage. Damage can also occur to the smooth-barked trunks of advanced specimens of trees such as *Angophora costata*, *Corymbia citriodora* and *C. maculata* that were previously grown in a sheltered nursery site and then planted in an exposed position where the sun can shine directly onto the trunk. Bark scorch shows up as a colour change – commonly brown, orange or reddish patches. Bark can even be killed on species with thin bark. Protection can be given by applications of whitewash (lime and water) or loosely wrapping shade cloth around the exposed areas until the natural cover is returned.

Sweating damage to the lower growth of Melaleuca bracteata. [D. Jones]

Sweating

Sweating is a common problem in tropical gardens, although it can occur elsewhere following long wet spells. In the tropics it is most noticeable during the wet season when conditions of high rainfall, high humidity and high temperatures are prevalent. Windless days when there is little or no air movement exacerbates the problem. Sweating is caused by high levels of moisture and humidity adversely affecting the normal transpiration process. Moisture levels build up in the plant and the leaves suffer accordingly. Symptoms include a rapid discolouration of leaves which become brown to black. If leaves remain moist they may even go slimy. Effects are worse in parts of the plant where the growths and leaves are crowded together. Affected leaves fall prematurely, leaving patches of bare stems. Under severe circumstances it is even possible for the stems to rot. Leaves and stems damaged from sweating are susceptible to attack by secondary fungi such as moulds and *Alternaria*.

Some plants are more susceptible to sweating than others. In general species with crowded small leaves or a dense compact growth habit suffer badly. Plants with hairy leaves can also be damaged by this condition. Mat plants and ground covers are particularly susceptible to sweating. In severe cases the whole plant can blacken, collapse and die, although sometimes only part of a plant is affected. Badly affected plants however, may eventually die. Sweating is worse when plants are crowded together or when overgrown by weeds.

Wind

Wind causes obvious physical damage to plants, but it can also have other effects depending on prevailing wind strength, temperature and humidity.

- Scorching summer winds dry out plants. They cause leaf dehydration and wilting of young growth. Leaves can fall prematurely, even while still green. Plant deaths can occur after prolonged exposure.

- Strong wind during cold weather can have a chilling effect and retard plant growth.

- Turbulent winds, as experienced during sudden storms, can cause structural damage. Damage also occurs when strong winds arrive from a direction that is not commonly encountered.

- Strong winds in combination with wet soils often results in root movement and loose plants which will need staking. Staked plants must be tied tightly enough to prevent excessive movement but also loose enough to allow some movement which will help strengthen the trunk and stimulate new root growth.

- Sandblasting occurs when sand particles carried by strong winds physically impact on plants. This phenomenon, which can occur in coastal and inland areas, causes significant damage to bark, foliage, flowers and fruit. Using temporary screens of hessian or shadecloth provides some protection.

- Salt-laden winds severely restrict plant growth (see later entry on salt damage).

- Cyclones cause tremendous damage and can annihilate extensive areas of vegetation. Even large trees can be wrenched from the ground. Studies have shown that some trees are able to withstand cyclonic winds by bending with the wind rather than resisting (palms are well known as cyclone survivors). Cyclones also include extremes of rain; also salt-laden winds travel further inland than is customary, resulting in damage to plants that are not used to having salt deposited on the foliage.

Strong wind and a weak crotch caused this Acacia elata *to break.* [D. JONES]

The point of fracture, water and fungus has weakened the crotch. [D. JONES]

OTHER AILMENTS

Burls

These are abnormal growths that are produced on the trunks and larger branches of some trees, particularly species of *Eucalyptus*. They appear as localised swellings with a roughened warty or bumpy surface and range in size from quite small to very large protuberances. They are often hemispherical in shape, darker than the rest of the trunk (brown or black) and sometimes covered with coarse or

flaky bark. The wood of a burl, which is different from that of the tree trunk, is usually very hard with a twisted grain. Although of common occurrence, most trees lack burls or only support a few, and they produce no apparent impact on the health of the tree. It is noticeable however, that some individual trees can be badly affected, with the whole trunk supporting numerous outgrowths of various shapes and sizes, even on relatively new growth. The causative agent of a burl is unknown, however it is likely to be some form of infection within the surface of the bark. Disease-producing agents such as bacteria, fungi (especially moulds) or viruses seem to be the probable cause. Mechanical injuries and insect damage may also perhaps result in burl formation. Burls are of more interest to woodworkers than to gardeners. Because the reason behind burl formation is unknown, there is no means of control.

Burls on a eucalypt trunk. [D. Jones]

Eucalypt Dieback

Since the 1940s the baffling problem known as eucalypt dieback (also called gum tree decline and native tree decline) has become more prominent and is on the increase. Dieback reports are not new (they date back to the 1850s), however the problem seems to have increased greatly in recent years. Dieback not only occurs in forests but is also prominent along roadsides, in partially cleared grazing country and even affects isolated trees in paddocks (this impact is sometimes termed 'rural dieback').

There are several different types of dieback which are often included under the one title. In general,

trees affected by dieback have sparse foliage cover and prominent secondary growths (termed epicormic growths) on the larger branches. These secondary growths impart a crowded or bushy appearance in the crown. The main branches die back from the tips and when the secondary growths become defoliated or die the whole tree becomes very sparse. Dieback of the main branches and secondary growths continues and, while the trees may linger for many years, they eventually succumb and die.

Advanced stage of dieback. [D. Jones]

Last stages of dieback in Eucalyptus blakelyi. [D. Jones]

Distribution of Dieback

Dieback achieved considerable publicity in the 1950s because of its dramatic effects on the eucalypts of the New England Tablelands area of northern New South Wales. In some parts of this region it is not possible to see healthy eucalypts during kilometres of travel. Dieback however, is not confined to this region but occurs over most of mainland Australia and much of Tasmania. Dieback of trees in paddocks has become very prominent in recent years and is now known to be widely distributed in many parts of Australia, affecting numerous species.

Affected Species

Most species of eucalypts in the subgenus *Monocalyptus* are affected or susceptible to decline, some much worse than others. Very sensitive species include *Eucalyptus blakelyi, E. drepanophylla, E. grandis, E. macrorrhyncha, E. maculata, E. marginata. E. melliodora, E. nova-anglica, E. saligna* and *E. tereticornis.* Many species, previously considered resistant to dieback, became sensitive after extended periods of exposure to the contributing factors. Species other than eucalypts include Cypress Pine (*Callitris columellaris*), Brush Box (*Lophostemon confertus*), Turpentine (*Syncarpia glomulifera*) and River Oak (*Casuarina cunninghamiana*).

Contributing Factors

Dieback is rarely caused by a single organism (an exception is the virulent soil root-rotting fungus, *Phytophthora cinnamomi*). The common view seems to be that most decline involves a complex of factors which mesh together, each contributing to some degree. Some of these factors may have been involved for decades, even centuries. These include:

- Forest clearing associated with pasture improvement. Factors include planting introduced pasture species and the increased use of fertilisers and pesticides. Often only a few trees are retained after clearing for stock shelter. It has also been shown that lush pastures provide a better habitat for Christmas beetle larvae (the adult beetles are a major cause of defoliation) than natural grassland.

- Clearing associated with urban developments results in similar problems. Isolated trees on suburban blocks commonly exhibit symptoms of dieback.

- An upsurge in the activities of insects which concentrate on the isolated trees. A whole host of pests is involved including leaf-eating beetles, stick insects, caterpillars, lerps, leaf miners and borers. Some of these may cause secondary attack after decline has begun. In the New England Tableland, New South Wales, it has been shown that different species of beetles attack all stages of shoot growth from the first flush in spring to the final growth in autumn. This constant defoliation year after year is an important factor in the decline of trees in this area.

- A reduction of insectivorous birds. Heavy clearing vastly changes the natural habitat and isolated trees in paddocks are less attractive to birds than are forests where a whole regime of species is present. Trees that are in decline flower poorly (if at all) and this further reduces the attraction they have for birds.

- Increase in soil salinity. After clearing there is often an increase in the salinity of shallow water tables in valley floors. Trees in contact with this saline water suffer dieback. Water tables move closer to the surface and impact over larger areas.

- Reduction in burning of forested areas allowing a build up of the shrub layer and invasion by colonisers such as *Pittosporum undulatum*. Resultant changes in soil moisture levels and humidity impact adversely on the established forest trees.

- Changes in soil water regimes including irrigation and water diversion.

- Increasing fungal attack, both soil borne and aerial.

Long-term droughts have caused serious dieback of Eucalyptus viminalis *on the Monaro Plain of New South Wales.* [S. Jones]

The devastating impact of long term drought on Eucalyptus viminalis *on the Monaro Plain of New South Wales.* [S. Jones]

- The impacts of grazing on soil compaction, and cropping resulting in changed soil horizons.

- The impacts of Bell Miners in some coastal forests.

- The impacts of adverse climatic conditions including droughts, fires and floods. Long-term drought particularly impacts species growing on the edges of their range as can be seen in the extensive dieback of *Eucalyptus viminalis* on the Monaro Plain of New South Wales.

Dieback Control

Recent studies suggest that certain primary factors are fundamental in most, if not all types of dieback. These are increased soil moisture levels from land clearing (promoting unhealthy roots), and reduced low intensity burns which increase nitrogen levels in the soil (resulting in highly nutritious leaves that are more attractive to insects). The solution seems to be changed fire regimes and altering the ecology to favour the trees, but these are very complex adjustments. Hopefully scientists can come up with some practical solutions.

Fasciation

Fasciation, sometimes also known as cresting, is a disfiguration that appears as if many stems have become fused together. It originates when the growth apex proliferates to form a series of growing points instead of a single growth. Fasciated stems become flattened and develop multiple leaves and ridges. Fasciation is usually found on young shoots which are growing rapidly. It is believed to be purely a physiological disorder developing with the very rapid growth, although similar symptoms occur occasionally following attacks by insects and mites; also some viruses. There is no method of control and if unsightly the fasciated stems should be pruned off (they are an interesting addition to floral arrangements). Fasciation does not seem to weaken a plant or stunt its growth. In some instances flowering and fruit set is inhibited on fasciated stems. Although generally uncommon, fasciation can occur on a wide range of native plants including species of *Allocasuarina, Banksia, Exocarpus, Santalum* and *Templetonia retusa.* Flowers can also become fasciated as is fairly commonly seen in *Xerochrysum bracteatum* and its cultivars.

Fasciated stem of Templetonia retusa. [D. Beardsell]

Fasciated flower stem of Stylidium armeria. [R. Elliot]

Lichens on Bark

Lichens are mutualistic organisms that consist of a species of fungus living in a symbiotic relationship with a species of green alga. They synthesise their own food using water and sunlight. Lichens are excellent indicators of good air quality, quickly dying out in the presence of pollution. They also help to remove carbon dioxide from the atmosphere and some species produce nitrogen which is washed into the soil. Lichens grow on rocks, stumps, fallen logs, wooden fences and directly on soil in natural grasslands. They also grow on tree bark and there is a common mistaken belief that lichens are detrimental to the health of the tree. Lichens, however, are not parasitic and as they pose no threat to trees, and their removal is totally unnecessary.

Lichens do not have deleterious effects on shrubs or trees. [D. Jones]

Mundulla Yellows

A significant growth problem, first brought to major attention as a disease in eucalypts from the vicinity of the South Australian town of Mundulla during the 1970s, has caused considerable interest to the scientific community and much anguish to landholders. Early literature on the subject detailed it (perhaps extravagantly) as a new highly contagious disease with the potential to spread into native vegetation causing irreparable damage and major changes. Its occurrence included areas in South Australia, the vicinity of Perth in Western Australia, parts of western Victoria and

Eucalypt leaves affected by Mundulla yellows. [S. Jones]

the Hunter Valley in New South Wales. A wide range of plant species is affected, with dramatic impacts on iconic large old trees highlighting the problem. Its effects are especially obvious along roadsides.

Although Mundulla yellows has been attributed to the various causal agents, including fungi, bacteria, nematodes and a range of virus-like organisms, detailed studies by several research groups failed to reveal any biotic causative agent. The common thread through all of the investigations is the link with alkaline soils and low concentrations of iron and manganese in affected plant tissue. The conclusion reached by some researchers is that the disease is most likely the result of lime-induced iron chlorosis, a problem well known to gardeners who live with alkaline soils (iron is extremely insoluble at high pH levels). Proof was provided when affected trees treated with iron responded positively and reverted to normal growth. Mundulla yellows has not been recorded in undisturbed native vegetation and its apparent spread can be explained by the widespread use of crushed limestone for making roads and the application of alkaline irrigation water in parks and gardens. Healthy trees growing in acidic soils can be severely impacted by such alkaline materials, resulting in decline and eventual death.

Symptoms: early symptoms show up as interveinal chlorosis in young leaves. Mature leaves become completely yellow with darker veins and fall prematurely. Leaf yellowing can be uneven within parts of the crown and it is not uncommon to see patches of bright yellow leaves among apparently normal green leaves. Sometimes whole trees have a yellow appearance, but more often the colouration is patchy in the crown. Limb dieback follows leaf fall, flowering and seed set declines and dieback progresses in stages through the tree. Once symptoms appear the trees decline in health steadily and eventually die.

Control: applications of iron are best applied in the early stages of deficiency (see chapter 20).

SALINITY

Plants can be damaged by salt in two main ways – firstly from salt carried in onshore coastal winds and secondly by salt building up in soils.

Wind-blown salt-damaged leaves of Calophyllum inophyllum. [D. Jones]

Wind-borne Salt

Salt is picked up from the sea by onshore winds and transported inland. Plants impacted by these salt-laden winds are commonly damaged by the deposition of salt on exposed foliage. Damage is more severe following strong winds in storms because more salt is transported and deposited further inland. The worst foliage damage follows strong onshore blows not accompanied by rain. Symptoms of wind-borne salt damage, include withering of new growth, stem dieback, browning of the leaves (especially the margins) which can blacken or become papery, and premature leaf fall. Plants grown in gardens exposed to salt-laden winds benefit from spraying with fresh water, but this action leads to a build-up of soil salt (next heading). Symptoms of wind-borne salt damage vary with the species. The most tolerant plants are those which grow naturally on exposed beaches and headlands.

Soil Salinity

Most, if not all Australian soils contain salt. Salt in soil originates from sources such as the type of parent rock, relictual deposits from previous inland seas and salt blown in from the ocean. Rainfall washes the salt through the soil profile where it remains in the groundwater. However, rising water tables bring the groundwater close to the soil surface and the salt becomes concentrated as a dense solution when the water evaporates during warm to hot weather. Salty water impedes the ability of plant roots to take up water and nutrients. The water need not ever appear on the soil surface for salting to occur, but if the water table rises near enough to the soil surface for evaporation to occur by capillary action then salt will build up and plants will be affected. Salting of this type is common in areas which have been extensively cleared of natural vegetation, because the deep-rooted shrubs and trees that occurred there naturally kept the groundwater at safe levels. Shallow-rooted crops, pasture and weeds are unable to stop the inexorable rise of the water table. When the water table is very high the salt appears as a white encrustation on the surface. Symptoms of plants grown in saline soil include stunting, reduced growth, reduced leaf size and an inward cupping of the leaves with associated brown or papery patches. Dieback and plant death is common. Soil salinity affects many areas of inland Australia and is a very difficult problem to solve. Lowering the water table is the most obvious answer but this involves large pumps and suitable sites where the salty water can be stored. Once the water table is sufficiently lowered, planting salt-tolerant shrubs and trees will help stabilise the situation.

The disastrous impact of increasing soil salinity. [C. French]

Salt damage – a stark reminder of its impact. [C. French]

SPRAY DAMAGE

Damage to plant tissues can result from the application of sprays. Herbicides (weed killers) are well known for causing plant damage which mostly results from wind moving the droplets further than anticipated. Herbicide damage, which is usually indicated by yellowing stems and leaves, may not become evident for several days. Spraying equipment used for weed killers should not be used to apply fungicides and pesticides. Many herbicides are highly residual and remnant material in spraying equipment may cause plant damage.

Pesticides can also cause plant damage. Plants with very soft or wilted growth are more likely to suffer spray damage than well-watered hardened plants. Plant sensitivity also varies with the species. Some ferns and ground orchids for example are very easily damaged by sprays. Some chemicals, such as oil-based sprays, are more likely to cause spray damage than other types. Spraying on hot days (above 25 °C) can also induce spray damage. White oil especially, should be used with care in hot weather. If in doubt apply white oil at lower than recommended strength. Pesticides applied as aerosols may also cause damage from the chemical carriers used to disperse the spray.

WATERLOGGING

Waterlogging occurs when water replaces air in the soil pores for long periods. Waterlogging is common in low-lying areas and also in heavy clay soils. Damage to plants grown in these soils occurs when the roots cannot extract sufficient oxygen for normal growth. Symptoms vary with the species but leaf cupping and irregular dead patches in the leaves are a useful guide. Wilting of new growth is also frequent because the plants have difficulty transporting water through their vascular system. Branch dieback and premature leaf-fall give affected plants an unthrifty appearance. Sensitive species die rapidly when waterlogged, whereas tolerant species exhibit few symptoms although generally make limited growth. Waterlogging is complicated by soil sulphur levels and the presence of disease organisms. Under waterlogged conditions sulphur becomes converted to rotten egg gas which is toxic to roots. Very few plants can survive waterlogged soil when vigorous root pathogens such as *Phytophthora cinnamomi* are present (see cinnamon fungus, chapter 19).

Waterlogging can also occur during floods when the soil is saturated for a period. Long-term flooding can have disastrous effects on gardens and natural communities. The strong flow of floodwaters can also cause physical damage to trees and shrubs.

Waterlogging damage to Eucalyptus conferruminata. [D. BEARDSELL]

Waterlogging damage in a natural habitat. [D. JONES]

A fast-moving flood ripped a large branch off this ironbark. [D. JONES]

MISCELLANEOUS PROBLEMS

- Mechanical damage is a common problem wherever machines interact with trees. Heavy impacts can result in the death of a tree but in lesser impacts large chunks of bark are usually torn off. Healthy trees can survive these impacts but often the affected area becomes the site of insect attack, especially borers.

- Graft incompatability sometimes occurs in grafted native plants. Often the development of numerous strong shoots from the rootstock is a sign of incompatability. Another common symptom is uneven size development at the junction between the rootstock and the grafted plant (scion). The rootstock can grow much larger than the scion or the reverse can occur.

- Cockatoo damage is common in trees where borers are active. After locating a borer the cockatoos tear open the bark and wood often causing considerable damage in the process. See also chapter 13.

Mechanical damage by vehicle. [D. Jones]

This site of mechanical damage has lead to invasion by borers. [D. Jones]

Overgrowth of scion compared with rootstock in Corymbia *graft.* [D. Jones]

Damage to Melaleuca armillaris *caused by a Yellow-tailed Black Cockatoo looking for borer grubs.* [D. Jones]

NATIVE PLANT SUSCEPTIBILITY AND RESISTANCE TO CINNAMON FUNGUS (*PHYTOPHTHORA CINNAMOMI*)

Australian native plants believed to be susceptible to *Phytophthora cinnamomi*

Plants of the same species can be sensitive to varying degrees and in some rare cases variants may even be resistant to *Phytopthora cinnamomi*.

Ongoing research and observation is most likely to increase the number of species known to be affected by this disease. ***Denotes very susceptible.**

Acacia aculeatissima
  aspera
  brownii
  buxifolia ssp.
    buxifolia
  campylophylla
  continua
  drummondii
  genistifolia
  glandulicarpa
  gracilifolia
  lanigera
  mitchellii
  myrtifolia
  oxycedrus
  pulchella
  rigens
  siculiformis
  spinescens
  stenoptera
  suaveolens
  terminalis
  triptera
  ulicifolia
  varia*
Actinostrobus
  pyramidalis
Actinotus helianthi
Acrotriche fasciculiflora
  halmaturina
  serrulata
Adenanthos barbiger
  cunninghamii
  cuneatus
  cygnorum
  detmoldii
  dobagii
  ellipticus
  filifolius
  ileticos
  macropidiana
  meisneri
  obovatus
  oreophilus

  pungens ssp. *effusus*
  pungens ssp. *pungens*
  sericeus
  terminalis
Agastachys odorata
Agrostocrinum scabrum
Allocasuarina crassa
  fraseriana
  humilis
  muelleriana
  nana
  paludosa
  pusilla
  rigida
  thuyoides
  verticillata
Amperea xiphoclada
Amphipogon
  amphipogonoides
Andersonia axilliflora
  caerulea
  echinocephala
  heterophylla
  lehmanniana
  pinaster
  simplex
Angophora costata
Anigoxanthos humilis
  ssp. *chrysantha*
Anopterus glandulosus
Aotus ericoides
  passerinoides
Argentipallium
  obtusifolium
Astartea astarteoides
Asterolasia asteriscophora
  phebalioides
Astroloma ciliatum
  conostephioides
  foliosum
  humifusum
  pinifolium
  xerophyllum

Austrostipa compressa

Babbingtonia behrii
Baeckea crassifolia
  leptocaulis
Banksia aculeata
  anatona
  ashbyi
  attenuata
  audax
  baueri
  baxteri
  brownii
  burdettii
  calyei
  candolleana
  coccinea
  concinna
  cuneata
  dryandroides
  elderiana
  gardneri
  gardneri var.
    brevidentata
  gardneri var.
    gardneri
  goodii
  grandis
  hirta
  hookeriana
  ilicifolia
  integrifolia
  laevigata
  laricina
  lehmanniana
  lindleyana
  littoralis
  luffitzi
  marginata
  media
  menziesii
  micrantha
  montana

  nutans
  occidentalis
  oligantha
  oreophila
  petiolaris
  pilostylis
  praemorsa
  prionotes
  pseudoplumosa
  pulchella
  quercifolia
  repens
  rufa ssp. *pumila*
  saxicola
  seminuda
  serrata
  solandri
  speciosa
  sphaerocarpa
  spinulosa ssp.
    cunninghamii
  telmatiaea
  tricuspis
  verticillata
  victoriae
Bauera sessiliflora
Beaufortia anisandra
  decussata
  heterophylla
  micrantha
  purpurea
  squarrosa
Blandfordia punicea
Boronia anemonifolia
  citriodora
  denticulata
  fastigiata
  heterophylla
  megastigma
  nana
  pilosa
  revoluta
  rhomboidea

Borya mirabilis
Bossiaea aquifolium
  cinerea
  eriocarpa
  obcordata
  ornata
  prostrata
Brachyloma ciliatum
  daphnoides
  depressum
Brachyscome uliginosa

Callitris preissii
  rhomboidea
Calothamnus gilesii
  villosus
Calytrix fraseri
  tetragona
Cassinia aculeata
Cenarrhenes nitida
Chamelaucium axillare
  erythrochlora
  griffinii
  roycei
  sp. *gingin*
  uncinatum
Conospermum mitchellii
  stoechadis
  toddii
  taxifolium
  triplinervium
Conostephium pendulum
Coronidium rupicola
Correa reflexa
Crowea angustifolia
  exalata
  saligna
Cyathodes
  (many species*)*

Dampiera alata
Darwinia collina
  lejostyla
  macrostegia

meeboldii
micropetala
oxylepis
squarrosa
wittwerorum
Dasypogon bromeliifolius
Daviesia brevifolia
  decurrens
  glossosema
  incrassata
  inflata
  leptophylla
  megacalyx
  mimosoides
  obovata
  physodes
  preissii
  pseudaphylla
  rhombifolia
  ulicifoli
  wyattiana
Dianella longifolia
  revoluta
Dillwynia cinerascens
  glaberrima
  phylicoides
  sericea
Diplarrena moroea
Dodonaea boroniifolia
  viscosa
Dryandra anatona
  arctotidis
  armata
  baxteri*
  bipinnatifida
  calophylla*
  cirsioides
  concinna
  conferta*
  falcata
  formosa*
  fraseri*
  ionthocarpa
  lindleyana
  mimica
  montana
  mucronulata
  nivea
  plumosa
  polycephala*
  pteridifolia
  quercifolia
  seneciifolia
  serra*
  serratuloides
  serratuloides ssp.
    perissa
  sessilis
  squarrosa
  squarrosa ssp.
    argillacea
  tenuifolia.

Epacris (most species,
  especially those
  from Tas.)
Eremaea beaufortioides
Eucalyptus baxteri
  caesia
  calycogona
  campaspe

consideniana
cosmophylla
crucis*
delegatensis
desmondensis
diversifolia*
dives
dumosa
eremophila
erythrocorys*
forrestiana
gracilis
grossa
kingsmillii*
macrandra
macrocarpa*
macrorhyncha
marginata S
moorei
muelleriana
nitens
obliqua
oleosa
orbifolia*
pauciflora ssp.
  pauciflora
pauciflora ssp.
  niphophila
playtpus
pleurocarpa
polyanthemos
preissiana
pyriformis
radiata
regnans
rhodantha*
salmonophloia
salubris
scabra
sepulcralis
sieberi
stricklandii
tetraptera
todtiana
torquata
viridis
websteriana*
willisii
woodwardii
Euryomyrtus ramosissima
Evandra aristata

Gastrolobium leakeanum
  luteifolium
  papilio
  pulchellum
  vestitum
Gompholobium
  confertum
  ecostatum
  knightianum
  polymorphum
Goodenia hederacea
  humilis
  lanata
Grevillea acanthifolia
  ssp. tenomera
  acerosa
  alpina
  althoferorum ssp.
    fragilis

aquifolium
asparagoides
baueri
bipinnatifda*
calliantha
capitellata
chrysophaea*
cirsiifolia
concinna
confertifolia
dielsiana*
dimorpha
dryophylla
endlicheriana
floribunda
granulifera
ilicifolia
insignis
irasa ssp. irasa
jephcottii
juniperina
lanigera
lavandulacea
leucopteris
linsmithii
miqueliana
mucronulata
pilulifera*
polybractea
quinquenervis
rogersii
rosmarinifolia
saccata
steiglitziana
stenomera
trifida
triloba
tripartita
victoriae

Hakea ambigua
  baxteri
  bakeriana
  bucculenta
  cucullata
  ferruginea
  flabellifolia
  francisiana
  lehmanniana
  marginata
  multilineata
  oleifolia
  pandanicarpa ssp.
    crassifolia
  prostrata
  subvaginata
  trifurcata
  ulicina
  undulata
Haloragodendron
  monospermum
Hibbbertia acerosa
  amplexicaulis
  calycina
  cistiflora
  commutata
  desmophylla
  furfuracea
  huegelli
  hypericoides
  inconspicua

lineata
montana
montana var. major
obtusifolia
pedunculata
prostrata
quadricolor
rhadinopoda
riparia
sericea
stellaris
villosa
virgata
Hovea elliptica
  linearis
  pungens
Hybanthus floribundus
Hypocalymma
  angustifolium
  robustum
  strictum

Isopogon attenuatus
  axillaris
  baxteri
  buxifolius var.
    obovatus
  ceratophyllus
  cucullata
  dubius
  formosus
  latifolius
  petiolaris
  sphaerocephalus
  teretifolius var.
    petrophiloides
  trilobus
  uncinatus

Jacksonia floribunda
  furcellata
  horrida
  spinosa
  sternbergiana

Kennedia prostrata
Kingia australis
Kunzea ambigua
  ericifolia
  montana
  sulphurea

Labichea punctata
Lambertia ericifolia
  echinata ssp. citrina
  echinata ssp.
    echinata
  echinata ssp.
    occidentalis
  ericifolia
  fairallii
  ilicifolia
  inermis var.
    drummondii
  inermis var. inermis
  multiflora var.
    multiflora
  occidentalis
  orbifolia
  rariflora ssp. lutea
  uniflora

Lasiopetalum
  floribundum
  glabratum
Latrobea genistoides
  hirtella
Laxmannia orientalis
Leionema phylicifolium
  ralstonii
Leptospermum coriaceum
  glaucescens
  juniperinum
  lanigerum
Leschenaultia formosa
Leucopogon australis
  capitellatus
  collinus
  concinnus
  conostephioides
  cymbiformis
  distans
  distans ssp.
    contractus
  elegans
  ericoides
  esquamatus
  flavescens
  gibbosus
  glacialis
  gnaphalioides
  gracillimus
  lanceolatus
  maccraei
  microphyllus var.
    pilibundus
  nutans
  obtectus
  oxycedrus
  parviflorus
  polymorphus
  propinquus
  pulchellus
  revolutus
  verticillatus
  virgatus
Lissanthe strigosa
Lomandra filiformis
  odora
  sonderi
Lomatia fraseri
Loxocarya cinerea
Lysinema conspicuum
  cillatum

Macrozamia riedlei
Macrozamia spp.
Melaleuca coccinea
  filifolia
  nematophylla
  scabra
  squamea
  subfalcata
  thymoides
  uncinata
Monotoca elliptica
  glauca
  scoparia
  tamariscina

Nematolepis squamea
Nothofagus
  cunninghamii

Olearia ciliata
  pannosa ssp.
    cardiophylla
Opercularia vaginata
Oxylobium spp.
Ozothamnus obcordatus
  ssp. *major*

Patersonia babianoides.
  occidentalis
  rudis
  sericea
  umbrosa
Pericalymma ellipticum
Persoonia elliptica
  juniperina
  longifolia
  micrantha
Petrophile biloba
  divaricata
  diversifolia
  drummondii
  ericifolia
  linearis
  longifolia
  media
  multisecta
  pulchella
  seminuda
  serruriae

  squamata
  stricta
Phebalium daviesii
Philotheca myoporoides
Phyllanthus hirtellus
Pimelea humilis
  pagophila
  suaveolens
Platylobium
  obtusangulum
Platysace compressa
  heterophylla
Podocarpus drouynianus
  lawrencei
Pomaderris intermedia
Prostanthera cuneata
  decussata
  lasianthos
  ovalifolia
  ringens
  saxicola var.
    montana
Pultenaea altissima
  benthamii
  daphnoides
  flexilis
  graveolens
  gunnii
  humilis
  involucrata

  mollis
  pedunculata
  procumbens
  pycnocephala
  scabra
  spinosa
  stricta
  subcapitata
  subalpine
  trifida
  villifera var. *villifera*

Richea pandanifolia
  scoparia

Scaevola calliptera
Scholtzia involucrata
Sphaerolobium acanthos
  medium
Sphenotoma
  dracophylloides
  drummondii
  gracile
  sp. Stirling Range
  squarrosum
Sprengelia incarnata
Stirlingia latifolia
  tenuifolia
Stylidium amoenum
  graminifolium

  junceum
  schoenoides
  spathulatum
Styphelia adscendens
  tenuiflora
Synaphea petiolaris
  polymorpha

Tasmannia lanceolata
  purpurascens
Taxandria linearifolia
  spathulata
Telopea mongaensis
  speciosissima
Tetrarrhena laevis
Tetratheca ciliata
  hirsuta
  pilosa
  setigera
  subaphylla
Themeda triandra
Thomasia grandifolia
Thryptomene calycina
  baeckeacea
  denticulata
  saxicola
Thysanotus multiflorus
  thyrsoideus
Tremandra stelligera
Trymalium ledifolium

Velleia foliosa
Verticordia densifolia
  grandis
  huegelii
  nitens

Westringia davidii
Wollemia nobilis
Woollsia pungens

Xanthorrhoea australis
  brevistyla
  drummondii
  glauca ssp. *glauca*
  gracilis
  minor
  nana
  platyphylla
  preissii
  resinifera
  semiplana ssp.
    semiplana
  semiplana ssp.
    tateana
Xanthosia dissecta
  tridentata
Xylomelum
  angustifolium
  occidentalis
Zieria laevigata

Australian Plants believed to show resistance to *Phytophthora cinnamomi*
*Denotes high resistance.

Acacia
  acinacea
  alata
  awestoniana
  baileyana
  barbinervis
  baxteri
  browniana
  browniana var.
    intermedia
  cyclops
  dealbata *
  decurrens
  drummondii
  extensa
  fimbriata
  floribunda
  howittii
  huegelii
  laterriticola
  longifolia
  mearnsii
  mitchelii
  mucronata
  nervosa
  pravissima
  preissiana
  prominens
  pulchella
  retinodes
  riceana *
  salicina

  saligna
  semitrullata
  urophylla
  verniciflua
  veronica
Acaena echinata
Adiantum aethiopicum
Agonis flexuosa
  juniperina
Allocasuarina
  lehmanniana
  microstachya
  pusilla
Anarthria gracilis
  prolifera
  scabra
Angophora costata
  floribunda
  hispida
  melanoxylon
Anigozanthos flavidus
  manglesii
  rufus
Astartea fascicularis
  heteranthera
Astroloma pallidum
Austrostipa mollis

Babingtonia
  camphorosmae
  virgata
Baeckea imbricata

  linifolia
  pachyphylla
Banksia integrifolia
  robur
  spinulosa var. *collina*
  spinulosa var.
    spinulosa
Billardiera
  drummondiana
  variifolia
Boronia crenulata
  spathulata
Bossiaea linophylla
  rufa
  webbii
Brachychiton acerifolius
  populneus
Burchardia multiflora
  umbellata
Bursaria spinosa

Callistemon acuminatus
  *citrinus**
  comboynensis
  linearis
  linearifolius
  *macropunctatus**
  pallidus
  paludosus
  *salignus**
  sieberi *
  viminalis *

Calothamnus affinis var.
  longistamineus
  crassus
  quadrifidus
  sanguineus
Calytrix alpestris
  asperula
  flavescens
  leschenaultii
  tenuiramea
Cassytha flava
  glabella
Castanospermum
  australe
Casuarina
  cunninghamiana
  glauca
  obesa
Caustis dioica
  pentandra
Ceratopetalum
  gummiferum
Chamaescilla corymbosa
Chamaexeros serra
Chorizema aciculare
Clematis pubescens
Comesperma calymega
  confertum
  virgatum
Conostylis aculeata
  pusilla
  serrulata

  setigera
  setosa
Corymbia calophylla
  citriodora
  eximia
  ficifolia
  gummifera
  intermedia
  maculata
  tessellaris
  torelliana
Cryptostylis ovata
Cyathochaeta avenacea
  clandestina
Cymbonotus preissianus

Dampiera linearis
Darwinia camptostylis
  citriodora
  leiostyla
  vestita
Desmocladus fasciculatus
  flexuosa
Deyeuxia drummondii
Dodonaea viscosa
Drosera erythrorhiza
  macrantha
  pallida
  peltata
Elaeocarpus angustifolius
  reticulatus
Epacris longiflora

microphylla
Eriochilus dilatatus
Eucalyptus acaciiformis
accedens
alba
alpina
angulosa
astringens
bancroftii
bosistoana
botryoides*
buprestium
camaldulensis
camphora*
cephalocarpa
cinerea
cladocalyx
coccifera*
conferruminata
cordata
cornuta
crenulata
deanei*
decurva
dunnii
dwyeri
falcata
forrestiana
gardneri
globulus*
gomphocephala*
gunnii*
haemastoma
incrassata
kitsoniana*
kruseana
laeliae
lansdowneana
lehmanniii
leucoxylon
ligulata ssp.
   stirlingica
maidenii
mannifera
megacarpa
melliodora
nicholii
occidentalis
ovata*
pachyloma
patens
perriniana*
polybractea
preissiana
pulverulenta
risdonii
robusta*
rubida
rudis
saligna
salmonophloia
sideroxylon*
spathulata
stellulata
tasmanica
tereticornis
tricarpa
uncinata
vernicosa

viminalis
wandoo
yarraensis
Eupomatia laurina

Gahnia radula
sieberiana
trifida
Gastrolobium modestum
Goodia lotifolia
Gompholobium
capitatum
tomentosum
Gonocarpus tetragynus
Goodenia caerulea
geniculata
ovata
scapigera
Grevillea acanthifolia
asplenifolia
banksii
buxifolia
'Clearview David'
fasciculata var.
   linearis
glabella
hookeriana
'Ivanhoe'
juniperina
lanigera
longifolia
mucronulata
pteridifolia
robusta
rosmarinifolia
shiressii
synaphea

Haemodorum
paniculatum
Hakea amplexicaulis
corymbosa
dactyloides
decurrens
drupacea
lissocarpha
nodosa*
petiolaris
salicifolia*
scoparia
ulicina
Hardenbergia
comptoniana
violacea
Harperia
confertospicatus
Hemiandra pungens
Hemigenia curvifolia
Hibbertia empetrifolia
racemosa
scandens
silvestris
vaginata
Hovea chorizemifolia
trisperma
Howittia trilocularis
Hydrocotyle hirta
Hypericum gramineum
Hymenosporum flavum

Hypocalymma
angustifolium
myrtifolium
speciosum
Hypolaena fastigiata

Indigofera australis
Isotoma
hypocrateriformis

Johnsonia lupulina
teretifolia
Juncus australis

Kennedia coccinea
prostrata
rubicunda

Kunzea ambigua
peduncularis
phylicoides
preissiana

Lambertia formosa
Lechenaultia biloba
Lepidosperma
brunonianum
concavum
laterale
scabrum
squamatum
tenue
tetraquetrum
viscidum
Leporella fimbriata
Leptocarpus tenax
Leptomeria
cunninghamii
eriocoides
Leptospermum
continentale
ellipticum
epacridoideum
erubescens
flavescens*
grandiflorum
lanigerum
myrsinoides
scoparium
squarrosum
Leucopogon glabellus
pendulus
Levenhookia pusilla
Lindsaea linearis
Lobelia gibbosa
rhytidosperma
Logania serpyllifolia
Lomandra confertifolia
integra
filiformis
longifolia
nigricans
pauciflora
preissii
sonderi
Lomatia myricoides
polymorpha
Lyginia barbata
imberbis

Macropidia fuliginosa
Meeboldina scariosa
Melaleuca armillaris
bracteata
cuticularis
decussata
diosmifolia
ericifolia
gibbosa
halmaturorum
holosericea
hypericifolia
lanceolata
laxiflora
leucadendron
linariifolia
macronychia
nesophila
parviflora
pentagona
preissiana
pulchella
spathulata
squamea
striata
styphelioides*
suberosa
thymifolia
viminea
violacea
wilsonii
Melia azedarach
Mesomelaena graciliceps
stygia
tetragona
Microseris scapigera
Millotia tenuifolia
Mirbelia dilatata
Myoporum insulare

Nuytsia floribunda

Olax phyllanthi
Opercularia
echinocephala
varia
Orthrosanthus laxus

Patersonia occidentalis
pygmaea
Pelargonium australe
rodneyanum
Pentapeltis peltigera
Persoonia aff. saccata
Philotheca freyciana
spicata
Phlebocarya ciliata
Phyllanthus calycinus
Phyllota humilis
Pimelea axiflora
hispida
Pittosporum
angustifolium*
revolutum*
rhombifolium*
undulatum*
Platysace tenuissima
Poa ensiformis
poiformis

sieberiana
Pteridium esculentum
Pterochaeta paniculata
Ptilotus manglesii
Pultenaea daphnoides
ericifolia

Rhtidosporum
procumbens

Scaevola striata
Schoenus curvifolius
pedicellatus
rigens
Sphaerolobium
vimineum
Stylidium brunonianum
imbricatum
piliferum ssp. minor
scandens
verticillatum
Swainsona galegifolia
Synaphea petiolaris
Syzygium oleosum
paniculatum
smithii

Taxandria floribunda
parviceps*
Tetraria capillaris
octandra
Thysanotus dichotomus
tenellus
Trachymene pilosa
Trichocline spathulata
Tricoryne elatior

Utricularia dichotoma
multifida

Veronica gracilis
Verticordia habrantha
Viminaria juncea

Xanthosia atkinsoniana
candida
huegelii
rotundifolia

GLOSSARY

abdomen The posterior of the three main body divisions of insects.

acaricide A chemical used to control mites and ticks.

aerosol Finely dispersed particles in air; a pressure pack.

alate Winged, often referring to winged and reproductive caste of ants and termites.

amino acid An organic compound which is a structural unit of protein.

apterous Wingless.

arachnida Class of arthropods that includes ticks and mites.

attractants Chemicals having an attraction for animals such as insects.

auxin A growth regulating compound controlling many growth processes such as bud-break, root development.

germination The process whereby seeds or spores sprout and begin to grow.

bacterial wilts Plant disease in which the causative bacteria produce slime that plugs the water-conducting tissue.

baculovirus A family of large rod-like viruses that commonly use larvae of moths as hosts.

biological control (biocontrol) The control of pests by encouraging natural enemies, predators, parasites or diseases.

byre Protective shelter of debris constructed by ants to cover a feeding colony of sap-sucking insects.

callus A hard formation of tissue, especially new tissue formed over a wound.

calyces (calyx) The sepals of a flower, typically forming a whorl that encloses the petals and forms a protective layer around a flower in bud.

cambium A cellular plant tissue from which phloem, xylem, or cork grows by division, resulting (in woody plants) in secondary thickening.

canopy The uppermost branches of the trees in a forest, forming a more or less continuous layer of foliage.

canker An area of tissue death on a stem, trunk or roots caused by fungal or nematode infection.

capitulum A structure similar to an elaiosome found on the eggs of some species of stick insects.

caterpillar The larvae of a moth, butterfly or sawfly.

chlorophyll The green pigment of leaves and other organs, important as a light-absorbing agent in photosynthesis.

chlorosis Yellowing of green plant tissue due to damage to chlorophyll, often caused by a virus (being chlorotic).

crawler The first active stage of a scale insect.

crown The part of the plant just above and below the ground from which the leaves and roots branch out.

croziers The coiled young fronds of ferns.

damping-off A condition in which young seedlings are attacked and killed by soil-borne fungi.

diapause A period of suspended development in an insect or other invertebrate, especially during unfavourable environmental conditions.

diptera Insect order with a single pair of wings, the other pair being replaced by small structures known as halteres, for example flies.

dormancy A state of quiescence or inactivity.

dorsal Top or uppermost, the back.

ecology The study of the interaction of plants and animals within their natural environment.

elytra Thickened leathery front wings of beetles, earwigs and such.

emulsion A suspension of fine droplets of one liquid in another, for example oil in water.

entomophagous Entomophagy is the consumption of insects as food.

entomopathogenic An organism which can kill arthropods by way of poisoning, either through its own toxins or those it harbours.

epicormic Growing from a previously dormant bud on the trunk or a limb of a tree.

exoskeleton A rigid external covering for the body in some invertebrate animals, especially arthropods.

exotic A plant or animal introduced from overseas.

filaments Slender thread-like objects or fibres, especially ones found in animal or plant structures.

forewings Either of the two front wings of a four-winged insect.

frass Refers to both the substance produced by stem borers which may conceal the borer hole and the waste of insect larvae such as caterpillars.

frond Leaf of a fern or palm.

fruit The seed bearing organ developed after fertilisation.

fungicide A chemical used to control fungus diseases.

gall An abnormal growth of plant tissues induced by the presence of a foreign animal or plant.

genus A taxonomic group of closely related species.

groundcover Low-growing, spreading plants that help to stop weeds growing.

grub A thick-bodied sluggish beetle larvae with prolegs and a well-developed head.

haustoria The appendange or portion of a parasitic plant (such as mistletoe) or parasitic fungus tip that penetrates the host plants and absorbs water and nutrients.

herbicide A chemical used to control weeds.

hindwing either of the two back wings of a four-winged insect.

honeydew A sugary liquid discharged from the rear end of some sucking insects.

hormone A chemical substance produced in one part of a plant and inducing a growth response when transferred to another part.

imago The adult stage of an insect.

insecticide A chemical used to control insect pests.

instar The form of an insect between successive moults.

interveinal Area of leaf or wing located between veins.

kino A resin-like exudate from trunks or branches which is especially common in eucalypts.

larva The immature insect from hatching to pupation.

LD50 Lethal dose to 50 per cent of test animals; usually expressed in milligrams of pesticide per kilograms of body weight of the test animal.

leaf miner An insect which lives in and feeds upon the leaf cells between the upper and lower surfaces of a leaf.

lesion A region in an organ or tissue which has suffered damage through injury or disease, such as a stem wound.

looper A caterpillar lacking two or more pairs of its prolegs; it crawls by looping its body.

maggot A larva without legs and without a well-developed head, typical of flies.

mandibles The anterior paired mouth-parts in insects.

mericlone A copy of an original plant made using the laboratory technique of meristem propagation or tissue culture.

meristems A region of plant tissue, found chiefly at the growing tips of roots and shoots and in the cambium, consisting of actively dividing cells forming new tissue.

metamorphosis The change in form during the development of an insect, i.e. from pupa to adult.

midstorey In a forest, those plants that form an intermediate height between the canopy or sub-canopy and the understorey.

miticide (also acaricide) A chemical used to control mites.

moult The process where an insect sheds its outer coat.

mycorrhiza A beneficial relationship between the roots of a plant and fungi or bacteria resulting in a nutrient exchange system. Some plants cannot grow without such a relationship.

native A plant or animal which is indigenous to a region.

natural control The reduction of populations of pests by natural means.

necrosis Death of plant tissue.

nematicide A chemical used to control nematodes.

nematodes Unsegmented worms with cylindrical elongated bodies.

nymph A young insect of a species with no metamorphosis; a smaller wingless version of an adult.

ootheca The egg case of cockroaches, mantises, and related insects.

osmosis Diffusion of water through a membrane caused by different concentrations of salts on either side of the membrane.

ovipositor A tubular structure of a female insect by which means eggs are deposited.

palps Each of a pair of elongated segmented appendages near the mouth of an arthropod, usually concerned with the senses of touch and taste.

parasite An organism growing or living on another, for example mistletoe.

parasitoid An organism that feeds on a host organism until larval development is complete.

parthenogenic The formation of fertile eggs without fertilisation by a male.

pesticide A chemical used to control pests. In this case the word pest can refer to a range of plant enemies including fungi, bacteria, nematodes, insects, mites and such.

petiole The stem or stalk of a leaf.

phloem That part of the vascular system concerned with the movement and storage of nutrients and hormones.

photosynthesis The conversion of carbon dioxide from the atmosphere into sugars within green parts of the plant, using chlorophyll and energy from the sun's rays.

phyllode A modified petiole acting as a leaf, often seen in acacias.

phyllodinous Having phyllodes.

phytotoxic Poisonous to plants.

pneumostome Snail breathinig tube.

pollarding Cut off the top and branches of (a tree) to encourage new growth at the top.

predator an animal that naturally preys on others.

proboscis The elongated tubular mouth-part of a sucking insect.

proleg A fleshy abdominal leg of an insect larvae.

prothallus The small flat structure that produces the germinating spores and reproductive organs of a plant, especially ferns.

prothorax The first or anterior segment of the thorax.

pubescent Covered with short soft downy hairs.

pupa The stage between the larva and the adult in metamorphic insects.

radula (in a mollusc) A rasp-like structure of tiny teeth used for scraping food particles off a surface and drawing them into the mouth.

resistence (pesticide) The decreased susceptibility of a pest population to a pesticide that was previously successful in control.

repellents Substances which induce avoidance.

rhizobia Soil bacteria that fix nitrogen

rhizosphere The region of soil that is directly influenced by root secretions and associated soil microorganisms.

sp. Species singular.

spp. Plural of species.

sporeling A young fern plant.

stage A distinct period in the development of an insect.

thorax Section of an animal's body between the head and the abdomen.

trade name A registered trade mark of the company manufacturing the chemical.

unthrifty (of livestock or plants) Not strong and healthy.

vascular tissues The fluid conducting tissues of a plant including phloem and xylem.

viviparous Bearing live young.

wettable powder Insecticidal dusts with wetting agents which make the product suitable for mixing with water.

woolly bear Common term used to describe a very hairy caterpillar.

xylem The water conducting tissue of a plant.

yellows A plant disease which causes yellowing and stunting.

BIBLIOGRAPHY

Barrett, S., Shearer, B.L., Crane, C.E., and Cochrane, A. (2008). An extinction-risk assessment tool for flora threatened by *Phytophthora cinnamomi*. *Australian Journal of Botany* 56: 477–486.

Berry, J.A., and Withers, T.M. (2002). New gall-inducing species of ormocerine pteromalid (Hymenoptera: Pteromalidae: Ormocerinae) described from New Zealand. *Australian Journal of Entomology* 41(1): 18–22.

Bodman, K., *et al.* (1996). *Ornamental Plant Pests, Diseases and Disorders*, Queensland Department of Primary Industries, Information Series Q196001.

Bowman, D. and Yeates, D. (2006). *A remarkable moment in Australian biogeography*. Wiley-Blackwell, Report Book No. 0028646X.

Brimblecombe, A.R. and Heather, N.W. (1965). Occurrence of the kauri coccid, Coniferococcus agathidis, Brimblecombe (Homoptera: Monophlebidae) in Queensland. *Journal of the Entomological Society of Queensland* 4: 83–85.

Brock, P.D. and Hasenpusch, J.W. (2009). *The Complete Guide to Stick and Leaf Insects of Australia*. CSIRO Publishing, Collingwood, Australia.

Büchen-Osmond, C., Crabtree, K., Gibbs, A. and Mclean, G. [Eds] (1988). *Viruses of Plants in Australia*. Australian National University Printing Service, Canberra.

Buffington, M.L. (2008). A revision of Australian Thrasorinae (Hymenoptera: Figitidae) with a description of a new genus and six new species. *Australian Journal of Entomology* 47(3): 203–212.

Cahill, D.M., Rookes, J.E., Wilson, B.A., Gibson, L. and McDougall, K.L. (2008). Turner Review no.17: *Pytophthora cinnamomi* and Australia's biodiversity: impacts, predictions and progress towards control. *Australian Journal of Botany* 56: 279–310.

Campbell, C. (1994). *Simple Pest Control*. Thomas C. Lothian Pty Ltd, Port Melbourne, Victoria.

Carnegie, A.J. (2002). *Field guide to common pests and diseases in eucalypt plantations in NSW*. State Forests of NSW, Sydney.

Carnegie, A.J. (2007). Forest health condition in New South Wales, Australia, 1996–2005. 1. Fungi recorded from eucalypt plantations during forest health surveys. *Australasian Plant Pathology* 36: 213–224.

Carver, M., Blüthgen, N., Grimshaw, J.F. and Bellis, G.A. (2003). Aphis clerodendri Matsumura (Hemiptera: Aphididae), attendant ants (Hymenoptera: Formicidae) and associates on Clerodendrum (Verbenaceae) in Australia. *Australian Journal of Entomology* 42(2): 109–113.

Carver, M. 1978. The Black Citrus Aphids, Toxoptera citricidus (Kirkaldy) and T. auranti (Boyer de Fonscolombe) (Homptera: Aphidiae) *Journal of Australian Entomolological Society* 17: 263–270.

Cho, J.J. (1983). Variability in susceptibility of some *Banksia* species to *Phytophthora cinnamomi* and their distribution in Australia. *Plant Disease* 67: 869–871.

Code of Practice for safe use of Pesticides, CSIRO Safety Booklet No. 3 (1976). CSIRO, Melbourne, Australia.

Connelly, D. (2012). *Honeybee pesticide poisoning; a risk management tool for Australian Farmers and Bee Keepers*. Rural Industry Research and Development Corporation. https://rirdc.infoservices.com.au/items/12-043

Cook L.G. and Gullan P.J. (2004). The gall-inducing habit has evolved multiple times among the eriococcid scale insects (Sternorrhyncha: Coccoidea: Eriococcidae). *Biological Journal of the Linnean Society* 83: 441–452.

Cook, L.G. (2003). *Apiomorpha gullanae* sp. n., an unusual new species of gall-inducing scale insect (Hemiptera: Eriococcidae). *Australian Journal of Entomology* 42(4): 327.

Cook L.G. and Gullan P.J. (2008). Insect, not plant, determines gall morphology in the *Apiomorpha pharetrata* species-group (Hemiptera: Coccoidea). *Australian Journal of Entomology* 47(1): 51–57.

Cook, L.G., Gullan, P.J. and Stewart, A.C. (2000). First-instar morphology and sexual dimorphism in the gall-inducing scale insect Apiomorpha Rubsaamen (Hemiptera: Coccoidea: Eriococcidae). *Journal of Natural History* 34(6): 879–894.

Cook, L.G. and Rowell, D.M. (2007). Genetic diversity, host-specificity and unusual phylogeography of a cryptic, host-associated species complex of gall-inducing scale insects. *Ecological Entomology* 32(5): 506–515.

Cranston, P. and Gullan, P. (1995). Gums and galls. *Australian Natural History* 24(12): 66.

Creighton, W. (1981). *The Australian Fungicide Handbook*. Plantection, Boonah, Queensland.

Crespi, B. and Worobey, M. (1998). Comparative Analysis of Gall Morphology in Australian Gall Thrips: The Evolution of Extended Phenotypes. *Evolution* 52(6): 1686.

Crespi, B.J. (1992). Behavioural ecology of Australian gall thrips (Insecta, Thysanoptera). *Journal of Natural History* 26(4): 769.

Crous, P.W., *et al.* (2011). Fungal pathogens of Proteaceae. *Persoonia* 27: 20–45.

Crous, P.W., *et al.* (2008). Host specificity and speciation of *Mycosphaerella* and *Tetraspaeria* associated with leaf spots. *Persoonia* 20: 59–86.

Crous, P.W., *et al.* (2009). Unravelling Mycosphaerella. *Persoonia* 23: 99–118.

CSIRO Entomologists (1970). *The Insects of Australia*. Dai Nippon, Hong Kong.

D'Souza, N.K., Colquhoun, I.J., Shearer, B.L. and Hardy, G.E.S.J. (2002). Assessing the potential for biological control of Phytophthora cinnamomi by fifteen native Western Australian jarrah-forest legume species. *Australasian Plant Pathology* 34: 533–540.

Dahms, E.C., Monteith, G. and Monteith, S. (1994). *Collecting, preserving and classifying insects*. Queensland Museum.

de Lima Detoni, M., Vasconcelos, E.G., Gomes Maia, A.C.R., do Nascimento Gusmão, M.A., dos Santos Isaias, R.M., Soares, G.L.G., Santos, J.C. and Fernandes, G.W. (2011). Protein content and electrophoretic profile of insect galls on susceptible and resistant host plants of *Bauhinia brevipes* Vogel (Fabaceae). *Australian Journal of Botany* 59(6): 509–514.

Dennill, G.B. and Gordon, A.J. (1990). Climate-related differences in the efficacy of the Australian gall wasp (Hymenoptera: Pteromalidae) released for the control of *Acacia longifolia* in South Africa. *Environmental Entomology* 19(1): 130.

Detoni, M.d.L., Faria-Pinto, P., *et al.* (2012). Galls from *Calliandra brevipes* BENTH (Fabaceae:Mimosoidae): evidence of apyrase activity contribution in a plant-insect interaction. *Australian Journal of Botany* 60(6): 559–567.

Dhileepan, K., Lockett, C.J. and McFadyen, R.E. (2005). Larval parasitism by native insects on the introduced stem-galling moth *Epiblema strenuana* Walker (Lepidoptera: Tortricidae) and its implications for biological control of *Parthenium hysterophorus* (Asteraceae). *Australian Journal of Entomology* 44(1): 83–88.

Dittrich-Schröder, G., Wingfield, M.J., Hurley, B.P. and Slippers, B. (2012). Diversity in *Eucalyptus* susceptibility to the gall-forming wasp *Leptocybe invasa*. *Agricultural & Forest Entomology* 14(4): 419–427.

Dorchin, N. and Adair, R.J. (2011). Two new *Dasineura* species (Diptera: Cecidomyiidae) from coastal tea tree, *Leptospermum laevigatum* (Myrtaceae) in Australia. *Australian Journal of Entomology* 50(1): 65–71.

Edmiston, R.J. (1989). *Plants Resistant to Dieback.* Department of Conservation and Land Management. 1–89.

Elliot, H.J. and deLittle, D.W. (1985). *Insect Pests of Trees and Timber in Tasmania.* Forestry Commission, Hobart.

Elliot, H.J., Ohmart, C.P. and Wylie, F.R. (1998). *Insect Pests of Australian Forests: ecology and management.* Inkata Press, Melbourne, Australia.

Environment Australia (2001). *Threat Abatement Plan for Dieback caused by the root-rot fungus* Phytophthora cinnamomi.

Farrow, R.A. (1996). *Insect pests of eucalypts on farms and in planations.* Identification Leaflet No. 11. Witches' broom scale. CSIRO Australia.

Freeman, T.P., Goolsby, J.A., Ozman, S.K. and Nelson, D.R. (2005). An ultrastructural study of the relationship between the mite *Floracarus perrepae* (Knihinicki & Boczek) (Acariformes: Eriophyidae) and the fern *Lygodium microphyllum* (Lygodiaceae). *Australian Journal of Entomology* 44(1): 57–61.

Fry, S., Caravan, H. and Chapman, T. (2011). Reproductive Caste Beats a Hasty Retreat. *Journal of Insect Behavior* 24(6): 413–422.

Garbou, S.S., Jaenson, T.G.T. and Pålsson, K. (2006). Repellency of MyggA® Natural spray (para-menthane-3,8-diol) and RB86 (neem oil) against the tick *Ixodes ricinus* (Acari: Ixodidae) in the field in east-central Sweden. *Experimental & Applied Acarology* 40(3/4): 271–277.

Gibbs, A.J., Mackenzie, A.M., Wei, K.J. and Gibbs, M.J. (2008). The potyviruses of Australia. *Arch Virol* 153:1411–1420.

Gibbs, A., Mackenzie, A., Blanchfield, A., Cross, P., Wilson, C., Kitajima, E., Nightingale, M. and Clements, M. (2000). Viruses of orchids in Australia; their identification, biology and control. *Austral. Orch. Rev.* June/July: 10–21.

Goolsby, J.A., Makinson, J. and Purcell, M. (2000). Seasonal phenology of the gall-making fly *Fergusonina* sp. (Diptera: Fergusoninidae) and its implications for biological control of *Melaleuca quinquenervia*. *Australian Journal of Entomology* 39(4): 336–343.

Groves, E., Hollick, P., Hardy, G. and McComb, J. (2009). *Native Plants Resistant to Dieback (Phytophora cinnamomi).* http://www.cpsm-phytophthora.org/downloads/CPSM_resistanceBrochures.pdf

Gullan, P.J. (1984a). A revision of the gall-forming coccoid genus *Apiomorpha* Rübsaaman (Homoptera: Eriococcidae: Apiomorphinae). *Australian Journal of Zoology Supplementary Series* 97: 1–203.

Gullan, P.J. (1984b). A revision of the gall-forming coccoid genus *Cylindrococcus* Maskell (Homoptera: Eriococcidae). *Australian Journal of Zoology* 32: 677–690.

Gullan, P.J. (1978). Male insects and galls of the genus *Cylindrococcus* Maskell (Homoptera: Coccoidea). *J. Aust. Ent. Soc.* 17: 53–61.

Gullan, P.J., Miller, D.R. and Cook, L.G. (2005). Gall-inducing scale insects (Hemiptera: Sternorrhyncha: Coccoidea). In: Raman, A., Schaefer, C.W. and Withers, T.M. (Eds). *Biology, Ecology and Evolution of Gall-inducing Arthropods.* pp. 159–229, Enfield, NH:Science Publishers.

Hadlington, P.W. and Johnston, J.A. (1977). *A Guide to the Care and Cure of Australian Trees.* New South Wales University Press, Kensington, New South Wales.

Hangay, G. and Zborowski, P. (2010). *A Guide to the Beetles of Australia.* CSIRO Publishing, Collingwood, Australia.

Hardy, N.B. and Gullan, P.J. (2007). A new genus and four new species of felt scales on Eucalyptus (Hemiptera: Coccoidea: Eriococcidae) in south-eastern Australia. *Australian Journal of Entomology* 46(2): 106–120.

Hassan, E. (1977). *Major Insect and Mite Pests of Australian Crops.* Ento Press, Gatton, Queensland.

Heather, N.W. and Schaumberg, J.B. (1966). Plantation problems of kauri pine in south-east Queensland. *Australian Forestry* 30: 12–19.

Hockings, F.D. (1980). *Friends and Foes of Australian Gardens.* A.H. & A.W. Reed Pty Ltd, Sydney.

Hockings, F.D. (2014). *Pests, Diseases and Beneficials.* CSIRO Publishing, Collingwood, Victoria.

Hoffman, H. (2009). *Common diseases of native plants in home gardens.* Garden note 357, Dept. of Agric. and Food, Western Australia.

Hoffmann, J.H., Impson, F.A.C., Moran, V.C. and Donnelly, D. (2002). Biological control of invasive golden wattle trees (*Acacia pycnantha*) by a gall wasp, *Trichilogaster* sp. (Hymenoptera: Pteromalidae), in South Africa. *Biological Control* 25(1): 64.

Horton, B.M. (2012). Mitigating the effects of forest eucalypt dieback associated with psyllids and bell miners in World Heritage Areas. *Australasian Plant Conservation* 20(4): 11–13.

House, S. (2010). *Kirramyces leaf diseases.* Primary Industries and Fisheries, Queensland Government.

Impson, F.A.C., Kleinjan, C.A., Hoffmann, J.H. and Post, J.A. (2008). *Dasineura rubiformis* (Diptera: Cecidomyiidae), a new biological control agent for *Acacia mearnsii* in South Africa. *South African Journal of Science* 104(7/8): 247–249.

Jurksis, V. (2005). Eucalypt decline in Australia, and a general concept of tree decline and dieback. *Forest ecology and management* 215: 1–20.

Knihinicki, D.K. and Boczek, J. (2003). Studies on eriophyoid mites (Acari: Eriophyoidea) of Australia: A new genus and seven new species associated with tea trees, *Melaleuca* spp. (Myrtaceae). *Australian Journal of Entomology* 42(3): 215–232.

Kolesik, P. (2000). Gall midges (Diptera: Cecidomyiidae) of Australian cypress-pines, *Callitris* spp. (Cupressaceae), with descriptions of three new genera and three new species. *Australian Journal of Entomology* 39(4): 244–255.

Kolesik, P. and Adair, R.J. (2012). A new genus of gall midge (Diptera: Cecidomyiidae) from Australian Acacia. *Australian Journal of Entomology* 51(2): 97–103.

Kolesik, P. and Baker, G. (2013). New species of *Resseliella* (Diptera: Cecidomyiidae), injurious to *Dianella* and *Phormium* (Xanthorrhoeaceae) in Australia and New Zealand. *Australian Journal of Entomology* 52(1): 67–72.

Kolesik, P., Brown, B.T., Purcell, M.F. and Taylor, G.S. (2012). A new genus and species of gall midge (Diptera: Cecidomyiidae) from *Casuarina* trees in Australia. *Australian Journal of Entomology* 51(4): 223–228.

Kolesik, P., Geyer, C. and Morawetz, W. (2006). First known gall midge (Diptera: Cecidomyiidae) from Arecaceae: *Normanbyomyia fructivora* gen. & sp. n. damaging fruit of the black palm, *Normanbya normanbyi*, in tropical Australia. *Australian Journal of Entomology* 45(1): 38–43.

Kolesik, P., Rice, A.D., Bellis, G.A. and Wirthensohn, M.G. (2009). *Procontarinia pustulata*, a new gall midge species (Diptera: Cecidomyiidae) feeding on mango, *Mangifera indica* (Anarcadiaceae), in northern Australia and Papua New Guinea. *Australian Journal of Entomology* 48(4): 310–316.

Kolesik, P., Taylor, G.S. and Kent, D.S. (2002). New genus and two new species of gall midge (Diptera: Cecidomyiidae) damaging buds on Eucalyptus in Australia. *Australian Journal of Entomology* 41(1): 23–29.

Kolesik, P. and Veenstra-Quah, A. (2008). New gall midge taxa (Diptera: Cecidomyiidae) from Australian Chenopodiaceae. *Australian Journal of Entomology* 47(3): 213–224.

Kolesik, P., Woods, B., Crowhurst, M. and Wirthensohn, M.G. (2007). *Dasineura banksiae*: a new species of gall midge (Diptera: Cecidomyiidae) feeding on *Banksia coccinea* (Proteaceae) in Australia. *Australian Journal of Entomology* 46(1): 40–44.

Kolesik, P. and Adair, R.J. (2002). A new genus of gall midge (Diptera: Cecidomyiidae) from Australian *Acacia*. *Australian Journal of Entomology* 51: 97–103.

Kolesik, P. and Barker, W.R. (2013*)*. A new species of gall midge (Diptera: Cecidomyiidae) feeding on *Hakea, Australian Journal of Entomology* 52 (3): 212–217.

Kranz, B., Schwarz, M., Wills, T., Chapman, T., Morris, D. and Crespi, B. (2001). A fully reproductive fighting morph in a soldier clade of gall-inducing thrips (*Oncothrips morrisi*). *Behavioral Ecology & Sociobiology* 50(2): 151–161.

Kranz, B.D., Schwarz, M.P., Morris, D.C., Crespi, B.J. (2002). Life history of *Kladothrips ellobus* and *Oncothrips rodwayi*: insight into the origin and loss of soldiers in gall-inducing thrips. *Ecological Entomology* 27(1): 49.

Kranz, B.D., Schwarz, M.P., Mound, L.A., Crespi, B.J. (1999). Social biology and sex ratios of the eusocial gall-inducing thrips *Kladothrips hamiltoni*. *Ecological Entomology* 24(4): 432–442.

Krimi, Z. (2001). Crown gall on *Eucalyptus occidentalis* seedlings. *EPPO Bulletin* 31(1): 114.

Laidlaw, W.S. and Wilson, B.A. (2003*)*. Floristic and structural characteristics of a coastal heathland exhibiting symptoms of *Phytophthora cinnamomi* infestation in the eastern Otway Ranges, Victoria. *Australian Journal of Botany* 51: 283–293.

Latz, P. (1995). *Bushfires and bush tucker. Aboriginal plant use in Central Australia*. IAD Press, Alice Springs, Australia.

Lee, T.C. and Wicks, T. J. (1977). *Phytophthora cinnamomi* in native vegetation in south Australia. *Australasian Plant Pathology Newsletter* 6: 22–23.

Leech, M. (2013). *Bee Friendly: a planting guide for European honeybees and Australian native pollinators*. Rural Industry Research and Development Corporation https://rirdc. infoservices.com.au/items/12-014

Llewellyn, R. [Ed.] (1995). *The Good Bug, Australasian Biological Control inc.* Department of primary Industries, Queensland and Rural Industries Research and Development Corporation.

Marshall, T. (2010). *Bug.* Harper Collins, Sydney.

McCredie, T.A., Dixon, K.W. and Sivasithamparam, K. (1985). Variability in the resistance of *Banksia* L.f. species to *Phytophthora cinnamomi*. *Australian Journal of Botany* 33: 629–637.

McDougall, K.L., Hardy, G.E.S.J. and Hobbs, R.J. (2001). Additions to the host range of *Phytophthora cinnamomi* in the jarrah (*Eucalyptus marginata*) forest of Western Australia. *Australian Journal of Botany* 49: 193–198.

McDougall, K.L., Hobbs, R.J. and Hardy, G.E.S.J. (2005). Distribution of understorey species in forest affected by *Phytophthora cinnamomi* in south-western Western Australia. *Australian Journal of Botany* 53: 813–819.

McDougall, K.L., Summerell, B.A., Coburn, D. and Newton, M. (2003). *Phytophthora cinnamomi* causing disease in subalpine vegetation in New South Wales. *Australasian Plant Pathology* 32: 113–115.

McDougall, K.L., Hobbs, R.J. and Hardy, G.E.S.J. (2002). Vegetation of *Pytophora cinnamomii*-infested and adjoining uninfested sited in the northern jarrah (*Eucalyptus marginate*) forest of adjoining uninfested sited in the northern jarrah (*Eucalyptus marginate*) forest of Western Australia. *Australian Journal of Botany* 50: 277–288.

McFadyen, R.E.C., Desmier de Chenon, R. and Sipayung, A. (2003). Biology and host specificity of the chromolaena stem gall fly, *Cecidochares connexa* (Macquart) (Diptera: Tephritidae). *Australian Journal of Entomology* 42(3): 294–297.

McLeish, M. (2011). Speciation in gall-inducing thrips on *Acacia* in arid and non-arid areas of Australia. *Journal of Arid Environments* 75(9): 793–801.

McLeish, M.J., Perry, S.P., Gruber, D. and Chapman, T.W. (2003). Dispersal patterns of an Australian gall-forming thrips and its host tree (*Oncothrips tepperi* and *Acacia oswaldii*). *Ecological Entomology* 28(2): 243–246.

McLeish, M.J., Schwarz, M.P. and Chapman, T.W. (2011). Gall inducers take a leap: host-range differences explain speciation opportunity (Thysanoptera: Phlaeothripidae). *Australian Journal of Entomology* 50(4): 405–417.

McMaugh, J. (1985). *What Garden Pest or Disease is That?* Lansdowne Press, Sydney, Australia.

Mendel, Z., Protasov, A., Fisher, N. and Salle, J.L. (2004). Taxonomy and biology of Leptocybe invasa gen. & sp. n. (Hymenoptera: Eulophidae), an invasive gall inducer on Eucalyptus. *Australian Journal of Entomology* 43(2)I 101–113.

Messing, R.H. and Wang, X-G. (2009). Competitor-free space mediates non-target impact of an introduced biological control agent. *Ecological Entomology* 34(1): 107–113.

Moulds, M.S. (1981). Larval food plants of hawk moths (Lepidoptera: Sphingidae) affecting commercial crops in Australia. *General and Applied Entomology* 13: 69–80.

Moulds, M.S. (1984). Larval food plants of hawk moths (Lepidoptera: Sphingidae) affecting garden ornamentals in Australia. *General and Applied Entomology* 16: 57–64.

Mound, L.A. (2004). Australian long-tailed gall thrips (Thysanoptera: Phlaeothripinae, Leeuweniini), with comments on related Old World taxa. *Australian Journal of Entomology* 43(1): 28–37.

Mound, L.A., Crespi, B.J. and Tucker, A. (1998). Polymorphism and kleptoparasitism in thrips (Thysanoptera: Phlaeothripidae) from woody galls on *Casuarina* trees. *Australian Journal of Entomology* 37(1): 8–16.

Mound, L.A. and Minaei, K. (2007). Australian thrips of the Haplothrips lineage (Insecta: Thysanoptera). *Journal of Natural History* 41(45–48): 2919–2978.

Mound, L.A. (1976). Thysanoptera of the genus *Dichromothrips* on old world Orchidaceae. *Biol. J. Linn. Soc.* 8(3): 245–265.

Naumann, I.D., Williams, M.A. and Schmidt, S. (2002). Synopsis of the Tenthredinidae (Hymenoptera) in Australia, including two newly recorded, introduced sawfly species associated with willows (*Salix* spp.). *Australian Journal of Entomology* 41(1): 1–6.

Nichol, A., Sivasithamparam, K. and Dixon, K.W. (1988). Rust infections of Western Australian orchids. *Lindleyana* 3(1): 1–8.

Nyeko, P., Mutitu, E.K. and Day, R.K. (2007). Farmers' knowledge, perceptions and management of the gall-forming wasp, *Leptocybe invasa* (Hymenoptera: Eulophidae), on *Eucalyptus* species in Uganda. *International Journal of Pest Management* 53(2): 111–119.

Ozman, S.K. and Goolsby, J.A. (2005). Biology and phenology of the eriophyid mite, *Floracarus perrepae*, on its native host in Australia, Old World climbing fern, *Lygodium microphyllum*. *Experimental & Applied Acarology* 35(3): 197–213.

Panjehkeh, N., Backhouse, D. and Taji, A. (2007). Flower colour is associated with susceptibility to disease in the legume *Swainsona formosa*. *Australasian Plant Pathology* 36: 341–346.

Parsons, R.F. and Uren, N.C. (2011). Is Mundulla Yellows really a threat to undisturbed native vegetation? A comment. *Ecological Management & Restoration* 12 (1): 72–74.

Perry, S.P., Chapman, T.W., Schwarz, M.P. and Crespi, B.J. (2004). Proclivity and effectiveness in gall defence by soldiers in five species of gall-inducing thrips: benefits of morphological caste dimorphism in two species (*Kladothrips intermedius* and *K. habrus*). *Behavioral Ecology & Sociobiology* 56(6): 602–610.

Podger, F.D. (1972). *Phytophthora cinnamomi*, a cause of lethal disease in indigenous plant communities in Western Australia. *Phytopathology* 62: 972–981.

Post, J.A., Kleinjan, C.A., Hoffmann, J.H. and Impson, F.A.C. (2010). Biological control of *Acacia cyclops* in South Africa: The fundamental and realized host range of *Dasineura dielsi* (Diptera: Cecidomyiidae). *Biological Control* 53(1): 68–75.

Prider, J., Watling, J. and Facelli, J.M. (2009). Impacts of a native parasitic plant on an introduced and a native host species: implications for the control of an invasive weed. *Annals of Botany* 103: 107–115.

Qin, T.K. and Gullan, P.J. (1992). A revision of the Australian soft scales *Pulvinaria. J. Nat. Hist.* 26: 103–164.

Qin, T.K. and Gullan, P.J. (1994). Taxonomy of the wax scales (Hemiptera: Coccidae: Ceroplastinae) in Australia. *Invertebrate Taxonomy* 8 (4): 923–959.

Radunz, L.A. and Allwood, A.J. (1981). *Insect Pests in the Home Garden and Recommendations for their Control,* Department of Primary Production, Northern Territory.

Rea, A.J., *et al.* (2011). Two novel and potentially endemic species of *Phytophthora* associated with episode dieback of Kwongan Vegetation in WA. *Plant Pathol.* 60: 1055–1068.

Reid, C.A.M (2006). A taxonomic revision of the Australian Chrysomelina, with a key to the genera (Coleoptera: Chrysomelidae). *Zootaxa* 1292: 1–119.

Reiter, N., Weste, G. and Guest, D. (2004). The risk of extinction resulting from disease caused by *Phytophthora cinnamomi* to endangered, vulnerable or rare plant species endemic to the Grampians, western Victoria. *Australian Journal of Botany* 52: 425–433.

Rentz, D.C.F. (1995). The Changa Mole Cricket, *Scapteriscus didactylus* (Latreille), a New World Pest Established in Australia (Orthoptera: Gryllotalpidae). *Australian Journal of Entomology* 34(4): 303–306.

Rentz, D. (1996). *Grasshopper Country.* University of New South Wales Press, Sydney, Australia.

Rentz, D. (2010) *A guide to the katydids of Australia.* CSIRO Publishing, Collingwood, Australia.

Scheffer, S.J., Nelson, L.A., Davies, K.A., Lewis, M.L., Giblin-Davis, R.M., Taylor, G.S. and Yeates, D.K. (2013). Sex-limited association of *Fergusobia* nematodes with female *Fergusonina* flies in a unique Australasian mutualism (Nematoda: Neotylenchidae; Diptera: Fergusoninidae). *Australian Journal of Entomology* 52(2): 125–128.

Scott, J.K., Yeoh, P.B. and Knihinicki, D.K. (2008). Redberry mite, *Acalitus essigi* (Hassan) (Acari: Eriophyidae), an additional biological control agent for Rubus species (blackberry) (Rosaceae) in Australia. *Australian Journal of Entomology* 47(3): 261–264.

Scott, P.M., *et al.* (2009). *Phytophthora multivora* sp. nov., a new species affecting *Eucalyptus, Banksia, Agonis* and other plant species in WA. *Persoonia* 22: 1–13.

Seymour, C.L. and Veldtman, R. (2010). Ecological role of control agent, and not just host-specificity, determine risks of biological control. *Austral Ecology* 35(6): 704–711.

Shearer, B.L. and Dillon, M. (1995). Susceptibility of plant species in *Eucalyptus marginata* forest to infection by *Phytophthora cinnamomi. Australian Journal of Botany* 43: 113–134.

Shearer, B.L. and Dillon, M. (1996). Susceptibility of plant species in *Banksia* woodlands on the Swan Coastal Plain, Western Australia, to infection by *Phytophthora cinnamomi. Australian Journal of Botany* 44: 433–445.

Shearer, B.L. and Fairman, R.G. (2007). Application of phosphite in a high-volume foliar spray delays and reduces the rate of mortality of four *Banksia* species infected with *Phytophthora cinnamomi. Australasian Plant Pathology* 36: 358–368.

Shearer, B.L., Crane, C. E. and Cochrane, A. (2004). Phosphite reduces disease extension of a *Phytophthora cinnamomi* front in *Banksia* woodland, even after fire. *Australasian Plant Pathology* 33: 249–254.

Shearer, B.L., Crane, C.E. and Cochrane, A. (2004). Quantification of the susceptibility of the native flora of the Southwest Botanical Province, Western Australia, to

Phytophthora cinnamomi. Australian Journal of Botany 52: 435–443.

Shearer, B.L., Crane, C.E., Barrett, S. and Cochrane, A. (2007). Assessment of threatened flora susceptibility to *Phytophthora cinnamomi* by analysis of disease progress curves in shadehouse and natural environments. *Australasian Plant Pathology* 36: 609–620.

Shepard, M., Lawn, R.J. and Schneider, M.A. (1983). *Insects on Grain Legumes in Northern Australia.* University of Queensland Press, St Lucia.

Stone, C. and Madden, J.L. (1984). Induction and formation of pouch and emergence galls in *Eucalyptus pulchella* leaves. *Australian Journal of Botany* 32(1): 33.

Struckley, M.C., *et al.* (2007). *Phytophthora inundata* from native vegetation in WA. *Aust. Plant Pathol.* 36: 606–608.

Suddaby, T., Alhussaen, K., Daniel, R. and Guest, D. (2008). Phosphonate alters the defense responses of *Lambertia* species challenged by *Phytophthora cinnamomi. Australian Journal of Botany* 56: 550–556.

Summerell, B.A. and Rugg, C.A. (1991). Wilt of *Helichrysum* species caused by *Fusarium oxysporum. Australasian Plant Pathology* 21: 18–19.

Summerell, B.A., *et al.* (1990). Crown and stem canker of Waratah caused by *Cylindrocarpon destructans. Australasian Plant Pathology* 19: 13–15.

Summerell, B.A., *et al.* (2011). *Fusarium* species associated with plants in Australia. *Fungal Diversity* 46: 1–27.

Taylor, G.S. and Davies, K.A. (2008). New species of gall flies (Diptera: Fergusoninidae) and an associated nematode (Tylenchida: Neotylenchidae) from flower bud galls on *Corymbia* (Myrtaceae). *Australian Journal of Entomology* 47(4): 336–349.

Taylor, S.G. (1990). Revision of the genus *Schedotrioza* Tuthill & Taylor (Homoptera: Psylloidea: Triozidae). *Invertebrate Taxonomy* 4(4): 721–751.

Tree, D.J. and Mound, L.A. (2009). Gall-induction by an Australian insect of the family Thripidae (Thysanoptera: Terebrantia). *Journal of Natural History* 43(19/20): 1147–1158.

Tree, D.J. and Walter, G.H. (2009). Diversity of host plant relationships and leaf galling behaviours within a small genus of thrips – *Gynaikothrips* and *Ficus* in south-east Queensland, Australia. *Australian Journal of Entomology* 48(4): 269–275.

Tynan, K.M., Scott, E.S. and Sedgley, M. (1998). Evaluation of *Banksia* species for response to *Phytophthora* infection. *Plant Pathology* 47: 446–455.

Tynan, K.M., Wilkinson, C.J., Holmes, J.M., Dell, B., Colquhoun, I.J., McComb, J.A. and Hardy, G.E.St.J. (2001). The long-term ability of phosphite to control *Phytophthora cinnamomi* in two native plant communities of Western Australia. *Australian Journal of Botany* 49: 761–770.

Veenstra, A.A., Michalczyk, A. and Kolesik, P. (2011). Taxonomy of two new species of gall midge (Diptera: Cecidomyiidae) infesting *Tecticornia arbuscula* (Salicornioideae: Chenopodiaceae) in Australian saltmarshes. *Australian Journal of Entomology* 50(4): 393–404.

Veenstra-Quah, A.A., Milne, J. and Kolesik, P. (2007). Taxonomy and biology of two new species of gall midge (Diptera: Cecidomyiidae) infesting *Sarcocornia quinqueflora*

(Chenopodiaceae) in Australian salt marshes. *Australian Journal of Entomology* 46(3): 198–206.

Veldtman, R., Lado, T.F., Botes, A., Proche , S., Timm, A.E., Geertsema, H. and Chown, S.L. (2011). Creating novel food webs on introduced Australian acacias: indirect effects of galling biological control agents. *Diversity & Distributions* 17(5): 958–967.

Wardell-Johnson, G., Stone, C., Recher, H. and Lynch, A.J.J. (2005). A review of eucalypt dieback associated with bell miner habitat in south-eastern Australia. *Australian Forestry* 68: 231.

Wasson, A.P., Ramsay, K., Jones, M.G.K. and Mathesius, U. (2009). Differing requirements for flavonoids during the formation of lateral roots, nodules and root knot nematode galls in *Medicago truncatula. New Phytologist* 183(1): 167–179.

Weste, G. (2003). The dieback cycle in Victorian forests: a 30-year study of changes caused by *Phytophthora cinnamomi* in Victorian open forests, woodlands and heathlands. *Australasian Plant Pathology* 32: 247–256.

Weste, G., Brown, K., Kennedy, J. and Walshe, T. (2002). *Phytophthora cinnamomi* infestation – a 24-year study of vegetation change in forests and woodlands of the Grampians, Western Victoria. *Australian Journal of Botany* 50: 247–274.

Wheeler, G.S., Taylor, G.S., Gaskin, J.F. and Purcell, M.F. (2011). Ecology and Management of Sheoak (*Casuarina* spp.), an Invader of Coastal Florida, USA. *Journal of Coastal Research* 27(3): 485–492.

Wills, T.E., Chapman, T.W., Mound, L.A., Kranz, B.D. and Schwarz, M.P. (2004). Natural history and description of *Oncothrips kinchega*, a new species of gall-inducing thrips with soldiers (Thysanoptera: Phlaeothripidae). *Australian Journal of Entomology* 43(2): 169–176.

Womersley, H. (1941). The red-legged earth-mite (Acarina, Penthaleidae) of Australia. *Transactions of the Royal Society of South Australia* 65(2): 292–294.

Worrall, R., Gollnow, B. and Taylor, G. (2005). *Management of psyllids in NSW Christmas bush plantations,* Primefact 17, NSW Department of Primary Industries.

Wright, S.A.W. and Center, T.D. (2008). Nonselective oviposition by a fastidious insect: the laboratory host range of the melaleuca gall midge *Lophodiplosis trifida* (Diptera: Cecidomyiidae). *Biocontrol Science & Technology* 18(8): 793–807.

Wylie, S.J., Li, H., Dixon, K.W., Richards, H. and Jones, M.G.K. (2013). Exotic and indigenous viruses infect wild populations and captive collections of temperate terrestrial orchids (*Diuris species*) in Australia. *Virus Research* 171: 22–32.

Wylie, S.J., Li, H. and Jones, M.G.K. (2013). Donkey Orchid Symptomless Virus: A Viral 'Platypus' from Australian Terrestrial Orchids. *PLoS One* 8(11):1:10.

Wylie, S.J., Tan, A., Li, H., Dixon, K.W. and Jones, M.G.K. (2012). *Caladenia* virus A, an unusual new member of the Potyviridae from terrestrial orchids in Western Australia. *Archives of Virology* 157: 2447–2452.

Zborowski, P. and Edwards, T. (2007). *A Guide to Australian Moths.* CSIRO Publishing, Collingwood, Australia.

Zborowski, P. and Storey, R. (1995) *A Field Guide to Insects of Australia.* Reed Books Australia, Chatswood, NSW.

INDEX

David L. Jones loves to write as shown by his authorship or co-authorship of more than 35 books. As a botanist at the CSIRO, Canberra, his studies led to contributions for the *Flora of Australia* and numerous scientific papers resulting in the description of more than 300 species of native orchids, and a life membership of the Australasian Native Orchid Society. Results of his enquiring mind can be seen in books on Australian native plants, orchids, ferns, palms, cycads, rainforest plants, climbers, bulbs, succulents, cacti and the insects and diseases that impact on Australian plants. He has collected and photographed extensively throughout Australia. David has been recognized by his peers, being awarded the Victorian College of Agriculture and Horticulture Medal (1991), Australian Plants Award (Professional Category) from the Australian Societies for Growing Australian Plants (1999), Award of Honour from the Australian Orchid Foundation (2001) and Certificate of Appreciation from the Australian Region of the International Plant Propagators' Society (2001). He was also a finalist in the Eureka Science Awards in 2004. David's most recent plant books published by New Holland, the well-received *Starting Out* series, stimulated him into rewriting this important book for Australian gardeners.

W. Rodger Elliot is a horticulturist, writer, lecturer, photographer, pseudo-botanist and gardener who has had a passion for Australian plants for a long time. Rodger and his wife Gwen have owned and operated retail and wholesale nurseries specialising in Australian plants. He has written for Australian and overseas garden journals and magazines since the 1970s and written a number of books on Australian plants. He is co-author with David Jones of the internationally renowned nine-volume *Encyclopaedia of Australian Plants Suitable for Cultivation* (illustrated by Trevor Blake). Rodger has lectured widely on Australian Plants in many parts of Australia as well as in the USA, Europe and Japan.

Rodger was a member of the Royal Botanic Gardens Board Victoria (which includes Melbourne and Cranbourne Gardens) for over 15 years. In 2001 Rodger and Gwen were appointed Members of the Order of Australia (AM) for their contribution to the horticulture of Australian plants. Rodger's other awards include Honorary Life Membership of Australian Plants Society (Vic), Australian Natural History Medallion, Australian Institute of Horticulture's Award for Excellence and the Gold Veitch Memorial Medal from the Royal Horticultural Society, London.

Sandra R. Jones is an ecologist whose 20-year professional career has spanned research, education, communication and management in both the private and public sectors. She has a PhD in applied ecology, and her research topics have included plant taxonomy, herpetology, conservation biology and habitat management. She has authored scientific and popular articles and considers herself to be an old-fashioned naturalist. Her passion for understanding and protecting the natural world has led her to explore Australia and seek to understand the interrelationships between the plants and animals she has observed. This interest has proven to be a perfect catalyst for investigating the role of pests and diseases of Australian plants, natural balances between pests and their hosts and methods for and implications of pest control.

Trevor L. Blake has been collecting, photographing and growing Australian plants for over 55 years. He lectures regularly to interested groups, is involved with workshops and writes for specialist publications. He is also strongly involved in the conservation movement and is especially opposed to the damage and destruction of natural habitats and landscape by totally inappropriate landuse. Trevor is also an illustrator whose line-drawings of Australian plants have been included in the nine volumes of the *Encyclopaedia of Australian Plants Suitable for Cultivation*, the *Identikit* series of books on Australian plants, *Flora of the Grampians* and *A Guide to Darwinia and Homoranthus*.